# Salt Tectonics

Geological Society Special Publications
*Series Editor* A. J. FLEET

GEOLOGICAL SOCIETY SPECIAL PUBLICATION NO. 100

# Salt Tectonics

EDITED BY

G. I. ALSOP, D. J. BLUNDELL & I. DAVISON
Department of Geology, Royal Holloway
University of London, UK

1996
Published by
The Geological Society
London

# THE GEOLOGICAL SOCIETY

The Society was founded in 1807 as The Geological Society of London and is the oldest geological society in the world. It received its Royal Charter in 1825 for the purpose of 'investigating the mineral structure of the Earth'. The Society is Britain's national society for geology with a membership of around 8000. It has countrywide coverage and approximately 1000 members reside overseas. The Society is responsible for all aspects of the geological sciences including professional matters. The Society has its own publishing house, which produces the Society's international journals, books and maps, and which acts as the European distributor for publications of the American Association of Petroleum Geologists, SEPM and the Geological Society of America

Fellowship is open to those holding a recognized honours degree in geology or cognate subject and who have at least two years' relevant postgraduate experience, or who have not less than six years' relevant experience in geology or a cognate subject. A Fellow who has not less than five years' relevant postgraduate experience in the practice of geology may apply for validation and, subject to approval, may be able to use the designatory letters C Geol (Chartered Geologist).

Further information about the Society is available from the Membership Manager, The Geological Society,

Published by the Geological Society from:
The Geological Society Publishing House
Unit 7, Brassmill Enterprise Centre
Brassmill Lane
Bath BA1 3JN
UK (*Orders*: Tel. 01225 445046
Fax 01225 442836)

First published 1996



**British Library Cataloguing in Publication Data**

A catalogue record for this book is available from the British Library.

ISBN 1–897799–44–6

Typeset by Type Study, Scarborough, UK

Printed in Great Britain by
Alden Press, Oxford

*Distributors*
*USA*
AAPG Bookstore
PO Box 979
Tulsa
OK 74101–0979
USA
(*Orders*: Tel. (918) 584–2555
Fax (918) 584–0469)

*Australia*
Australian Mineral Foundation
63 Conyngham Street
Glenside
South Australia 5065
Australia
(*Orders*: Tel (08) 379–0444
Fax (08) 379–4634)

*India*
Affiliated East–West Press PVT Ltd
G–1/16 Ansari Road
New Delhi 110 002
India
(*Orders*: Tel. (11) 327–9113
Fax (11) 326–0538)

*Japan*
Kanda Book Trading Co.
Tanikawa Building
3–2 Kanda Surugadai
Chiyoda-Ku
Tokyo 101
Japan
(*Orders*: Tel. (03) 3255–3497
Fax (03) 3255–3495)

# Contents

## Acknowledgements

We would like to thank the following companies that sponsored this meeting: BP, Conoco (UK), Enterprise Oil, Esso (UK), and Ranger Oil (UK). This meeting was convened and supported by the Petroleum and Tectonic Studies Group of the Geological Society. The following people acted as referees for the papers submitted to this volume, and we thank them for their considerable efforts which have improved this publication.

C. J. Banks
D. J. W. Bosence
M. Brooks
J. P. Brun
J. A. Cartwright
H. Cohen
J. Dixon
C. F. Elders
R. Evans
O. Graverson
R. Heaton
R. Holmes
J. R. Hossack
M. Hudec
M. Insley
M. P. A. Jackson
M. Jenyon
G. D. Karner
H. Koyi
D. McGuinness
J. Milson
T. Nalpas
A. Poliakov
J. Price
Gerald Roberts
Gordon Roberts
M. Sans
S. Stewart
P. Szatmari
C. J. Talbot
D. A. Waltham
M. O. Withjack
S. Wu

*Ian Alsop, Derek Blundell and Ian Davison, London, March 1995*

# Salt tectonics: some aspects of deformation mechanics

IAN DAVISON, IAN ALSOP & DEREK BLUNDELL

*Department of Geology, Royal Holloway, University of London, Egham, TW20 OEX, UK*

This volume is dedicated to studies of the deformation of evaporite rocks in pillows and diapirs, and the surrounding sedimentary overburden rocks and sediments. Salt diapirs have become the focus of attention in the last forty years, because of their strategic importance in controlling hydrocarbon reserves, and their unique physical properties enable storage of hydrocarbons and toxic waste. Their economic importance is unique on the Earth's surface, as evaporites in the Middle East are responsible for trapping up to 60% of its hydrocarbon reserves (**Edgell**).

Salt also produces some of the most complex and beautiful deformation features on the Earth's surface, although few of these surface exposures have been examined in detail. The first section of this volume consists of analyses of outcrop, cave, mine and borehole information which add to our general understanding of the internal diapir deformation patterns and overburden tectonics. This is followed by papers mainly based on seismic reflection profile interpretation, which provide accurate documentation of salt tectonics in the NW Europe Zechstein Basin (**Buchanan *et al.***, **Stewart *et al.***, Zirngast), Persian Gulf (**Edgell**), and the Angolan Margin (**Spathopoulos**).

Salt diapirism is principally controlled by the rheology of the overburden rocks and physical models have given us important new insights into the deformation of overburden. Deformation experiments are presented on overburden around salt structures (**Alsop**), salt sheet segmentation (**Koyi**), and development of giant counter-regional faults on the Brazilian continental margin (**Szatmari *et al.***). **Petersen & Lerche** investigate the influence of salt sheets and diapirs on thermal history. Salt has a thermal conductivity ($K$) which is up to three times greater than the surrounding sediments, but $K$ decreases by up to a third downwards through a 5 km high salt diapir. In contrast, increasing compaction produces higher $K$ in the sedimentary rocks with increasing depth. Variable conductivity models with compaction and temperature indicate that thermal effects are much larger at the top of the salt structure than at the base.

The following sections summarize some of the fundamental aspects relating to salt tectonics covered by this thematic set of papers, but it is not intended to be a full review.

## Deformation mechanisms

**Burliga, Davison *et al.*, Sans *et al.*, Smith** and **Talbot & Alavi** provide new insights into the internal deformation patterns in salt from mesoscopic observations in deformed bodies. Shear zones are commonly developed parallel to bedding in potassic horizons (**Sans *et al.***), where the salt becomes gneissose with X:Z axial ratios of crystals reaching commonly around 4:1 in Zechstein salt (**Smith**), and Red Sea diapirs (**Davison *et al.***, Fig. 1). Relatively undeformed halite layers are carried laterally (**Smith**) and upwards as rafts between shear zones into diapirs (**Burliga**), and undeformed halite rafts are often transported in a highly-sheared sylvinite matrix (**Sans *et al.***).

Average halite grain sizes generally range between 5–10 mm in North Sea Zechstein halite (our own observations) and 15–20 mm in deformed Gulf Coast Louann Salt (Talbot & Jackson 1987). At this size, both pressure solution and dislocation creep will be active mechanisms of deformation (Van Keken *et al.* 1993) and characteristics of both creep mechanisms have been described in natural salt (Spiers et al. 1990).

Salt can fault in a brittle mode at high strain rates (Fig. 2, **Smith**; **Davison *et al.***). Long range methane gas migration occurs through fractures in salt in the Algarve Basin, Portugal (Terrinha *et al.* 1994) and in the Sergipe–Alagoas Basin, NE Brazil (Fig. 2). This has important implications for both hydrocarbon potential and toxic waste disposal.

## Flow patterns inside the source layer

Fold wavelengths of salt layers that have original thicknesses greater than 1 km with one kilometre of overburden will typically be about 20–30 km in the North Sea (Fig. 3). The initial wavelength of the perturbations controls the maximum amount of salt that is available to flow into an individual salt structure. Assuming a circular drainage radius of 10 km for North Sea examples implies that for a 1.5 km thick salt layer approximately 470 km$^3$ of salt could be transported into a single diapir. This is much greater than the amount of salt contained in most North Sea diapirs which will be around 25–75 km$^3$. The rate of radial flow of this amount of salt will increase as the diapir is approached, so that there is a mass acceleration and constriction of flow

*From* ALSOP, G. I., BLUNDELL, D. J. & DAVISON, I. (eds), 1996, *Salt Tectonics*, Geological Society Special Publication No. 100, pp. 1–10.

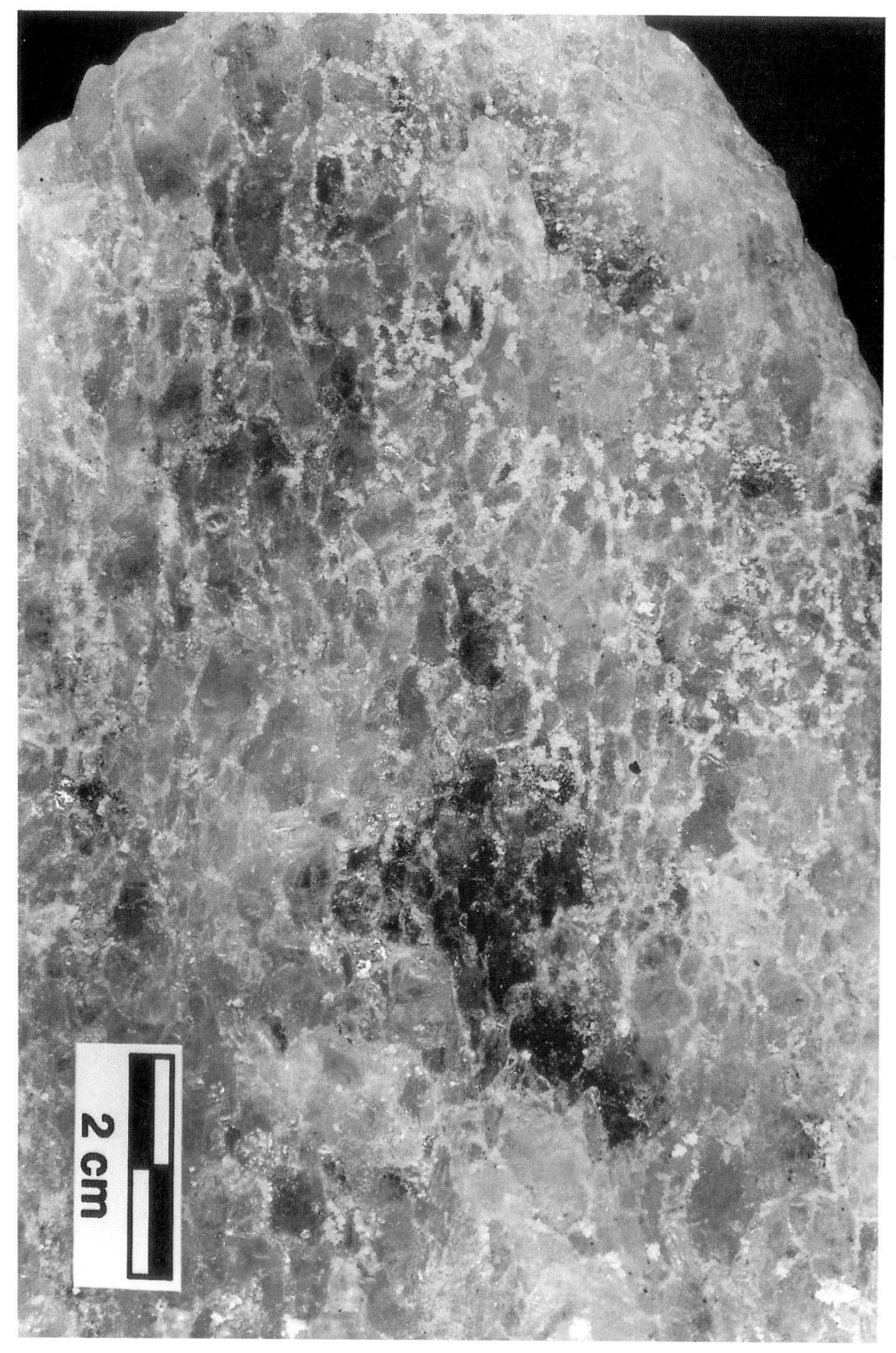


**Fig. 1.** Photograph of halite fabric in the centre of a mature Red Sea diapir in NW Yemen (**Davison *et al.***). This coarse grain size (5–10 mm) is typical of many major diapir provinces and suggests that diapirs deform by both pressure solution and dislocation creep. Note the vertical extension lineation in the salt produced by vertical flow in the diapir neck.

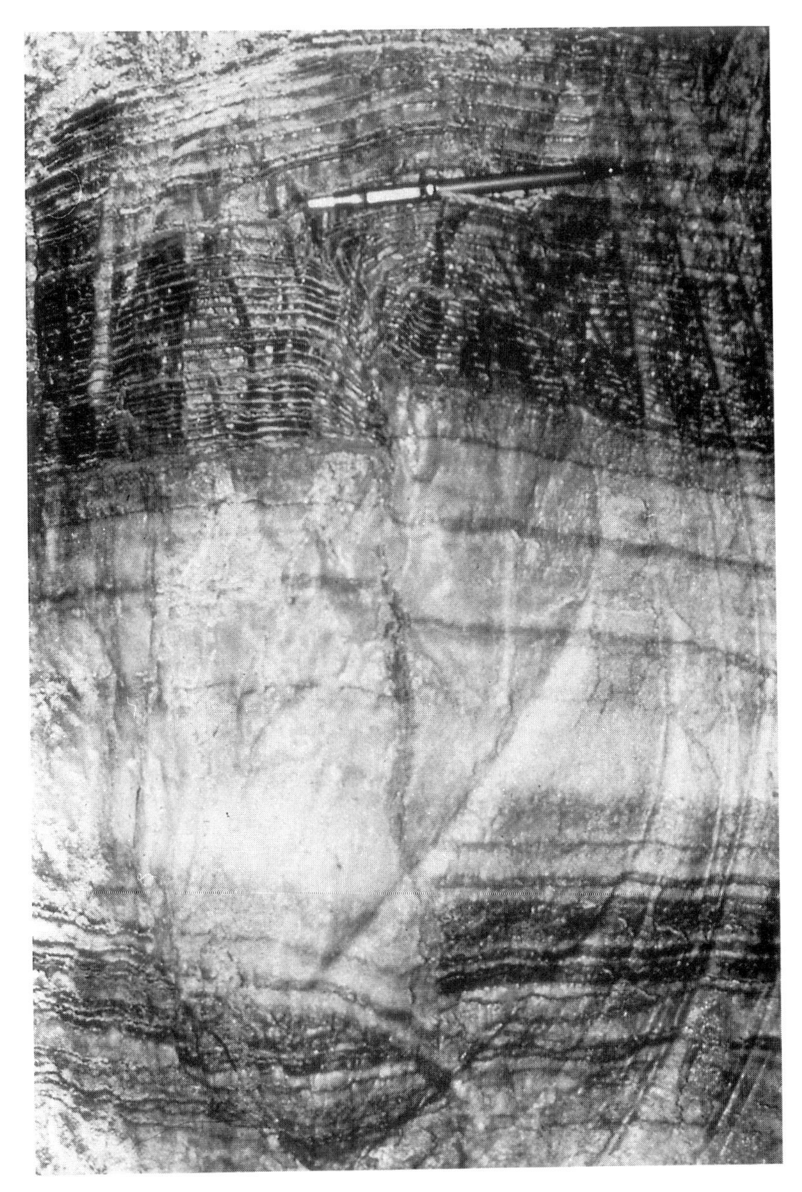

**Fig. 2.** Fault in Aptian age halite and sylvinite from the PETROMISA Mine near Aracaju, Sergipe-Alagoas Basin, NE Brazil. The salt lies at a present depth of approximately 300 m with overburden of Cretacous age. Coarse-grained halite has recrystallized along the fault plane. Open fractures filled with methane are also present in this mine.

towards the diapir (Fig. 4). Strain rates in the diapir neck could be up to 80 times larger than the average flow rate in the source layer (our own calculations from experimantal models using SGM polymer, Fig. 4). Vertical flow in the centre of the diapiric neck produces extreme prolate strains (**Davison *et al.***). These are nullified at the top of the diapir where vertical flattening takes place; this may leave apparently undeformed surfaces in diapiric heads (Figs 4 & 5).

Flow of viscous salt in a confined layer exhibits a Poiseuille velocity profile, with the highest flow velocity in the centre of the layer (Fig. 6a). Intervening sedimentary layers within the evaporite sequence produce a relative decrease in the rate of lateral flow due to increased drag effects (Fig. 6d).

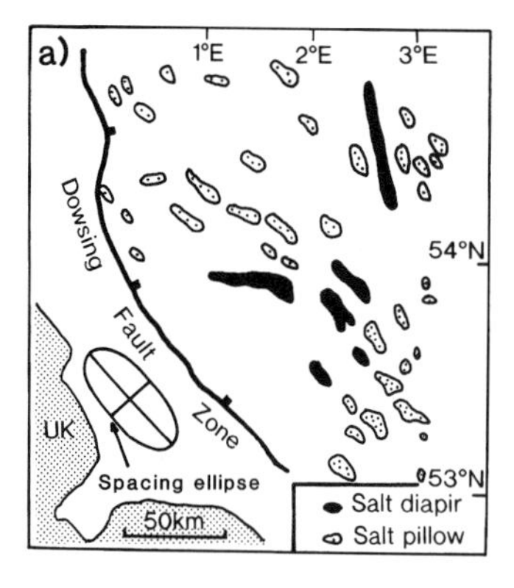


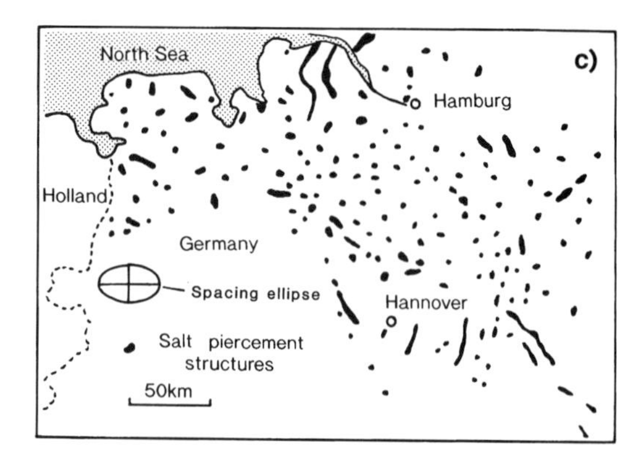


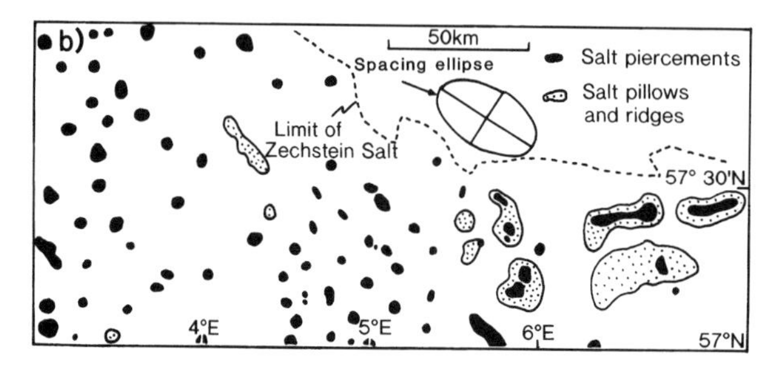


**Fig. 3.** Summary of salt structures in various provinces of the North Sea Zechstein Basin: (a) southern North Sea (Hughes & Davison 1993). Fry plots indicate average minimum spacing between structures in a NE–SW direction is 25 km, and 58 km in a NW–SE direction. (b) Norwegian-Danish basin (Hospers *et al.* 1988). Average spacing between structures is 23 km in a NE–SW direction, and 39 km in an NW– SE direction. (c) Northern Germany (Trusheim 1960). Average spacing between structures is 31 km in an E–W direction and 26 km in a N–S direction.

Each layer behaves individually when controlling lateral flow, but the layers must act together during vertical flow perpendicular to the initially horizontal layering (Fig. 6e). Couette plate type flow (Fig. 6b) will occur in the source layer where overburden immediately above is shortening or extending. Hence, salt at the top of the layer is dragged by the overburden and appreciable mass movement may occur, with salt being dragged along detachments to thicken downslope (Fig. 6f). Downslope thickening of salt on the western and eastern platforms of the Central Graben may have been produced by flow of salt into the Graben despite the natural buoyancy of the salt.

## Effective viscosity of salt and overburden and anisotropy

Laboratory deformation experiments on evaporite rocks have measured viscosity variations greater than ten orders of magnitude. Wet carnallite may have a viscosity as low as $10^8$, whereas dry halite may have a viscosity around $10^{18}$ Pa s (Van Keken *et al.* 1993). It is difficult to make direct estimates of large-scale effective viscosities of diapirs over long time periods. Different parts of the diapir deform at different velocities and by different mechanisms (Van Keken *et al.* 1993). Analogies with physical or mathematical models give estimates of effective viscosities of diapir source layers which vary between $10^{16}$ and $10^{18}$ Pa s (e.g. Koyi 1988, 1991; Nalpas & Brun 1993). However, the viscosity of the salt has little to do with the triggering of diapirism; which is primarily dependent on the strength and stress state of the overburden. The few examples of viscosity contrast estimates of salt and overburden in natural examples have been based on initial wavelength-thickness estimates using Ramberg theory (Rönnlund 1989; Hughes & Davison 1993). These estimates give sedimentary rock:evaporite rock viscosity ratios of 50–$10^4$. To think of sedimentary rocks as purely viscous (Renzhiglov & Pavlishcheva 1970) or brittle end members is an oversimplification (**Poliakov *et al.***). Besides the obvious evidence of brittle faulting

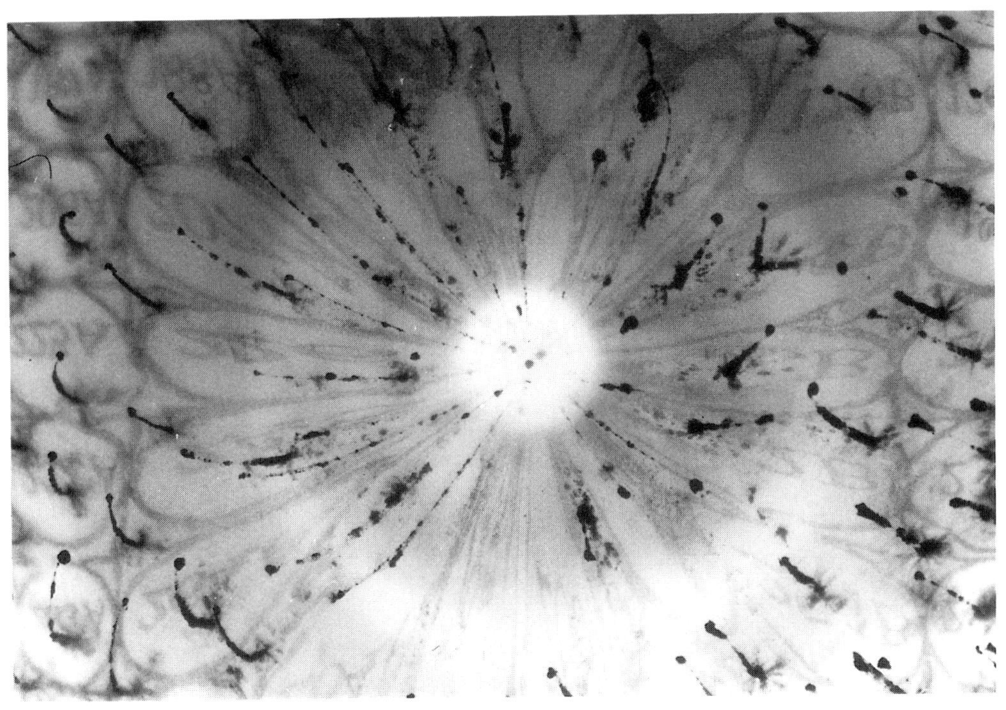

**Fig. 4.** View of physical model looking up a diapir neck (1 cm wide lightest zone in centre). SGM-36 polymer diapir ($\rho$ = 965 kg m$^{-3}$, $\mu$ = 5 x 10$^4$ Pas) intruding into Johnson's Baby Oil ($\rho$ = 1100 kg m$^{-3}$, $\mu$ = 160 Pa s). Strain reversal of X:Z of much greater than 5:1 occurs in the polymer diapir. Numbered strain marker circles (originally 1 cm in diameter) on the top of the polymer layer, which are highly stretched going up the neck of the diapir, become readable again once the polymer spreads out in a diapir head (lateral extent of the diapir mushroom head is shown by the lighter annulus around the neck, 3 cm in radius). Original vertical coloured powder streaks and horizontal strain circles have been used to calculate the strains. The horizontal component of shear strain in the model is represented by the projected horizontal length of the streaks.

there are several lines of evidence which also support viscous behaviour of the salt and overburden system, when considered over long time periods and large scales.

- Folded rocks are present at the Earth's surface, with little evidence of surface faulting (e.g. Zagros mountains, **Talbot & Alavi**; Mexican Fold Ridges, Gulf Coast of Mexico, Tharp & Scarborough 1994)
- As a first approximation, many salt diapirs have elongated droplet or finger shapes (e.g. Jackson & Seni 1984, Ratcliffe *et al.* 1992);
- Smooth drag profiles of sediments near diapirs are observed on seismic sections (e.g. Ratcliffe *et al.* 1992), suggesting the sedimentary rocks approximate to a viscous or power law fluid behaviour.
- Overpressured mud diapirs form sinusoidal waves (e.g. Collier & White 1990).
- Salt pillows and diapir spacings appear at consistent wavelengths over large areas (Fig. 3) in contrast to the non-periodic spacing of major brittle structures in the basement.

Hence, the best approach to modelling of diapirs should be to simulate overburden as a visco-elastic, or power-law elastic combination where the effective deformation mechanism will depend on strain rate and environmental conditions.

The presence of interbedded sedimentary rocks will greatly stengthen the salt unit and produce a large strength anisotropy. Experimental deformation of multilaminates indicates that degree of anisotropy can depend on strain rate (**Koyi**). Strength anisotropy is often inherent throughout the massive salt layers which contain rhythmic layering of anhydrite-rich beds at centimetre scales (Fig. 2). This will be sufficient to generate pervasive small wavelength sheath folds (metric scale) throughout large areas of the weakly-deformed diapir. Each salt layer will flow at a rate dependent on its thickness, as effective viscosity depends on thickness (**Cohen & Hardy**). Once diapirism is well advanced intervening sedimentary layers, up to at least 40 m thick (**Burliga**, **Davison *et al.***) and probably much greater, will be boudinaged so that salt can flow between the layers, at which point an interbedded

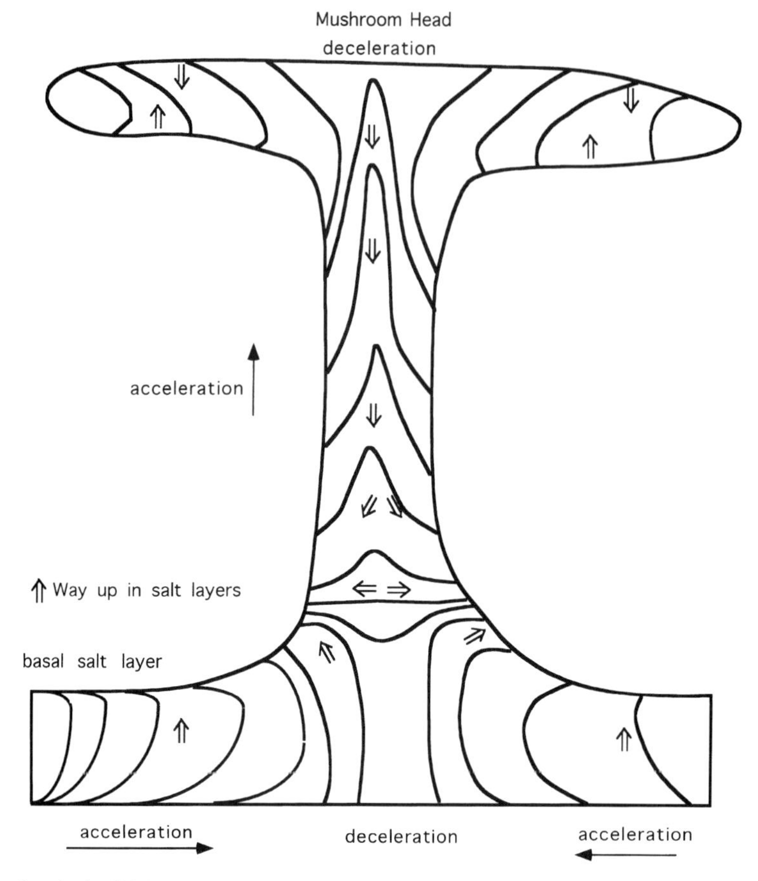


**Fig. 5.** Schematic velocity fields in a salt diapir, based on physical models using viscous fluids. The approximate deformed shapes of original linear vertical markers are shown.

unit will behave essentially like pure salt. Hence, during the pillow stage the source layer will have a relatively high strength and high anisotropy, but during the diapir stage these will be drastically reduced due to boudinage and separation of stronger sedimentary rock layers.

## Strain rates

Stress, temperature (Carter & Heard 1970), water content (Spiers *et al.* 1982), and grain size (van Keken *et al.* 1993) are the main controlling factors on local strain rates in salt. However, the time-averaged vertical strain rate of the salt is usually equal to the overburden aggradation rate in pure downbuilding situations. Occasionally, the salt extrudes to the surface, and namakier strain rates can dramatically increase to several metres per annum (Talbot & Jarvis 1984).

## Active versus passive diapirism

There has been a long-standing debate on the relative importance of active diapirism, and whether it is possible for diapirs to pierce overburden thicknesses of 1 km or more by buoyancy alone. Vendeville & Jackson (1992) indicate that this is highly unlikely as the strength of the sediments is too great to allow diapir-related faulting; and they suggest that many diapirs are initiated by thin-skinned extension or erosion of the overburden. However, once diapirs become established, with vertical reliefs of 4 km or more, then theoretically, an appreciable buoyancy force is developed (5–20 MPa, **Davison *et al.***). The role of active buoyant rise of salt in extensional provinces has been clearly demonstrated in Israel (**Frumkin**) and Yemen (**Davison *et al.***). Both of these studies measured diapir uplift rates of 1–5 mm $a^{-1}$. Sedimentary rocks deposited at sea level have been uplifted at a rate of

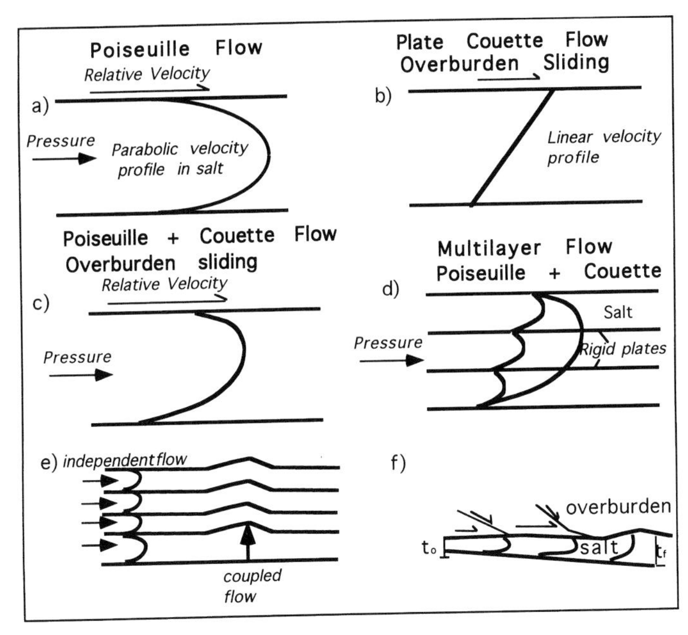


**Fig. 6.** Schematic velocity profiles. (a) Poiseuille flow in a single thick layer. (b) Plate Couette flow produced by overburden sliding. (c) Combination of Poiseuille and Couette flow, which probably occurs in most salt layers where overburden is extended or contracted. (d) Poiseuille and Couette flow in a multilayer of equivalent thickness. (e) Layers react together during movement perpendicular to layering. (f) Drag of salt (original thickness $t_0$) down a detachment horizon to produce thicker salt (final thickness $t_f$) at the down-dip end of the slide.

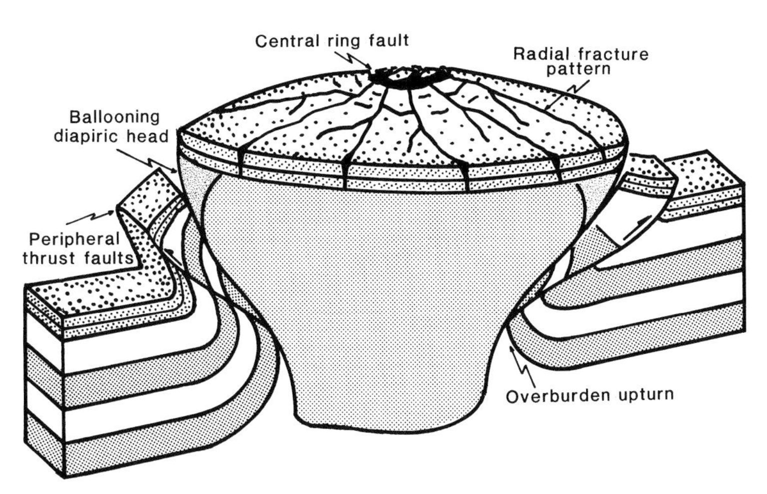


**Fig. 7.** Schematic fault and fracture patterns produced in overburden layers above a diapir. Based on observations of physical models and seismic interpretation.

6 mm a$^{-1}$ in Israel (**Frumkin**) and up to 17 m above sea level in the last 3750 years in Yemen (**Davison *et al.***). Salt diapirs act as tectonic pressure gauges (**Talbot & Alavi**). The effective buoyancy force can be measured by the heights of diapiric freeboards above a fixed reference level (e.g. sea level). Salt buoyancy is usually only capable of uplifting sediment to a maximum of 300 m above a regional

datum level in most extensional settings (see table 1 in **Davison *et al.***). This indicates a minimum buoyancy force of approximately 7.5 MPa (for an overburden sediment density of 2500 kg $m^{-3}$). Hence, buoyancy forces may be strong enough to break through thin (approximately 50 m) covers of weak sediments and sedimentary rocks, by inducing faulting or pervasive grain flow. In the active contractional setting of the Zagros mountains diapirs produce salt fountains up to 1.5 km above regional level, indicating there is a large component of horizontal tectonic stress of at least 25 MPa which is driving the salt upward. This amount of stress could induce appreciable tectonic thinning by viscous deformation of the overburden. The main mechanisms of overburden creep would be grain boundary sliding, pressure solution, and compaction. This may allow salt to move up through overburden over long time scales, by physically pushing material aside or by volume reduction, without major faulting occurring. For example, Langheinrich & Plessman (1968) indicate that Late Cretaceous chalk overlying a diapir in Germany has been thinned by up to 28% by pressure solution with development of stylolites.

## Overburden deformation

The exact mechanism by which pillows become intrusive (diapiric) is poorly documented, but usually involves faulting and tectonic attentuation of the overburden. Recent modelling experiments (**Alsop, Szatmari *et al.***) indicate that pillow crests are often cusp-shaped and intervening synclines are lobe-shaped due to the high ductility contrast at the folding salt interface (Biot 1964) and also due to listric faulting. These cusps may be present on the top of many diapirs but they are not visible on seismic data due to their steep sides. Cusps produce intense extensional deformation in the surrounding overburden, which can be used by salt to inject upwards into faults. Breaking the lid above a cusp allows the beds to flex upwards in a broken beam, where the shape of the bending depends on the flexural rigidity of the overburden (Schultz-Ela *et al.* 1993). Tighter bends will be produced by weaker overburden. Once the overburden is broken by faults and tensional fractures it is possible for the salt to migrate upwards. However, the exact mechanism of upward movement of the salt is still poorly documented. Normal faulting is produced in thin overburdens above natural salt diapirs indicating that extension is dominant over diapir roofs and salt may move upward by pushing the overburden aside (**Davison *et al.***). Modelling studies predict that extension in the overburden at the diapir crest will produce a radial fault pattern with subordinate ring faults (Fig. 7, **Alsop**). Thrust faults can be developed at the outside edge of the diapiric head (Davison *et al.* 1993, Schultz-Ela *et al.* 1993, **Alsop**, and Fig. 7).

## Salt welds and dissolution

Fluid dynamics indicates that flow in a salt layer is inversely proportional to the thickness of the layer raised to the third power (e.g. **Hardy & Cohen**). Hence, it is theoretically impossible to completely remove a salt layer (salt weld) by deformation alone, as the drag resistance exerted on the layer is much greater than any known tectonic or buoyancy driving force. Dissolution must take place to produce total removal of the salt.

A typical example of a salt weld observed on reflection seismic data was penetrated by the Phillips well 47/4a-3 in the southern North Sea. It encountered 60m of Zechstein rock which was predominantly competent carbonate and anhydrite of all four cycles recognized in this area. Two thin halites are present. In the 4th cycle 7m of Roter Salzton was found, and a further 4m was discovered in a deeper cycle (M. Lappin, pers. comm). Estimates of salt dissolution in cross sectional reconstructions of the central North Sea indicate values varying between 50% to 0% of the original volume depending on the assumptions made in the balancing procedure (cf. Allan *et al.* 1994, and **Buchanan *et al.*** respectively). This serves to show the inherent problems which are involved in balancing sections of salt tectonics.

## Salt triggering mechanisms

Although there is empirical evidence that some salt dome provinces have not started to deform until an appreciable amount of sediment has been deposited (e.g. Jenyon 1986), it is now accepted that a density inversion is not necessary to promote vertical movement of salt (Jackson & Vendeville 1994, **Cohen & Hardy**). Overburden can be less dense than the salt horizon and upward vertical movement of salt can still take place (Vendeville & Jackson 1993, Jackson & Vendeville 1994). **Hoyos *et al.*** indicate that vertical pillow growth initiated with no overburden present at the crest of pillows in the Madrid basin, and there is no minimum thickness of sediment required to initiate halokinesis. Halokinesis may be triggered by either compression, as has been suggested for the Tertiary pillow growth in the southern North Sea (Stewart & Coward 1995), by regional extension of overburden (Jackson & Vendeville 1994), by erosion (Evans 1993), or differential loading (**Cohen & Hardy**). Basement faults do not necessarily trigger diapirs even when they offset the salt layer with a throw greater than the salt thickness (Koyi 1991; Jackson & Vendeville 1994).

## Physical modelling of salt diapirism

Recently, analogue modelling studies, usually using sand and viscous polymer, have provided important new insights into the geometry of salt diapirs and the conditions under which diapirism will occur, especially in extensional situations (e.g. Vendeville & Jackson 1992; Jackson & Vendeville 1994). Modelling has been extremely successful in producing a range of geometrical solutions to brittle-viscous deformation problems. However, sedimentary rocks (up to 3000 m thick) with extremely variable lithology will not deform by the same mechanisms as a 1–2 cm thick sandpack, typically used in experimental modelling. Over time periods of 250 Ma (e.g. since deposition of NW European Zechstein salt) overburden sediments may deform in a non-brittle way so that diapirs may slowly develop, without regional tectonics, erosion, or differential loading. It is apparent that more work is required on the rates and mechanisms of creep deformation of overburden sedimentary rocks and its relationships to salt tectonics.

Since the pioneering work of Parker & McDowell (1955), the majority of experiments involving granular overburdens have used sand to simulate the sedimentary overburden, despite the sand scaling up to be too strong to represent shallowly-buried sediment (Parker & McDowell 1955, page 2401). Hence, many modelled diapirs fail to display the viscous folding geometries characteristic of overburden deformation surrounding natural diapirs (Ratcliffe *et al.* 1992). The introduction of new synthetic modelling materials such as glass beads, may simulate deformation of overburden more closely, resulting in more realistic diapiric models.

## Future research

Great advances have been made in the understanding of salt tectonics in recent years principally using improved seismic data and modelling. In areas of good seismic velocity control, even salt overhangs can be imaged using turning wave migration (Ratcliffe *et al.* 1992). 3D seismic data acquisition over exploration propects is now routine and has led to a much more detailed understanding of how salt behaves in sedimentary basins. More detailed 3D data acquisition from new areas of salt tectonics will hopefully increase the level of understanding and reveal new unexpected tectonic features (such as the Sigsbee Scarp discoverd in the 1970s). Recent physical modeling of salt diapirs has been mainly limited to isotropic sand overburdens and isotropic viscous polymers and putties. More modelling work is required using a variable range of substances to test the effects of differing amounts of anisotropy in the salt and the overburden layers. Numerical modelling is reaching the stage of sophistication where it is now possible to model brittle and viscous deformation in parallel, so that the notional material can effectively 'decide' depending on the conditions, by which mechanism it deforms. This will lead to more realistic numerical models of salt tectonics, which are much easier to run than physical analogue models.

Very few detailed outcrop studies have been undertaken on salt structures. This is mainly due to salt exposures being normally restricted to hot and cold desert climates, in remote and often politically unstable countries. However, future work on salt outcrop studies will enable more detailed analysis of the nature of overburden and salt deformation and answer questions such as how does salt actually penetrate through overburden and how much viscous and/or plastic deformation occurs in overburden rocks, as compared to the more familiar brittle faulting. To conclude, there are a great deal of unknown aspects of salt tectonics. Fortunately, there are now new tools which can be employed to increase knowledge, and a strong driving force from the energy and waste disposal industry which requires this better understanding of salt tectonics.

Most of the papers in this volume were presented at a meeting of the Petroleum and Tectonics Groups of the Geological Society on 14–15th September 1994, and we would like to thank authors for the use of some of their ideas in this introduction. We would also like to thank the following companies for sponsoring salt research at Royal Holloway: Amoco (UK), Conoco (UK), Enterprise Oil, Hardy Oil and Gas, Phillips (UK), Ranger Oil (UK), Santos (Europe), & Saxon Oil. Max Harper ran the model shown in Fig. 4.

## References

ALLAN, P., ANDERTON, R., DAVIES, M., MARSHALL, A., POOLER, J. & VAUGHAN, O. 1994. Structural development of the ETAP diapirs, central North Sea. *Salt Tectonics meeting, Geological Society, London, 14–15th September 1994*, abstract.

BIOT, M. A. 1964. Theory of viscous buckling and gravity instability of multilayers with large deformation. *Bulletin of the Geological Society of America*, **76**, 371–378.

CARTER, N. L. & HEARD, H. L. 1970. Temperature and rate-dependent deformation of halite. *American Journal of Science*, **269**, 193–249.

COLLIER, J. S. & WHITE, R. S., 1990. Mud diapirism within Indus Fan sediments: Murray Ridge, Gulf of Oman. *Geophysical Journal International*, **101**, 345–353.

Davison, I., Insley, M., Harper, M., Weston, P., Blundell, D., McClay, K. & Quallington, A. 1993. Physical modelling of overburden deformation around salt diapirs. *Tectonophysics*, **228**, 255–274.

Evans, R. 1993. Lowering of sea level as a cause of initiation of diapirism of salt. *American Association of Petroleum Geologists, Hedberg Research Conference on Salt Tectonics, Bath, September 13th–17th 1993*, abstract.

Hospers, J., Rathore, J. S., Jianhua, F., Finnstrøm, E. G. & Holthe, J. 1988. Salt tectonics in the Norwegian-Danish Basin. *Tectonophysics*, **149**, 35–60.

Hughes, M. & Davison, I. 1993. Geometry and growth kinematics of salt pillows in the southern North Sea. *Tectonophysics*, **228**, 239–254.

Jackson, M. P. A. & Seni, S. J. 1984. *Atlas of Salt Domes*. Report of Investigations no. 140. Bureau of Economic Geology, University of Texas at Austin.

—— & Vendeville, B. C. 1994. Regional extension as a geologic trigger for diapirism. *Bulletin of the Geological Society of America*, **106**, 57–73.

Jenyon, M. K. 1986. *Salt Tectonics*. Elsevier, Barking.

Koyi, H. 1988. Experimental modelling of the role of gravity and lateral shortening in Zagros Mountain Belt. *AAPG Bulletin*, **72**, 1381–1394.

—— 1991. Gravity overturns, extension, and basement fault reactivation. *Journal of Petroleum Geology*, **14**, 117–142.

Langheinrich, G. & Plessman, W. 1968. Zue Enstehungsweise von Schieferungs-Flächen in Kalkstein (turon-kalke, eines Salzauftreibs-Sattels im Harz-Vorland). *Geologische Mitteilungen*, **8**, 111–142.

Nalpas, T. & Brun, J. P. 1993. Salt flow and diapirism related to extension at crustal scale. *Tectonophysics*, **228**, 349–362.

Parker, J. M. & McDowell, A. N. 1955. Model studies of salt dome tectonics. *AAPG Bulletin*, **39**, 2384–2470.

Ratcliffe, D., Gray, S. H. & Whitmore, N. D., Jr 1992. Seismic imaging of salt structures in the Gulf of Mexico. *The Leading Edge*, **11**, 15–33.

Renzhiglov, N. F. & Pavlishcheva, T. V. 1970. On the viscosity of rocks. *Soviet Mineral Science*, **5**, 582–585.

Rönnlund, P. 1989. Viscosity ratio estimates from natural Rayleigh-Taylor instabilities. *Terra Nova*. **1**, 344–348.

Schultz-Ela, D., Jackson, M.P.A. & Vendeville, B.C. 1993. Mechanics of active salt diapirism. *Tectonophysics*, **228**, 275–312.

Spiers, C. J., Lister, G. S. & Zwart, H. J. 1982. *The Influence of Fluid-Rock Interaction on the Rheology of Salt Rock, and on Ionic Transport in the Salt: First Results*. European Atomic Energy Community Publication, WAS-153-80-7N, 268–280.

——, Schutjens, P. M. T. M., Brzesowsky, R. H., Peach, C. J., Liezenberg, J. L. & Zwart, H. J. 1990. Experimental deformation of constitutive parameters governing creep of rocksalt by pressure solution. *In*: Knipe, R. J. & Rutter, E. H. *Deformation Mechanisms, Rheology and Tectonics*. Geological Society, London, Special Publication, **54**, 215–227.

Stewart, S. & Coward, M. P. 1995. A synthesis of salt tectonics in the southern North Sea. *Marine and Petroleum Geology*, **12**, 457–476.

Talbot, C. J. & Jackson, M. P. A. 1987. Internal kinematics of salt diapirs. *AAPG Bulletin*, **71**, 1068–1093.

—— & Jarvis, R. J. 1984. Age, budget and dynamics of an active salt extrusion in Iran. *Journal of Structural Geology*, **6**, 521–533.

Terrinha, P., Ribeiro, A. & Coward, M. P. 1994. Mesoscopic structures in a salt wall. The Loule salt wall diapir, Algarve Basin, south Portugal. *Salt Tectonics meeting, Geological Society, London, 14–15th September 1994*, abstract.

Tharp, T. M. & Scarborough, M. G. 1994. Application of hyperbolic stress-strain models for sandstone and shale to fold wavelength in the Mexican Ridges Foldbelt. *Journal of Structural Geology*, **16**, 1603–1619.

Trusheim, F. 1960. Mechanism of salt migration in northern Germany. *AAPG Bulletin*, **44**, 1519–1540.

Van Keken, P. E., Spiers, C. J., Van den Berg, A. P., & Muyzert, E. J. 1993. The effective viscosity of rocksalt: implementation of steady-state creep laws in numerical models of salt diapirism. *Tectonophysics*, **225**, 457–476.

Vendeville, B. C. & Jackson, M. P. A. 1992. The rise of diapirs during thin-skinned extension. *Marine & Petroleum Geology*, **9**, 331-353.

—— & —— 1993. Some dogmas in salt tectonics challenged by modelling. *American Association of Petroleum Geologists, Hedberg Research Conference, Salt Tectonics, Bath, September 13-17th 1993*, abstract.

# Kinematics within the Kłodawa salt diapir, central Poland

STANISŁAW BURLIGA

*Institute of Geology, Adam Mickiewicz University, ul. Maków Polnych 16, 61–686 Poznań Poland*

**Abstract**: The Kłodawa salt diapir consists of deformed Zechstein rock salts, with subordinate anhydrite, dolomite and shales. Upward movement of these rocks and the structural evolution of the diapir started in the Triassic and continued at least until the Tertiary. Lithological, and thus rheological variation resulted in the development of a variety of tectonic structures on different scales, which can be observed in mine galleries intersecting the diapir at several levels. The rocks are folded, faulted, fractured and boudinaged, and their behaviour is strictly related to the competence contrasts between layers of different lithology. There is a range of competence from dolomite through anhydrite, shale, clayey salt and rock salt, to the least competent, potassium salt. Kinematic indications at level 600 are provided by boudins, book-shelf structures, asymmetric folds, small-scale faults, tension gashes and veins. These were produced during different stages in the evolution of the diapir and provide evidence for various mechanisms of deformation, ranging from ductile flow to brittle fracturing. The small-scale structures allow the visualization and identification of larger folds and the depiction of their development. They record changes of shear directions along the fold limbs, and also indicate a change in folding mechanism from flexural-slip folding to shear folding. High strain is localized in bedding-parallel zones and within areas dominated by low competence rocks. These zones usually form discrete horizons separating domains of more competent rocks, which show little evidence of deformation and, occasionally, undeformed synsedimentary structures (dessication polygons). The displacements were zonal, accommodated by low competence rocks, with more competent rocks passively transported as rafts within the former. The zones of intense transport are localized within beds of rock salt and potassium salt.

Geophysical prospecting in the northwestern part of Poland revealed a great number of Zechstein salt structures occurring within or under thick (up to several thousand metres) sequences of Mesozoic and Cenozoic deposits (e.g. Dadlez & Marek 1969; Pożaryski 1977). They have different geometries, corresponding to different stages of evolution – from salt pillows to diapirs and salt walls with associated gypsum–clayey caps reaching the surface. One of the largest structures is the Izbica Kujawska–Kłodawa–Łęczyca salt ridge, located within the Danish–Polish Trough in central Poland (Fig. 1). This is a northwest – trending salt ridge more than 60 km long and 0.5–2 km wide (width of caprock). It can be subdivided into three parts – northern, middle and southern – of different heights and different types of contacts with the surrounding deposits. The northern and southern parts of the ridge do not pierce through the Mesozoic sequence but only buckle it into large anticlines separated by the middle part, about 30 km long which does pierce through the Mesozoic cover. The middle part is also the highest part of the ridge and is referred to as the 'Kłodawa salt diapir' (Fig. 1a). The upper part of the diapir is geologically best known, being accessible because of active salt mine excavations which provide basic information about the structure.

This paper presents an attempt at kinematic analysis of the diapir, based on small-scale structures. This is a preliminary study as data was collected only from one level of the mine (level 600).

## Geological setting

The general architecture of the Kłodawa diapir was described by Poborski (1955, 1957) and Werner *et al.* (1960). Their interpretation, based on observations from the mine, was also furnished with subsurface well-log and geophysical data. According to them, the diapir rises from a depth of more than 6000 m up to several tens of metres below ground level in the highest places. It is an asymmetrical feature with the eastern side dipping moderately eastward and the western side almost vertical and locally inverted, dipping in the uppermost part towards the NE (Fig. 2). The shape of the western side of the ridge is probably due to a fault occurring in the basement on this side of the ridge (Figs. 1a, 2; Dadlez & Marek 1969, 1974; Pożaryski 1977; Dadlez 1994), which was active during the development of the ridge. The Permian–Mesozoic complex is downthrown along the western side of the ridge, relative to its eastern side, and the stratigraphic thicknesses of particular formations differ on both sides of the ridge (Fig. 2). The diapir consists of Zechstein evaporites and terrigeneous

*From* ALSOP, G. I., BLUNDELL, D. J. & DAVISON, I. (eds), 1996, *Salt Tectonics*, Geological Society Special Publication No. 100, pp. 11–21.

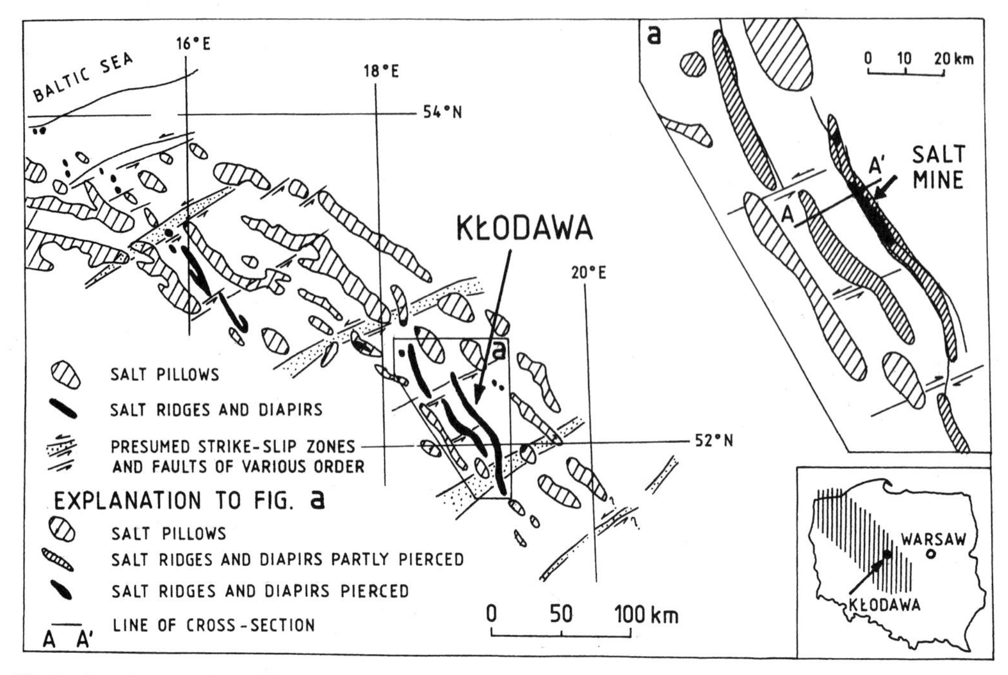


**Fig. 1.** Location map of the Kłodawa salt diapir showing general tectonic features of the Polish Trough (based on Dadlez 1994). A–A', location of Fig. 2.

rocks, topped by a gypsum–clayey cap of varying thickness (on average 100–150 m), overlain in turn by Tertiary and Quaternary deposits The detailed lithostratigraphic column of the Zechstein sediments is shown in Fig. 3. It is very similar to the typical Zechstein profile of Central Europe, with distinguishable counterparts of cyclothems from Z2 to Z4 and probably also Z1 (Charysz 1973). The older stratigraphical units occur in the central part of the diapir and they pierce through the younger ones.

The diapir is composed of bedded rock salt, clayey salt and subordinate beds of anhydrite, dolomite, shale and potassium salt. All rocks are laminated. Stratigraphical thicknesses and alternations of the rock types vary considerably (see Figs. 4, 5) and this variation was additionally modified by tectonics. The rocks became deformed during the evolution of the diapir (from the Triassic until at least the Tertiary) and at present they are steeply dipping, being folded, faulted or boudinaged. The type of deformation depends on the competence contrast of a particular rock with respect to its immediate surroundings. Based on the competence, the rocks can be arranged from the most competent dolomite, through anhydrite, shale, clayey salt and

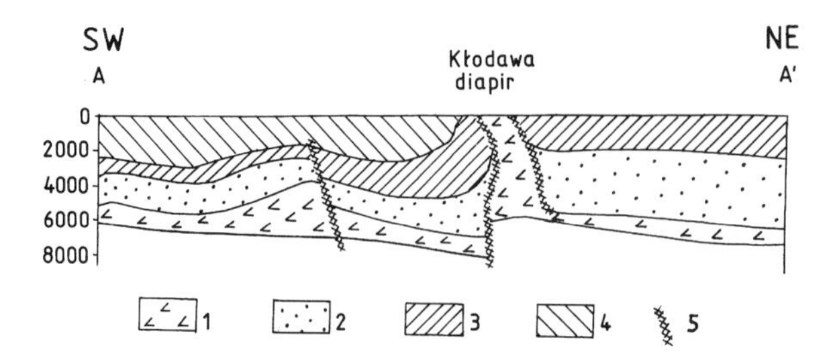


**Fig. 2.** Generalized geological cross-section through the Polish Trough in the vicinity of Kłodawa. See Fig. 1 for location (based on Dadlez & Marek 1969). 1, Zechstein, 2, Triassic, 3, Jurassic, 4, Cretaceous, 5, major faults.

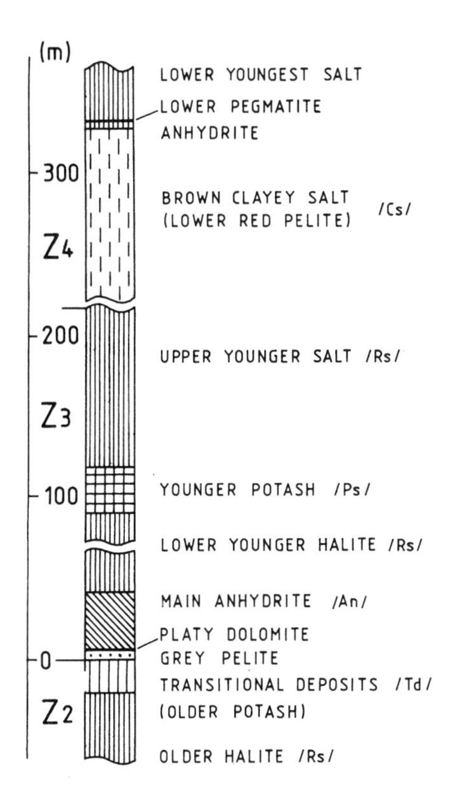


**Fig. 3.** Zechstein lithostratigraphical column for the Kłodawa salt diapir (based on Charysz 1973).

rock salt, to the least competent potassium salt. Alternations and combinations of lithological variants modify the competence of various multilayer sets, which is especially important for rock salt. Any admixture of clay or anhydrite within rock salt raises its competence relative to pure rock salt.

As a result of these lithological and rheological variations, a lot of tectonic structures were formed during the development of the diapir and are now preserved and observable within it. They developed at different stages of its evolution and have different dimensions, ranging from microscopic to hundreds of metres across (e.g. Figs 5, 6 and 7). These structures allow the reconstruction, at least in part, of the kinematics of the diapir.

## Methods

Kinematic analysis of the diapir in question is based on small-scale structures. Data were collected from level 600 of the mine, which has the most extensive net of galleries. All the small-scale structures occurring at this level have been mapped; They consist of boudins, book-shelf structures, asymmetric folds, small-scale faults, veins and tension gashes. Their presence, geometry and dimensions are determined by the thickness and competence contrast of beds in a given multilayer set together with the overall tectonics. As a result of the varying thicknesses and compositions of the strained layers, and the limited width and height of galleries (on average 2–4 m and 2–3 m, respectively), it is difficult to use the structures for measurement of the spatial distribution of the state of strain throughout the diapir. However, the sense and mode of displacements and their succession in time may be estimated, although this is possible only in the case of small-scale structures developed during late deformational increments. The older structures have been preserved only in exceptional cases.

As particular lithological varieties show different deformation features, each of them will be discussed separately. It is also necessary to divide the analysed part of the diapir into two domains, northern and southern, characterized by different tectonic styles (Fig. 5).

## The Northern domain

The northern domain of the diapir (Fig. 5) is characterized by beds striking parallel to the ridge boundaries, typically NW–SE, and by only slight folding. The dip of the beds is 75° (and steeper), and dominantly towards the NE. The beds generally represent a complete stratigraphic column, except in a very few places, with older units occurring in the central part of the diapir and younger units in its outer parts. All the lithological varieties form distinct horizons separated by sharp sedimentary boundaries and their characteristics are given below.

One of the most prominent features in the northern area is the presence of a boudinaged anhydrite bed together with adjacent beds of dolomite and shale. These rocks are of the highest competence. Although they display similar amounts of strain, judged from similarities of boudinage style in

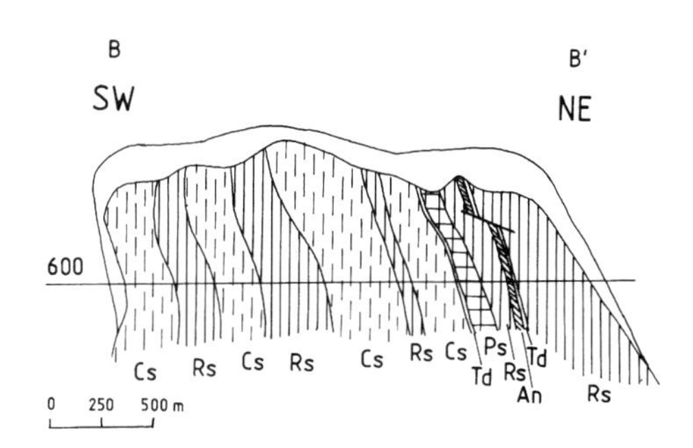


**Fig. 4.** Geological cross-section through the uppermost part of the diapir (based on Tarka 1992). See Fig. 5 for location. Rs, rock salt; Cs, clayey salt; Ps, potassium salt; An, anhydrite, dolomite, shale; Td, transitional deposits.

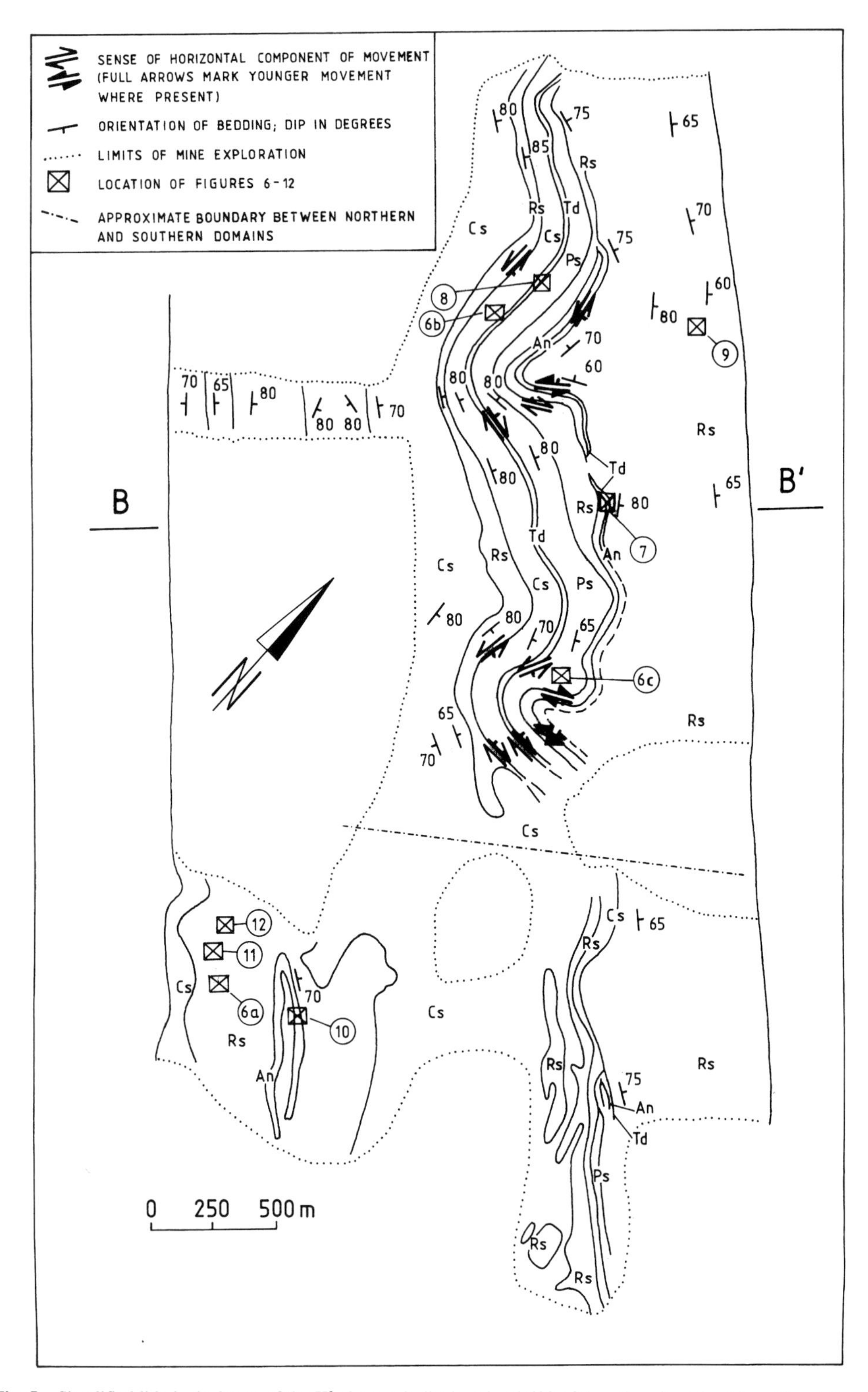


**Fig. 5.** Simplified lithological map of the Kłodawa salt diapir at level 600 of the mine. See Fig. 4 for abbreviations of rocks. Major vertical component of displacement is marked on the arrows showing the sense of horizontal component of movement with a tick on the down-side. B–B', location of Fig.4.

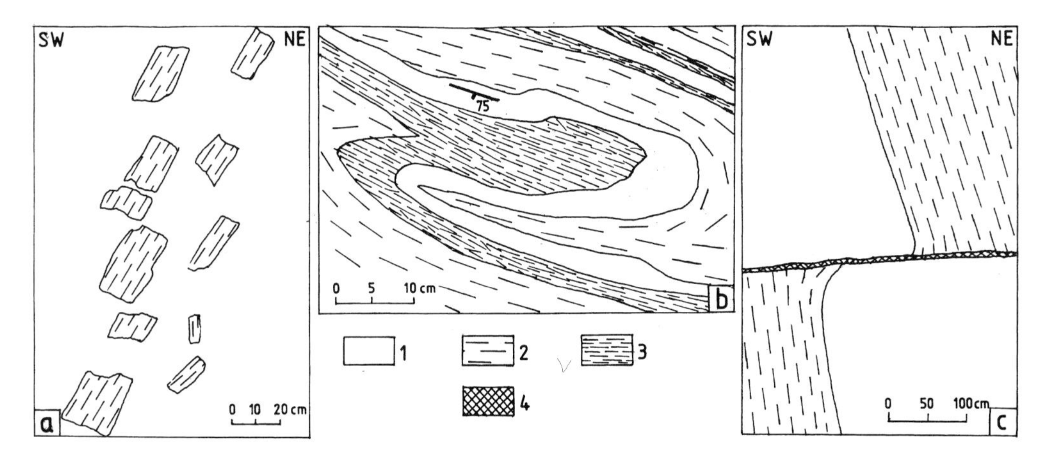


**Fig. 6.** Examples of small-scale structures used for kinematic analysis. See Fig. 5 for location. (**a**) Boudinaged (book-shelf structure) layer of clayey salt within rock salt; (**b**) small parasitic fold marked by relatively competent clayey salt layer; (**c**) small-scale fault shifting bed of laminated clayey salt. 1, rock salt; 2, clayey salt; 3, clays/shale; 4, potassium salt.

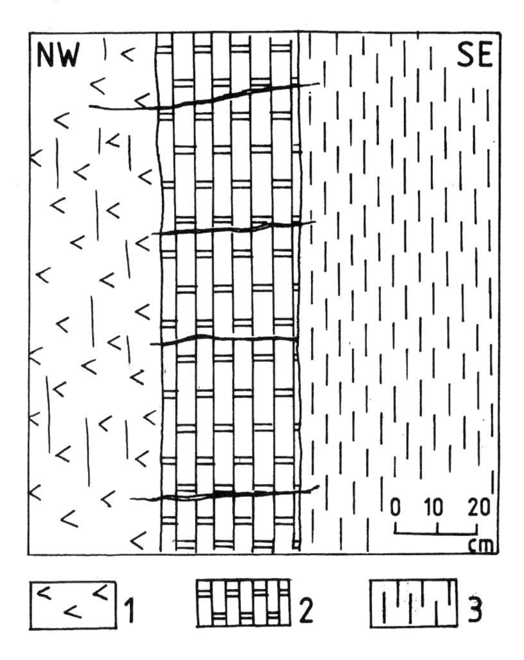


**Fig. 7.** Boundary zone of anhydrite, dolomite and shale beds in the northern domain. Further description is given in the text. See Fig. 5 for location. 1, Anhydrite; 2, dolomite; 3, shale.

competent beds and considered at a map scale, the deformation styles clearly differ between these rocks. The thickest (40–50 m) bed of anhydrite developed only large-scale boudins (see Fig. 5). Almost no deformation is visible inside these boudins. The strain is demonstrated by rare tension fractures, and by veins filled with epigenetic salts of different types. The anhydrite is finely laminated and this lamination is occasionally gently buckled at the boundary zone with the dolomite. The dolomite bed is about 40 cm thick, and is also boudinaged on a smaller scale (Fig. 7). It appears in the diapir as an array of boudins, less than 1 m long in their greatest dimensions. The boudin neck zones are narrower than 10 cm and are filled with epigenetic salts (halite or potassium salts). The geometry of the boudins and their orientation is not known precisely, but their longest dimensions are observed on the walls of the galleries. The calculated value of extension $e$ is in the range of 0.02 to 0.05. The 5 m thick bed of shale is deformed only in a zone up to 1m thick adjacent to the dolomite. There the shale is strongly fractured into rhombohedral fragments (less than 15 cm long), while the rest of the bed is finely laminated and undisturbed. Some of the fractures are filled with epigenetic salts. There are no other kinematic indicators in these lithologies except boudins or veins. These indicate stretching parallel to the bedding, but it is not possible to interpret precisely the stretching direction, because of the lack of striations and the difficulties in determining the three-dimensional geometry of the boudins. Judging from the veins (fibres of salt filling the veins are occasionally buckled, suggesting changing orientations of the strain ellipsoids), the principal stretching direction seems to be concordant with the dip of the beds, thus, indicating subvertical stretching.

Similar slight tectonic deformation is observed in clayey salt and in deposits transitional to pure rock salt or potassium salt. The term 'transitional deposits' is used for numerous alternations of relatively thin layers of shale, clayey salt and rock salt with an admixture of anhydrite or clay and, in some layers, with fine laminae of potassium salt. The

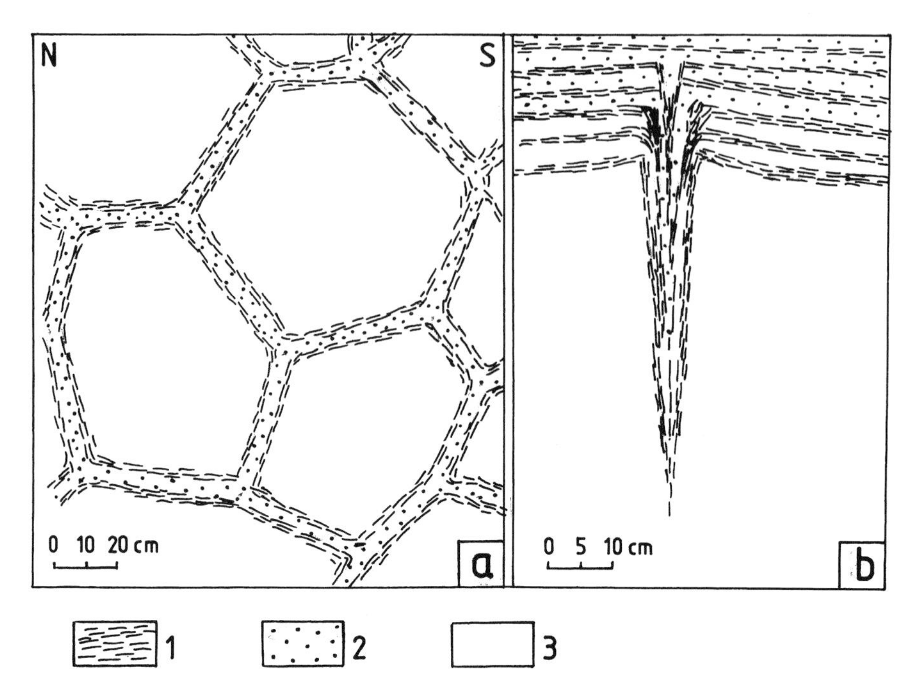


**Fig. 8.** Desiccation polygons in rock salt. (**a**) planar view; (**b**) bedding-perpendicular cross-section of individual desiccation fissure. See Fig. 5 for location. 1, clays; 2, pinkish rock salt; 3, greyish rock salt.

thicknesses of these layers vary: layers of rock salt and clayey salt have thicknesses ranging from 3 cm to more than 1 m, while other variants are in the range of only a few centimetres. The best evidence for there being only a weak tectonic effect on the transitional deposits comes from the presence of continuous horizons of desiccation polygons. These polygons are most often hexagonal or pentagonal in the bedding plane and their sides are usually in the range of 0.7–1.5 m (Fig. 8). In vertical sections (relative to the bedding plane) they apear to be V-shaped wedge fissures, opening towards the stratigraphic top. They are generally perpendicular to the bedding, their length ranges between 0.7 and 1.2 m, and they are filled with clay or laminated clay and salt, the laminae of which are parallel to the walls of these fissures.

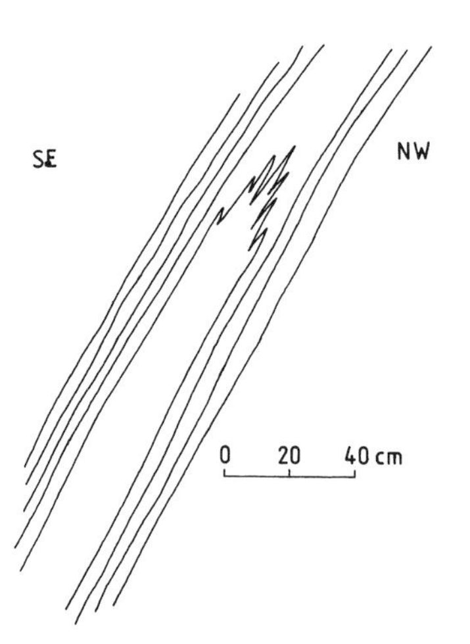


**Fig. 9.** Cross-sectional geometry of isoclinal fold typical of thick rock salt units in the northern domain. See Fig. 5 for location.

Clayey salt is almost homogeneous in composition, although layering occurs infrequently due to varying content of rock salt and clays. Strain in the clayey salt and transitional deposits is occasionally demonstrated by small parasitic folds, boudins, veins and small-scale faults (e.g. Fig. 6b). The dimensions of these structures differ and are determined, as was stated above, by the thicknesses of layers and the competence contrast between them. Their orientation depends on their position in relation to larger fold structures in the diapir (Fig. 5). Common features of the parasitic folds are steeply plunging axes and steeply dipping axial planes (65 –90°), with the trend depending on the geometry of large-scale folds containing them. The dip of the faults vary from subhorizontal to subvertical, depending on the location in the diapir. These

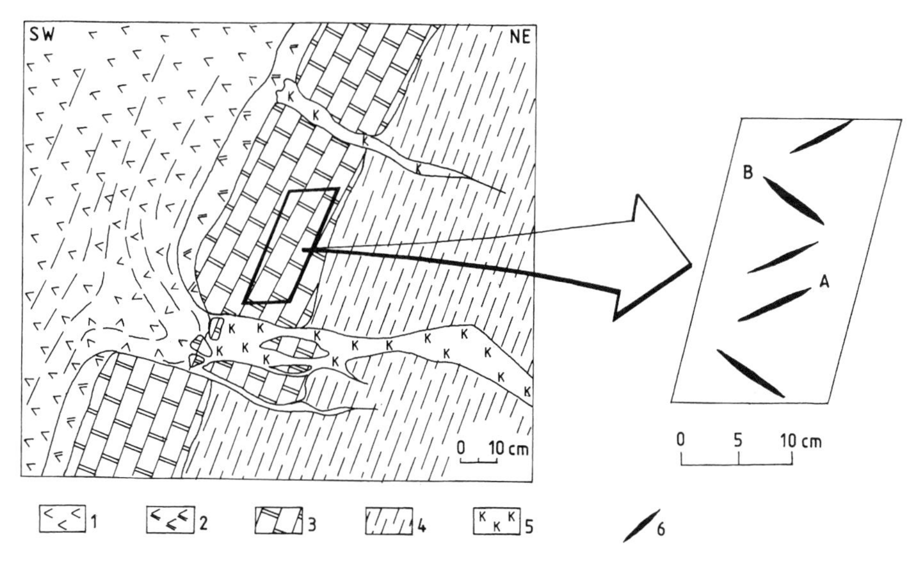


**Fig. 10.** Deformed dolomite bed in the limb of the isolated syncline in the northern domain containing sets of tension gashes. Details are given in the text. See Fig. 5 for location. 1, anhydrite; 2, anhydrite with admixture of dolomite and clays; 3, dolomite; 4, shale; 5, potassium salt; 6, tension gashes.

mesostructures occur only in some layers adjacent to the boundaries of the units and they indicate, as in the previous case, steep (60–90°), bedding-parallel, upward displacement. The portions closer to the centre of the diapir apparently moved more quickly towards the top than the outer portions of these lithological varieties.

Small-scale structures of all types are abundant in potassium salt. The unit of potassium salt is 50 m thick and contains interlayers of rock salt. The rock salt layers, being competent, became strongly folded, faulted, boudinaged or even distorted into chaotically dispersed fragments, reminiscent of melange. The attitude of the mesostructures varies in a similar way to the units of clayey salt and transitional deposits described above, which indicates the same type of bedding-parallel upward displacement.

Although the rock salt occurs in the diapir in very thick units (Fig. 5) and is faintly layered, it does not appear to contain any small-scale structures that could be used as kinematic indicators. The reason for this is its high mineralogical purity and, thus, its lack of markers (except for a faint lamination). The rock salt is, however, strongly folded. Isoclinal, folds with very high amplitude and small wavelength are the dominant fold type (Fig. 9). In these folds, layering is almost parallel in both limbs, even immediately adjacent to the hinge zones. The axial planes of these folds are parallel to the bedding and their dips, as well as the plunges of their axes, are steep. The occurrence of the folds resulted in an increase of the total thickness of the rock salt units, and thus the observed thicknesses are not true stratigraphic thicknesses.

There are two large asymmetrical anticlines in the northern domain (Fig. 5). They have nearly vertical axes and axial planes, and two sets of small-scale structures occur in their limbs. One of the sets suggests bedding-parallel, hingeward displacement along the outer portions of the folds (open arrows on Fig. 5), while the other set indicates similar movement along their inner portions (solid arrows on Fig. 5). Such kinematics would indicate strain partitioning and/or a change of folding mechanism within a fold stucture during progressive deformation from flexural-slip folding to shear folding. Parasitic folds in the limbs of these anticlines suggesting displacement by a shear folding mechanism, obliterate parasitic folds related to flexural-slip folding. This implies that shear folding is the younger phenomenon in these folds, and also suggests a short-lived 'kinematic memory' of the rocks forming the diapir.

## Southern domain

The southern domain (Fig. 5) of the diapir, in contrast to the northern one, is characterized by intensely folded beds on a mesoscopic scale. Another prominent difference is the lack of thick beds of potassium salt which occur as bedding-parallel veins rather than beds. The discrete beds of anhydrite, dolomite and shale form an isolated

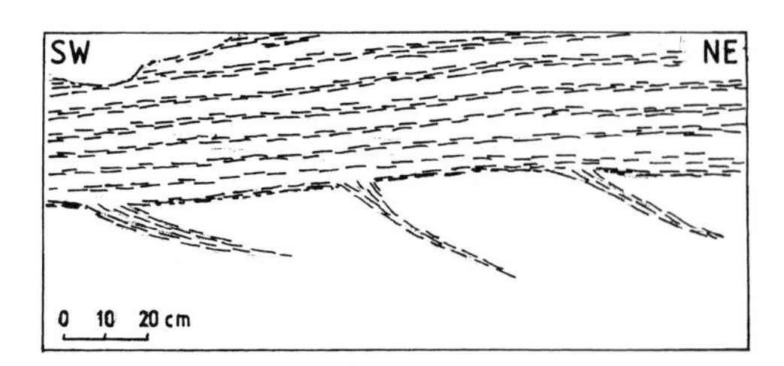


**Fig. 11.** Rotated desiccation fissures from the southern domain. See Fig. 8 for explanation of rock types, and Fig. 5 for location. Details are given in the text.

syncline (Fig. 5). Thicknesses of these beds are tectonically reduced in places but, in general, they are boudinaged, as in the northern domain. However, more intense strain is observable in the dolomite bed. The dolomite boudins have the same dimensions but their separation is wider, up to 0.5 m, and the largest of them are filled with anhydrite that has been plastically squeezed into the necks (Fig. 10). As in the northern domain, it is impossible to depict the exact geometry and orientation of the boudines but their longest dimensions are observable on the walls of the galleries. The calculated value of extension $e$ is in the range of 0.28–0.33, indicating on average 30% extension of the dolomite bed. Additionally, there are two sets of tension gashes in some parts of the dolomite bed filled with anhydrite, which form *en echelon* arrays (Fig. 10b). The gashes are straight, with sharp boundaries and opposite dips, which suggests that they formed in differently oriented shear conditions (Burliga 1994). The set labelled A on Fig. 10b suggests upward displacement along the inner side of the dolomite bed (i.e. the outer portion of the syncline) and in the set labelled B reverse movement. As in the case of the large folds in the northern domain, this situation implies a change of folding mechanism from a phase of dominant buckling and flexural-slip folding to a phase of dominant shear folding.

Small-scale structures are infrequently observed in clayey salt and transitional deposits. They do not form regular horizons as in the northern domain but are folded on a large scale and probably boudinaged. The desiccation polygons are strained in this part of the diapir: the fissures are rotated and are not perpendicular to the bedding planes but inclined at low angles (Fig. 11). Shear strain was calculated for these rotationally deformed fissures and the obtained value of shear strain $\gamma$ (gamma) varies from –1.8 to –4.1. However, these results were, obtained from only five structures, and thus there are too few to give full statistical characteristics of the shear strain.

Beds of anhydrite, shale and dolomite also occur as strongly boudinaged, brecciated or distorted irregular bodies within the rock salt. Their appearance and occurrence is linked with the refolding that affected the rock salt in this part of the diapir. The isoclinal folds which are like those typical of the rock salt in the northern domain, here have been refolded at least once around steeply plunging axes (Fig. 12). The plunge angle is in the range 55–75°. The dimensions of these folds are larger than the width of the mine galleries, and thus it is impossible to map them in detail. There are, however, a few places where their cross-sections are visible in three planes. The folds are oval or zig-zag in horizontal cross-section (normal to their axes), with diameters of the closed outcrop patterns in the range 10–30 m and amplitudes of at least several tens of metres (Fig. 12). These features are typical of sheath folds (Cobbold & Quinquis 1980; Ramsay & Huber 1993).

## Discussion

The presented description of mesostructures and their occurrences in different lithological units is very general, but it justifies the adopted approach of dividing the analysed portion of the diapir into two domains: the northern part, characterized by slight tectonic deformation, and the southern part, with intense deformation. Unfortunately, the net of galleries is too limited and the nature of the transition between the domains remains unknown. Therefore, it is impossible to decide whether the transition is continuous, or whether the two domains are separated by a fault. Another possibility is that the strongly strained southern area is typical of the innermost part of the diapir, which is not accessible in the northern domain and which is much thinner there. Whatever the nature of the

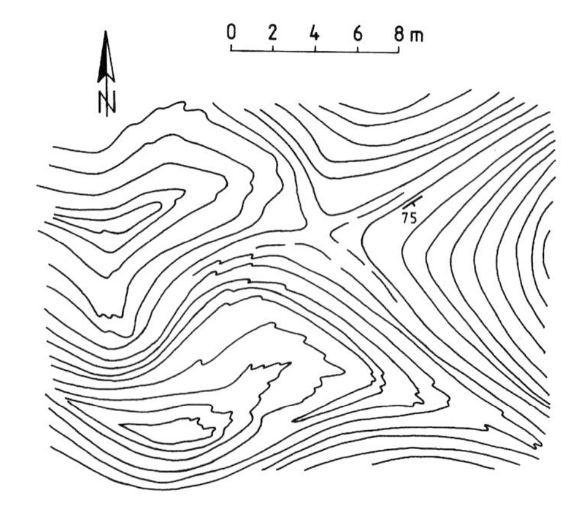


**Fig. 12.** Horizontal cross-section through folded rock salt beds; note closed outcrop pattern and earlier folding in later folds. See Fig. 5 for location.

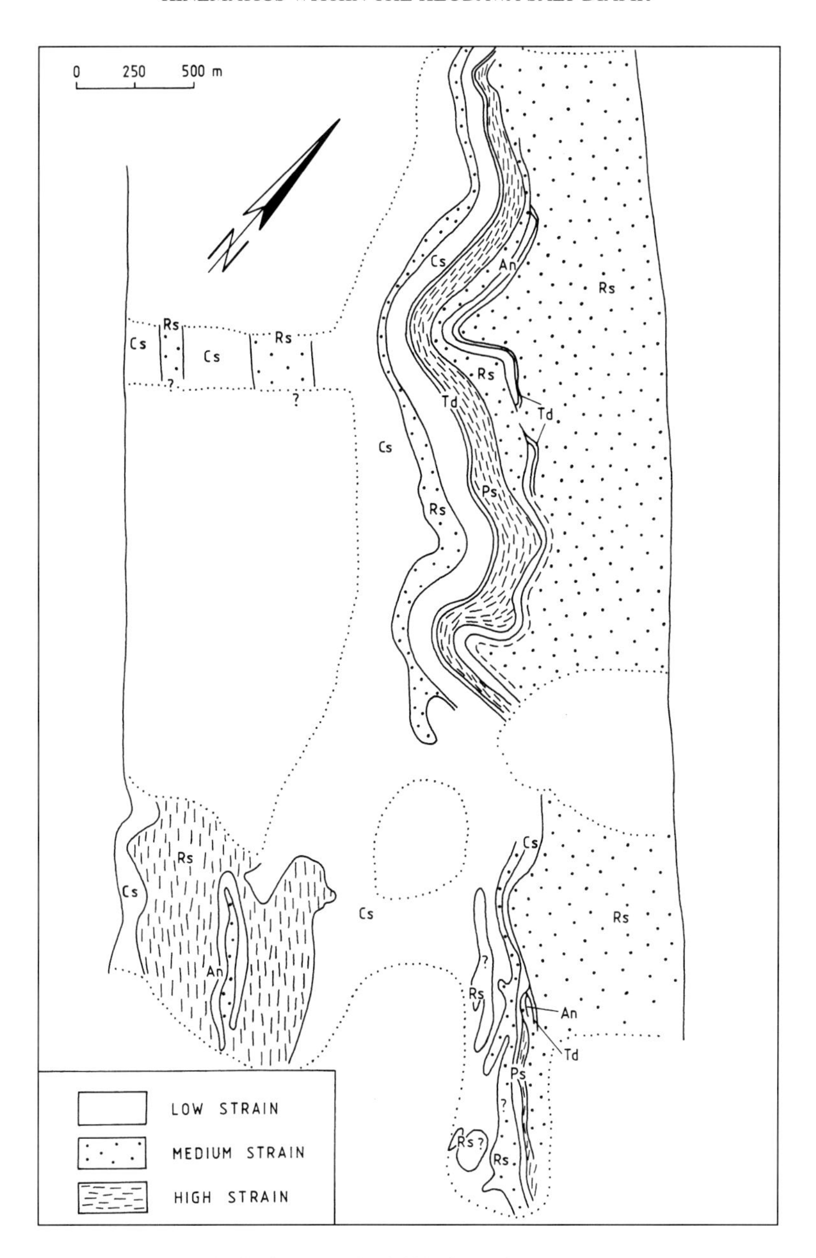


**Fig. 13.** Map of the qualitative strain distribution at level 600 of the salt mine.

transition between the two areas, the evident relation between strain and lithology is well pronounced in both domains. Numerous tectonic structures recording displacements occur within the potassium salt, and rather few appear in the transitional deposits, clayey salt, anhydrite, dolomite and shale. The problem with the clayey salt is that this lithological variety is almost homogeneous in composition and contains no sedimentary structures that could mark deformation. Thus, it is possible that these units are strained but because of the paucity of markers, this is not visible. The same is true for the rock salt; it suffered intense folding but the high mineralogical purity results in rare occurrences of megascopically recognizable internal structures (except for the folds).

The abundant occurrences of mesostructures in the potassium salt can not be explained and justified only on the grounds that the numerous alternations of rock salt and potassium salt layers favour development of these structures. Similar alternations of different lithological varieties occur in the transitional deposits and they demonstrate weak deformation and scarce mesostructures. The transitional deposits are, however, richer in more competent components than the potassium salt units, which are composed of the least competent rocks. Thus, the difference in bulk competence is the most likely reason for strain distribution.

On the basis of the given qualitative analysis, a map of strain distribution has been made (Fig. 13). It is only a schematic depiction of strain within the diapir but it clearly shows the occurrence of more and less strained discrete zones. The most peculiar are the zones containing very few tectonic structures, such as those with the desiccation polygons. The presence of portions in the diapir, despite nearly 6000 m of upward transport (the source layers are at a depth of about 6000 m), points to a passive mode of transport as rafts within the surrounding rocks. Thus, the movement was zonal, mainly parallel to the bedding planes, with the portions closer to the diapir centre moving faster than the outer parts. It was localized and accommodated in the most strained rocks, i.e. the potassium salt and rock salt, resulting in a system of variously spaced shear zones.

The calculated values of extension of the dolomite bed, based on boudins, show pronounced differences in the amount of strain between beds in the northern and southern domains. In the northern domain extension is relatively small, while in the southern domain up to 30% extension is indicated. These strain values, however, relate only to the bedding-parallel vertical direction in which the longest boudin dimensions were observable, and are obtained from 2–3 m long sectors of the beds, thus, they cannot be regarded as fully representative for the dolomite bed throughout the diapir.

It is difficult to establish the sequence of events for the whole diapir. It is possible to refer to particular structures within it, but without precise time correlation. The most spectacular features are the large folds. Each type seems to relate to stages in the development of the diapir. The large-scale isoclinal folds, which are typical of the rock salt in the northern part of the analysed area are interpreted as having developed according to the recumbent sheath fold model proposed by Talbot & Jackson (1987), and resemble the overlapping folds described by Richter-Bernburg (1980). They could have formed in the initial stage of diapirism, possibly in rocks that were still horizontally layered. Their development might progress during steepening of beds and upward movement. These folds were subsequently refolded, which resulted in a complex interference pattern, typical of the southern area of the diapir. Their closed or zig-zag outcrop patterns, and the very steep orientation of fold axes and sheath geometry indicate that their development may be linked with the upward movement of the salt.

## Conclusions

The differentially strained zones indicate that the Kłodawa salt diapir is tectonically heterogeneous. There are weakly and strongly strained zones with the amount of strain being dependent on the lithology. Rock salt and potassium salt are the least competent and the most strained, while dolomite and anhydrite are the most competent and the least strained rock types. The occurrence of undeformed synsedimentary structures (desiccation polygons) in some horizons indicates strain partitioning and the operation of passive flow of more rigid rock masses within rock salt and potassium salt. In the weakly deformed northern domain there is bedding-parallel subvertical displacement towards the surface, while in the southern domain, which is more strongly strained, complementary to this upward movement is the rotation of earlier formed structures around steeply plunging axes.

The collection of data for this work in the Kłodawa salt mine was made possible by the cooperation and assistance of the mine geologists; considerable thanks are due to Grzegorz Misiek, Piotr Kolonko, Andrzej Sobiś and Janek Chojnacki. Thanks are also due to Andrzej Żelaźniewicz for useful comments and stimulating discussions, the two referees, Maura Sans and Ian Alsop, who helped to improve the original version of this paper, and Derek Handley who kindly corrected the English.

## References

BURLIGA, S. 1994. Tension gashes in the Platy Dolomite (Zechstein) of the S-W part of the Kłodawa Diapir. *Przegld Geologiczny*, **2**, 99–102 (in Polish with English summary).

CHARYSZ, W. 1973. Cechsztyńskie piętro soli młodszych (Z3) w regionie kujawskim. *Prace Geologiczne Polskiej Akademii Nauk*, **75**, 7–60

COBBOLD, P. R., QUINQUIS, H. 1980. Development of sheath folds in shear regimes. *Journal of Structural Geology*, **2**, 119–126

DADLEZ, R. 1994. Strike-slip movements in the Polish Lowlands. *Geological Quarterly*, **38**, 307–318

—— & MAREK, S. 1969. Styl strukturalny kompleksu cechsztyńsko – mezozoicznego na niektórych obszarach Niżu Polskiego. *Geological Quarterly*, **3**, 543–561

—— & —— 1974. General outline of the tectonics of the Zechstein–Mesosoic complex in the Polish Lowlands. *Biuletyn Instytutu Geologicznego*, **274**, 111–140.

POBORSKI, J. 1955. Złoże solne w Kłodawie. *Biuletyn Instytutu Geologicznego,* 6–52.

—— 1957. Cechsztyńska struktura solna Izbica – Łęczyca. *Przegld Geologiczny,* **1**, 31–32.

POŻARYSKI , W. 1977. *Geology of Poland, Vol. 4, Tectonics.* Wydawnictwa Geologiczne, Warsaw.

RAMSAY J. G. & HUBER M. I. 1993. *The Techniques of Modern Structural Geology. Vol. 2: Folds and Fractures.* Academic, London

RICHTER-BERNBURG, G. 1980. *Salt Tectonics, Interior Structures of Salt Bodies.* Bulletin des Centres de Recherches Exploration – Production Elf – Aquitaine, **4**, 373–393.

TALBOT, C. J. & JACKSON M. P. A. 1987. Internal kinematics of salt diapirs. *AAPG Bulletin,* **71**, 1068–1091.

TARKA, R. 1992. Tektonika wybranych złóż soli w Polsce ma podstawie badań mezostrukturalnych (Tectonics of some salt deposits in Poland based on mesostructural analysis). *Prace Państwowego Instyutu Geologicznego,* **137**, 5–39 (in Polish with English summary).

WERNER, Z., POBORSKI, J. *et al.* 1960. Złoże solne w Kłodawie w zarysie geologiczno – górniczym. *Prace Państwowego Instytutu Geologicznego,* **7/3,** 467–497.

# Deformation and sedimentation around active Miocene salt diapirs on the Tihama Plain, northwest Yemen

IAN DAVISON[1], DAN BOSENCE[1], G. IAN ALSOP[1] & MOHAMMED H. AL-AAWAH[2]

[1]*Department of Geology, Royal Holloway, University of London, Egham, Surrey, TW20 0EX, UK*

[2]*Department of Geology, Sana'a University, Sana'a, Republic of Yemen*

**Abstract:** The Al Salif and Jabal al Milh salt diapirs of Miocene age cut through a 4–5 km thick overburden of Miocene to Recent sedimentary rocks in the southern Red Sea. The Al Salif diapir is part of a north–south oriented diapiric wall which has caused updoming of the overburden and active extensional faulting. The halite is folded into tight to isoclinal subvertical folds, with large-scale rafts of interbedded gypsum-anhydrite boudinaged on the fold limbs. There is no evidence for large-scale overburden stoping or injection of salt into overlying faults and fractures, and the upward movement of the salt dome is probably caused by the overburden being forced aside and sliding off the dome. The upper surface of the salt is an undulating smooth surface which was exposed at the sea-floor during the Quaternary. Coral reefs with $^{14}$C ages of 3700 a have been raised up to an elevation of 17 m above present-day sea-level, giving a surface uplift rate of 4.6 mm a$^{-1}$. At the Jabal Al Milh diapir, recumbent folding and thrusting are the main deformation features observed in the siltstones and gypsum layers of the overburden. The overburden has been rotated to the vertical at least 200 m from the diapir walls. The folding and thrusting is interpreted to be caused by overflow of salt in a namakier. This study shows that overburden sediments above the Al Salif diapir extended to allow sediments to slide from the salt dome crest by listric faulting, whereas sediments underlying the Jabal al Milh salt glacier were sheared and shortened in the horizontal direction.

Although the crests of many salt diapirs reach tantalizingly close to the Earth's surface, the geometry, nature and evolution of diapir roofs are poorly understood. A number of structural studies of diapiric roofs have been undertaken using seismic data (e.g. Jenyon 1988a, b), but the methods by which diapirs apparently intrude through overburden sediments is still poorly known. In order to further understanding of overburden deformation above diapirs, a case study of two well exposed structures in northwest Yemen was undertaken. The Al Salif diapir was intruded to the sea-bed and has updomed late Quaternary marine sediments by up to 45 m above sea-level so that they are now exposed onshore. The study documents and compares the range of structures produced by diapir upbuilding and by surface extrusion of salt. Improved seismic resolution has resulted in surface salt extrusions being more widely recognized, e.g. the Espirito Santo Basin, Brazil (P. Szatmari, pers. comm. 1993), Great Kavir, Iran (Jackson *et al.* 1990) and north Yemen (Heaton *et al.* 1995). However, no detailed information is available on the nature of the deformation induced in the sediments which are overridden by a surface salt extrusion or namakier. The Jabal al Milh structure is interpreted to be a diapir with an extrusive outflow and sedimentary rocks are well preserved around the diapir neck beneath the eroded namakier. Hence, the two salt structures described here provide outcrop data on two contrasting structural styles produced from (a) deformation around an upbuilding diapir breaking to the sea-bed, and (b) deformation underneath a spreading namakier. This study also provides a rare opportunity to examine the relationship between salt dome uplift, relative sea-level fluctuations and sedimentation patterns.

## Stratigraphic setting

The southern Red Sea rift initiated soon after the main phase of Afar plume magmatism between 25 and 30 Ma BP (Davison *et al.* 1994). Hughes & Beydoun (1992) suggest that brackish and marginal marine environments would have characterized the early Red Sea in this area during the late Oligocene. Synrift deposits do not outcrop onshore in western Yemen and have not been cored, but are probably clastic rocks, volcanic lavas and pyroclastic rocks (Davison *et al.* 1994). The main halite deposits are thought to be mid- to late Miocene in age (El Anbaawy *et al.* 1992) and were precipitated soon after rifting of continental crust ceased (Heaton *et al.* 1995), when extension became localized at the

*From* Alsop, G. I., Blundell, D. J. & Davison, I. (eds), 1996, *Salt Tectonics*, Geological Society Special Publication No. 100, pp. 23–39.

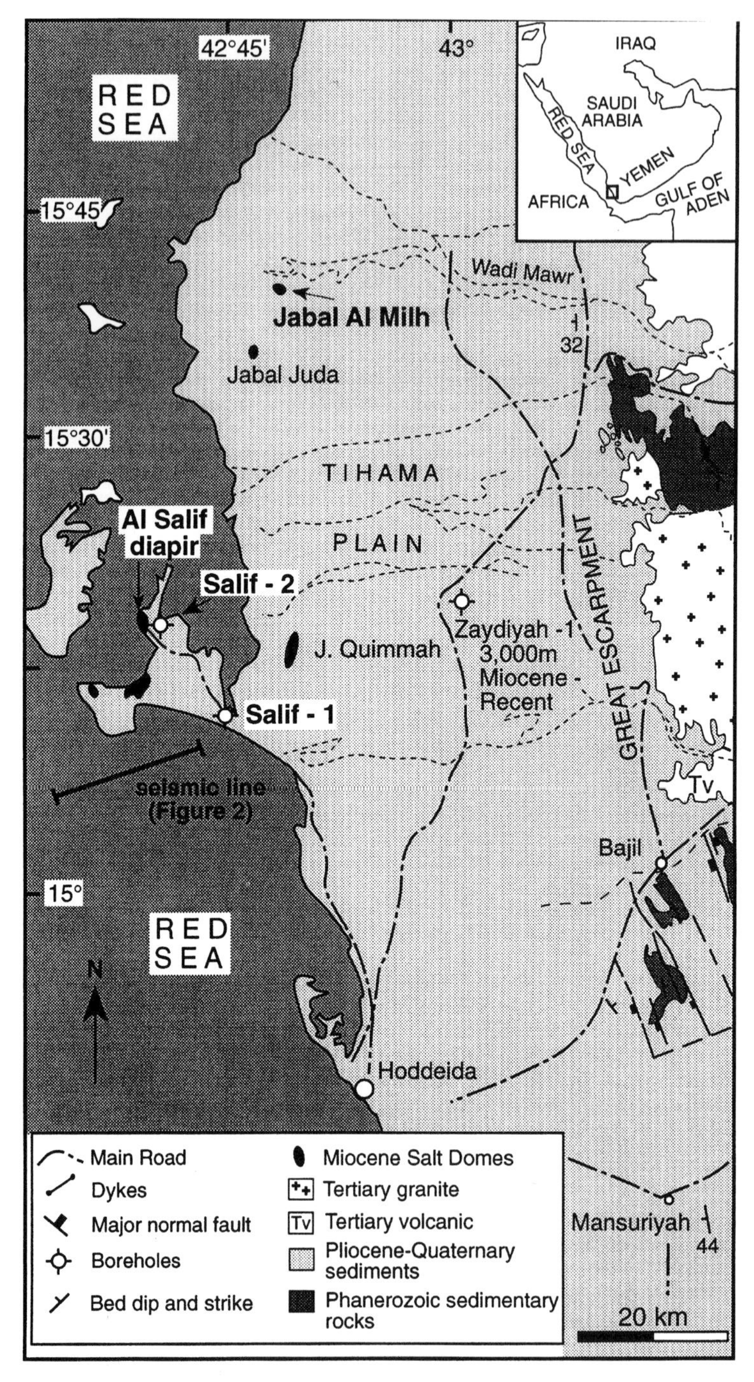


**Fig. 1.** Location map of the salt diapirs in northwest Yemen with regional geology (after Davison *et al.* 1994).

mid-ocean ridge spreading centre. The main halite unit may have reached up to 2 km in thickness (Heaton *et al.* 1995). Interbedded evaporite deposits and shallow marine and continental clastic sediments were deposited through to the present day (Heaton *et al.* 1995) Present-day deposits of the Tihama Plain are alluvial fans derived from the Yemen Highlands, aeolian dunes, and coastal

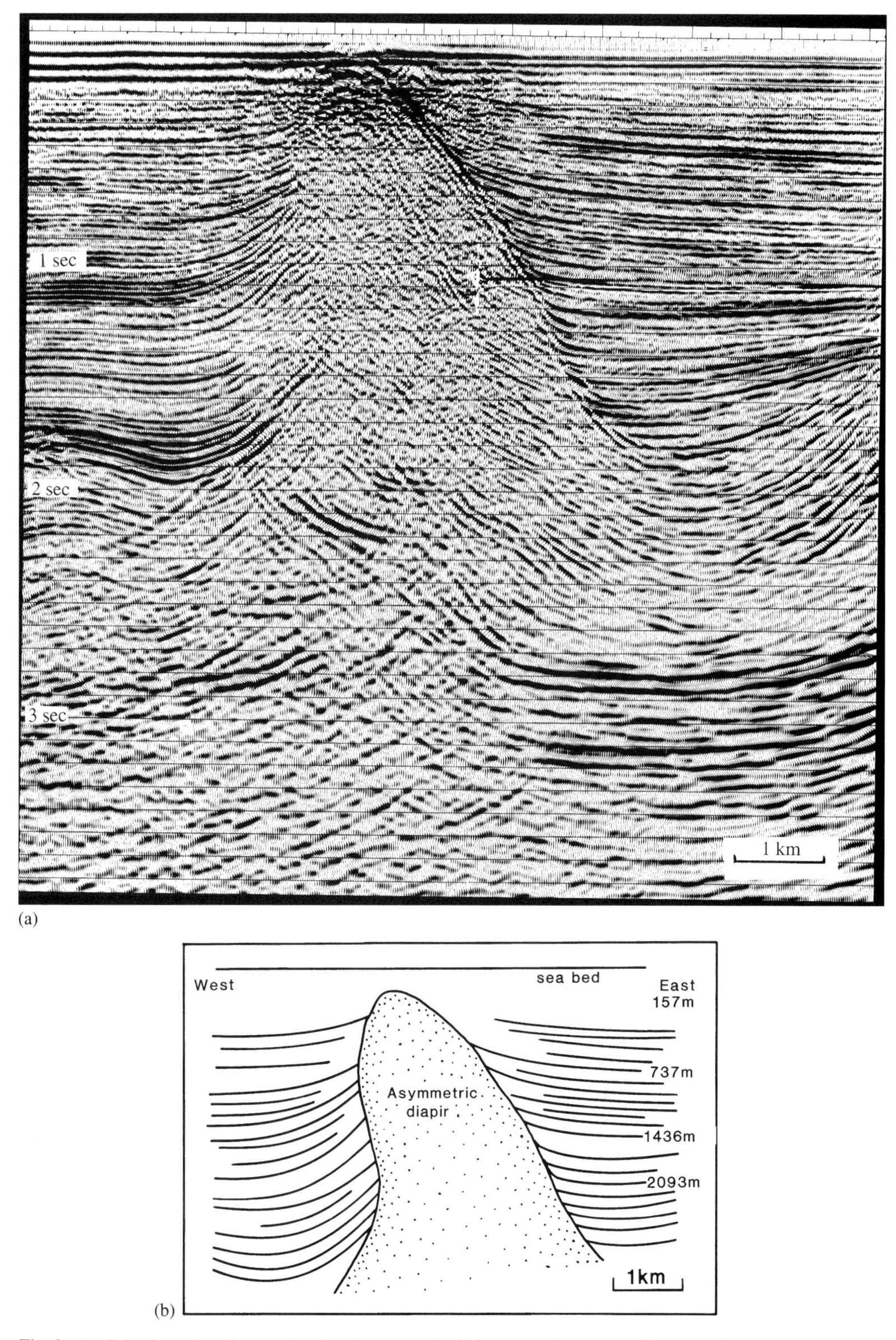


**Fig. 2.** (a) Seismic section through the diapiric wall which is interpreted to be the offshore continuation of the Al Salif diapir; location of line shown in Fig. 1. (b) Depth-converted seismic interpretation of line in (a).

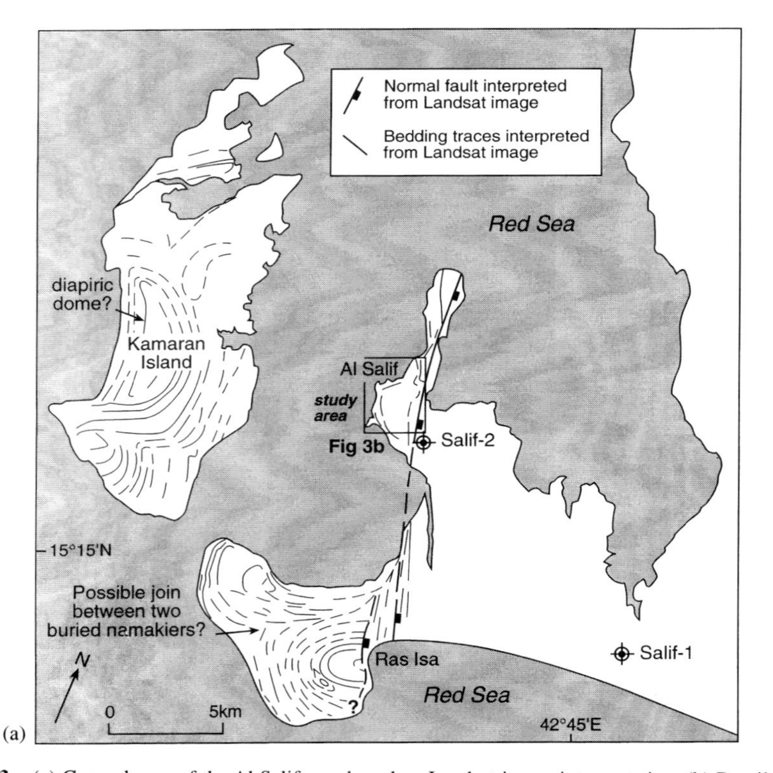


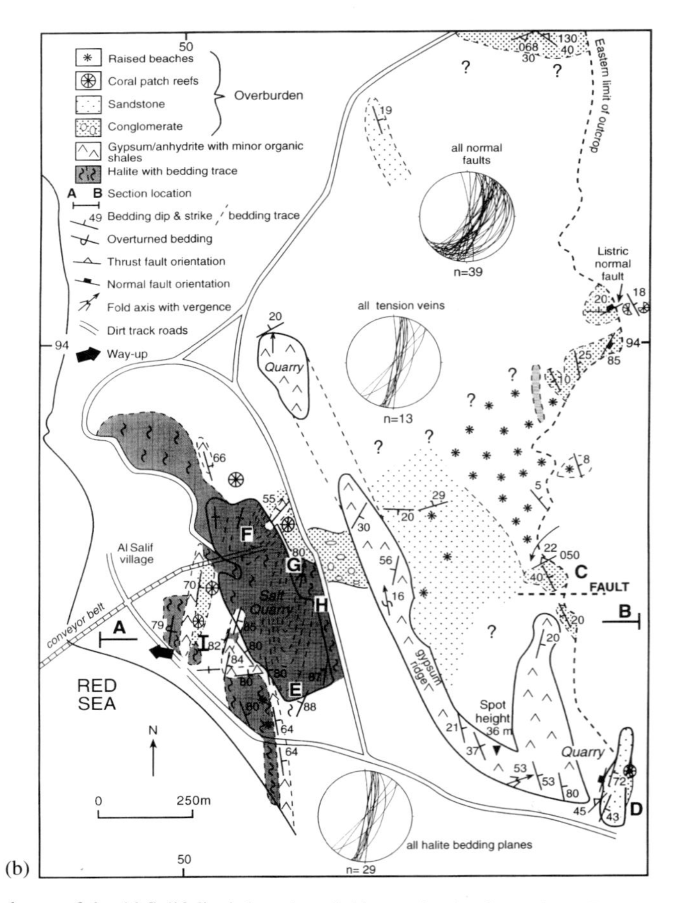


**Fig. 3.** (a) General map of the Al Salif area based on Landsat image interpretation. (b) Detailed map of the Al Salif diapir based on field mapping by the authors. Equal-area lower-hemisphere projections of fault and tensional gypsum vein orientations throughout the whole area are also shown.

siliciclastic beach sands, sabkhas and muddy, mangrove shorelines. The evaporite deposits overlying the main halite unit are dominated by sabkha and layered gypsum deposits (El-Anbaawy *et al.* 1992), with individual layers reaching at least 140 m in thickness at outcrop (this study, Jabal Al Milh), and in offshore wells (Heaton *et al.* 1995). Two diapirs (Al Salif and Jabal Al Milh) break to the surface near to the present-day Red Sea coastline in Yemen (Fig. 1), and salt is quarried down to a depth of 30 m below surface. Six other buried salt structures reach close to the surface in Yemen (Fig. 1), whilst salt domes are also known in southern Saudi Arabia at Jizan and the Farasan Islands (Wade 1931; Heybroek 1965).

## Al Salif diapir – an example of active diaprism

### *Geological setting*

The Al Salif diapir is an asymmetric structure which has risen up to 4 km above the main Miocene halite source layer (Fig. 2). The Salif-2 well (Fig. 3a) penetrated 1981 m of massive halite before drilling was suspended (Heaton *et al.* 1995). The diapir lies very close to the present-day Red Sea coastline and our mapping (Fig. 3) indicates that the Al Salif peninsula is produced by a linear N–S trending salt diapir bounded by a normal growth fault along the eastern margin (Figs 2 and 3). The diapiric wall continues offshore for several kilometres; where it is visible on seismic sections (Fig. 2). Diapirism began soon after deposition of the main salt interval, as can be inferred from the seismic section (Fig. 2), and thickening of sedimentary strata into the eastern flank can be observed. The diapir is asymmetric and overturned towards the west, with an overhang on the western flank of up to 1 km in width (Fig. 2b). Overburden reflectors break up close to the top of the diapir (above 0.5 s TWT), suggesting that rapid facies variations or pervasive faulting has occurred. This is consistent with our field observations that the vertical salt within the diapir is overlain unconformably by gently folded overburden of variable lithologies, which are cut by numerous small faults (Figs 3 and 4). Conglomerates, sands and reefs of late Quaternary age are now raised above sea-level, indicating that the diapir is still actively growing and that salt is still available for supply into the diapir. We see no evidence to support the views of El-Anbaawy *et al.* (1992) that the Al Salif structure is a ‘gently folded anticlinal fault block’.

Two separate partial ring structures defined by overburden bedding structures with radii of 2–5 km have been identified on Landsat images on the Ras Isa peninsula (Fig. 3a, see also Heaton *et al.* 1995). These are interpreted to be caused by surface salt flows which have coalesced along an E–W striking contact. Both structures appear to be truncated by a large N–S trending extensional fault system which bounds the eastern margin of the Al Salif diapir. A similar deformation style is seen on Kamaran Island (Fig. 3), where reconnaissance field work verified that coarse carbonate grainstones and rudstones of the overburden have produced an elongate domal structure which is 7 km long in the N–S direction and up to 5 km wide. It is probable that the whole island is a carbonate platform uplifted on a large salt dome. Very saline waters have been encountered in wells on the island that are probably produced by dissolution of underlying salt, and many other offshore islands in the Red Sea are produced in a similar manner, e.g. Farasan Islands, Saudi Arabia (Wade 1931). Similar salt canopies have been described in the offshore area by Heaton *et al.* (1995).

### *Internal deformation fabrics within the Al Salif diapir*

The diapir is best exposed in the main salt quarry at Al Salif (Figs 3 and 4) and is also known from shallow boreholes to extend in the subsurface at least 1 km to the east of the present-day quarry (J.C. Sproule and Associates 1981, unpublished report; and Fig. 4). Bedding within the salt is subvertical and a thickness of 1981 m of salt was drilled in the Salif-2 borehole (Fig. 3a). The horizontal width measured in an E–W direction is at least 1.5 km, but this is not a stratigraphic thickness as the salt is tightly folded with fold wavelengths up to 200 m traced out by resistant anhydrite/gypsum ridges (Fig. 3b). The N–S exposed length of salt along strike at Salif is at least 1 km, but is inferred to extend much further in this orientation in the subsurface. The top salt surface reaches a maximum height of 15 m above sea-level, and is quarried to 10 m below sea-level.

The Al Salif diapir consists predominantly of halite, with fine laminae of anhydrite 1–5 mm thick, which are regularly spaced at 2–15 cm intervals with an average spacing of 10 cm (Fig. 4b). The bedding in the diapir generally strikes NNE–SSW, is sub-vertical and is tightly folded with fold amplitudes up to two hundred metres, with minor parasitic folds developed in the hinges of the folds (Fig. 4b). Few minor folds were observed on the limbs of the major folds; their absence is considered to be due to extreme stretching on the fold limbs (Fig. 4b). Minor folds with subvertical axes display consistent vergence patterns in relation to major folds of presumed similar generation. However, patterns of vergence may be ambiguous where

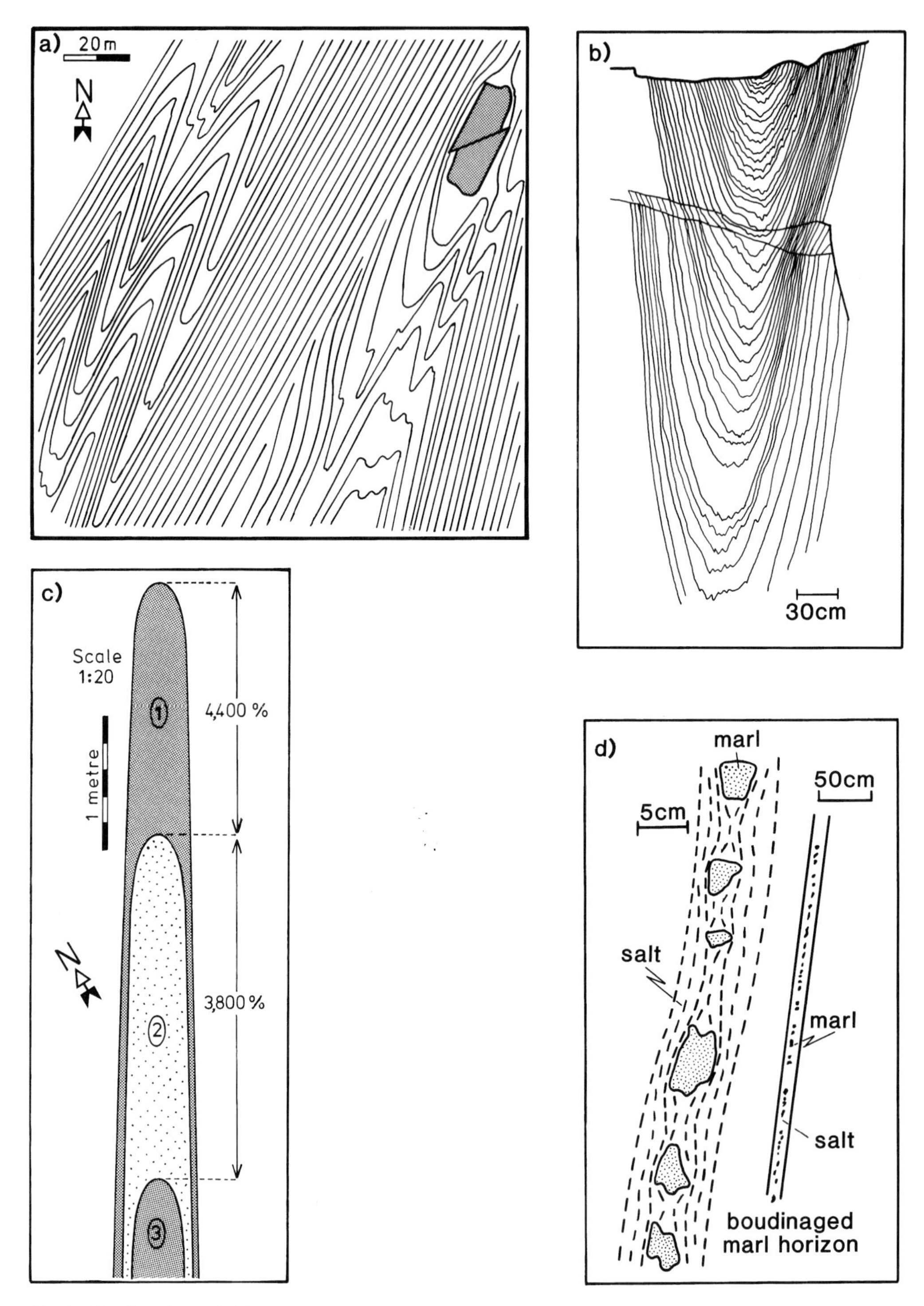


**Fig. 4.** Details of internal folding and boudinage of sedimentary layers with the Al Salif diapir. (a) Horizontal plan view of folding and boudinaged gypsum/anhydrite units (shaded) in the northern part of the main Salif quarry. (b) Folding style in vertical section along the southern termination of the quarry in January 1994. Location E in Fig. 3b. (c) Folding in salt with extreme changes in thickness of salt layers from limb to hinge, from location F in Fig. 3b. (d) Details of boudinaged anhydrite and marl horizons from location E in Fig. 3b.

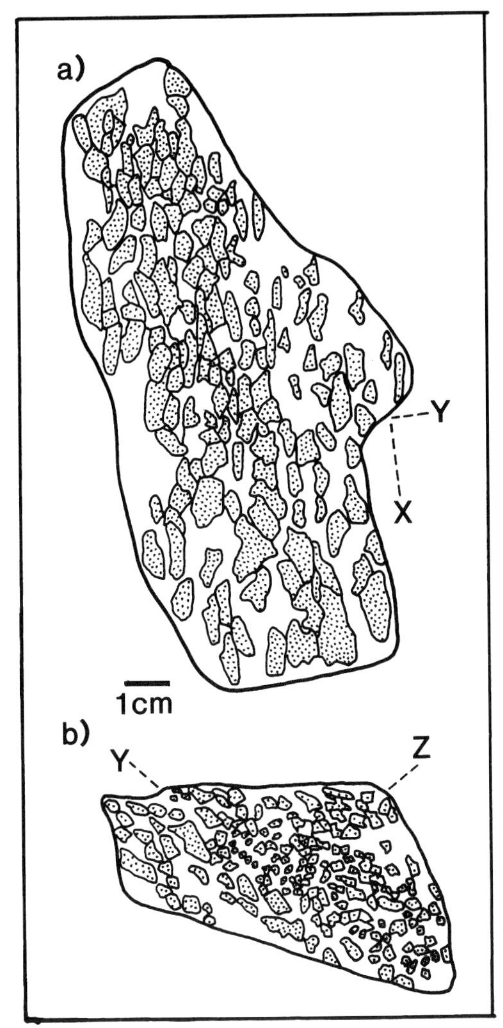


**Fig. 5.** Fabric shape of halite crystals in the neck of the Al Salif diapir. White areas are where crystal faces were obliterated and it was not possible to evaluate grain shape. X, vertical and parallel to stretching lineation; Y and Z, horizontal. Sample from location G in Fig. 3b.

curvilinear sheath folds are developed. A subvertical axial planar foliation of flattened halite crystals is developed, and a subvertical stretching lineation defined by elongate halite crystals is observed parallel to fold axes (Fig. 5, and fig. 1 in Davison *et al.* 1996). The individual halite grains vary in size up to 100 mm diameter, with an average grain size varying between 5 and 10 mm. An LS fabric is developed in the diapir, with an average halite crystal shape fabric of X:Y:Z = 4:1.5:1 (where X is vertical and parallel to the stretching lineation and Z is perpendicular to the foliation, Fig. 5). This plots well within the prolate field of the strain ellipse, as expected in the centre of a constricted diapiric neck. Deformation of halite results in boudinaged anhydrite beds elongated in the vertical direction parallel to the stretching lineation, with maximum extension ranging between 65 and 220% (Fig. 4d). Beds in fold hinges are between 3800 and 4400% thicker than on the fold limbs (Fig. 4c). The thicker anhydrite/gypsum, dolomite and shale interbeds occur as rafts up to 15 m thick and 250 m across, with horizontal separations of 150 m produced by large-scale boudinage (Fig. 3b) in the fold limbs. Three such rafts occur in an en-echelon arrangement to the southwest of the quarried area, with bedding overturned (determined from graded and resedimented dolomite beds) and younging to the west (Fig. 3b). These boudins have a rectangular shape and are composed of laminated and nodular anhydrite, organic-rich shales, 2–10 cm bedded fine sands and silts, and laminated dolomites (for detailed descriptions see El-Anbaawy *et al.* (1992)). The large boudins show little deformation of internal sedimentary structures, suggesting a strong viscosity constrast between them and the halite. Coarse halite crystals, up to several tens of centimetres long, are preserved in the necks of these boudins. The structures and fabrics reported here indicate intense vertical constrictional flow within the salt diapir (cf. El-Anbaawy *et al.* (1992, p. 70) who described the structure as 'gently folded' and with a 'lack of strong deformation structures'). The outcrops to the east of the quarry comprise about 30 m of tightly folded and sheared nodular, mylonitic and brecciated brown gypsum which forms the resistant N–S trending ridge to the Salif peninsular (Figs 3b and 4). The contact between this gypsum and the halite is not exposed, but because it is mylonitized and tightly folded it is assumed to be contained within the halite neck. A borehole to the east of the ridge penetrated at least 2 km of salt, which indicates that this gypsum ridge is within the diapir neck and is an elongate raft (Fig. 8).

Surface precipitation (80 mm $a^{-1}$) and water run-off, since mining operations began over the last ten years, have produced a spectacular karst and pot-hole system on top of the Al Salif diapir (Fig. 6b), with one crevasse proved by drilling to reach 20 m depth (J. C. Sproule and Associates 1981, unpublished report). The presence of large cavities which are linked to the surface would permit appreciable volumes of clastic sediment or water to be incorporated in the diapir head. These observations suggest a new mechanism for salt diapirism, which would be triggered by dissolution of the salt and collapse of overburden blocks into subterranean caverns, allowing salt to break the overburden. Rezak *et al.* (1985) have suggested that a similar dissolution system has developed above a diapir in the Gulf Coast to Mexico.

**Fig. 6.** (a) Photograph looking northwest at the upper dissolution surface of the Al Salif diapir from location H in Fig. 3b. (b) Dissolved crevasse and pot-hole system caused by surface water run-off since excavation of the salt quarry began in about 1930. Location H in Fig. 3b. Hammer is 35 cm long. Also note the regular upstanding layers of anhydrite in the halite which define the bedding in the diapir.

## *Stratigraphy of the sediments over the Al Salif diapir*

*Description* The subvertical and deformed evaporites of the diapir are unconformably overlain by around 40 m of strata which are subhorizontal to moderately dipping (up to 45°), with occasional steeper dips close to faults (Figs 3 & 7). The unconformity is a smooth, undulating (approximately 1 m of relief) but sharp surface of halite (Fig. 6a). These undulations may be due to wave action or river-channel scouring. The overlying sediments are varied and have a complex relationship with the underlying salt and the topographic unconformity. Their occurrence and sedimentology will be described in detail in a future publication.

To the southeast of the gypsum ridge, a roadside quarry (location D, Fig. 3b) exposes a similar succession but with basal coarse quartz sand (>1.8 m thick) overlying sheared gypsum seen on the hill-side above the quarry. The sandstone and siltstone

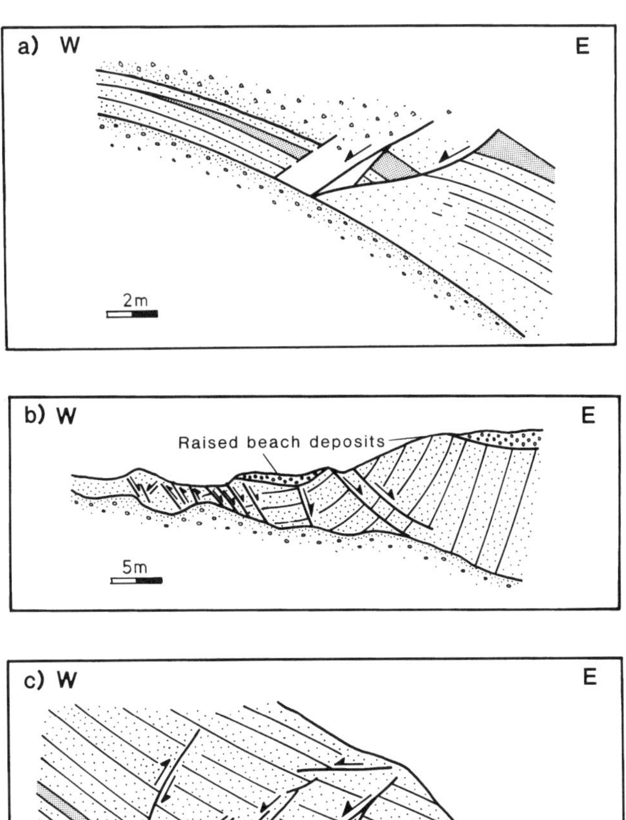


**Fig. 7.** Details of faulting deformation within the sedimentary overburden of the Al Salif diapir. (a) Growth faulting in sandstones and siltstones. Location C in Fig. 3b. (b) Faulting in bioclastic sandstone raised beaches with bedding rotated up to 50°. Location D in Fig. 3b. (c) Faulting in sandstone and siltstone overburden at locality C in Fig. 3b.

dip at 43–72° and contain scattered pebbles of Precambrian basement, and are overlain by >2.2 m of grey to white marls and micritic limestone. These are truncated with an irregular erosion surface covered by conglomerate, and then overlain by 4 m of reefal limestone. The latter is divided into a lower unit dipping at 28° due east and an upper 1 m unit which is subhorizontal. $^{14}C$ dating of a *Tridacna* shell (sample DBSY25) from the upper unit, which is 7–8 m above sea-level, gives an age of 12 200 ± 400 a (Vita-Finzi pers comm. 1993).

Exposures to the west of the quarry are smaller, but display a spectacular unconformity surface where the boudins or rafts of gypsum/anyhdrite, dolomite and shale lithologies stand up as ridges, with a relief above the salt of at least 6 m (Fig. 8). These ridges are partially exhumed by present-day weathering. Around the ridges are basal conglomerates (<2.5 m) with gypsum/anhydrite and dolomite pebbles, which are overlain by coarse bioclastic gravels and coral patch reefs built mainly by the branching coral *Galaxea fascicularis*. Two *Tridacna* shells from an altitude of 17–18 m (from location I in Fig. 3b) give dates of 3700 ± 250 and

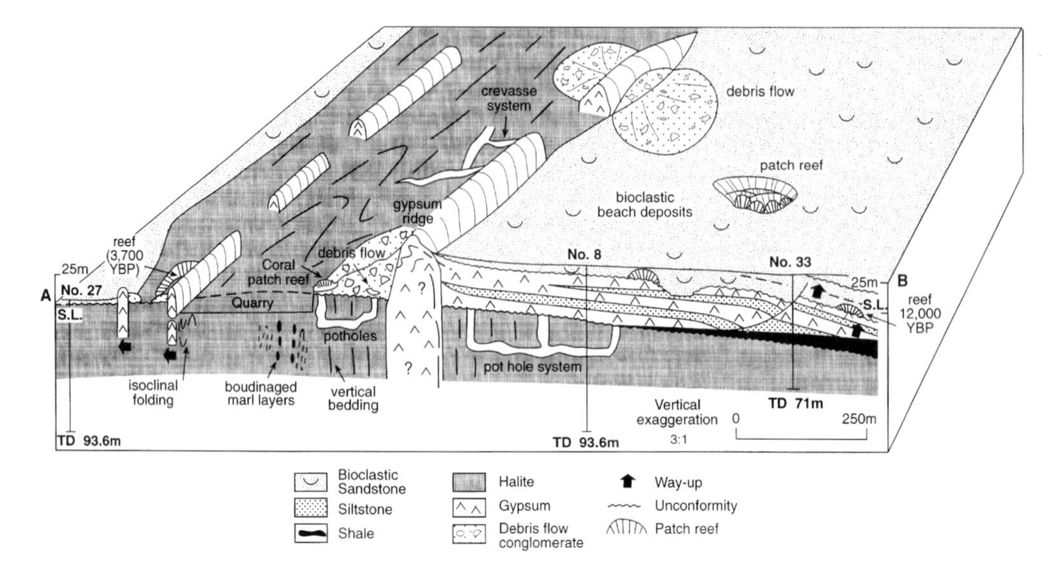


**Fig. 8.** E–W cross-section and top surface of the Al Salif diapir, based on our surface observation and shallow borehole data (from J. C. Sproule and Associates, unpublished reports, 1984). Location of section is marked A–B in Fig. 3b.

3850 ± 250 a BP (Vita-Finzi pers comm. 1993). Many archaeological remains have been unearthed in the Ras Isa peninsula, and $^{14}C$ dating of shell piles (middens) at the surface has also given dates of 3700 a BP (C. Phillips pers. comm. 1994). This suggests that the first preserved human colonization in this area occurred at about the same time that the beach deposits became emergent on the crest of the diapir.

*Environmental interpretation* Clues to the environment in which the upper surface of the salt was dissolved come from the immediately overlying deposits. The eastern outcrops of conglomerates around the gypsum ridge are interpreted as scree deposits. The sands to the southeast are unfossiliferous and the scattered basement pebbles indicate a source from the Yemen Highlands to the east, and are interpreted to be alluvial in origin with lacustrine or marginal marine ponding for the silts. The dissolution of the upper surface of the salt in this area is thus interpreted to occur in a subaerial environment. These subaerial deposits were subsequently transgressed by the margins of the Red Sea in the late Quaternary to produce shoreline conglomerates and then shallow-marine patch reefs. During growth of reefs in the east of the diapir, they were being tilted towards the east and were finally overlain by subhorizontal reefs at around 12 200 a BP.

Dissolution in the western outcrops is likely to have occurred in a submarine environment, as the first deposits over the salt are marine beach deposits and coral reefs. The protruding ridges of less-soluble gypsum and anhydrite control local sediment deposition, with banking-up of predominantly alluvial sediments on their eastern flanks, as this area was presumably protected from wave action by the ridges (Fig. 8).

The western localities are more centrally located over the diapir and the reefs are younger (3700–3850 a BP) than the tilted reefs to the east. In the simplest case it would be expected that the central top to the diapir would impinge on the seafloor first and preserve the oldest reefs, with progressively younger reefs forming on the margins of the diapir in an offlapping fashion. This is not the case in Salif, where an early stage of uplift and tilting of reefs on the margin of the diapir is recorded by easterly dipping reefs and unconformities (Fig. 8). During the earlier period (12000 a BP), the active crest of the diapir may have been moving up too quickly and the water was too salty to allow reefs to colonize, but it was easier for them to grow on the flanks of the diapir where clastic sediments onlapped onto the salt surface. Later subsidence of the diapir top surface by at least 10 m over 8450 a (the height and age difference between the two dated reefs) must have occurred, bringing the upper levels of the salt ridge into the shallow marine environment where the younger reefs accumulated. Since 3700–3850 a BP the diapir has moved upwards by 17 to 18 m if it is assumed that the reefs formed at or just below sea level.

**Fig. 9.** General view of the Jabal Al Milh salt diapir looking west from the extended mosque, showing isolated mounds of overburden sediments which have been pushed approximately 45 m above the regional land surface.

## *Deformation and structure of the overburden around the Al Salif diapir*

The diapir has reached the surface and is actively lifting overburden of marine sediments aged 3700 a to 45 m above sea-level. Hence it is an example of active diapirism rather than downbuilding and differential compaction, as sea-level changes in the last 4000 a have been less than 2 m. Quaternary raised beaches lie unconformably and directly on the dissolved halite surface around the main quarry area at the crest of the diapir, and there is no indication of a cap-rock development. This indicates that any Miocene to Pliocene overburden was removed by erosion or was never deposited. Sandstones and siltstones are exposed on the eastern part of the diapir, slightly down-flank from the crest of the dome, and are mainly deformed by extensional NNE to NE striking faults with a maximum throw of 5 m (stereogram in Figs 3b & 7). The faults are planar and listric with sedimentary growth present in the hangingwall (Fig. 7b). The minimum horizontal extension produced by faulting in the lowermost overburden siltstones, which have a regional dip of 20°E, is approximately 25%. Extensional faults generally dip inwards towards the centre of the diapiric ridge, and the layering in the overburden dips away from the diapiric crest. Some normal faults deform the most recent raised carbonate platform and also displace the top of the salt surface by 4 m. One fault plane was recognized within the exposed salt body, and fault slickensides within the halite were observed on the southern edge of the quarry.

Many vertical extensional veins cut through the late Quaternary marine sediments and consolidated wadi deposits. The veins generally strike NNE to NE and have a similar orientation to the normal faults (see stereogram in Fig. 3b). They reach up to 1 m in thickness and are filled with fibrous gypsum, indicating that vein opening was incremental and there was abundant circulation of fluids during deformation. Some gypsum veins with subvertical fibres are orientated parallel to bedding in the oldest overburden siltstones. The veins reach up to 100–200 mm thick, indicating that fluid overpressures must have been temporarily and locally greater than the vertical pressure exerted by the overburden. Minor, high-angle (>45°) reverse faults which have bedding cut-off angles of 60° are also present, with up to 50 cm of throw (Figs 7b & c). The high bedding cut-off angles suggest that bedding was rotated before faults were formed.

The diapir has intruded to the surface and is actively lifting overburden sediments above their regional level and beginning to push them sideways.

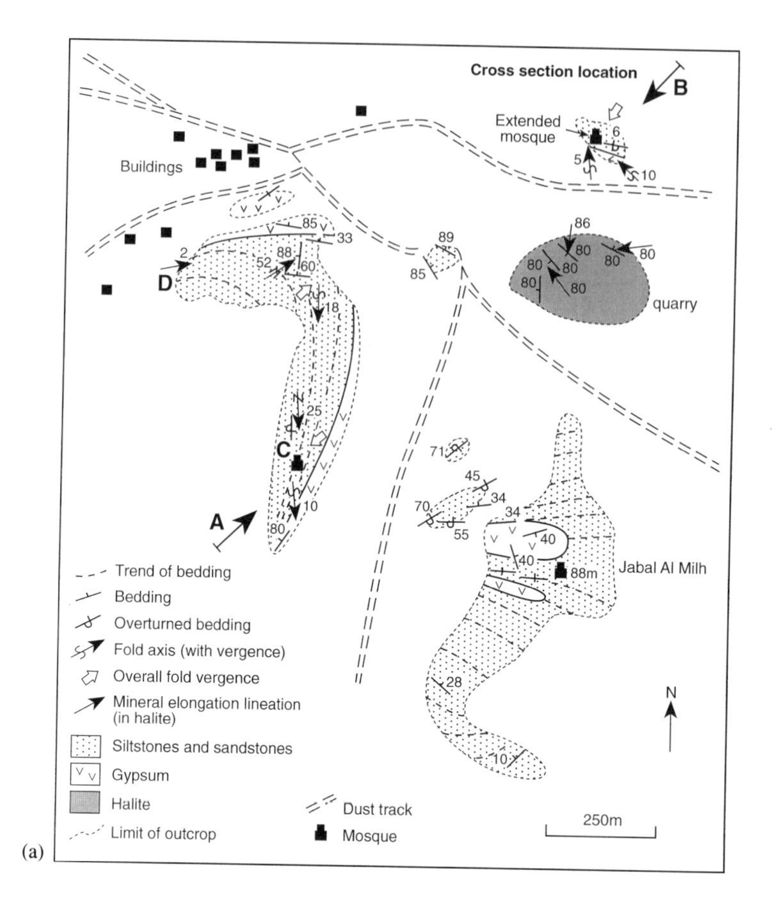


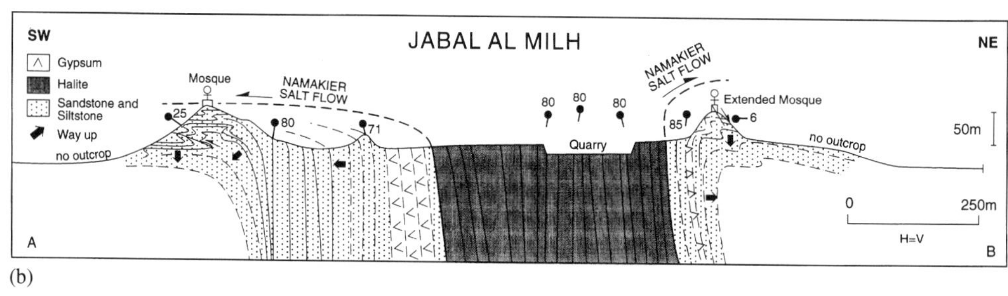


**Fig. 10.** (a) Map of the Jabal al Milh diapir based on field mapping by the authors. (b) NE–SW cross-section of the Jabal al Milh diapir.

The current rate of surface uplift may be calculated from the height of the youngest reefs (see above) and is 4.6 mm $a^{-1}$. This compares with the average vertical growth rate of 0.32 mm $a^{-1}$, assuming that the salt has a mid- to late Miocene depositional age (14 Ma) and the diapir has a height of 4500 m. The rapid growth of the salt diapirs is probably due to the high heat flow and the weak overburden containing layered evaporites, which reach up to 40% of the total overburden thickness (Heaton *et al.* 1995) and the rapid sedimentation rate.

## Jabal al Milh – an example of sedimentary rock deformation beneath a namakier

### *Regional setting and structure*

This diapir lies approximately 12 km inland from the Red Sea shoreline, at 45° 4′N (Fig. 1), and the regional land surface stands around 30m above sea-level. There are no nearby outcrops on this flat-lying alluvial plain. A small open-cast salt mine in the centre of the diapiric neck (Figs 9–11) shows

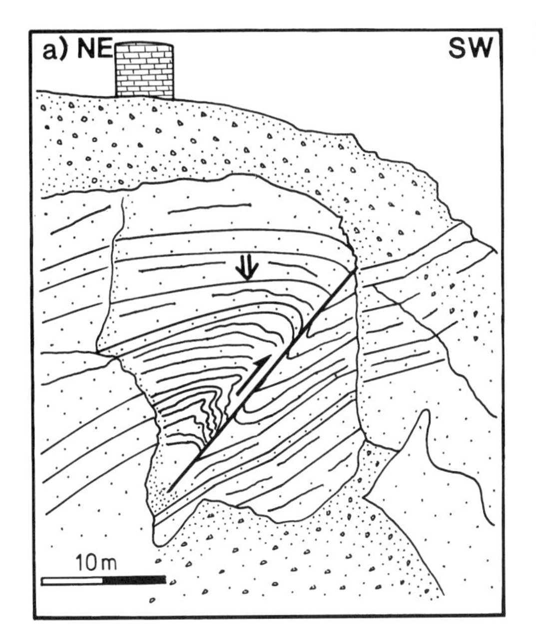


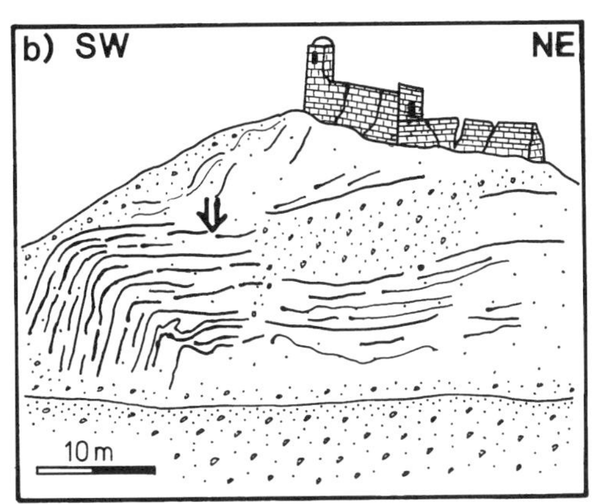


**Fig. 11.** Details of recumbent fold deformation of the overturned siltstones and anhydrite layers below the extrusive salt flow of the Jabal al Milh diapir. (a) Folding and thrusting with thrust displacement of several metres. Location C on Fig. 10a. (b) Folding at the edge of the diapir wall, interpreted to have been caused by surface extrusion of salt. Note that beds were inverted before folding occurred. Location is at the extended mosque on Fig. 10a.

subvertical NW–SE trending bedding within the laminated halite and anhydrite, and a vertical stretching lineation is preserved (Fig. 10a) which is visually similar to that at Al Salif. The overburden sediments immediately adjacent to the diapir form a distinct group of hills, with an elevation of up to 85 m above sea level (Fig. 9).

The largest section of overburden examined

| Description | Interpretation |
|---|---|
| Stratigraphic top of section (i.e. base of the outcrop) not seen | |
| 2.5 m – Interbedded medium sands and laminated and cross laminated fine sands, current rippled and with bed surfaces with plant debris. Slumped sand bed with truncated top gives way-up evidence. | Alluvial flood plain deposits |
| 1.5 m – Interlaminated gypsum and shale and occasional sandy partings. | Hypersaline lake on floodplain |
| 6.5 m – Dark grey silts with scattered plant fragments in 1-10 cm parallel bedding. | Overbank deposits |
| 0.5 m – Grey/brown paper shales. | Lacustrine |
| 3.0 m – Nodular gypsum with enterolithic folds. | Sabkha |
| 8.0 m – Brown siltstone with plant fragments. | Overbank deposits |
| 2.0 m – Nodular gypsum. | Sabkha |
| 5.0 m – Brown siltstone with plant fragments. | Overbank deposits |
| Stratigraphic base of section (i.e. top of the exposure) not seen | |

occurs beneath the extended mosque (Fig. 10b), where the rocks are inverted and the top and base of the section are not observed. The 29 m of stratigraphic section (above) is recorded here as there are no previous stratigraphic logs of these Miocene or younger rocks in Yemen:

The outcrop to the west of the diapir contains an overturned sequence of fining-upward cycles interpreted to be of alluvial origin. Erosively based large-scale, crossbedded, coarse-grained sandstones are overlain by plane-bedded, medium-grained sandstones passing up into interlaminated fine-grained sandstones and siltstones. These may pass up into laminated gypsum or crosslaminated fine-grained sandstones.

These overburden sedimentary rocks are estimated to reach at least 140 m in thickness and clearly represent alluvial plain and sabkha environments. We have no data on the age of these deposits, and they appear to be devoid of any fossil remains apart from unidentified plant fragments.

### *Sedimentary overburden deformation*

The bedding either side of the halite mine dips vertically and the exposed halite appears to have a strike width of 500 m in a NE–SW direction. The overburden is rotated to the vertical up to 200 m from the vertical halite contact, and further away from the diapir the rocks are folded so that the sequence is inverted (Fig. 10b). Recumbent folding ($A$ = 10 m, $\lambda$ = 10–20 m) took place after the

**Fig. 12.** Photograph of the northeastern flank of the Jabal Al Milh salt diapir, where the bedding is overturned, folded and thrusted, and later gravity-driven extension has caused cracking of the 100-year-old mosque walls. Location shown in Fig. 10a. The measured section in the text is through the overturned sequence measured from left to right of the cliff in the foreground at the base of the hill.

overburden had been inverted, resulting in downward-facing folds with a sense of fold vergence directed away from the diapir neck. The folding is interpreted to be produced by salt extrusion, which has now been removed. However, the lack of outcrop makes it impossible to determine whether the clastic sedimentary rocks are huge rafts within the diapir or the overburden.

Asymmetric folds immediately adjacent to the neck at either side of the halite core preserve an opposite sense of vergence to that expected for outward flow of the salt (Fig. 10b). This is due to bending of the layers into a monoclinal fold where the salt neck terminated and surface extrusion commenced. Many minor folds, with axes trending N to NW, are developed in the multilayered siliciclastic–gypsum sequences, and thrusts with outward vergence are developed with up to 10 m of throw (Fig. 11). A large-scale ($\lambda$ = 200 m) open upright fold, trending E–W with a horizontal axis, represents a further phase of folding at location D in Fig. 10a, which probably occurred after inversion of the sedimentary cover. Besides the major folds and thrusts, many small-scale normal faults are present, with throws of 5–10 cm, which trend approximately ENE in the region of the extended mosque (Fig. 10a). Extension fractures are perpendicular and parallel to the bedding planes and are filled with fibrous gypsum. The veins perpendicular to bedding trend E–W at the mosque near to locality C on Fig. 10a. These fractures were subsequently folded during contractional deformation of the sedimentary rocks around the diapirs. This suggests that high pore pressures were developed above the Jabal al Milh diapir before folding occurred.

The overburden sediments are now undergoing active extension, and there is an excellent modern-day strain marker on top of one of the overburden outcrops in the form of a mosque, 80–100 years old, which has been extended by up to 14% by normal faulting and extensional cracking, measured by the gaps in the walls (Fig. 12). This implies a local extensional strain rate of $5.5 \times 10^{-11} s^{-1}$. The normal faults have produced pull-aparts where they steepen up through more rigid layers and leave gaping holes in the sedimentary sequence.

## Discussion

There has been a long-standing debate about the role of active upward intrusion of salt into overburden and how important this is in relation to diapiric growth by passive downbuilding (e.g. Trusheim 1960). Active intrusion is controlled by the density difference between the salt and overburden, and the vertical relief of the salt structure. However, the effectiveness of the buoyancy force is inhibited by the viscous drag force exerted by the walls of the diapiric neck. This viscous drag force is dependent on the strength of the overburden and the salt, and on the cross-sectional area of the orifice through which the salt is flowing. The rate of salt flow $q$ may be expressed as

$$q = \frac{\sigma t^3}{2\mu\kappa L} \quad (1)$$

where $\sigma$ is the hydraulic pressure, $t$ is salt thickness, $\mu$ is the salt viscosity, $\kappa$ is the roughness constant and $L$ is the length of the flowing layer (Price & Cosgrove 1990, p. 94). The most direct way to determine whether there is a vertical buoyancy force driving diapirism is by observation of the amount of uplift of a salt body and its overburden above a regional marker elevation. A survey of the amount of sea-bed relief around diapirs in several different provinces is summarized in Table 1. These observations indicate that the buoyancy force is capable of pushing sediments up to 260 m above regional in the Gulf Coast. If we assume that near-surface sediments have a zero flexural rigidity and there is a density contrast of 800 kg $m^{-3}$ between the overburden and the sea-water, then the calculated buoyancy force is 2 MPa. This will be greater if the sediments have finite flexural rigidity. The predicted buoyancy force (with no viscous drag) is approximately 15 MPa for Gulf Coast of Mexico structures with up to 8 km of relief. In the smaller Yemen diapirs, uplift of the overburden is not as great, and the uppermost overburden of stronger siliciclastics and carbonates may be responsible for this. The crests of diapirs in the Great Kavir Province of Iran reach up to 1.5 km above the desert surface (Jackson *et al.* 1990; C. J. Talbot, pers. comm. 1994). These diapirs are intruded in an active compressional regime, which suggests that the horizontal compressive stress is large enough to hold a salt column 1300 m high (considering that the buoyancy of the salt alone probably supports 200 m of the column). This suggests that a horizontal tectonic stress of approximately 25 MPa is supporting the salt fountain.

Salt diapirs may appear as intrusive bodies penetrating into sedimentary rocks on seismic sections because of several processes: (a) evaporite dissolution with development of infilled caverns which apparently look liked stoped blocks; (b) dissolution at the sea-bed leaving a flat upper surface with no deposition of sediments above the diapir; (c) extensional faulting which thins sediments until salt breaks through to produce stoping, or sediments slide off the dome to expose salt at the surface (Schulz-Ela *et al.* 1993); or (d) development of a very thin sequence above the diapir that has been thinned by pervasive brittle or viscous deformation (Fig. 13).

The Al Salif diapir indicates that active upbuilding produces extensional faulting as observed in physical and numerical models of diapirs (Schulz-Ela *et al.* 1993; Vendeville & Jackson 1992; Alsop 1996; Poliakov *et al.* 1996). There is no evidence of large-stoped blocks within the two halite diapirs, nor is there evidence for salt injecting upwards into the overburden along faults or fractures. The late Quaternary raised beaches at Al Salif are deposited directly on salt, indicating non-deposition or removal of older overburden; the former is more likely as the overburden sediments on the eastern flank of the diapir suggest that there was little relief

**Table 1.** *A summary of vertical relief (freeboard) produced at the sea-bed or above regional desert surface by diapirs from different provinces*

| Diapir location | Vertical relief (m) | Reference |
|---|---|---|
| Gulf Coast off Mississippi delta | 100–240 | Jackson *et al.* (1990, fig. 1.77) |
| Gulf Coast of Mexico | 190 | Liro (1992, fig. 10) |
| Gulf Coast of Mexico | 255 | Wu *et al.* (1990, fig. 14) |
| Gulf Coast of Mexico | 260 | Wu *et al.* (1990, fig. 15) |
| NW Gulf Coast of Mexico | 130 | Rezak *et al.* (1985) |
| Gulf Coast of Mexico | 144 | Seni (1992, fig. 11) |
| Cabo St Tome, Brazil | 97 | Demercian *et al.* (1993, fig. 7) |
| Cabo St Tome, Brazil | 131 | Demercian *et al.* (1993, fig. 6) |
| Al Salif, NW Yemen | 45 | This study |
| Jabal al Milh, Red Sea | 45 | This study |
| J. Quimmah, Red Sea | 45 | This study |
| Saharan Atlas, Djelfa, Algeria | 300 | Kulke (1972) |
| Great Kavir Diapirs, Iran | 300–1500 | Kent (1979, Talbot & Alavi (1995) |
| Mt Sedom diapir, Dead Sea | 200 | Frumkin (1995) |

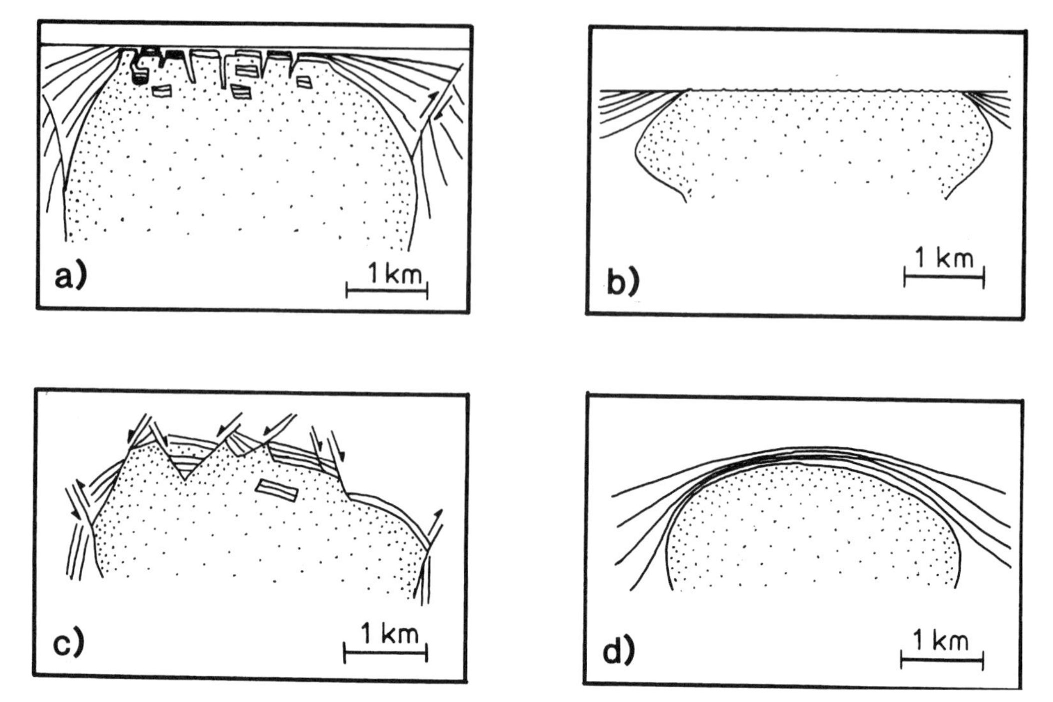


**Fig. 13.** Summary of how diapirs can 'apparently pierce' overburden sediments on seismic reflection profiles. (a) Dissolution above seal-level with surface run-off creates pot-hole crevasse system with stoping of sediments into dissolved channels and caverns. (b) Non-deposition of sediment and active erosion/dissolution at the sea-bed. (c) Extensional faulting and condensed sequence. (d) Condensed sequence and/or attenuation by ductile creep.

when these sediments were deposited. This suggests that the diapir breaks through the cover rocks by actively lifting them to produce normal faults and pushing them aside to produce thrusts. Extensional faulting aids this process by attenuating the overburden which makes it weaker, allowing material to slide off from the crest of the diapir.

## Conclusions

1. Salt diapirs in northwest Yemen are actively deforming Quaternary marine overburden sediments and lifting them up to 45 m above sea-level, indicating that active diapirism has occurred rather than differential compaction and downbuilding. These Quaternary raised-beach deposits were deposited directly on the salt, indicating that the Miocene–Pliocene overburden has been removed or was never deposited in the last 14 Ma.
2. The Al Salif diapir has produced mainly small-scale normal faulting in the overburden, with 25% extension recorded in the overburden on the east flank of the diapir crest. Minor contractional faulting also occurs in the overburden, showing active lateral pushing.
3. Surface uplift rate over the emergent diapir crest at Al Salif is currently estimated to be approximately 4.6 mm $a^{-1}$, compared to the time-averaged growth rate since late Miocene times of 0.32 mm $a^{-1}$. The drastic acceleration of salt movement is common when diapirs break to the surface (our own observations on physical models).
4. The Jabal al Milh structure is interpreted to have been produced by a former surface salt extrusion, with deformation dominated by recumbent folding and occasional thrusting observed in the sedimentary rocks beneath the namakier.
5. Minor reverse faulting occurs above the Al Salif diapir, and minor extensional faulting and gypsum tensional veins, some of which are folded around the Jabal al Milh diapir, indicate complex local changes in stress regimes around diapirs.
6. Both diapirs have tensional veins developed parallel to bedding filled with fibrous gypsum, which attests to very high pore pressures developed during faulting. These pore pressures probably developed due to impermeable thin evaporite layers, which occur throughout the Miocene Pliocene overburden sequence in a

domal structure above the diapir crest. This would have promoted faulting and facilitated removal of overburden by extension.

We would like to thank BP Exploration (Yemen) for providing seismic lines and access to Landsat imagery, and the Ministry of Oil and Natural Resource in Yemen for permission to publish the seismic line. We acknowledge funding from the Royal Society (1991 Yemen Expedition) and the European Union Red Sea grant awarded to B. Purser, Université de Paris, with Bosence as UK Coordinator. B. Rosen helped to identify the corals. C. Vita-Finzi kindly dated the raised reefs. The Governor of Al Hodeida is thanked for permission to work in the Al Salif and Jabal al Milh areas. We thank the Yemen Salt Company for permission to visit the mines and Mohammed Al Ghaleb (mining engineer) assisted our field work with logistical support and informative discussions on the Salif salt. Gareth Jenkins performed the depth conversion used in the cross-section of the Salif diapir in Fig. 2b. Helpful reviews by Christopher Talbot and Richard Heaton are gratefully acknowledged.

## References

ALSOP, G. I. 1996. Physical modelling of fold and fracture geometries associated with salt diaprism, *this volume*.

DAVISON, I., BAKER, J., BLAKEY, S. *et al.* 1994. Geological Evolution of the south-eastern Red Sea Rift margin: Republic of Yemen. *Bulletin of the Geological Society of America*, ***106***, 1471–1492.

——, ALSOP, I. & BLUNDELL, D. 1996. Salt tectonics: some aspects of deformation mechanics, *this volume*.

DEMERCIAN, S., SZATMARI, P. & COBBOLD, P. 1993. Style and pattern of salt diapirs due to thin-skinned gravitational gliding, Campos and Santos Basins, Offshore Brazil. *Tectonophysics*, **228**, 393–433.

EL-ANBAAWY, M. I. H., AL-AAWAH, M. A. H., AL-THOUR, K. A. & TUCKER, M. 1992. Miocene evaporites of the Red Sea Rift, Yemen Republic: sedimentology of the Salif halite. *Sedimentary Geology*, **81**, 61–71.

FRUMKIN, A. 1995. Uplift rate relative to base levels of a salt diapir (Dead Sea basin, Israel) as indicated by cave levels, *this volume*.

HEATON, R., JACKSON, M. P. A., BAMAHMOUD, M. & NANI, A. S. O. 1995. Superposed Neogene extension, contraction and salt canopy emplacement in the Yemeni Red Sea. *In*: JACKSON, M. P. A., ROBERTS, D. G. & SNELSON, S. (eds) *Salt Tectonics: A Global Perspective for Exploration*. American Association of Petroleum Geologists, Memoir.

HEYBROEK, F. 1965. The Red Sea Miocene evaporite basin. *In*: *Salt Basins around Africa. Journal of the Institute of Petroleum Technology, London*, 17–40.

HUGHES, G. W. & BEYDOUN, Z. R. 1992. The Red-Sea-Gulf of Aden: biostratigraphy, lithostratigraphy and paleoenvironments. *Journal of Petroleum Geology*, **15**, 157–172.

JACKSON, M. P. A., CORNELIUS, R. R., CRAIG, C. H., GANSSER, A., STÖCKLIN, J. & TALBOT, C. J. 1990. *Salt Diapirs of the Great Kavir, Central Iran*. Geological Society of America, Memoir, **177**.

JENYON, M. K. 1988*a*. Some deformation effects in a clastic overburden resulting from salt mobility. *Journal of Petroleum Geology*, **11**, 309–324.

—— 1988*b*. Overburden deformation related to prepiercement development of salt structures in the North Sea. *Journal of the Geological Society, London*, **145**, 445–454.

KENT, P. E. 1979. The emergent Hormuz salt plugs of southern Iran. *Journal of Petroleum Geology*, **2**, 117–114.

KULKE, H. 1972. Le rocher de sel de Djelfa (atlas saharien algérien). Géologie et morphologie. *Societé Histoire Naturelle, Afrique du Nord Bulletin*, **9**, 433–451.

LIRO, L. M. 1992. Distribution of shallow salt structures, lower slope of the northern Gulf of Mexico, USA. *Marine and Petroleum Geology*, **9**, 433–451.

POLIAKOV, A. N. B., PODLADCHIKOV, Y. Y., DAWSON, E. C. & TALBOT, C. J. 1996. Salt diapirism with simultaneous brittle faulting and viscous flow. *this volume*.

PRICE, N. J. & COSGROVE, J. 1990. *The Analysis of Geological Structures*. Cambridge University Press.

REZAK, R., BRIGHT, T. & MCGRAIL, C. 1985. *Reefs and Banks of NW Gulf of Mexico – Geological, biological and Physical Dynamics*. Wiley Interscience, New York.

SCHULTZ-ELA, D., JACKSON, M. P. A. & VENDEVILLE, B. 1993. Mechanics of active salt diapirism. *Tectonophysics*, **228**, 275–312.

SENI, S. J. 1992. Evolution of salt structures during burial of salt sheets on the slope, northern Gulf of Mexico. *Marine and Petroleum Geology*, **9**, 452–468.

TALBOT, C. J. & ALAVI, M. 1996. The past of a future syntaxis across the Zagros, *this volume*.

TRUSHEIM, F. 1960. Mechanism of salt migration in N. Germany. *AAPG Bulletin*, **44**, 1519–1540.

VENDEVILLE, B. C. & JACKSON, M. P. A. 1992. The rise of diapirs during thin-skinned extension. *Marine and Petroleum Geology*, **9**, 331–353.

WADE, A. 1931. Intrusive salt bodies in coastal Asir, southwestern Arabia. Symposium of salt domes. *Journal of the Institute of Petroleum Technology, London*, 321–330.

WU, S., BALLY, A. W. & CRAMEZ, C. 1990. Allochthonous salt, structure, and stratigraphy of the north-eastern Gulf of Mexico, Part II: Structure. *Marine and Petroleum Geology*, **7**, 334–370.

# Uplift rate relative to base-levels of a salt diapir (Dead Sea Basin, Israel) as indicated by cave levels

AMOS FRUMKIN

*Israel Cave Research Center, Department of Geography, The Hebrew University of Jerusalem, Jerusalem 91905, Israel*

**Abstract**: Rapid downcutting rates in the extremely soluble salt of the Sedom diapir, Dead Sea basin, Israel, allow cave channels to become rapidly graded with respect to base level. Diapir uplift leaves the older passages high and dry above present base level. Dating these passages by $^{14}C$ allows us to estimate diapir uplift rates, taking into account previous Dead Sea levels. Maximum mean Holocene uplift rates are 6-7 mm $a^{-1}$ along the eastern fault of Mount Sedom.

Several salt diapirs occur in the Dead Sea Basin (Neev & Hall 1979) within the the Levant transform zone (Kovach & Ben Avraham 1990). Of these, Mount Sedom is the best exposed (Fig. 1). The diapir has an elongated map expression measuring 11 km (N–S) by 1–1.5 km (E–W), and rising 250 m above the Dead Sea. At present (1994) the Dead Sea has receded to 410 m below sea-level, leaving its south basin, east of Mount Sedom, as a dry plain serving as the local base-level of erosion. The salt exposed in Mount Sedom is of marine origin, and commonly assigned a Neogene–Pleistocene age (Zak and Bentor, 1968; Horowitz 1992, p. 332; Stein *et al.* 1994). The salt beds in the upper part of the diapir are steeply to vertically inclined (Fig. 2a) and slightly deformed, mainly along the diapir borders.

The salt extends to around 200 m above the Dead Sea base-level. It is mostly covered by a 5–50 m thick anhydrite cap-rock. The extremely arid climate, with mean precipitation of 50 mm $a^{-1}$, inhibits rapid surface dissolution (Gerson & Inbar 1974). Dissolution by rare run-off events is concentrated within an extensive karst system which drains 57% of Mount Sedom surface area (Frumkin 1992). The cave conduits of this karst system offer exceptional opportunities to study salt tectonic features.

The Sedom Diapir rises along several subparallel faults (Fig. 2a). Most of the studied caves, such as Mishqafaim Cave, drain towards the eastern margin of the mountain. The steep topography of the eastern escarpment (Fig. 3), its salt strain features (Zak and Freund 1980) and tectonic structures within cave passages (e.g. Fig. 4) indicate that much of the recent diapir rise has occurred along this border fault, while other faults are less active.

Caves begin to develop when a surface channel is captured by a fissure passing through the relatively insoluble cap-rock into the salt. Young unequilibrated passages grow downwards by dissolution into the rock salt until they reach the minimal gradient allowing transportation of their coarse sediment load (Frumkin 1994). The active alluviated channels are apparently adjusted to the local base-level of erosion with a profile below which the channel cannot downcut and at which neither net erosion nor deposition occurs. The downcutting process prior to reaching equilibrium may take some 100 years up to a few thousand years for the measured downcutting rates of around 1 cm $a^{-1}$, (Frumkin & Ford 1995). The equilibrium profile is a time-independent configuration, maintained as long as boundary conditions do not change.

The caves near the diapir boundary are often multi-storied (Fig. 2b), with high and dry levels hanging above the present base-level (Fig. 3). Assuming that a hanging passage became adjusted to base-level during its development, its present position is explained by a gradual uplift in respect to base-level of erosion after becoming inactive. The components contributing to this 'uplift' are discussed later. The estimated uplift of the top of the diapir is probably associated with salt flowage at depth.

Base-level of erosion changes, relief development and palaeosea-level position have been estimated previously by U series dating of speleothems in relict limestone caves (Ford *et al.* 1981; Gascoyne *et al.* 1983; Mylroie & Carew 1988). This study is the first attempt to estimate rising rates of a salt diapir by dating its cave passages.

## Methods

The elevations of cave outlets were measured to a precision of ±1 cm using an automatic surveying level and a theodolite. Cave passages were surveyed with a tape measure, inclinometer and compass to a precision of ±10 cm.

*From* Alsop, G. I., Blundell, D. J. & Davison, I. (eds), 1996, *Salt Tectonics*, Geological Society Special Publication No. 100, pp. 41–47.

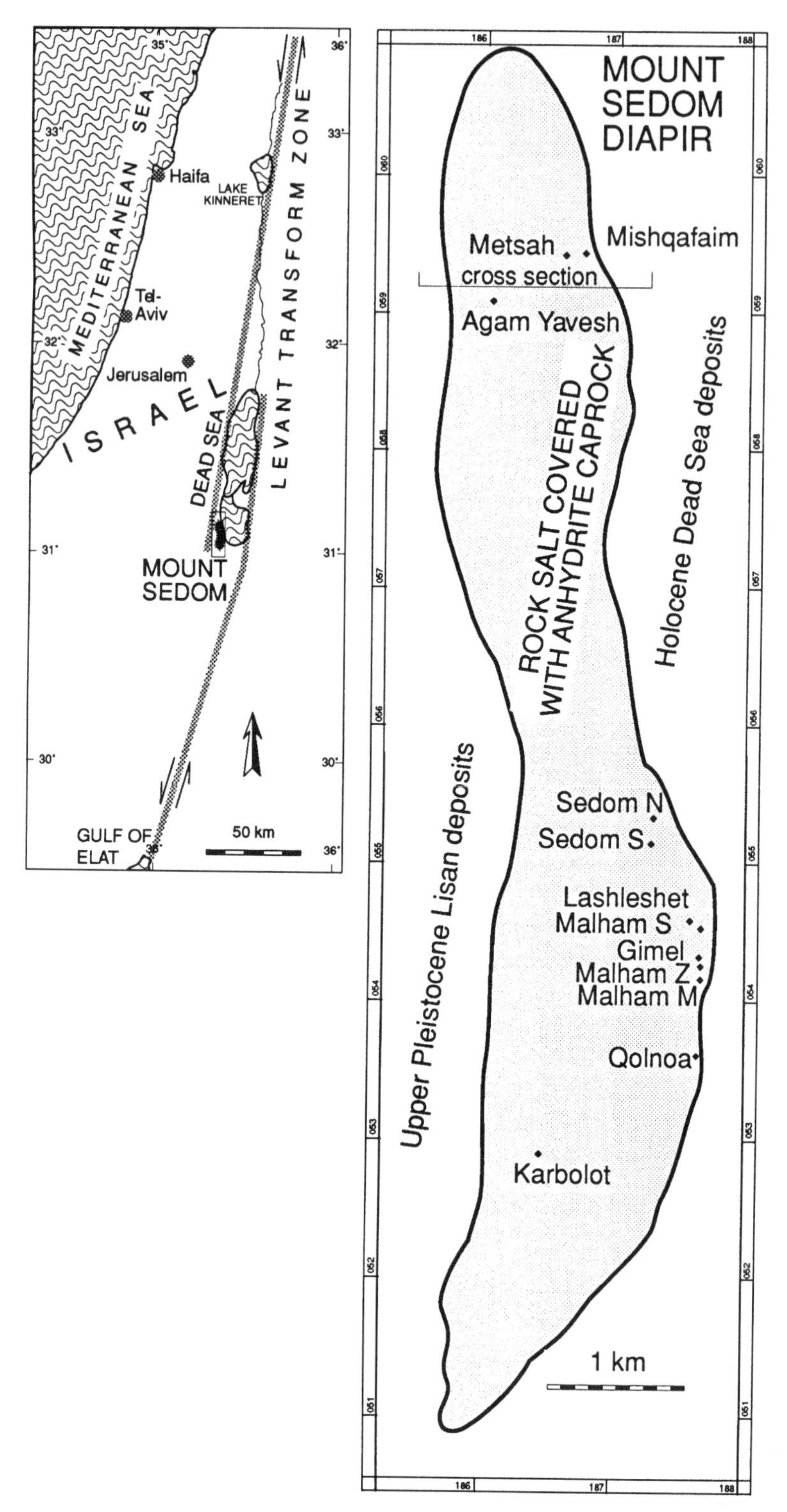


**Fig. 1**. Mount Sedom study area. Main Dead Sea rift faults are drawn schematically (after Garfunkel *et al.* 1981). Cross-section line (Fig. 2) and $^{14}C$ dated caves are shown.

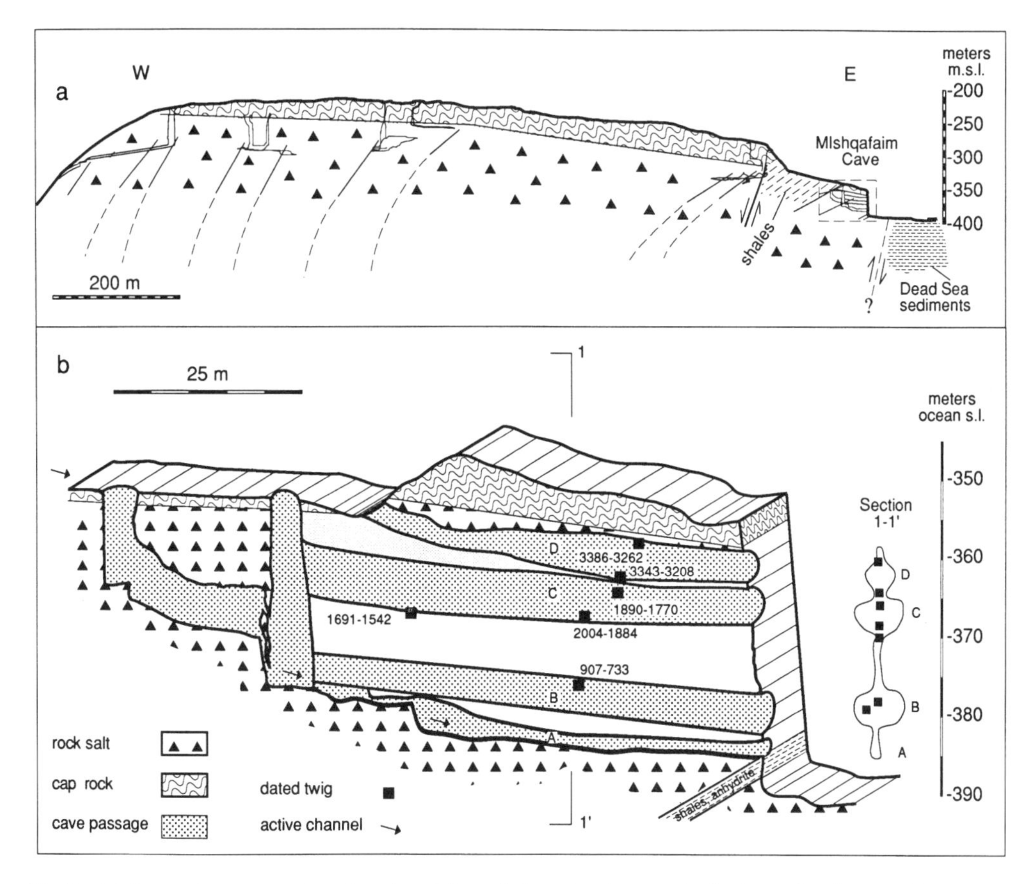


**Fig. 2.** (**a**) Composite cross-section of northern Mount Sedom, showing steeply dipping rock salt covered by anhydrite cap-rock. Broken line shows dip in depth, as indicated by boreholes (J. Charrach, pers. comm.). (**b**) Block diagram and cross-section (1–1′) of Mishqafaim Cave. Black squares indicate dated wood twigs. One standard deviation of calibrated radiocarbon ages is given. The modern passage (A) occurs above a relatively insoluble layer. B, C, D are relict passages from which diapir uplift rates are deduced. Present Dead Sea level is –410 m, having receded from the eastern escarpment of Mount Sedom during the last 80 years.

Wood fragments embedded in alluvial flood sediments within the caves were $^{14}C$ dated (for details see Frumkin *et al.* 1994). The wood is well preserved owing to the arid climate, and is suitable for $^{14}C$ dating. The dates obtained are believed to be close to the actual age of the cave passage (Frumkin & Zak 1991). This assumption is supported by the internal consistency of several dates in each cave (Fig. 2b). The dates are calibrated to calendar ages, using the program CalibETH (Niklaus 1991) which is based on dendrochronological calibration curves (Pearson *et al.* 1986; Stuiver & Pearson 1986). Ages are presented as one standard deviaton range of calendar years BP (before 1950).

As a first-degree approximation, subsequent equilibrated cave levels are assumed to be subparallel (Fig. 2b). The elevation of a dated twig above the currently active channel can therefore be used as a proxy for the respective elevations of the cave levels at the outlet (e.g. the relative elevation of Mishqafaim level D above level A is considered similar to the relative elevations of the outlets of these two levels; Fig. 2b).

## Results

The radiocarbon dates from Mount Sedom Caves were published by Frumkin *et al.* (1991, 1994). Figure 5 summarizes the data for 12 dated caves (for locations see Fig. 1), showing the elevation of 29 wood samples above the active channel versus age.

The oldest passage is 46 m above the currently active channel of the cave and dated to ~8000 a BP. The youngest levels (aged <1400 a) are <10 m

**Fig. 3.** The steep eastern fault escarpment of Mount Sedom at Sedom Cave. The active passage is graded to base level (bottom). A relict, partly collapsed passage is hanging 60 m above base level (top). On its left (centre of picture) is a pillar called 'Lot's wife' by early explorers. The arrow indicates the salt table to the right of the hanging passage – the highest rock salt outcrop in the escarpment, 75 m above base-level. Above the arrow the escarpment is built of cap-rock.

above the active channel. Most dated passages are between the two extremes, with a clear correlation between age and elevation, suggesting a mean rate of 6.2 mm $a^{-1}$ for Mount Sedom uplift with respect to base-level of erosion, during the past 8000 a.

Mishqafaim Cave is shown as a case study of a multi-level cave where the uplift can be observed in detail. The dry passages B, C, D are dated to 900–700, 2000–1600 and 3400–3200 a BP, respectively. Each level was dated by at least two wood samples whose locations are shown on Fig. 2b. The three distinct dry levels may correspond to stable periods in the evolution of the diapir. Relatively rapid uplift seems to have occurred during the periods 3200–2000, 1600–900 and 700–present. Assuming this time frame for rapid uplift, the elevation difference between successive levels suggests uplift rates with respect to base-level of 5, 18 and 4 mm $a^{-1}$ in each period respectively. Today down-cutting is inhibited by a relatively insoluble layer of shales and anhydrite, causing the currently active channel to hang above base level (Fig. 2). Such cases of lithologic constraint may lead to underestimation of diapir uplift rates.

The calculated mean rate of uplift relative to base-level during the last 3300 a at Mishqafaim Cave is 7 mm $a^{-1}$. This agrees well with the mean value of 6.2 mm $a^{-1}$ for all 12 dated caves (Fig. 5).

## Discussion

The mean 6–7 mm $a^{-1}$ rate of uplift with respect to base level is actually composed of two components: uplift of the exposed portions of the diapir, and Dead Sea fluctuations. If the Dead Sea level fluctuated around a constant value during the

**Fig. 4.** A tectonically deformed salt cave passage near the eastern escarpment. A vertical fault plane in the centre (behind and above the person) exhibits vertical movement. Note the 2 m offset of the dissolution horizontal plane (above the head of the person, left of fault, and to the right of his chest). Vertical salt beds trend from left to right, across the horizontal dissolution notches and the fault.

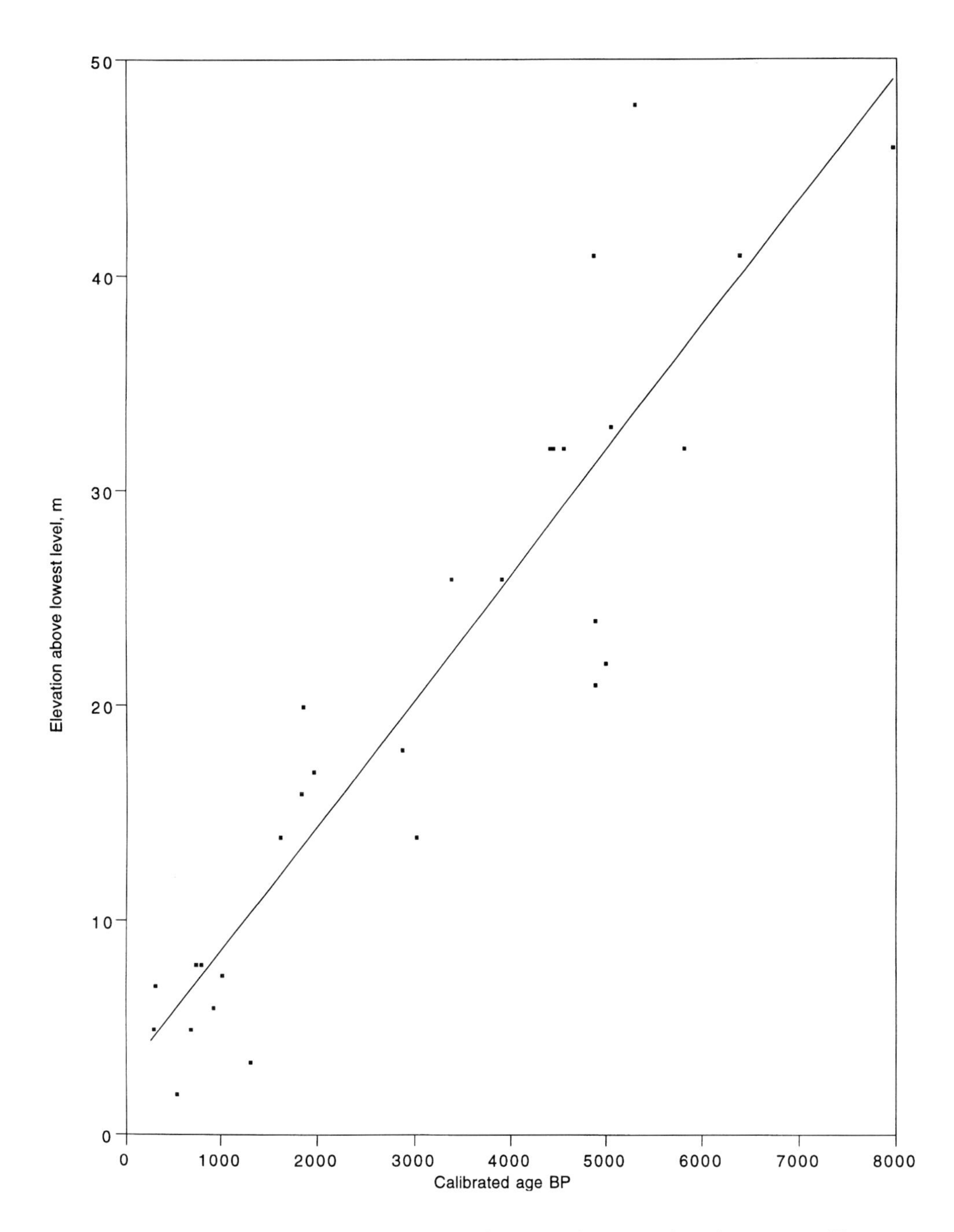


**Fig. 5.** Elevation (above the lowest level in a given cave) of the Mount Sedom wood samples versus age. The correlation suggests a mean downcutting rate of 6.2 mm $a^{-1}$ during the last 8000 a. This is attributed mainly to base-level lowering, resulting from diapir uplift and Dead Sea fluctuations.

Holocene, then it would not have contributed to the measured value of 6–7 mm $a^{-1}$. This value would then be attributable entirely to diapir uplift, with the scatter around the regression line of Fig. 5 attributed to Dead Sea level fluctuations. However, other evidence suggests higher Dead Sea levels during the early Holocene (for example, a salt crust covering land snails, indicating a –280 m level at *c.* 6700 a

BP; (Goodfriend *et al.* 1986), and gradual desiccation with fluctuations during the mid- to late Holocene (Frumkin *et al.* 1991). This suggests that the actual uplift rate of the diapir was somewhat less than 6–7 mm $a^{-1}$.

The dependence of inferred uplift rates on assumptions of Dead Sea levels can be demonstrated in Mishqafaim Cave (Fig. 2b). During the formation of level D in Mishqafaim Cave, the Dead Sea may have reached a level of –375 m (Klein 1982). This would result in a diapir rising rate of ~3.5 mm $a^{-1}$ during the last 3300 a. A similar value was calculated by Zak (1967) for southwestern Mount Sedom, and was adopted by Frumkin *et al.* (1991).

The distinct levels seen in Mishqafaim Cave (Fig. 2) and others may be a result of stability with respect to base-level, during which either the diapir was not rising significantly, or its uplift rate was comparable to the Dead Sea rise rate. However, distinct cave levels may also occur as a result of climatic change or channel piracy by a lower fissure, without distinct base-level control (Ford and Williams 1989; Palmer 1987). If we assume base-level control for each level in Mishqafaim Cave, then the rates of 4–18 mm $a^{-1}$ for the rapid uplift periods between formation of the successive levels indicate highly fluctuating uplift rates.

Fluctuating uplift rates suggest that the upward movement of the exposed portion of the diapir is being retarded by salt viscosity and resistance of sedimentary rocks breached by the diapir. The interplay between these forces and the buoyancy force causing the diapiric motion determine the rate of uplift at any one moment in geological time. The time scale for major changes in uplift rates of the Mount Sedom diapir appears to be 1000–1500 a. This does not exclude, however, short-term high-rate uplift events, whose timing may not be detected by the method used here.

Differential fault movement within the diapir and rotation of its upper portion (Frumkin 1992) may introduce variable uplift rates to different parts of the diapir. The results presented here should therefore be considered as an approximation representing regions with dated caves, specifically the eastern margin of Mount Sedom.

## Conclusions

The maximum mean uplift rate relative to base-level for the diapir along its eastern fault is 6–7 mm $a^{-1}$ during the last 8000 a. Lower mean values are possible, depending on exact Dead Sea levels during cave development. The uplift rate probably fluctuated widely in response to interplay between buoyant pressure release and inhibiting forces. This behaviour should be accounted for in diapirism models (Lerche & O'Brien 1987). The recurring interval of major uplift events in Mount Sedom is approximately 1000–1500 a.

A different approach for estimating short-term diapir uplift rates is currently applied by direct geodetic measurement across the diapir border, conducted by the author repeatedly since 1990. As this method requires a long observation period, the results will be discussed in the future.

Funding for this research was provided by the Dead Sea Works. The Natural Sciences and Engineering Research Council of Canada supported the author during writing of the paper. The Israel Cave Research Center team helped the field work. $^{14}C$ dating was done by Mr Israel Carmi in the Weizmann Institute of Science. Mr J. Charrach provided encouragement and help and Dr D. C. Ford provided useful remarks.

## References

FORD, D. C. & WILLIAMS, P. W. 1989. *Karst Geomorphology and Hydrology*. Unwin Hyman, London.

——, SCHWARCZ, H. P., DRAKE, J. J., GASCOYNE, M., HARMON, R. S. & LATHAM, A. G. 1981. Estimates of the age of the existing relief within the southern Rocky Mountains of Canada. *Arctic and Alpine Research*, **13**, 1–10.

FRUMKIN, A. 1992. *The Karst System of the Mount Sedom Salt Diapir*. PhD thesis (in Hebrew, English abstract), The Hebrew University.

—— 1994. Morphological features and development of salt caves. *Bulletin of the National Speleological Society*, **56**, 82–95.

—— & FORD, D. C. 1995. Rapid entrenchment of stream profiles in the salt caves of Mount Sedom, Israel, *Earth Surface Processes and Landforms*, **20**, 139–152.

—— & ZAK, I. 1991. Holocene evolution of Mount Sedom Diapir based on karst evidence (abstract). *In*: WEINBERGER, G. (ed.) *Annual Meeting, Akko, Israel Geological Society*, 35.

——, CARMI, I., ZAK, I. & MAGARITZ, M. 1994. Middle Holocene environmental change determined from the salt caves of Mount Sedom, Israel. *In:* BAR-YOSEF, O. & KRA, R. (eds) *Late Quaternary Chronology and Paleoclimates of the Eastern Mediterranean*. Tucson, Radiocarbon, The University of Arizona, 315–332.

——, MAGARITZ, M., CARMI, I. & ZAK, I. 1991. The Holocene climatic record of the salt caves of Mount Sedom, Israel. *The Holocene*, **1**, 191–200.

GARFUNKEL, Z., ZAK, I. & FREUND, R. 1981. Active faulting in the Dead Sea rift. *Tectonophysics*, **80**, 81–108.

GASCOYNE, M., FORD, D. C. & SCHWARCZ, H. P. 1983. Rates of cave and landform development in the Yorkshire Dales from speleothem age data. *Earth Surface Processes and Landforms*, **8**, 557–568.

GERSON, R. & INBAR, M. 1974. The field study program of the Jerusalem-Elat symposium, 1974. Reviews and summaries of Israeli research projects. *Zeitschrift für Geomorphologie Supplementband*, **20**, 7–11.

GOODFRIEND, G. A., MAGARITZ, M. & CARMI, I. 1986. A high stand of the Dead Sea at the end of the Neolithic period: paleoclimatic and archeological implications. *Climatic Changes*, **9**, 349–356.

HOROWITZ, A. 1992. *Palynology of Arid Lands*, Elsevier, Amsterdam.

KLEIN, C. 1982. Morphological evidence of lake level changes, western shore of the Dead Sea. *Israel Journal of Earth-Sciences*, **31**, 67–94.

KOVACH, R. L. & BEN AVRAHAM, Z. 1990. Geologic and tectonic processes of the Dead Sea Rift zone. *Tectonophysics*, **80**, 1–137.

LERCHE, I. & O'BRIEN, J. J. 1987. Modelling of buoyant salt diapirism, *In*: LERCHE, I. & O'BRIEN, J. J. (eds) *Dynamical Geology of Salt and Related Structures*. Academic Press, Orlando, 129–162.

MYLROIE, J. E. & CAREW, J. 1988. Solution conduits as indicators of Late Quaternary sea level position. *Quaternary Science Reviews*, **7**, 55–64.

NEEV, D. & HALL, J. K. 1979. Geophysical investigations in the Dead Sea. *Sedimentary Geology*, **23**, 209–238.

NIKLAUS, T. R. 1991. *CalibETH User's Manual*. ETH, Zürich.

PALMER, A. N. 1987. Cave levels and their interpretation. *NSS Bulletin*, **49**, 50–66.

PEARSON, G. W., PILCHER, J. R., BAILLIE, M. G. L., CORBETT, D. M. & QUA, F. 1986. High-precision $^{14}C$ measurement of Irish Oaks to show the natural $^{14}C$ variations from A.D. 1840 to 5210 B.C. *Radiocarbon*, **28**, 911–934.

STEIN, M., AGNON, A., STARINSKY, A., RAAB, M., KATZ, A. & ZAK, I. 1994. What is the "age" of the Sedom Formation? *Annual meeting, Israel Geological Society*, 108 (abstract).

STUIVER, M. & PEARSON, G. W. 1986. High-precision calibration of the radiocarbon time scale, A.D. 1950–500 B.C. *Radiocarbon*, **28**, 805–838.

ZAK, I. 1967. *The Geology of Mount Sedom*. PhD thesis, (in Hebrew, English abstract), The Hebrew University.

—— & BENTOR, Y. K. 1968. Some new data on the salt deposits of the Dead Sea area, Israel. *Symposium on the Geology of Saline Deposits*. Unesco, Hanover, 137–146.

—— & FREUND, R., 1980. Strain measurements in eastern marginal shear zone of Mount Sedom salt diapir, Israel. *AAPG Bulletin*, **64**, 568–581.

# Hydration diapirism: a climate-related initiation of evaporite mounds in two continental Neogene basins of central Spain

MANUEL HOYOS[1], MIGUEL DOBLAS[1], SERGIO SÁNCHEZ-MORAL[1], JUAN CARLOS CAÑAVERAS[1], SALVADOR ORDOÑEZ[2], CARMEN SESÉ[1], ENRIQUE SANZ[1] & VICENTE MAHECHA[1].

[1]*Museo Nacional de Ciencias Naturales, Consejo Superior de Investigaciones Científicas (C.S.I.C.), José Gutierrez Abascal 2, 28006 Madrid, Spain*

[2]*Departamento de Ciencias Ambientales y Recursos Naturales, División de Geología, Universidad de Alicante, 03690 Alicante, Spain*

**Abstract**: While many halokinetic models consider that a thick overburden is necessary before diapirism develops, other studies show that salt might begin to move soon after deposition in response to different factors (e.g. regional extension). We describe two continental Neogene basins in central Spain, characterized by three Miocene sedimentary units, where unusually early initiation of evaporite mounds or 'incipient pillows' occurs, with no overburden on top of them. The arguments suggesting the early bulging of the evaporites of the Lower Unit, forming a palaeorelief before and during the deposition of the Intermediate Unit, rely on certain characteristics of the palaeokarsts developed on the roof of the Lower Unit, as well as in the nature, disposition, thickness, and synsedimentary processes of the Intermediate Unit. Folds, faults, tension gashes, and collapse structures develop within the evaporite mounds, and these deformation features might be related to their uplift history. The agent of initiation suggested in this paper for the evaporite mounds 'hydration diapirism', or a volume increase related to the hydration of anhydrite to gypsum has not been previously recognized in the literature. A series of sedimentological and palaeontological arguments further suggest that this chemical reaction was triggered by a climatic change at the boundary between the Lower and the Intermediate Units.

Timing and agents of initiation of diapirism are two of the most interesting topics in salt tectonics. The halokinetic model of Trusheim (1960) for gravity-driven salt tectonics considered that a thick overburden was essential for the generation of pillows, and their subsequent evolution into diapirs. Even if this early notion has been maintained by some authors (Jenyon 1986), a growing body of evidence from seismic studies (Hughes 1968) and quantitative reconstructions of cross-sections (Wu *et al.* 1990) tends to indicate that salt might begin to move soon after deposition, even below thin overburdens (Jackson & Talbot 1994; Jackson & Vendeville 1994). According to Jackson & Vendeville (1994), 'early' initiation 'refers to the time soon after salt deposition when overburden was thinner than 1 km'. In this sense, the concept is linked to the existence of at least some overburden, normally between 350 and 1000 m (Jackson & Vendeville 1994).

The possible agent of initiation of salt pillows is the other debatable subject, and it is one of the least understood aspects of salt tectonics (Jackson & Vendeville 1994). Several processes have been suggested to initiate salt movement: erosional unloading (Evans 1993), sedimentary differential loading, large-scale thermal convection of salt, local heat sources, gravity gliding and spreading (Jackson & Talbot 1986), sub-salt deformation, and thin-skinned extension and contraction (Woodbury *et al.* 1980; Jackson & Talbot 1994).

In this paper, we study two continental Neogene basins (Madrid and Calatayud), characterized by three sedimentary units: a 'Lower Unit', comprising up to 500 m of Agenian and Aragonian sediments with detrital deposits in the borders, passing to evaporites in the centre; and up to 150 m of detrital and carbonate sediments, constituting the 'Intermediate and Upper Units' of the Miocene in central Spain (Hoyos & López-Martínez 1985; Calvo *et al.* 1990).

These evaporite mounds show unique features in their evolution in relation to the timing and agents of initiation of diapirism. In the first place, these mounds began to move unusually early in their history, i.e. when no overburden whatsoever existed on top of them. Several authors have previously recognized the existence of palaeokarsts, upwelling phenomena, and the occurrence of internal highs in the Madrid Basin (Alberdi *et al.* 1983; Martín Escorza 1983; Hoyos *et al.* 1985; Calvo *et al.* 1984; Ordoñez & García del Cura 1992).

*From* ALSOP, G. I., BLUNDELL, D. J. & DAVISON, I. (eds), 1996, *Salt Tectonics*, Geological Society Special Publication No. 100, pp. 49–63.

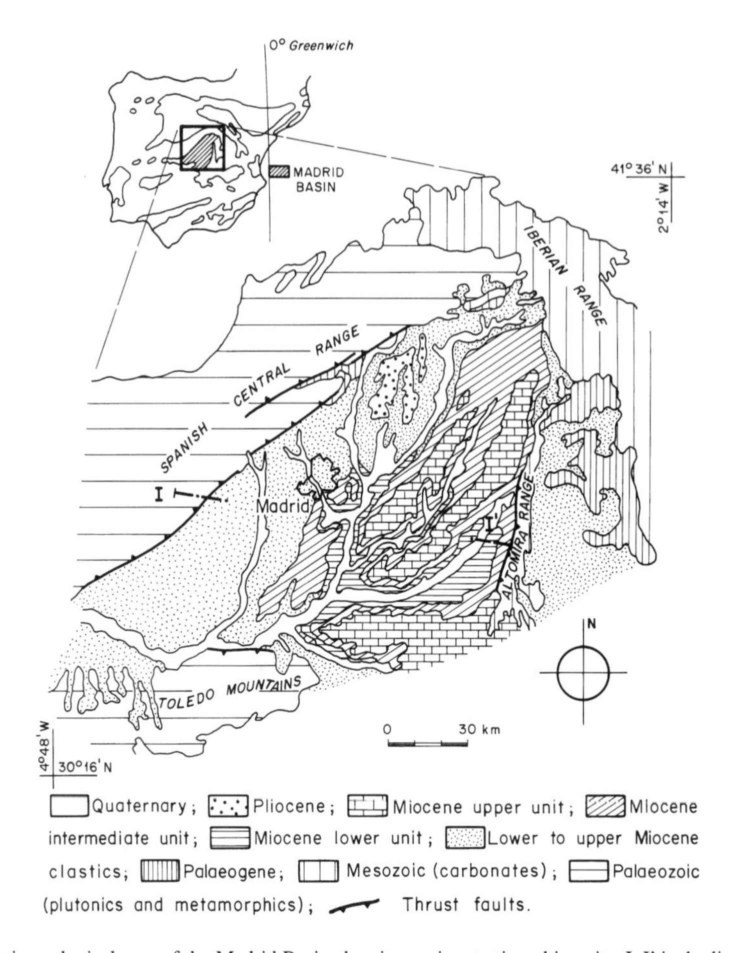


**Fig. 1**. Schematic geological map of the Madrid Basin showing major stratigraphic units. I–I' is the line of the cross-section in Fig. 8.

It is suggested that the agent of initiation of the evaporite mounds is related to an increase in volume associated with the chemical reaction anhydrite + water —> gypsum. This chemically induced disturbance is termed 'hydration diapirism', and, to our knowledge, it has not been identified previously as a trigger in the initiation of diapirism. We further suggest that hydration diapirism is the result of a climatic change in the study area.

Tsui & Cruden (1984) described the effect of hydration of anhydrite to gypsum in the genesis of anticlinal structures. One mechanism suggested for the formation of the intestine-like enterolithic folds is related to volume changes, such as the hydration of anhydrite to gypsum (Jackson & Talbot 1994). Oedometer tests carried out by Nuesch *et al.* (1984) have shown that the hydration of samples with shale and anhydrite might yield dramatic volume increases of up to 120%. Geological hazard assessment in the northeastern part of Spain (Ebro Basin), suggested that the hydration of anhydrite to gypsum was responsible for rates up to 1 cm $a^{-1}$ of ground elevation change (Ordoñez *et al.* 1990). Climate-related salt tectonics has been described in Iranian salt glaciers, which have been shown to flow after rainfall at rates much faster than predicted by experiments (Talbot & Jarvis 1984).

## The Madrid Basin

The Neogene Madrid Basin occupies more than 10 000 $km^2$ in the centre of the Iberian Peninsula (Fig. 1). It is bounded by different geological units: the Iberian Variscan Massif to the north, west, and south, with the NE–SW orientated Spanish Central Range and the E–W trending Toledo Mountains (Palaeozoic metamorphics and granites). To the east, the Madrid Basin is bounded by the NW–SE orientated Iberian Range, and the N–S trending

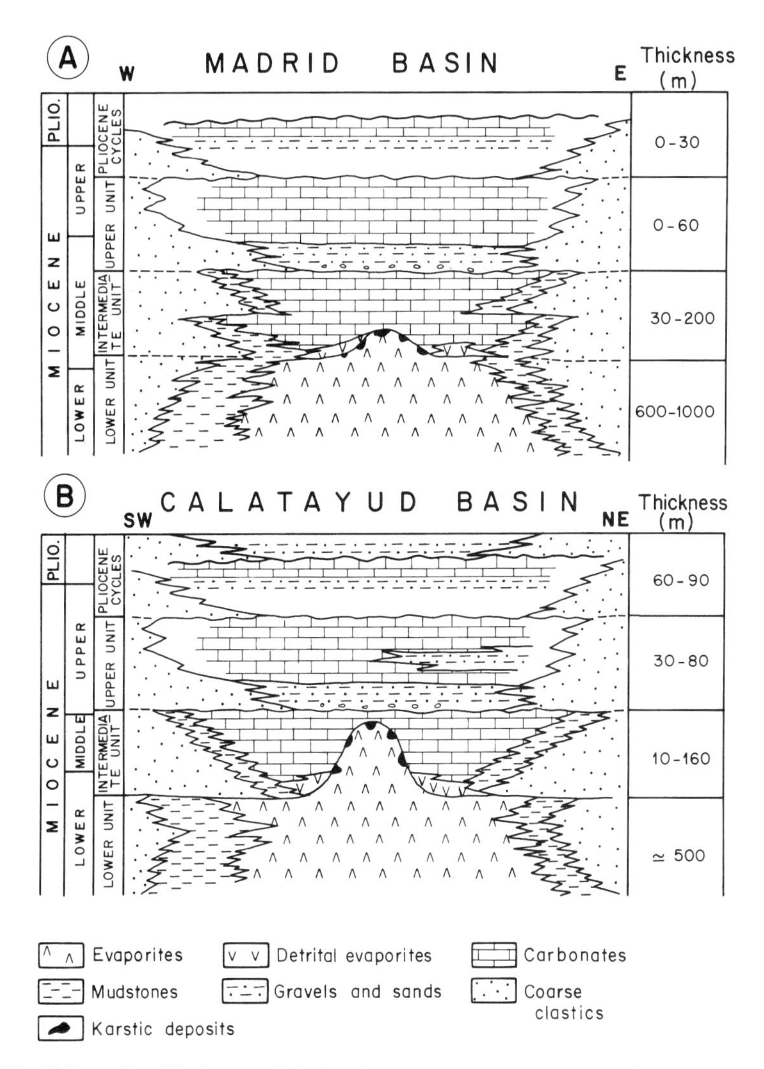


**Fig. 2.** Simplified lithostratigraphic sketches depicting the sedimentary series of the Madrid and Calatayud Basins. Three major units are recognized during the Miocene, separated by discontinuities: Lower Unit, Intermediate Unit, and Upper Unit.

Altomira Range, both characterized by Mesozoic carbonates and detrital sediments (Martín Escorza 1983; Calvo *et al.* 1990).

The Madrid Basin is filled with a sequence of Tertiary sediments, showing a marked thickness asymmetry: 3000 m in the west and 1500 m in the east. The basal part to the west and in the centre of the basin displays Cretaceous carbonates, while in the east these are in places underlain by Triassic evaporites (Querol 1989). The marine Mesozoic basin occupied the eastern half of the Iberian Peninsula, and the Madrid area corresponded to the western margin of this basin. Mesozoic deposits are overlain by a thick succession of Palaeogene clastics and evaporites. The Palaeogene continental basin occupied a much wider area in central Spain as compared with the present shape of the Madrid Basin. The sedimentary infilling of the Madrid Basin, as a true and separate tectonic trough, is Miocene in age, and it was related to the Alpine uplift of its surrounding ranges. The Miocene sediments occupy most of the areal extent of the Madrid Basin. The thickness of the complete Neogene series ranges between 800 and 1200 m.

Three main Miocene sedimentary units have been distinguished in the Madrid Basin (Fig. 2A; Alberdi *et al.* 1983; Cañaveras 1994): Lower Unit Intermediate Unit, and Upper Unit. These units are

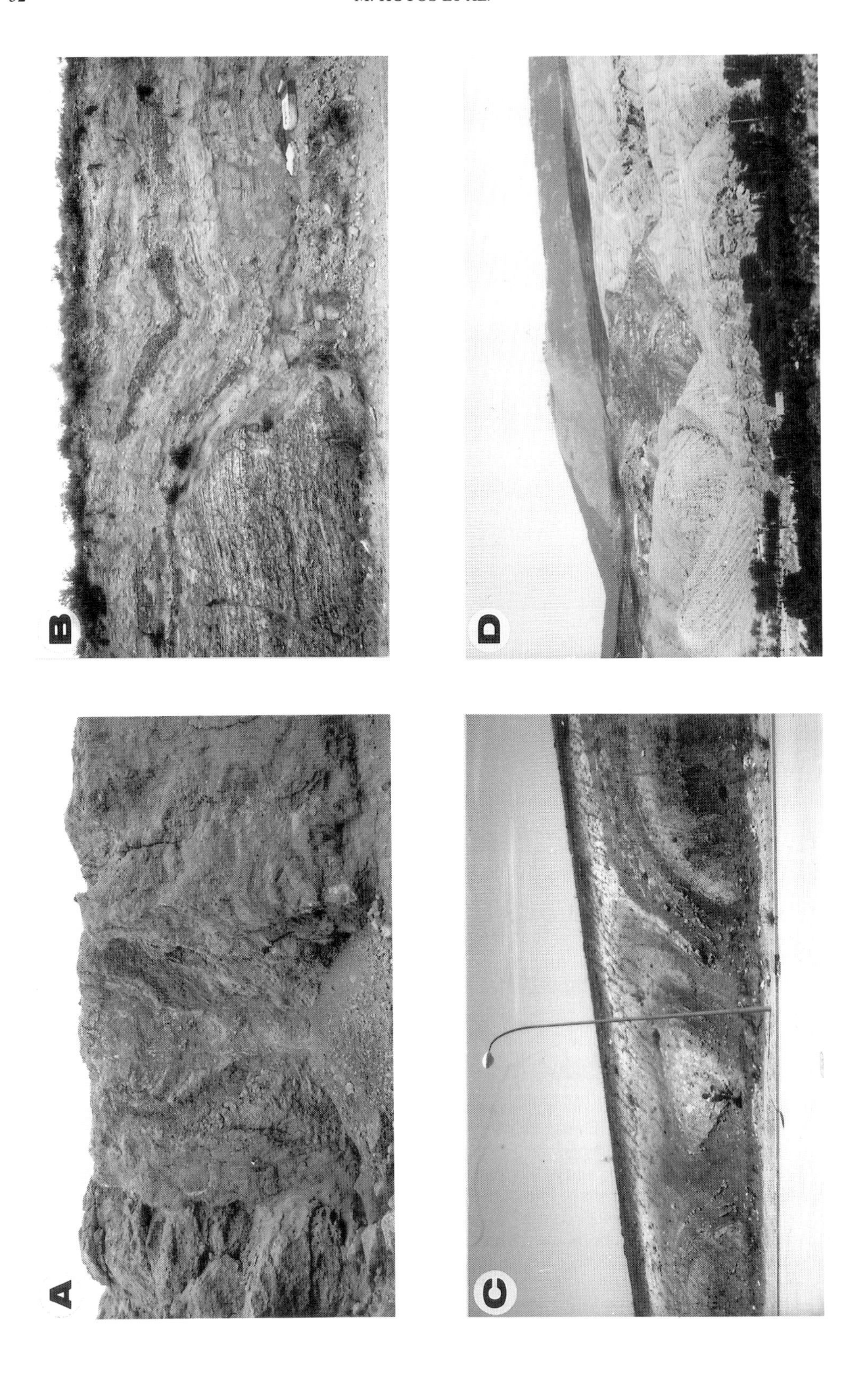

B
D
A
C

separated by two sedimentary discontinuities. The Lower Unit (600–1000 m of Agenian to the Middle Aragonian deposits) is characterized by the following elements from the borders to the centre (Ordoñez *et al.* 1991; Ordoñez & García del Cura 1992; Sanchez-Moral *et al.* 1993): alluvial fans with a maximum width of 50 km in the northwestern border (these fans are only 3–4 km wide in the southern and eastern borders); saline mudflats and playa lakes with transitional facies; central evaporitic facies, deposited in saline lakes with shallow waters. The evaporitic domain has an approximate thickness of 500 m and, from the base to the top, consists of the following sequence (García del Cura 1979; Ordoñez *et al.* 1991): (1) a lowermost section with alternating anhydrite, halite, glauberite, polyhalite and magnesite, enriched towards the borders of the basin with early diagenetic glauberite, massive thenardite, and occasional reddish muds with halite and/or glauberite (Ordoñez & García del Cura 1994); and (2) an upper part where anhydrite and gypsum are the most abundant phases, with occasional halite. Anhydrite and halite rarely outcrop, but they have been observed in boreholes. Anhydrite presents nodular and laminar fabrics, with many varieties of habit (Ordoñez *et al.* 1991). Secondary gypsum derived from the hydration of anhydrite is the dominant evaporite and is the only type exposed at the surface, with a thickness of up to 120 m (primary gypsum is negligible): this secondary gypsum commonly displays pseudomorphic fabrics and relicts of anhydrite.

The Intermediate Unit spans the middle/late Aragonian to the early Vallesian, with thicknesses between 30 and 200 m. Alluvial fans outcrop along the borders, while clastics and lacustrine and palustrine carbonates occur in the centre. Finally, the Upper Unit (late Vallesian to the late Turolian) displays fluvial and shallow-water carbonates, with a maximum thickness of 50 m. The Miocene sediments are locally overlain by Pliocene clastics and carbonates, and alluvial and fluvial Quaternary sediments (0–30 m). A series of palaeokarsts and collapse 'dolines' (or sinkholes) are found in this basin on top of the Lower Unit, as well as higher in the series (Figs 3A and 3C; Hoyos *et al.* 1985; Hoyos & López-Martínez 1985; López-Martínez *et al.* 1987).

The triangular shape of the Madrid Basin is conditioned by a series of bounding thrust faults (NE–SW, Spanish Central Range; E–W, Toledo Mountains; and N–S, Altomira Range) of late Variscan age (Doblas 1991), which were reactivated during the Alpine formation of the Madrid Basin (Fig. 1; Martín Escorza 1983). The lateral facies variations of the Miocene sediments, and the present orientation of the fluvial network were controlled by a series of NE–SW and E–W striking faults (Alía 1960; Martín Escorza 1976, 1983). In contrast to the bounding faults, the ones found within the Madrid Basin are mostly normal. Block tilting phenomena and large-scale gentle folds are typically associated with the NE–SW fault trends within the basin.

Folds, faults, tension gashes, tilted blocks and collapse structures are found within the evaporites of the Madrid Basin, and three major hypotheses have been suggested to explain these deformation features (Martín Escorza 1983): (1) halokinesis acting upon the gypsum layers (Alía 1960); (2) deformation events related to external tectonics (Capote & Fernández Casals 1978); (3) a mixed scheme involving basement faults acting upon weaker gypsum layers (Martín Escorza 1983). In this paper, we propose that the deformation can be explained by the uplift of the evaporite mound.

The possible Alpine tectonic phases which gave rise to this Neogene trough are poorly constrained (Martín Escorza 1983) as: (1) the pre-Miocene sediments are restricted to the borders, and they show strong facies variations; (2) most of the continental sediments display a poor faunal content; and (3) most of the discordances are local in character, and only two major sedimentary discontinuities bounding the Miocene units are of regional extent (Antunes *et al.* 1987). In this sense, recent attempts to unravel the tectonic history of the Madrid Basin using fault population analyses (Vicente *et al.* 1990, in press; Calvo *et al.* 1991) are difficult to accept. Moreover, the data used by these authors are critically unconstrained in several ways: (1) most of the faults they measure are found in the surrounding

**Fig. 3.** Some examples of palaeokarsts and collapse dolines in the two spanish continental basins. (A) Palaeokarst with an exokarstic depression filled with in situ sediments, developed on the roof of the gypsum deposits of the Lower Unit (Madrid Basin; Arganda; Long: 3°26′W; Lat: 40°18′N ). The width of the section is 13 m. (B) Collapse doline affecting the gypsum of the Lower Unit (horizontal sediments to the left), filled with finely banded carbonates of the Intermediate Unit (folded sediments to the right), in the Calatayud Basin (Munebrega; Long: 1°41′W; Lat: 41°15′N). Note the synsedimentary character of these later deposits, with respect to the collapse event (the thickness of the banded carbonates increases towards the right); the width of the section is 5 m. (C) Collapse doline affecting a series of lutites and palustrine carbonates of the Intermediate Unit in the Madrid Basin (Vallecas; Long: 3°33′W; Lat: 40°23′N). (D) Large-scale collapse doline within the Calatayud Basin (Munebrega; Long: 1°41′W; Lat: 41°15′N). The subhorizontal light-grey sediments surrounding the doline are gypsum deposits of the Lower Unit. The doline is filled with detrital and carbonate sediments of the Upper Unit.

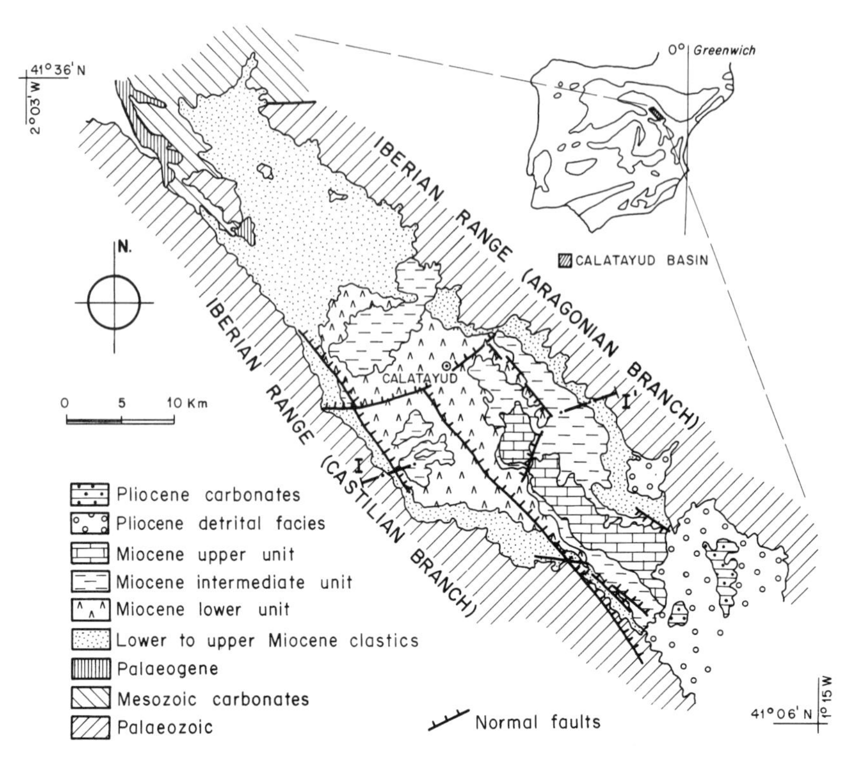


**Fig. 4.** Schematic geological map of the Calatayud Basin showing major stratigraphic units. I–I' is the line of the cross-section shown in Fig. 9.

ranges (and not within the basin), and they are often of late Variscan age, with "supposedly" Alpine reactivations (only relative ages are available); (2) the number and senses of movement since the late Variscan are unknown, as no detailed kinematic analyses of slickenside planes have been undertaken; (3) these authors use stress inversion methods which have been found to be based on 'invalid assumptions' (Pollard *et al.* 1993; Cashman & Ellis 1994); and (4) these fault analyses only account for the classic horizontal stresses, ignoring the relevance of vertical stresses associated with halokinesis and the upwelling of the salt mound (particularly in the case of the faults found within the basin). Using these unconstrained data, Vicente *et al.* (1990, in press) and Calvo *et al.* (1991) propose three deformation episodes for the Madrid Basin, including a final extensional phase following the deposition of the Intermediate Unit.

However, several arguments suggest that compressional tectonics acted during the whole Miocene: (1) the bounding reverse faults of the Spanish Central Range and the Toledo mountains were active until at least the upper Miocene/Pliocene (Martín Escorza 1990); (2) the Iberian Range to the east displays a series of Miocene to Pliocene folds, overriding the Tertiary sediments of the Madrid Basin; (3) a series of compressional structures are found in the roof of the Miocene Upper Unit, i.e. contractional karstic tubes, small-scale folds, etc.; (4) the normal faults, which are restricted to the evaporites of the Madrid Basin and do not affect the substratum, only reflect localized extensional stresses related to the uplift of the evaporite mound; (5) the present-day geotectonic setting of central Spain is compressional – the Madrid Basin constitutes an intraplate trough subject to the convergence of Africa and Eurasia during the Miocene, with NW directed indentation in the southeastern corner of the Iberian Peninsula (Doblas *et al.* 1991). Compression was transmitted towards central Spain through thin-skinned thrusting, involving a fundamental detachment at mid-crustal level, with the Spanish Central Range and the Toledo Mountains acting as 'pop-ups' (Banks & Warburton 1991; Doblas *et al.* 1991).

## The Calatayud Basin

The Calatayud Basin constitutes an intermontane basin elongated parallel to the NW–SE orientated Iberian Range (Fig. 4). The Iberian Range is an

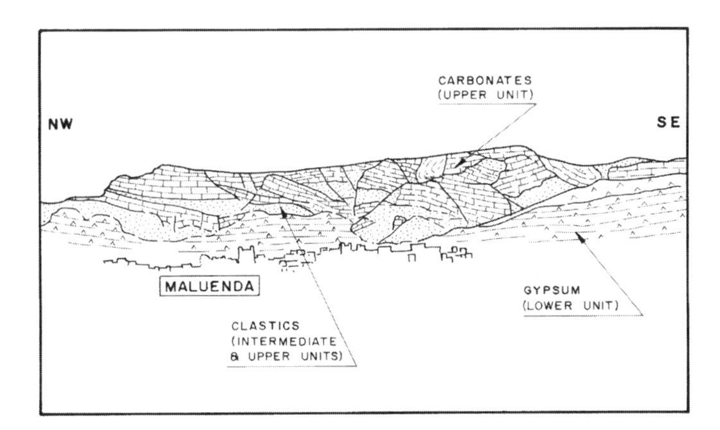

**Fig. 5.** Complex collapse/chaotic structure in the Calatayud Basin (Maluenda; Long: 1°37′W; Lat: 41°17′N). The detrital and carbonate deposits of the Intermediate and Upper Units collapsed on top of the gypsum of the Lower Unit.

intraplate fold-and-thrust belt composed of Palaeozoic and Mesozoic rocks, which underwent a three-stage evolution: (1) rifting and sedimentation during the Mesozoic; (2) crustal thickening and thin-skinned thrusting during the Palaeogene compression; and (3) post-orogenic transtension and crustal thinning, from the Neogene to the present (Capote 1983; Simón 1990).

The depth of the basin exceeds 1500 m and the lower part is unknown, except for the Palaeogene sediments. The Palaeogene only outcrops in the northwest of the basin, and consists of alluvial fan deposits with conglomerates, lutites and marls. As in the Madrid Basin, the Miocene series consists of three units (Fig. 2B). The Lower Unit is composed of more than 500 m of Agenian age sediments with clastics in the borders, progressively passing to evaporites in the centre with palaeokarsts in their roof (Hoyos & López Martínez 1985). The evaporitic facies consist essentially of gypsum and associated soluble saline phases (anhydrite and halite), with occasional intercalations of lutites and dolomitic marls which become more abundant towards the margins of the basin. Secondary gypsum is the dominant exposed evaporite, with a thickness of up to 200 m. Secondary gypsum is derived from the hydration of anhydrite, as revealed by petrography (Ortí *et al.* 1994). Boreholes in the central part of the basin reveal thick deposits of anhydrite (more than 400 m) and some halite, underneath the secondary gypsum. Both the Intermediate Unit (Agenian to Turolian age, with maximum thickness around 160 m), and the Upper Unit (35–75 m of Turolian to Ruscinian age deposits), consist of clastic sediments and lacustrine and palustrine carbonate rocks. The Miocene is covered

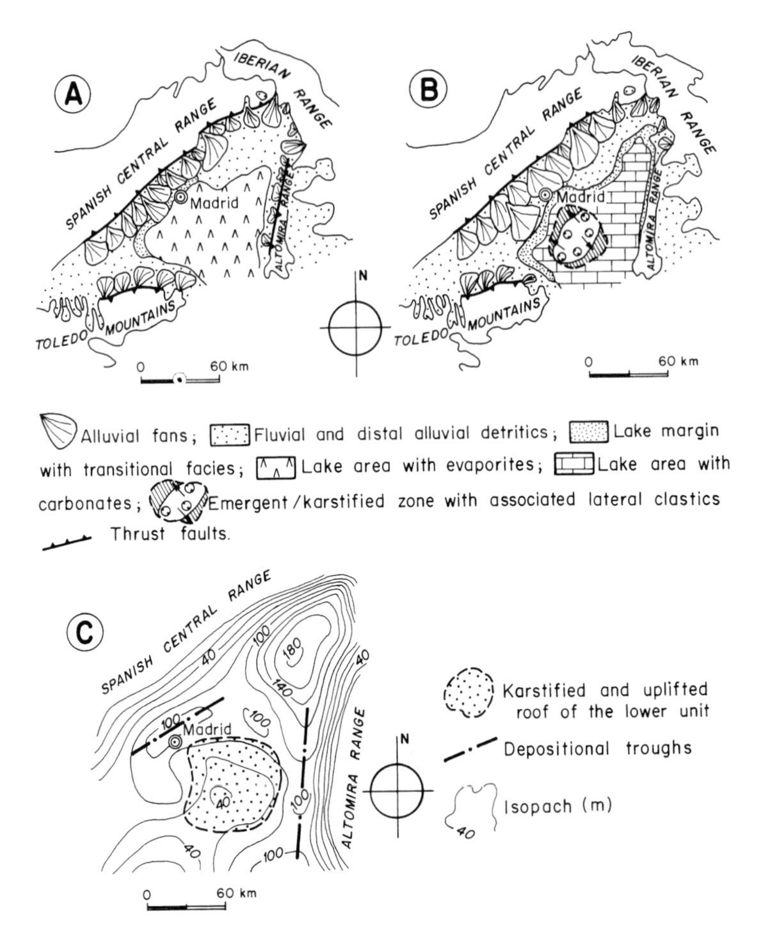


**Fig. 6.** The Madrid Basin during the lower and middle Miocene. (**A**) Palaeogeographic map, showing the facies distribution at the end of the Lower Unit. (**B**) Palaeogeographic map, showing the facies distribution during the middle part of the Intermediate Unit; an uplifted evaporite mound is shown in the central sector of the basin. (**C**) Isopach map of the Intermediate Unit (modified from Vicente *et al.* in press), highlighting the position of the evaporite mound.

by Plio/Quaternary clastics and carbonates. As in the Madrid Basin, dolines, palaeokarsts and collapse structures are found during the whole Miocene evolution of Calatayud (Figs 3B, 3D, and 5).

During the Miocene, the Calatayud Basin underwent extensional tectonics, together with a dextral strike-slip component along NW–SE bounding faults. The geotectonic setting of the Calatayud Basin is rather complex, being the result of three factors: (1) southward-directed compression from the Pyrenees (Simón 1990); (2) north- to northwest-directed indentation from the southeastern corner of the Betic cordilleras (Doblas *et al.* 1991; López-Ruiz *et al.* 1993); and (3) generalized low-angle extension disrupting the western Mediterranean (Valencia and Alborán basins; (Doblas & Oyarzun, 1989a, b, 1990).

Gravitational folds and normal faults are commonly found in the central evaporitic sectors of the basin, associated with the uplift of the mound. Abundant sets of fractures and tension gashes filled with fibrous gypsum only exist in the dry mudflat areas surrounding the evaporite mound.

## Early initiation of evaporite mounds in central Spain

In the Madrid and Calatayud Basins, internal highs developed at the site of the previous evaporitic depocentres of the Lower Unit, coevally with the initial pulses of the Intermediate Unit (Figs 6 and 7). Four different arguments clearly indicate the early initiation of these evaporite mounds when no overburden covered the deposits.

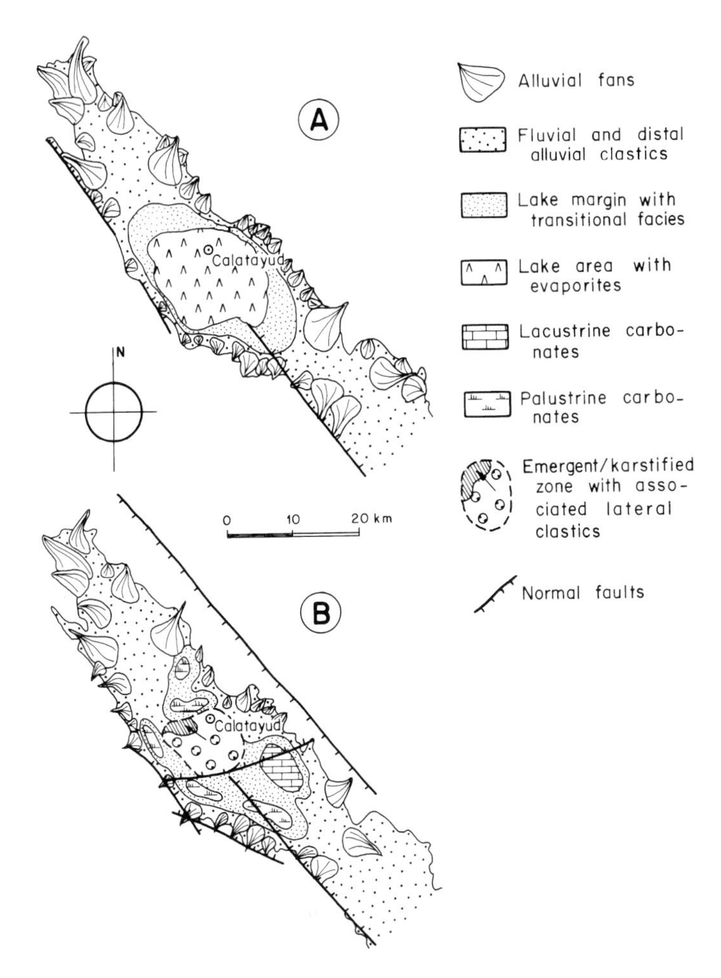


**Fig. 7.** The Calatayud Basin during the lower and middle Miocene. Palaeogeographic maps showing the facies distribution at the end of the Lower Unit (**A**) and the beginning of the Intermediate Unit (**B**). (**B**) shows an uplifted evaporite mound in the central part of the basin.

Firstly, a series of palaeokarsts with exokarstic depressions is developed on top of the Lower Unit gypsum (Figs 3A and 8). These dolines are filled with autochthonous karstic deposits (clay, marl, dolomite, and secondary gypsum), coming from the alteration of gypsum.

Secondly, a series of detrital gypsum sediments, which constitute the first elements of the Intermediate Unit, were deposited away from the central evaporitic palaeoreliefs, in the surrounding troughs (Fig. 8). This direction of sedimentation is opposed to the one which prevails in the borders of the basin i.e. directed towards the centre of the basin. These two arguments clearly indicate emergence and weathering of the gypsum, accompanied by a relative drop of the phreatic level, thus implying the presence of a palaeorelief in the centre of the basin.

Thirdly, synsedimentary structures are observed (slumping structures, localized increases in thickness, and carbonatic intraclasts), affecting the basal carbonates of the Intermediate Unit, which accommodate themselves within the karstic gypsum depressions (Fig. 3B). This clearly reveals that subsidence and collapse, associated with the karstification of the gypsum, continued within the dolines, until the complete fossilization of the evaporite upwellings.

The final argument is related to the variation in thickness of the Intermediate Unit (Figs 6C, 9 and 10), which is a minimum on top of the evaporitic mounds (0–15 m) and increases away from them (60–150 m in its deepest parts).

The above observations reveal that the evaporite masses in central Spain had been uplifted very early in their history, and it is argued that these upwellings or mounds might constitute 'incipient

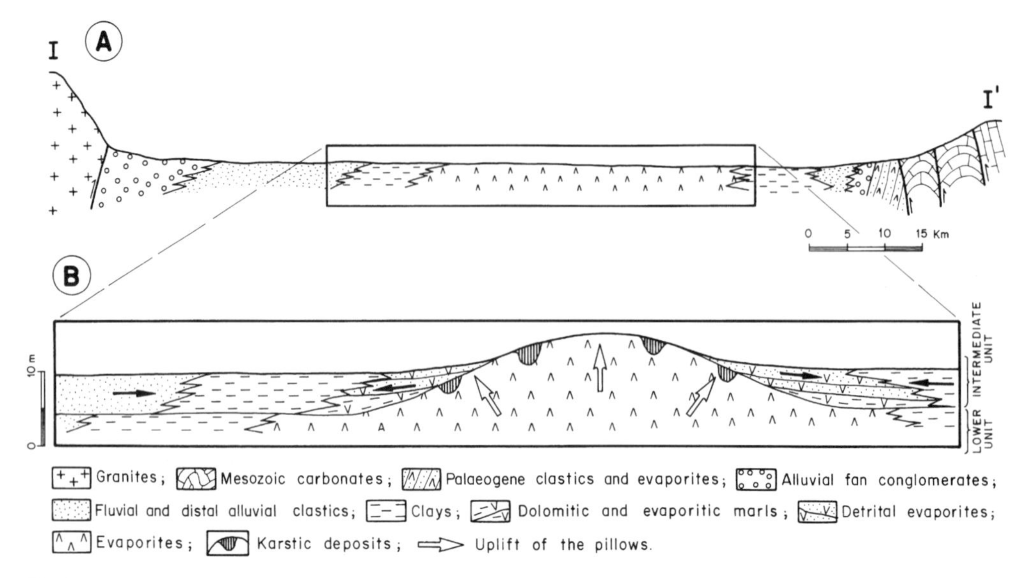


**Fig. 8.** Schematic cross-sections showing the evolution of the Madrid Basin from the end of the Lower Unit (**A**) to the first stages of the Intermediate Unit (**B**).

pillows', with maximum diameters of 30 and 10 km, and minimum amplitudes of 110 and 60 m for Madrid and Calatayud, respectively.

## Hydration diapirism as the agent of initiation of the evaporite mounds

We suggest that 'Hydration Diapirism', or the transition of anhydrite to gypsum by the chemical reaction anhydrite + water —> gypsum, may be the agent of initiation of the evaporite mounds in central Spain.

We further propose that the unusual influx of water responsible for the 'hydration', which occurred in both Spanish basins during the transition from the Lower Unit to the Intermediate Unit (Ramblian/Aragonian), may be related to a climatic change from arid conditions towards more humidity. Three independent sedimentological and palaeontological arguments support this interpretation. (1) There are changes in the lacustrine sedimentation, with evaporites of the Lower Unit replaced by more extensive carbonate deposits of the Intermediate Unit (both units are separated by a sedimentary discontinuity). This can only be explained by an increased influx of meteoric waters. (2) Changes in the fossil record are present, with 'more humid' associations, including a series of Upper Ramblian small mammals in the MN3 zone of Calatayud (mainly *Gliridae* and *Eomyidae*; (De Bruijn 1967)); and, the first arrival of large mammals (*Mastodontidae*, *Proboscidea* (De Bruijn 1967; Boné *et al.* 1980; Junco *et al.* 1985), together

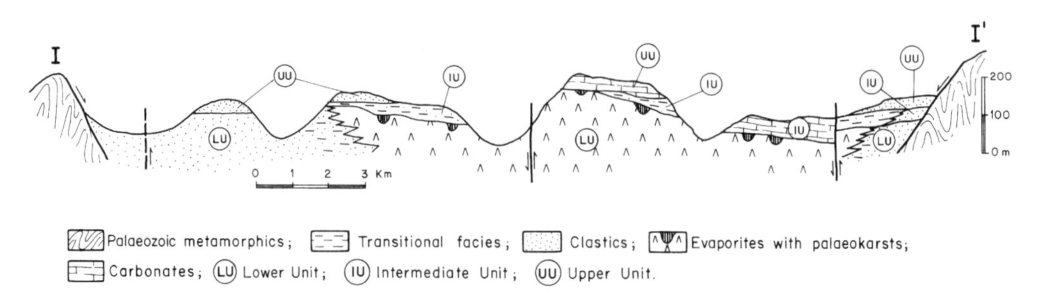


**Fig. 9.** Schematic transverse cross-section of the Calatayud Basin, showing the present arrangement of the Miocene depositional systems, and highlighting the variation in thickness of the Intermediate Unit, i.e. a minimum on top of the evaporite upwelling, and increasing away from it.

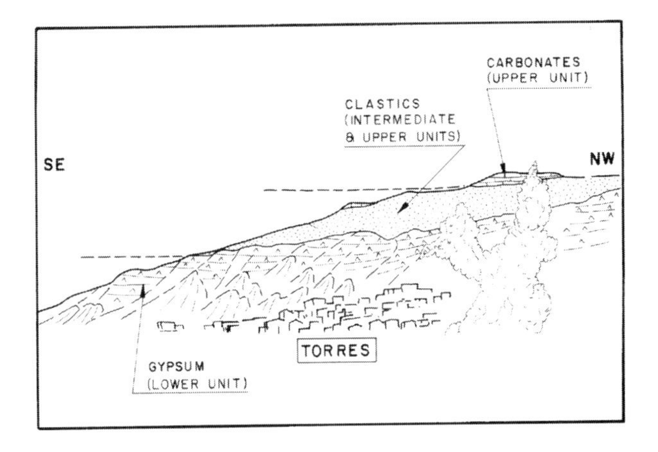

**Fig. 10.** Palaeorelief developed on the roof of the gypsum of the Lower Unit in the Calatayud Basin (Torres; Long: 1°35′W; Lat: 41°20′N). The photograph shows the variation in thickness of the Intermediate Unit. which is a minimum above the gypsum uplift (to the right) and increases away from it (to the left).

with the abundance in this last zone of the small mammal *Lagopsis* (a riparian *Ochotonidae* (Alberdi *et al.* 1985), in the Ramblian/Aragonian transition between the MN3 and MN4 zones. (3) Van der Meulen & Daams (1992) have shown that the rodent assemblages of the Calatayud Basin define a maximum in the humidity during the Ramblian/Aragonian time span.

We will now describe the three-stage evolution model involved in the 'Hydration Diapirism' hypothesis.

The first stage corresponds to the 'Deposition Stage (Fig. 11A), involving the deposition of the evaporites within the saline lakes in the centre of the basins, characterized by an internal drainage and dry conditions (Figs 6A and 7A).

The beginning of the second stage, or 'Upwelling Stage' (Fig. 11B), is marked by a major change in the sedimentation, with an augmented influx of meteoric waters, during the first pulses of deposition of the Intermediate Unit. This change produced a drop in the salinity, which would have triggered an augmented water activity in the lake, thus favouring the stability of the hydrated saline phases over the anhydrous ones. According to Plummer *et al.* (1988) and Sánchez-Moral (1994), a water activity of 0.777 marks the boundary between the stability of anhydrite (below it) and gypsum (above it). In this sense, the fresh waters, subsaturated in salt, that reached the lacustrine area induced a rehydration of the surficial anhydrite deposits, which would be transformed to gypsum (gypsification). This process usually induces a 61% volume increase (Sonnenfeld 1984). The crystallization pressure triggered by the hydration of anhydrite to gypsum depends on the saturation of the system in gypsum (Winkler 1975), and it might range between 282 and 334 atmospheres, within the expected temperatures of continental sedimentary environments (0 and 50°C). The volume increase is further accelerated by the presence of alkaline sulphates in dissolution (Conley & Bundy, 1958), which exist in both

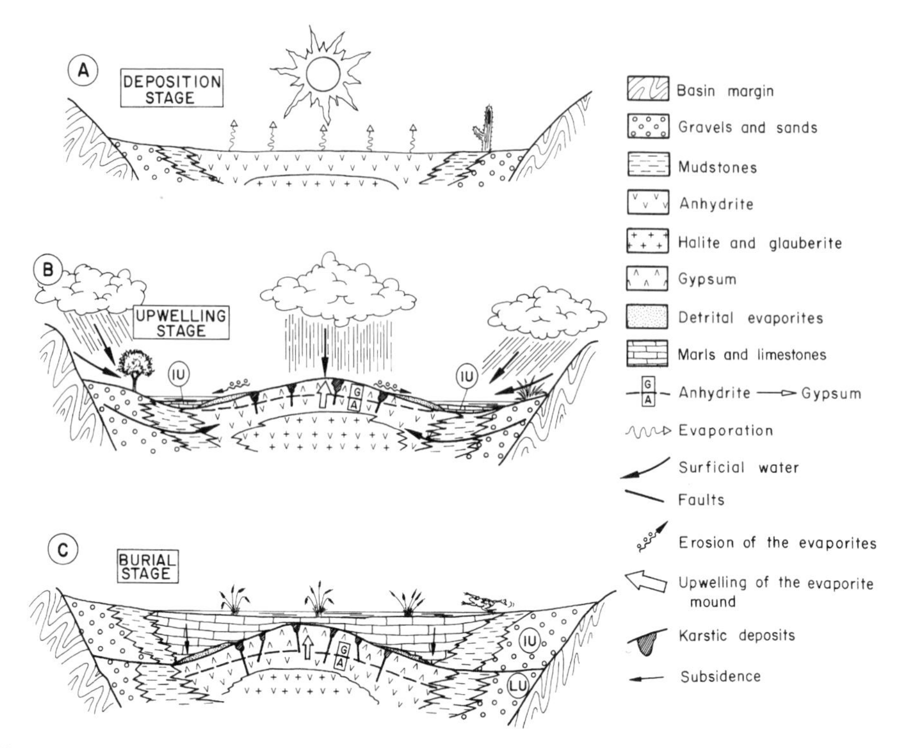


**Fig. 11.** Highly schematic diagrams showing the three-stage evolution of the Madrid and Calatayud Basins, with their associated evaporite mounds. A: Lower Unit. B: "early/middle" Intermediate Unit. C: "late" Intermediate Unit.

Spanish basins (Ordoñez *et al.* 1991; Sánchez-Moral *et al.* 1993; Ordoñez & García del Cura 1994). It can be shown that the hydration of anhydrite to gypsum occurred just after the deposition of the Lower Unit; this is revealed by detrital secondary gypsum pebbles with anhydrite nuclei, which are found in the base of the Intermediate Unit deposited in the troughs surrounding the evaporite upwellings, as well as by the climatic arguments which imply an influx of meteoric water. This increase in volume of the most surficial anhydrite layers changing to secondary gypsum would trigger a first upwelling of the previous evaporitic depocentre, as well as a depositional trough surrounding it (Figs 6B, 7B and 8). The emerged secondary gypsum would be weathered and the eroded sediments would be washed away from the central palaeorelief, while karstification, subsidence, collapse and dissolution would have taken place. In this sense, the chemical transformation of anhydrite to gypsum would continue, reaching progressively deeper evaporite layers and hence increasing the upwelling. Two major arguments explain why the layers of anhydrite did not expand by the same amount in all the basin i.e. why the uplift was greater in the middle part of the evaporite basin (see also Fig. 11B). (1) The anhydrite was not uniformly distributed areally in the saline lake, as abundant clay intercalations (distal facies of alluvial fans) existed towards the borders of the saline lake. In this sense, a differential volume increase was established between the central part (higher) and the borders (lower) of the evaporite basin. (2) The uplift was enhanced in the centre of the basin by a series of events taking place in the troughs surrounding the evaporite mounds – the partial dissolution of the anhydrite, and the deposition of the initial sediments of the Intermediate Unit – which exerted a differential loading in these areas, and buried the external parts of the saline lake, thus increasing the effect of the meteoric waters in the central and uncovered section of the evaporite domain.

These uplifting processes would go on until the progressive fossilization of both the surrounding depositional troughs and the evaporitic palaeorelief during the last stage ('Burial Stage'; Fig. 11C). This occurred in the upper Aragonian in Madrid and in the Turolian in Calatayud, with the final depositional pulses of the Intermediate Unit.

Although most of the upwelling activity was concentrated in both basins during the Intermediate Unit, the evaporitic mounds were partially reactivated later (Hoyos *et al.* 1979; Martín Escorza 1983; Cañaveras *et al.* 1991), particularly in the

Madrid Basin, where palaeokarsts developed on top of the Intermediate Unit, and a generalized dome-shaped topography was established in the centre of the basin during the Pliocene/Quaternary period. These reactivations might be related to differential erosion accompanying the fluvial downcutting which occurred during the Pliocene/Quaternary period, and to the partial dissolution of the evaporites in the borders of the mounds.

Some additional arguments that further support the hydration diapirism model are the following. (1) The climatic change responsible for the influx of fresh water in the sedimentary environment, and hence the hydration of anhydrite, is the only common factor which exists between Calatayud and Madrid during the Miocene. (2) According to Tsui & Cruden (1984), water bodies such as lakes and streams near the anhydritic strata are prerequisites for hydration to occur, and this is precisely the situation in the two Spanish basins.

We should also mention the importance of 'per descensum' meteoric waters in relation to the halite horizons, which are found in the lower parts of the evaporitic series at Calatayud and Madrid. In this sense, it is well known that the introduction of brines may lead to an immediate and large acceleration of the creep rate of halite (Jackson & Talbot 1994), hence facilitating the possible flow and upwelling of the salt.

## Conclusions

Early initiation of evaporite mounds occurred in two continental basins in central Spain, immediately following the deposition of the evaporites, and when no overburden had been deposited on top of them. The arguments that allow us to identify this incipient diapirism rely on the presence of palaeokarsts developed in the roof of the Lower Unit, as well as in the synsedimentary structures, disposition, thickness, and sedimentological characteristics of the Intermediate Unit. It is interesting to note that decreasing amounts of overburden have been advocated to initiate diapirism. While Trusheim (1960) considered that a thick overburden was necessary, Jackson & Vendeville (1994) recognize that salt might begin to well up with a thin overburden (350–1000 m). In this paper, we suggest an extreme situation, where no overburden whatsoever is required for halokinesis to begin.

To our knowledge, the agent of initiation suggested for the Madrid and Calatayud Basins has not been cited previously. Hydration has resulted from the chemical transition of anhydrite to gypsum. Several sedimentological and palaeontological arguments further suggest that this chemical reaction was triggered by a climatic change at the boundary between the Lower and the Intermediate Units.

These mounds or 'incipient pillows' developed within continental-derived evaporitic series (mostly anhydrite and gypsum) with no hydrocarbon potential, in contrast to the typical thick, marine-derived series (mostly halite, with minor anhydrite and argillaceous and potash-rich beds) classically associated with diapirism. In this sense, the Madrid and Calatayud Basins might provide unique clues to the incipient stages of diapirism, offering alternative mechanisms for these uplifts. As pointed out by Bishop (1978): 'There is no method currently known to describe the formation of incipient diapirs because they are without boundary conditions'.

The different deformation features found within the two Spanish basins (collapse structures, faults and folds), which were previously ascribed to a complex variety of changing tectonic stresses (Calvo *et al.* 1991), can now be explained as related to a simple cause: the upwelling of evaporite mounds. In this sense, this paper highlights the importance of vertical-related stresses in the evolution of the two Spanish Neogene continental basins, originally proposed for the Madrid Basin by Alía (1960, 1978).

Financial support was provided by the Dirección General de Investigación Científica y Técnica through Project PB-89-0047 awarded to M. Hoyos. We thank I. Davison, I. Alsop, O. Graversen, R. Evans, C. Talbot and M. Jackson for their reviews of an earlier version of this manuscript. Special thanks are due to José Arroyo for the drafting of the figures.

## References

ALBERDI, M. T., HOYOS, M., JUNCO, F., LÓPEZ MARTÍNEZ, N., MORALES J., SESÉ, C. & SORIA, D. 1983. Biostratigraphie et évolution sédimentaire de l'aire de Madrid. *Abstract International Colloquium on Mediterranean Neogene continental Paleoenvironments and Paleoclimatic Evolution*. Montpellier, 18–23.

——, ——, MAZO, A. V., MORALES, J., SESÉ, C. & SORIA, D. 1985. Bioestratigrafía, paleoecología, y biogeografía del Terciario de la Provincia de Madrid. *In*: ALBERDI, M. T. (ed.) *Geología y Paleontología del Terciario continental de la provincia de Madrid*. Consejo Superior de Investigaciones Científicas (C.S.I.C.), Madrid, 99–105.

ALÍA, M. 1960. Sobre la Tectónica profunda de la fosa del Tajo. *Notas y Comunicaciones del Instituto Geológico y Minero de España*, **58**, 125–162.

—— 1978. Geodinámica de la Meseta Ibérica. *Urania*, **289/290**, 25–50.

ANTUNES, M. T., CALVO, J. P., HOYOS, M., MORALES, J., ORDOÑEZ, S., PAIS, J. & SESÉ, C. 1987. Ensayo de correlación entre el Neógeno de las areas de Madrid y Lisboa (Cuencas Alta y Baja del Rio Tajo). *Comunicaciones do Servicio Geológico do Portugal*, **73**, 85–102.

BANKS, C. J. & WARBURTON, J. 1991. Mid-crustal detachment in the Betic system of southeast Spain. *Tectonophysics*, **191**, 275–289.

BISHOP, R. S. 1978. Mechanism for emplacement of piercement diapirs. *AAPG Bulletin*, **62**, 1561–1583.

BONÉ, E., ALBERDI, M. T., HOYOS, M. & LÓPEZ MARTÍNEZ, N. 1980. Prospéction paléontologique de la région de Torralba de Ribota (Burdigalien du bassin de Calatayud, province de Zaragoza, Espagne). *Paleovertebrata, Mémoire Jubilaire R. Lavocat*, 233–247.

CALVO, J. P., DE VICENTE, G. & ALONSO ZARZA, A. M. 1991. Correlación entre las deformaciones alpinas y la evolución del relleno sedimentario de la Cuenca de Madrid durante el Mioceno. *Comunicaciones del I Congreso del Grupo Español del Terciario, Vic*, 55–58.

——, HOYOS, M., MORALES, J. & ORDOÑEZ, S. 1990. Neogene stratigraphy, sedimentology, and raw materials of the Madrid basin. *Paleontologia y Evolución*, **2**, 63–95.

——, ORDOÑEZ, S., HOYOS, M. & GARCÍA DEL CURA, M. A. 1984. Caracterización sedimentológica de la Unidad Intermedia del Mioceno de la zona sur de Madrid. *Revista de Materiales y Procesos Geológicos*, **2**, 145–176.

CAÑAVERAS, J. C. 1994. *El paleokarst del techo de la Unidad Intermedia del Mioceno de la Cuenca de Madrid.* PhD Thesis, Complutense University, Madrid.

——, HOYOS, M., ORDOÑEZ, S. & CALVO, J. P. 1991. Características morfológicas y sedimentológicas, e implicaciones paleogeográficas de un paleokarst en una cuenca endorreica: el karst del techo de la Unidad Intermedia de la cuenca de Madrid. *Comunicaciones del I Congreso del Grupo Español del Terciario, Vic.* 67–70.

CAPOTE, R. 1983. La tectónica de la Cordillera Ibérica. *In*: Instituto Geológica y Minero de España (ed.) *Geología de España, Tomo II*, Instituto Geológico y Minero de España, Madrid, 108–120.

—— & FERNÁNDEZ-CASALS, M. J. 1978. La tectónica Postmiocena del sector central de la Depresión del Tajo. *Boletín Geológico y Minero*, **89**, 114–122.

CASHMAN, P. H. & ELLIS, M. A. 1994. Fault interaction may generate multiple slip vectors on a single fault surface. *Geology*, **22**, 1123–1126.

CONLEY, R. F. & BUNDY, W. M. 1958. Mechanism of gypsification. *Geochim. Cosmochim. Acta*, **15** (1–2), 57–72.

DE BRUIJN, H. 1967. Gliridae, Sciuridae, y Eomyiidae (Rodentia, Mammalia) Miocenos de Calatayud (provincia de Zaragoza, España) y su relación con la bioestratigrafía del área. *Boletín Instituto Geológico y Minero de España*, **78**, 189–365.

DOBLAS, M. 1991. Late Hercynian extensional and transcurrent tectonics in central Iberia. *Tectonophysics*, **191**, 325–334.

——, LÓPEZ-RUIZ, J., HOYOS, M., MARTÍN, C. & CEBRIÁ, J. M. 1991. Late Cenozoic indentation/escape tectonics in the eastern Betic Cordilleras and its consequences on the Iberian foreland. *Estudios Geológicos*, **47**, 193–205.

—— & OYARZUN, R. 1989a. Neogene extensional collapse in the western Mediterranean (Betic-Rif Alpine orogenic belt): Implications for the genesis of the Gibraltar Arc and magmatic activity. *Geology*, **17**, 430–433.

—— & —— 1989b. 'Mantle Core Complexes' and Neogene extensional detachment tectonics in the western Betic Cordilleras, Spain: an alternative model for the emplacement of the Ronda peridotites. *Earth and Planetary Science Letters*, **93**, 76–84.

—— & —— 1990. The late Oligocene–Miocene opening of the North Balearic Sea (Valencia Basin, western Mediterranean): A working hypothesis involving mantle upwelling and extensional detachment tectonics. *Marine Geology*, **94**, 155–163

EVANS, R. 1993. Lowering of sea-level as cause of the initiation of diapirism of salt. *American Association of Petroleum Geologists Hedberg Research Conference on Salt Tectonics, Bath, England, September 13–17th.*

GARCÍA DEL CURA, M. A. 1979. *Las Sales Sódicas, Calcosódicas y Magnésicas de la Cuenca del Tajo*, Fundación Juan March, Serie Universitaria.

HOYOS, M., JUNCO, F., PLAZA, J. M., RAMÍREZ, A. & RUIZ SANCHEZ-PORRO, J. 1985. El Mioceno de Madrid. *In*: ALBERDI, M. T. (ed.): *Geología y Paleontología del Terciacio continental de la provincia de Madrid.* Consejo Superior de Investigaciones Científicas, Madrid, 9–16.

—— & LÓPEZ MARTÍNEZ, N. 1985. Iberic Depression. *In*: STEININGER, *et al.* (eds) *Neogene of the Mediterranean Tethys and Paratethys: Stratigraphic correlations tables and sediment distribution maps.* International Geological Correlation Program, Project 25, Vol. 1.

——, ZAZO, C., GOY, J. L. & AGUIRRE, E. 1979. Estudio geomorfológico de los alrededores de Calatayud. *III Reunión del Grupo Español del Trabajo del Cuaternario*, Madrid, 149–160.

HUGHES, D. J. 1968. Salt tectonics as related to several Smackover fields along the northeast rim of the Gulf of Mexico basin. *Gulf Coast Association of Geological Societies Transactions*, **18**, 320–330.

JACKSON, M. P. A. & TALBOT, C. J. 1986. External shapes, strain rates, and dynamics of salt structures. *Bulletin of the Geological Society of America*, **97**, 305–323.

—— & —— 1994. Advances in salt tectonics. *In*: HANCOCK, P. L. (ed.) *Continental Deformation.* Pergamon, 159–179.

—— & VENDEVILLE, B. C. 1994. Regional extension as a geologic trigger for diapirism. *Bulletin of the Geological Society of America*, **106**, 57–73.

JENYON, M. K. 1986. *Salt Tectonics.* Elsevier, Barking.

JUNCO, F., ALBERDI, M. T. & HOYOS, M. 1985. Considerations on the biostratigraphy and paleoecology of the Aragonian age in the Madrid basin. *VIIIth Congress of the Regional Committee on Mediterranean Neogene Stratigraphy, Budapest*, 290–292.

LÓPEZ MARTÍNEZ, N., AGUSTÍ, J., CALVO, J. P. *et al.* 1987. Approach to the Spanish continental Neogene synthesis and palaeoclimatic interpretation. *Annales Instituti Geologici Publici Hungarici*, **70**, 383–391.

LÓPEZ-RUIZ, J., CEBRIÁ, J. M., DOBLAS, M., OYARZUN, R., HOYOS, M. & MARTÍN, C. 1993. Cenozoic intra-plate volcanism related to extensional tectonics at

Calatrava, central Iberia. *Journal of the Geological Society, London*, **150**, 915–922.

MARTÍN ESCORZA, C. 1976. Actividad tectónica durante el Mioceno de las fracturas del basamento de la Fosa del Tajo. *Estudios Geológicos*, **32,** 509–522.

—— 1983. Neotectónica de la cuenca de Madrid. *In*: COMBA, J. A. *et al.* (eds) *Geología de España, Tomo II*, Instituto Geológico y Minero de España, Madrid, 543–553.

—— 1990. Distensión-compresión en la cuenca de Campo Arañuelo. Implicación cortical. *Geogaceta*, **8**, 39–42.

NUESCH, R., MADSEN, F. T. & FONYO, I. 1984. Influence of shale interlayers in anhydrite on swelling behaviour. *9th International Clay Conference*, Strasbourg, 123–129.

ORDÓÑEZ, S. & GARCÍA DEL CURA, M. A. 1992. El sulfato sódico natural: Las sales sódicas de la Cuenca de Madrid. *In*: GARCÍA GUINEA, F. & MARTÍNEZ FRÍAS, J. (eds) *Recursos minerales de España*, Consejo Superior de Investigaciones Científicas, Madrid, 1229–1250.

—— & —— 1994. Deposition and diagenesis of sodium-calcium sulphate salts in the Tertiary saline lakes of the Madrid Basin, Spain. *In*: RENAUT, R. W. & LAST, W. M. (eds) *Sedimentology and Geochemistry of Modern and Ancient Saline Lakes*. Society of Economic Paleontologists and Mineralogists, Special Publication, **50**, 229–238.

——, CALVO, J. P., GARCÍA DEL CURA, M. A., ALONSO ZARZA, A. M. & HOYOS, M. 1991. *Sedimentology of Sodium Sulphate Deposits and Special Clays from the Tertiary Madrid Basin (Spain)*. International Association of Sedimentologists, Special Publication, **13,** 39–55.

——, SORIANO, A., GARCÍA DEL CURA, M. A. & ESTEBAN, F. 1990. Swelling mechanisms of Tertiary anhydritic–dolomitic shales. *6th Congress of the International Association of Economic Geologists*, Rotterdam, 1963–1971.

ORTÍ, F., ROSELL, L., PALLICK, A. E. & UTRILLA, R. 1994. Yesos de Calatayud: Aplicación del estudio de facies y geoquímica de sulfatos al conocimiento de un sistema evaporítico. *Geogaceta*, **15**, 74–77.

PLUMMER, L. N., PARKHURST, D. L., FLEMING, G. W. & DUNCKLE, S. A. 1988. *A Computer Program Incorporating Pitzer's Equations for Calculation of Geochemical Reactions in Brines*. US Geological Survey Water-Resources, Investigation Report, **88–4153**.

POLLARD, D. D., SALTZER, S. D. & RUBIN, A. M. 1993. Stress inversion methods: are they based on faulty assumptions? *Journal of Structural Geology*, **15**, 1045–1054.

QUEROL, R. 1989. *Geología del Subsuelo de la Cuenca del Tajo*. Escuela Técnica Superior de Ingenieros de Minas, Departamento de Investigación Geológica, Madrid.

SÁNCHEZ-MORAL, S. 1994. *Sedimentación Salina Actual en un Lago Continental (Laguna de Quero, Toledo). Aplicación de la Modelización Termodinámica al Estudio de Secuencias de Precipitación Salina*. PhD Thesis, Complutense University, Madrid.

——, HOYOS, M., ORDÓÑEZ, S., GARCÍA DEL CURA, M. A. & CAÑAVERAS, J. C. 1993. Génesis de epsomita infiltracional por dedolomitización en un ambiente sulfatado árido. Eflorescencias en la Unidad Inferior evaporítica de la Cuenca de Calatayud. *V Congreso de Geoquímica de España*, Soria, 24–29.

SIMÓN, J. L. 1990. Algunas reflexiones sobre los modelos tectónicos aplicados a la cordillera Ibérica. *Geogaceta*, **8**, 123–129.

SONNENFELD, P. 1984. *Brines and Evaporites*. Academic, London.

TALBOT, C. J. & JARVIS, R. J. 1984. Age, budget and dynamics of an active salt extrusion in Iran. *Journal of Structural Geology*, **6**, 521–533.

TRUSHEIM, F. 1960. Mechanism of salt migration in northern Germany. *AAPG Bulletin*, **44**, 1519–1540.

TSUI, P. C. & CRUDEN, D. M. 1984. Deformation associated with gypsum karst in the Salt River escarpment, northeastern Alberta. *Canadian Journal of Earth Sciences*, **21**, 949–959.

VAN DER MEULEN, A. J. & DAAMS, R. 1992. Evolution of Early–Middle Miocene rodent faunas in relation to long-term palaeoenvironmental changes. *Palaeogeography, Palaeoclimatology, Palaeoecology*, **93**, 227–253.

VICENTE, G. D., CALVO, J. P. & ALONSO ZARZA, A. M. 1990. Main sedimentary units and related strain fields of the Madrid basin (central Spain) during the Neogene. *IX Regional Committee on Mediterranean Neogene Stratigraphy*. Barcelona, 121–122.

——, —— & MUÑOZ-MARTÍN, A. (in press). Neogene tectono-sedimentary review of the Madrid Basin. *In*: FRIEND, P. & DABRIO C. J. (eds) *Tertiary Iberian Basins*. Cambridge University Press, World and Regional Series.

WINKLER, E. M. 1975. *Stone Properties and Durability in Man's Environment*. Springer, Berlin.

WOODBURY, H. O., MURRAY, I. B. & OSBORNE, R. E. 1980. Diapirs and their relation to hydrocarbon accumulation. *In*: MIALL, A. D. (ed.) *Facts and Principles of World Petroleum Occurrence*, Canadian Society of Petroleum Geologists, Memoir **6**, 119–142

WU, S., BALLY, A. W. & CRAMEZ, C. 1990. Allochtonous salt structure and stratigraphy of the north-eastern Gulf of Mexico, Part II: Structure. *Marine and Petroleum Geology*, **7**, 334–371.

# Internal structure of a detachment horizon in the most external part of the Pyrenean fold and thrust belt (northern Spain)

M. SANS[1,2], A. L. SÁNCHEZ[2] & P. SANTANACH[1]

[1]*Departament de Geologia Dinàmica, Geofísica i Paleontologia. Universitat de Barcelona, Barcelona 08071, Spain*

[2]*Súria K SA, Sales y Potasas, Súria, Barcelona 08260, Spain*

**Abstract:** The thick Eocene Cardona salt forms the detachment horizon of the most external part of the Pyrenean fold and thrust belt. This salt horizon forms the cores of anticlines and is intensively mined for potash. The study of the structure of the salt layer at three sites (Cabanasses, Súria-Shaft IV and Balsareny underground mines) reveals that the metric and decametric scale structures are consistent with deformation in a shear zone (detachment horizon) with a shear sense of top-to-the-south and not related to compressional or diapiric folding. A further study shows that these structures are later modified by larger wavelength folds and thrusts which have a surface expression. The total horizontal shortening in the two anticlines studied (Súria and El Guix) is a minimum of 750 m. All the folds developed in the detachment horizon have horizontal axes which are normal to the shear direction. This contrasts with folds developed in detachment horizons with intense shearing, where folds with axes parallel to shear direction are common.

Thick thrust sheets are emplaced almost without internal deformation, over salt and other evaporites, which are the most suitable rocks of a sedimentary pile to develop a detachment. Deformation is typically concentrated in the evaporitic layers located in the lower part of the thrust sheet as shown by Jordan & Nüesch (1989); Malavielle & Ritz (1989) and Marcoux *et al.* (1987). In many fold-and-thrust belts, thick salt sequences are also responsible for the transmission of the deformation to wide areas of the foreland (Davis & Engelder 1985). These thick salt sequences form the detachment horizon for all the folds and thrusts developed above. Although there are several works on the internal structure of diapirs and namakiers (Kupfer 1968; Talbot 1979; Woods 1979; Richter-Bernburg 1980; Talbot & Jackson 1987), little work has been published on the internal structure of salt detachment levels.

In the frontal-most portion of the Pyrenean fold-and-thrust belt, shortening has been slight and salt has only been involved in fold and thrust structures, without any further transport or modification by diapiric development. These characteristics together with intensive underground mining for potash make this area highly suitable to study the internal structure of an only moderately deformed detachment level. The purpose of this paper is to describe the structure of the Cardona salt detachment horizon from centimetre to kilometre scale and determine the relationship between structures at different scales. The relationship between salt behaviour during deformation and the kinematics of the formation of the anticlines will also be described.

Data from three underground potash mines have been examined. In each mine the structure of the potash layers is surveyed daily. The data collection is different in each mine and depends on the complexity in each case. In Súria IV and Cabanasses the geological monitoring is in the form of 1:500 cross sections completed with 1:10 diagrams of the mining fronts every 25 m. In Balsareny a topographic leveling of the potash layers is the structural monitoring. These data compiled over many years by the companies Súria K SA and Potasas del Llobregat SA, together with *in situ* observations in the mines and field work form the basis of this study.

## Geological setting

In the southern Pyrenees, basement and cover thrust sheets were emplaced to the south from late Cretaceous to Miocene times. In front of the main emergent frontal thrust which separates the allochthonous thrust sheets from the authochthonous deformed foreland, the deformation propagates southwards due to the presence of thick Tertiary evaporitic horizons at depth in the foreland (Fig. 1) (Vergés *et al.* 1992; Muñoz *et al.* 1994). The deformed foreland is characterised at the surface by the presence of three sets of folds with different orientations. A northeastern region of WNW-trending folds, a central region (study area, Fig. 1) with northeast trending folds and a southwestern region with east-trending folds. Each of these regions is located on top of a different evaporitic level that are

*From* ALSOP, G. I., BLUNDELL, D. J. & DAVISON, I. (eds), 1996, *Salt Tectonics*, Geological Society Special Publication No. 100, pp. 65–76.

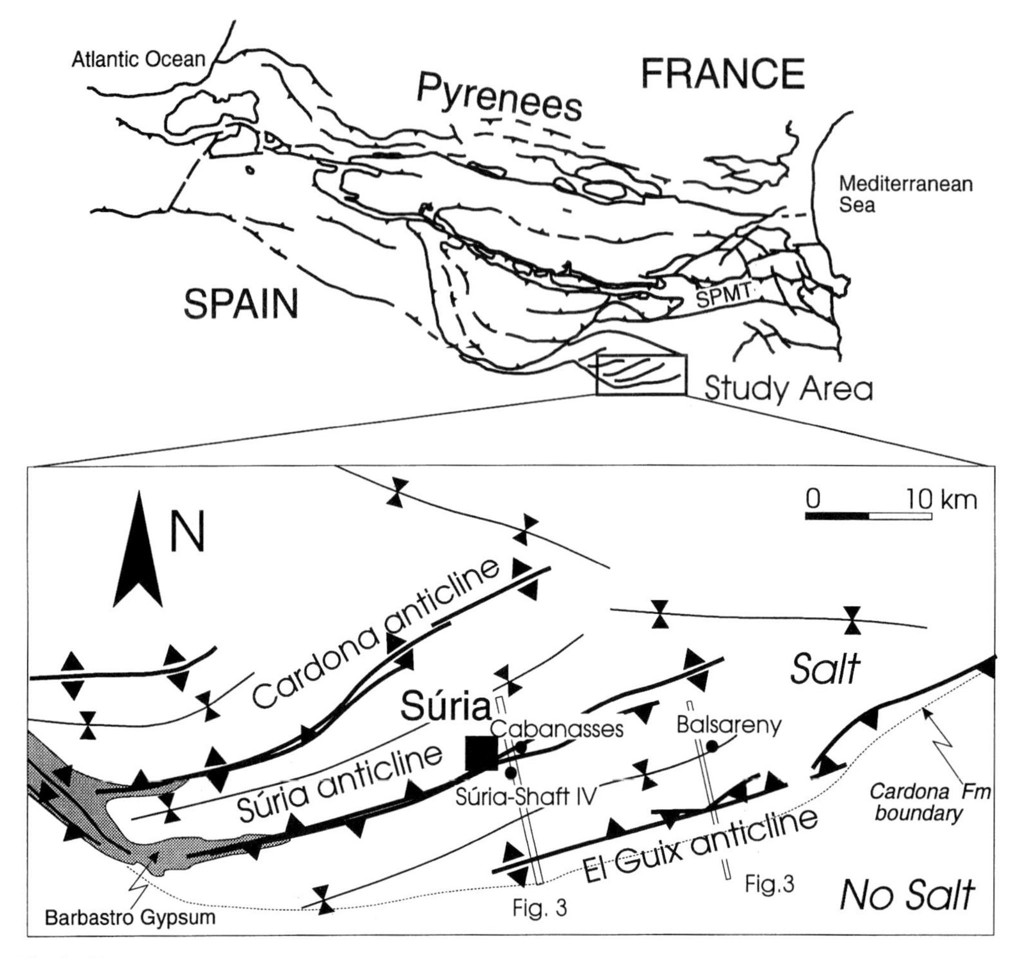


**Fig. 1.** The study area is located in the deformed foreland of the Pyrenees. The South Pyrenean Main Thrust (SPMT) separates the allochtonous thrust sheets from the deformed foreland. The northeast–southwest Súria and El Guix anticlines are the frontal structures of the Pyrenean fold-and-thrust system. They develop at the southern boundary of the Cardona detachment level.

from northeast to southwest: Beuda, Cardona and Barbastro formations (Vergés *et al.* 1992). In the study area, folds (Els Guix and Súria anticlines) detach above the thick Cardona Formation which acts as a very effective decoupling horizon.

The Cardona Formation represents a marine evaporitic macrocycle (Pueyo & San Miguel 1974; Pueyo 1975; Busquets *et al.* 1985) that started with the sedimentation of euxinic marls (Igualada marls Formation, Bartonian) and ended with non-marine detrital sediments (Sanaüja lacustrine complex and Solsona and Artés formations) which are late Eocene–Oligocene in age (Fig. 2).

The evaporitic sediments of this macrocycle are mainly halite with interbedded potash levels on top of a 5–10 m thick basal anhydrite (Fig.2). Two members can be differentiated: (1) the lower rocksalt member, and (2) the upper potash and rocksalt member (Pueyo 1975). The lower rocksalt member is 200–250 m thick and is formed by alternating lutite and halite layers. Halite lamination is 6 cm thick and grey in colour when it contains diffuse lutitic material and large cubic crystals of halite characterized by sunken depressions on the crystal face (hopper crystals). The halite is lighter grey in colour when it has scarce hoppers and no lutitic material (Busquets *et al.* 1985).

The upper member has some potash layers interbedded in the rocksalt. The total thickness is 100 m. The potash layers are an alternating sequence of sylvinite (rock consisting of halite and sylvite (KCl)), carnallite (K $MgCl_3$ $6H_2O$), halite and lutitic laminae. The halite laminae are 4 cm thick in this upper member, are fine grained

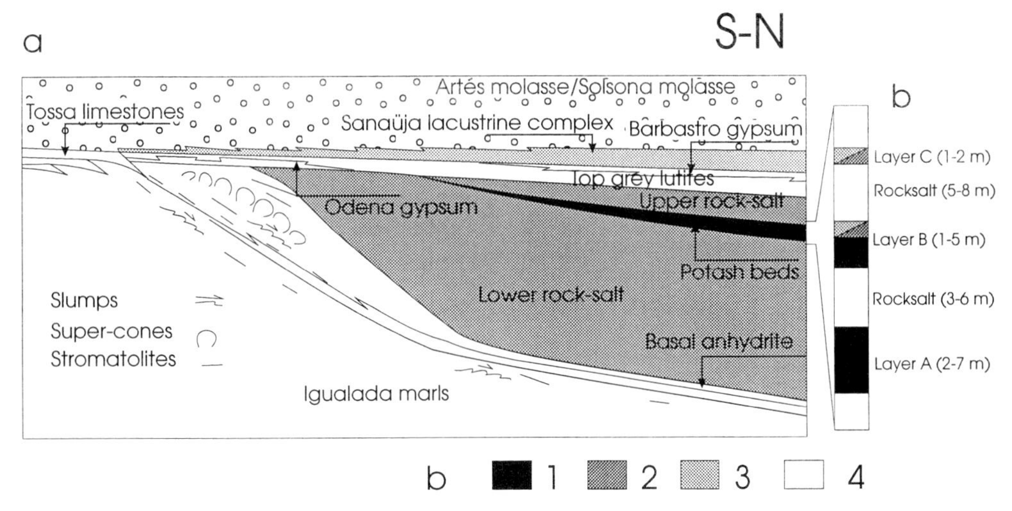


**Fig. 2.** (**a**) Facies distribution in the centre and eastern margin of the Catalan Potash basin (modified from Ortí *et al.* 1984). (**b**) Stratigraphic column of the potash layers at the Súria and the El Guix anticlines. 1, sylvinite; 2, 'transformed sylvite'; 3, carnallite; 4, halite.

(millimetre scale) and frequently show graded bedding. They are orange and contain small amounts of sulphates (anhydrite and polyhalite). Each lamina is a succession of hopper crystals and clear halite which are 5 mm thick on average (Busquets *et al.* 1985). The average thickness of the halite-sylvinite laminae in the potash member is 8 cm with a maximum of 30 cm. Banded sylvinite shows a laminated lithofacies similar to the one described for the halite in the potash member, but containing lutite, halite and sylvinite layers. Sylvite is present as polygonal grains larger than the halite ones. Secondary sylvite ('transformed sylvite') after carnallite is also found in some carnallite levels in the upper part of the potash beds (Pueyo 1975). It shows a coarser-grained texture with large euhedral and clean sylvite crystals set in an older generation of red sylvite, halite and insoluble material (clay, hematite, sulphates). In this member the contact between the evaporitic beds and the clay laminae is abrupt with the clay laminae being 1 mm thick on average.

In the upper member three potash layers were exploited. From bottom to top these are labelled A, B and C layers (Fig. 2). At present only layer B is mined by Súria K S A, and layers A and B are mined by Potasas del Llobregat SA. Layer B consists of sylvinite at the bottom and transformed sylvite or carnallite at the top in Súria K SA mines and of sylvinite in Potasas del Llobregat SA. Layer A consists of sylvinite and layer C, where it was exploited, consisted of transformed sylvite (Fig. 2).

The basal anhydrite is considered equivalent to stromatolite and gypsum layers on the border of the basin (Fig. 2). The extent of the deposition of the lower salt member led to a progressive shallowing of the basin, and it was probably at the end of the potash stage that the marine basin became filled up (Busquets et al. 1985). After the deposition of these marine evaporites, continental, alluvial, fluvial and lacustrine deposits (Fig. 2) covered most of the Ebro basin (Sáez 1987; Vergés & Burbank 1995; Puigdefàbregas *et al.* 1986, 1992).

## Structure

### *Macrostructure*

The study of the internal structure of the Cardona salt detachment horizon is based on data collected at three sites, corresponding to three underground potash mines. From north to south the mines are Cabanasses and Súria (Shaft IV), owned by the Súria K SA and Balsareny, owned by Potasas del Llobregat SA (Fig. 3). Súria K SA exploits the Súria 'anticline' and Potasas del Llobregat SA mines the El Guix anticline and its northern syncline (Fig. 1 & 3). These two anticlines are the frontal-most structures of the Pyrenean fold-and-thrust-belt.

The Súria 'anticline' is a compound double verging structure, with a south-vergent anticline in the north and a north-directed thrust in the south with a related hangingwall anticline, the Súria southern anticline (Fig. 3). The northern anticline of the Súria structure is mined at Cabanasses. The mine extends 4 km to the east from the main shaft in a N070°E direction following the fold axes, and it exploits the southern limb of the anticline, never

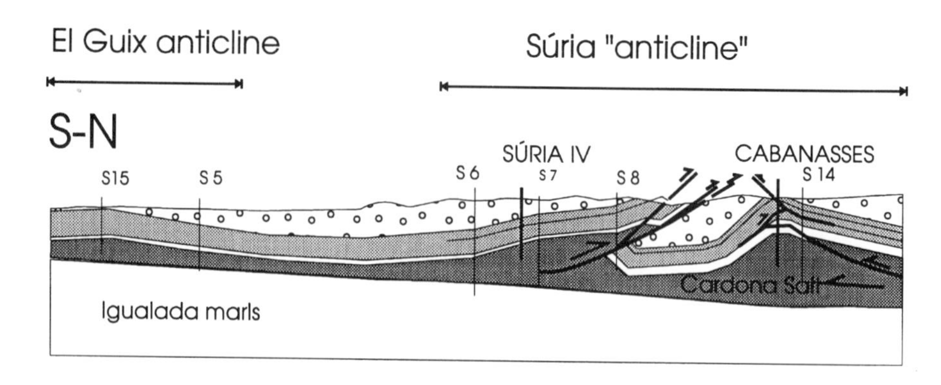


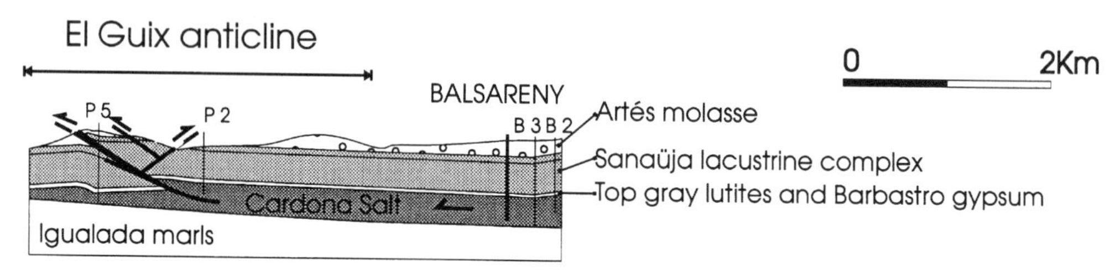


**Fig. 3.** Cross sections of the El Guix and Súria anticlines and location of the studied mines. See location in Fig. 1. From north to south the studied mines are Cabanasses, located in the southern limb of the northern Súria anticline, Súria (Shaft IV) located in the hangingwall of the backthrust, and to the south, Balsareny located in the southern limb of the syncline. These cross sections are based on potash exploration wells and seismic lines. Lithological ornament as in Fig. 2.

reaching the northern one. At the surface this anticline is south-vergent near the town of Súria and it opens and becomes symmetrical and neutral verging further east (Fig. 1). This change along strike of the fold surface expression can be followed in the subsurface by the interpretation of the seismic profiles and the potash exploration wells that cut this structure (Sans & Vergés in press). The fold is cored in the east by small and simple north-directed thrusts whereas towards the west it is cored by a complex array of thrusts which display a thrust-wedge geometry (Fig. 3). The Súria southern anticline is related to the north-directed backthrust and it is mined at Súria (Shaft IV). This mine extends 4.5 km east of the main shaft in a N070ºE direction and in a north–south direction it extends approximately 1.5 km, from 500 m south of the backthrust plane to aproximately 1000 m north of the southern syncline axes. The displacement of this thrust increases form east to west, to a maximum of 350 m in the central part, close to the main shaft.

The southern mine (Balsareny) is located in the northern limb of the El Guix anticline (Fig. 1). This anticline is a faulted south-vergent detachment fold (Fig. 3). At the surface, thrusts outcrop as a single thrust, or as a narrow band of thrusts which cut the south dipping limbs of the folds (Sans & Vergés in press). The thrusts are both north and south-directed, and their dip is similar (25°–40°). South-directed thrusts accommodate the major displacement in this fold (80% of the total). At present the mine extends, from the northern syncline axes to 1 km north of the main south-directed thrust (Fig.1).

The total shortening accumulated in the anticlines located in the southern foreland of the Pyrenees is 2 km (Vergés *et al.* 1992), of which a minimum of 750 m corresponds to the Súria and the El Guix anticlines. Deformation decreases towards the south, where the El Guix south-directed thrusts represent a total shortening of 200 m. Therefore the detached units located north of the Súria 'anticline' have been displaced a minimum of 750 m to the south, whereas those located in the northern limb of the El Guix anticline have only been displaced by 200 m.

### *Structures developed within the detachment level*

The studied structures within the Cardona evaporitic formation are all located in the potash layers or closely above or below it. The potash layers are located in the upper third of the Cardona evaporitic formation and therefore close to the detached overlying sediments. There are no observations at the

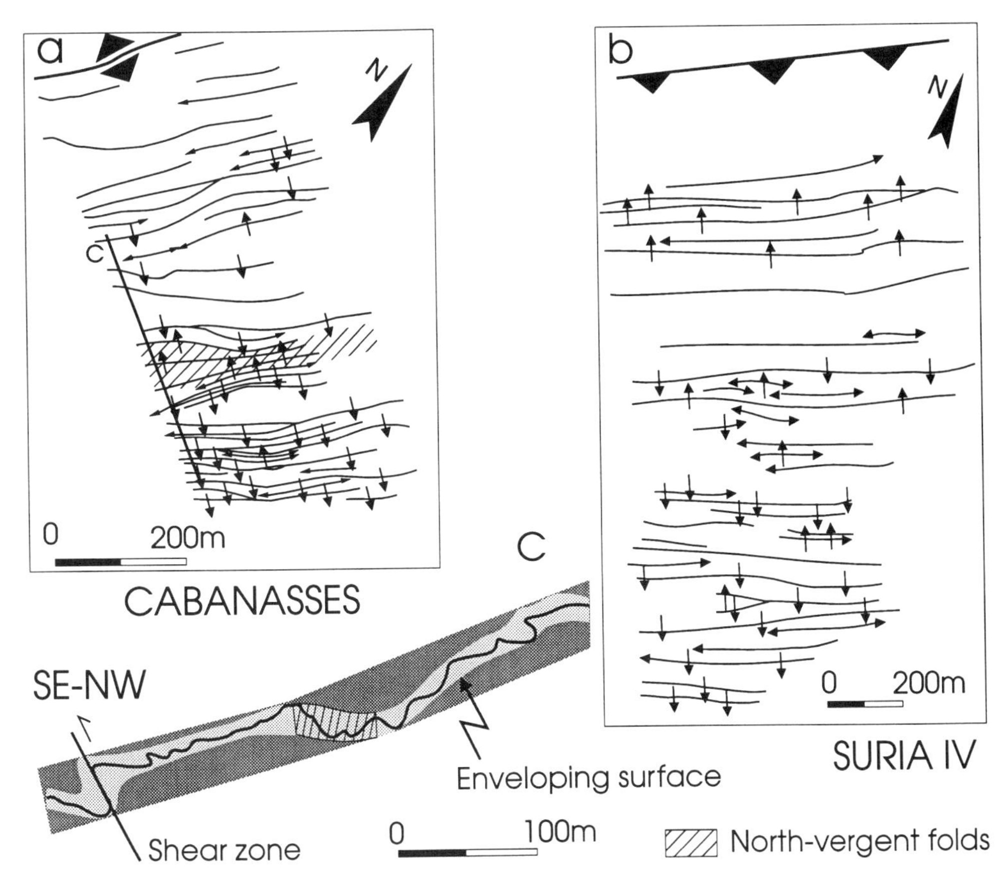


**Fig. 4.** Anticline traces map of Súria (Shaft IV) and Cabanasses mine, arrows indicate fold vergence. (**a**) At Cabanasses, folds are mostly south-directed except in the hatched area, where they are slightly north-vergent. In this area folds have been rotated as a consequence of the later, larger wavelength folding. C indicates location of Fig. 4*c*. (**b**) At Súria Shaft IV the anticlines are generally south-directed except close to the north-directed backthrust, where they are north-vergent. (**c**) Enveloping surfaces of the decametric folds at Cabanasses; different grey shades indicate the first and second order enveloping surfaces. Rotated south-vergent folds are located in the north-dipping short limb of the second-order folds (hatched area). See relation with map (4*a*).

bottom of the evaporite formation or at the top in the contact with the non-evaporitic materials.

*Folds.* Folds of a wide variety of scales are the most common structures in the salt layers. Metre-size folds can be directly observed in the mines whereas decametric folds are reconstructed from the gallery cross sections (Fig. 4a–c). Almost all of these folds have horizontal axes with a direction varying from N040°E to N080°E, mostly N070°E-N080°E, parallel to the surface trace of the major anticlines and thrusts. They show a dominant southerly directed vergence in all the structural positions examined (Figs 4*a* and 4*b*) except in the proximity of the north-directed backthrust (500 m from the inferred thrust plane) where they are mostly north-vergent folds (Fig. 4b). South-vergent folds are found in south-dipping limbs (Cabanasses and Súria), north-dipping limbs (Balsareny) and even in the hinge zone of the syncline (Balsareny-Fig. 5a). There is a reduction of the typical fold size and fold density from north to south (from Cabanasses to Balsareny) and in Balsareny there are wide areas which do not show any folding. The folds are usually of subsimilar class (Class 1C; Ramsay 1967). A single fold usually shows a thickening of the south-dipping limb and a thinning of the north-dipping limb (Fig. 5b). North-dipping overturned limbs are highly thinned in all scale folds (shear zones). The folds can change along strike from inclined to recumbent.

The enveloping surface of the metric and decametric scale folds described above is gently folded. The asymmetry of these larger wavelength folds is consistent with the location of these folds with

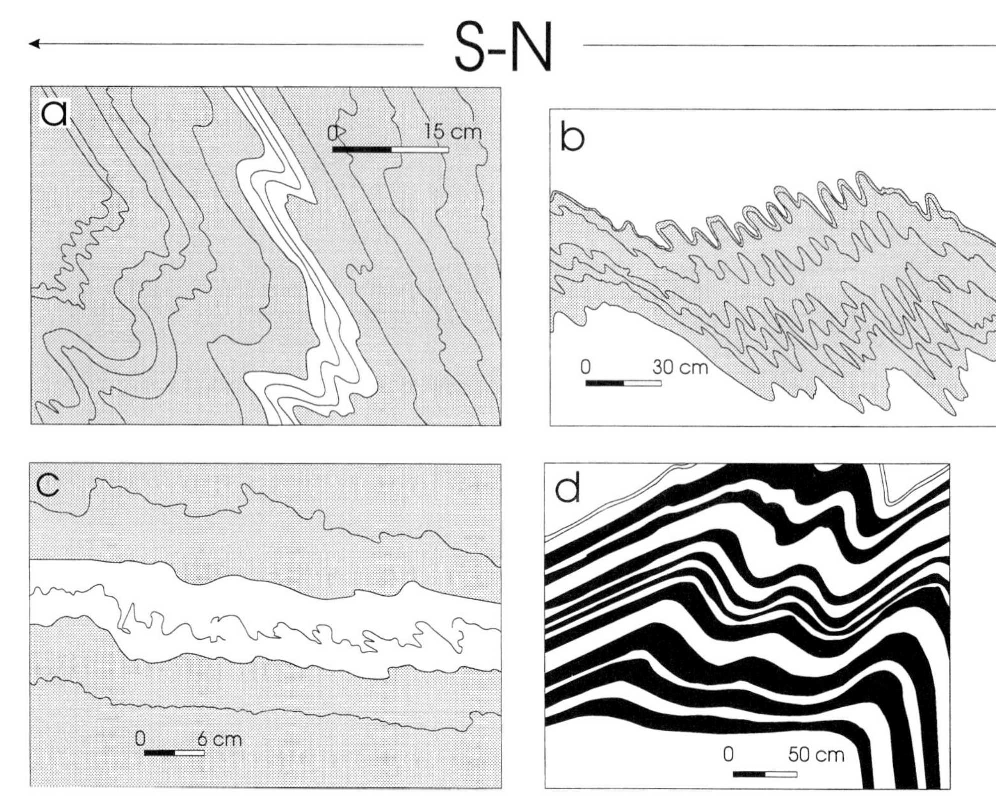


**Fig. 5.** Fold examples (vertical section). Black represents red sylvinite and grey and white represent different halite layers. (a) Geometry of the minor folds located at the syncline hinge zone between the Súria and El Guix anticlines (Balsareny). The north-dipping overturned limbs are thinned and stretched. (b) The south-dipping limbs of the anticlines are thickened whereas the north-dipping ones are thinned (Súria-Shaft IV). (c) Clay layers form ptygmatic folds that sometimes break and form small south-directed thrusts, which pass upward and downward to the salt layers as ductile deformation. (d) Geometry of north-vergent folds at the proximity of the north-directed thrust (Súria-Shaft IV).

respect to the kilometric scale folds (Súria and El Guix anticlines). Cabanasses is the best example of enveloping surfaces analysis (Fig. 4c). The enveloping surface displays a long limb-short limb geometry consistent with its location in the southern limb of the northern Súria anticline (Fig. 4c) although sometimes the enveloping surface is offset by reverse shear zones as in the southern fold of Fig. 4c (shear zones). The map of anticline traces (Fig. 4a) and the cross sections (Fig. 4c) show a band of north-vergent decametric folds located in the north-dipping limb of an enveloping surface fold. The decametric scale folds have an opposite vergence with respect to the widespread southern vergence of the folds of this scale. Although some of these folds could be related to the enveloping surface folds, unfolding of the enveloping surface rotates the north-vergent folds to a south-vergent position which gives an overall coherent view of the decametric scale folds. The north-vergent decametric scale folds located in the north-dipping limb of the enveloping surface folds have been interpreted as previous south-vergent folds rotated as a consequence of the new larger wavelength folding.

Centimetre-scale folds are mainly seen in the lutitic laminae. They generally show a vergence consistent with the major metric fold to which they are presumably related. The clay laminae can also break and form small thrusts that pass to a ductile deformation in the salt layers (Fig. 5c). Due to the difference in competence and in thickness between the clay layers and the salt around them, the folds in the clay laminae are ptygmatic folds in the more deformed stages (Fig. 5c).

*Lineations.* A common feature of the three sites is the presence of a well developed crenulation lineation in the clay layers. The orientation of the lineation coincides with the direction of the fold axes N60°E. No mineral stretching lineations are observed at mesoscopic scale.

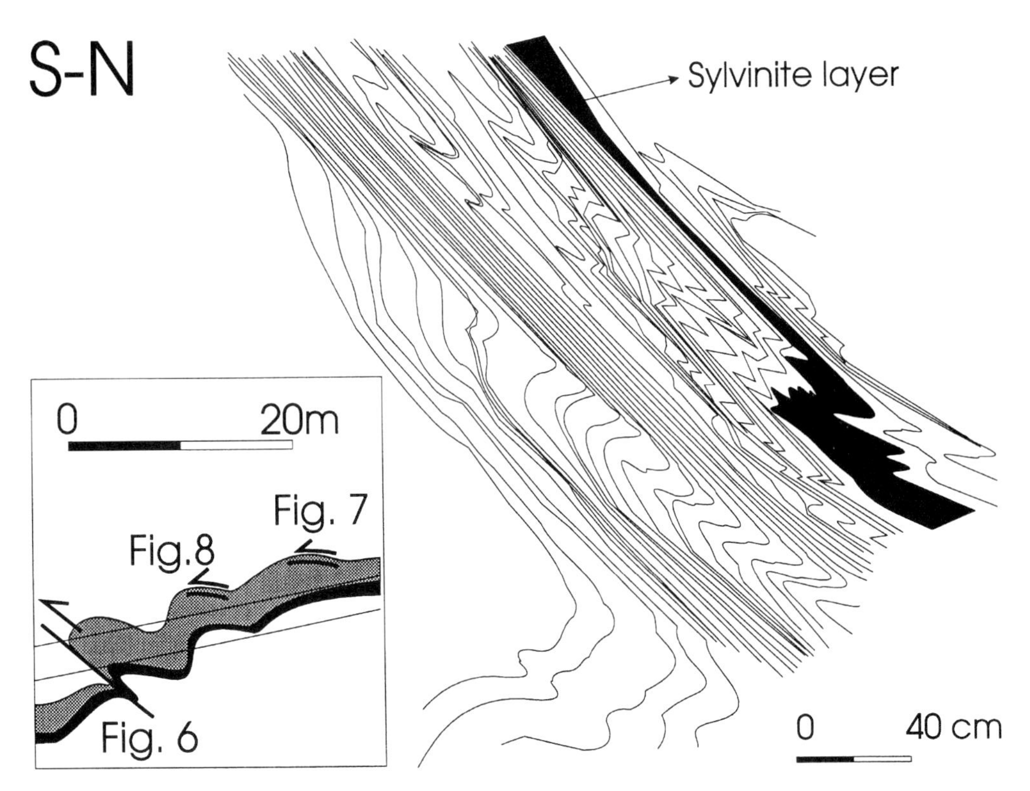


**Fig. 6.** Inhomogeneous distribution of shear in the overturned limb of a south-vergent fold. A marked banding, due to the stretching of the salt beds, develops at low angle to the shear band boundaries. Black represents red sylvinite layers and white represents halite layers.

*Shear zones.* Shear zones have been mainly observed in relation to two different structural settings, (1) overturned limbs of folds and (2) lithologic boundaries which represent a competence contrast. In the first case, the shear zones are located in the overturned and north-dipping limbs of inclined folds (Fig. 6). In these limbs, there is an inhomogeneous concentration of the deformation which results in distinct smaller shear zones. Each shear zone has abrupt and straight limits and a maximum width of 30 cm (Fig. 6). In the shear zone the layers are completely transposed and form a banding sub-parallel to the shear zone limit. The reduction in thickness of layers in the shear zones is from 6 cm (undeformed thickness, Pueyo 1970 ) to 0.5 cm (Fig. 6). In the hinges of the folds, the layer thickens to 12–15 cm. Shear zones terminate by a reduction to zero displacement, which results in no thickness reduction in the banding and no folding. These shear zones have a displacement of up to several metres (Fig. 6) and are related to the stretching of the overturned limbs during the folding process. The enveloping surface of the decametric scale folds is also displaced by these north-dipping shear zones which display a reverse displacement. This shear sense is consistent with a transport shear direction in the whole salt sequence of top-to-the-south.

The other common location for shear zones is parallel to bedding in the proximity of lithologic changes which represent a competence contrast (Figs 7 & 8). One of the most suitable contacts for shear zone development is the B layer-carnallite limit. In the observed locations the bedding dip is slight and the shear sense is always top-to-the-south, consistent with the general sense of shear in the salt sequence (Fig. 7). These shear zones have been observed in all the studied sites (Cabanasses, Súria, Balsareny). The shear zone limits are also sharp and the maximum shear is observed in the banded areas where there is great thickness reduction (Fig. 8). Other structures developed in relation to the shear zones are drag-folds. These folds nucleate where there is a small obliquity between bedding (south dipping) and shear (subhorizontal). The evolution of folds from a local bed thickening to a well developed fold have been used to infer the sense of shear (Figs 7 & 8) (Hanmer & Passchier 1991).

The folds related to the metric and decametric

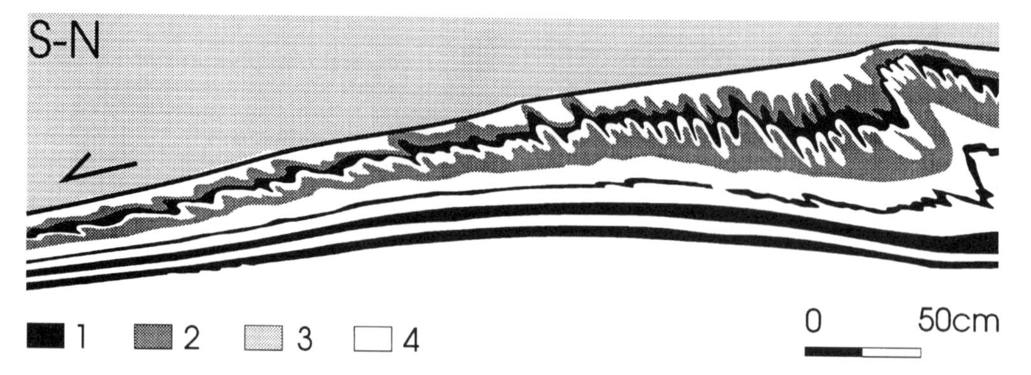


**Fig. 7.** Train of folds located in a shear zone parallel to a lithologic limit (carnallite-layer B). Several small detachments are localized in the whole sequence separating deformed and undeformed layers. In the fold train deformation increases towards the north and from open and slightly assymetrical folds in the southern part it evolves to asymmetrical and to isoclinal folding. 1, red sylvinite; 2, dark halite; 3, carnallite; 4, clear halite. Arrow indicates shear sense.

scale shear zones have similar orientations to those observed throughout the whole salt sequence. Both sets of folds have parallel fold axes, which are also perpendicular to the shear transport direction inferred to be top towards the south. These relations may indicate that the evaporitic horizons behave like a top-to-the-south shear zone.

*Boudinage.* Boudinage was observed in layered series of sylvinite and halite in the mines at several locations, mostly in Súria (Shaft IV). The sylvinite layers are boudinaged between the halite layers. Initial stages of boudinage show pinch and swell morphologies in continuous sylvinite layers (Fig. 9a). More evolved boudins form elongated lenses of sylvinite into the halite matrix (Fig. 9b), indicating a low viscosity contrast. The boudinage of sylvinite between halite layers contradicts the classification by Borchert & Muir (1964) and Kern & Richter (1985) which gives sylvinite as weaker than halite. In some locations it is possible to date boudinage with respect to the folding. All the potash layers show thinning and thickening in Súria (Shaft IV) which comprises several fold wavelengths indicating that boudinage does not have any relationship to folding. Furthermore, smaller folds develop in the thinner segments of the layers and larger folds where the layers are thicker. The observation that the wavelength of the folds depends on the thickness of the potash layer (Fig. 9c) indicates that boudinage formed prior to folding. At a smaller scale, boudins are present in limbs as well as in hinges of the ten-metre scale folds.

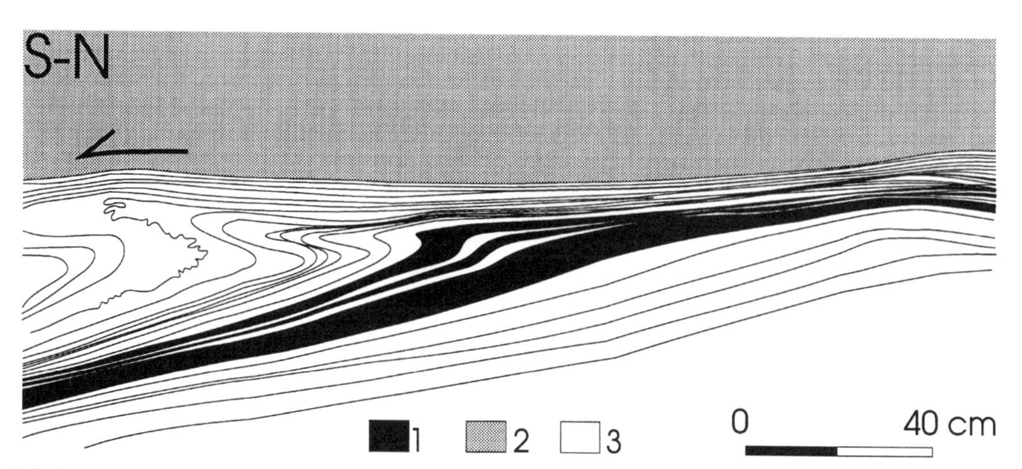


**Fig. 8.** Drag folds developed in a shear zone. The progressive deformation of the fold from a localized thickening to a well developed fold enables the shear sense direction to be inferred. Top to the south shear sense. 1, red sylvinite; 2, carnallite; 3, halite.

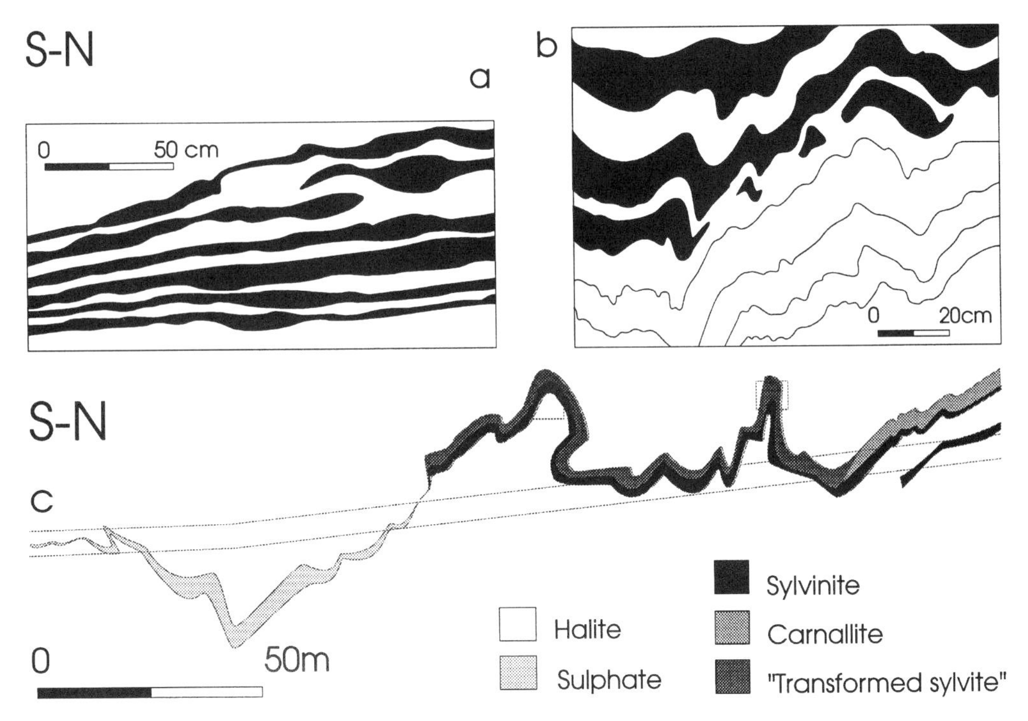


**Fig. 9.** Boudins can be observed in different outcrops in the mine. (a) Pinch and swell structures reflect a moderated boudinage of the sylvinite layers between the halite ones. (b) Sylvinite layers can form lenses in a halite matrix. Boudins can be folded after their formation. (c) The potash layers thin and thicken without any relation with folding, in contrast; their thickness controls the wavelength of the folds. Simplified cross section from mining survey.

*'Sterils'.* Sterils' are defined as zones where there are no potash layers. 'Sterils' can be of various origins; some of them show a relationship with the structure of the potash layers, whereas others are of unknown origin. Their sizes are variable. Most of the large 'sterils' are located in the central area of Súria (Shaft IV) (Fig. 4) and in Balsareny, and they coincide with the most boudinaged area on a large scale. This coincidence is interpreted as evidence that they are genetically linked. Absence of potash layers could be due to complete attenuation and boudinage of these layers. The 'steril' areas are elliptical in plan, with the longer axes oriented parallel to the fold axes. The internal structure of the "sterils" is not well known but it affects various anticlines and synclines. Small 'sterils' are more widespread and are also present in Cabanasses. Most of these small 'sterils' seem related to shear zones developed in overturned limbs of folds.

## Discussion and conclusions

In the study area the thick Cardona salt forms the detachment level of the fold-and-thrust system. The salt cores the anticlines and its internal structure has been described to date as chaotic or random (internal reports of Súria K SA; Richter-Bernburg 1980). The study of the internal structure of the salt-cored anticlines has shown that they are neither chaotic nor random, but that they record the deformation history of the detachment and associated folds and thrusts.

In the studied salt layers there are: (1) north-vergent folds and (2) south-vergent folds. North-vergent folds are located in the proximity of the Súria backthrust (less than 500 m from the inferred fault plane) and are not present to the south of this region. This has been interpreted as indicating that these structures are genetically related to one another. These folds and thrust show no geometric relation with the other structures and it is not possible to relatively date them. South-vergent folds, on the contrary, are widespread in the salt layers. The asymmetry of the south-vergent minor folds in relation to their location in the major folds (Súria and El Guix anticlines) cannot be interpreted as linked to a compressional or diapiric origin of the major folds (Figs 10a and b). The southern vergence of the minor folds in the southern limbs of the southern and northern Súria anticlines is incompatible with a compressional folding (Figs 5b and 5c). The southern vergence of the minor folds and the top-to-the-south shear sense inferred in the shear zones located in the north-dipping limb of the El Guix anticline

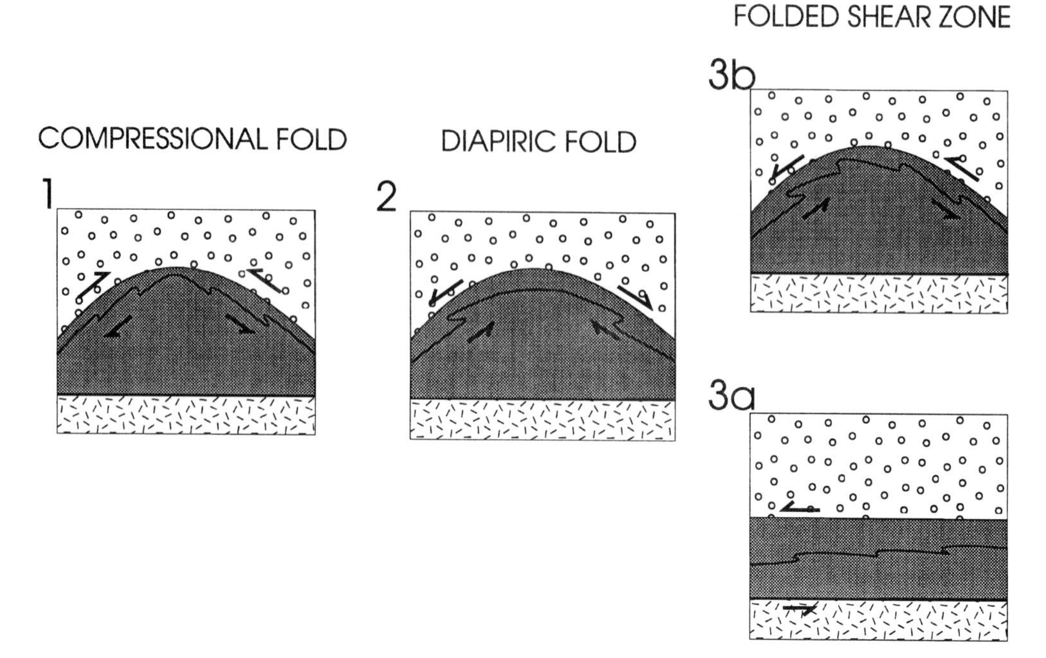


**Fig. 10.** Attitude of minor folding in relation to major folds of different origin. Compressional and diapiric folds have a symmetrical distribution of the minor folds asymmetry whereas a folded shear zone shows a generalized unique vergence in both limbs of the fold.

(Balsareny) is inconsistent with diapiric folding. Furthermore, the presence of south-vergent folds in the hinge zone of the syncline (Balsareny, Fig. 5a) is inconsistent with both diapiric and compressional folding. The generalized southern vergence of the folds and the direction of the shear zones are consistent with a top-to-the-south shear movement related to the detachment of the upper brittle units (Sanaüja lacustrine complex and Artés and Solsona molassic formations) above the Cardona detachment level (Fig. 10c).

In the initial stages, shearing results in three different types of mesoscopic structures in the salt layers: (1) folds, (2) boudins and (3) no-deformation-structures at mesoscopic scale, depending on the relative orientation of bedding with respect to the shear direction. In the study area, shear is subhorizontal and with a top-to-the-south displacement. This arrangement results in shortening and folding of the gently south-dipping beds because they are in the compressional quadrant together with stretching and boudinaging of the gently north-dipping beds because they are in the extensional quadrant (Fig. 11). Parallelism between shearing and bedding results in no deformation structures at mesoscopic scale in the salt layer. The increasing deformation led to the formation of (1) kilometre-scale salt-cored anticlines and (2) thrusts in the units located above the salt (Fig. 11). The detachment salt-cored anticlines fold the enveloping surface of the minor folds previously developed in the detachment layer. In Cabanasses (southern limb of the northern Súria anticline) the long limb-short limb asymmetry of the enveloping surface is opposite to that observed in the minor folds and consistent with the geometry of the compressional anticline which outcrops at the surface as a kilometric scale south-vergent asymmetric anticline (Figs 4a and c). The thrusts and their related folds penetrated at least to a certain depth in the detachment level as evidenced by the development of the north-vergent folds close to the Súria backthrust (Súria-Shaft IV). The lack of superimposition of these north-vergent folds on previous structures may indicate that in this region bedding and shearing were parallel before the development of the north-directed backthrust (Fig. 11). South of this region, the folded boudins observed in the mine (Fig. 9b) may indicate the change from a north-dipping limb to a south-dipping limb due to the northward displacement of the thrust.

In the study area there is a general deformation gradient from north to south, reflected in the fold size and density. Although deformation generally diminishes form north to south, deformation within the salt increases in the proximity of a major fault. The deformation evolves from small steeply-dipping sterils in the fault zone, to shear zones with

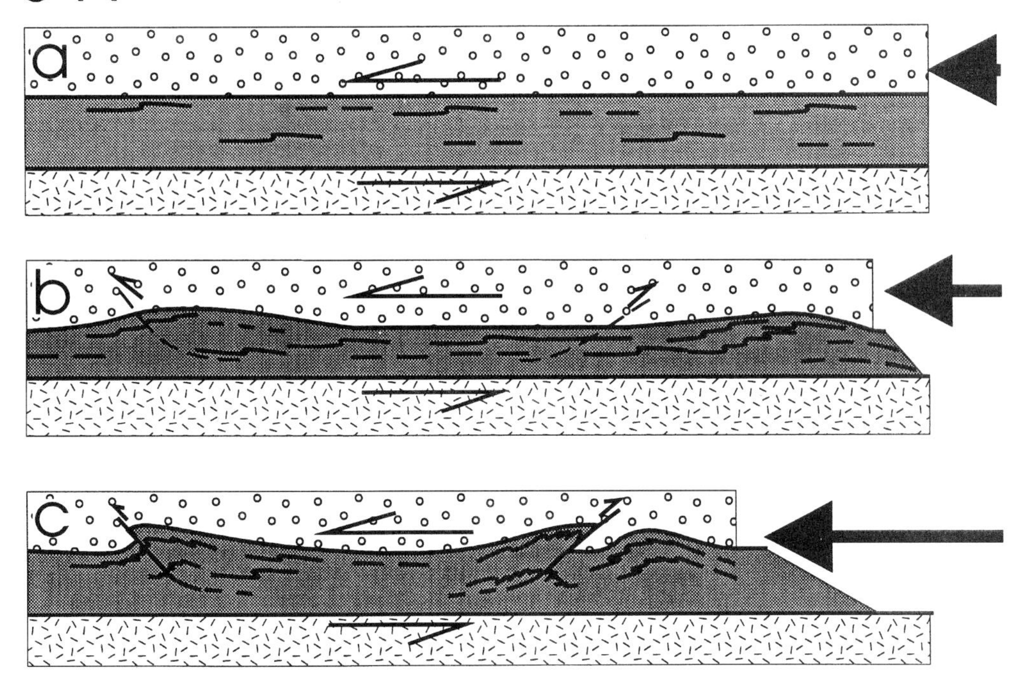


**Fig. 11.** Cartoon of the progressive deformation of the detachment level. (**a**) Initial shearing forms small scale structures such as folds and boudins in the salt layer. (**b**) The overburden deforms by thrusts and detachment folds at the same time that shearing in the salt layers causes the formation of new folds and boudins. (**c**) In the salt layers, deformation increases near the thrusts. The folds relating to the thrusts face the same direction as the thrust. The detachment folds modify the arrangement of the previous structures, rotating and folding them.

marked banding and finally to tight folding that gradually open and diminish in size away from the fault. The general deformation gradient is consistent with the shear displacement gradient towards the south.

The displacement of the units located north of the Súria 'anticline' is a minimum of 750 m whereas, in the northern limb of the El Guix anticline, the southward displacement is only 200 m. Thus, the example studied shows the internal structure of a detachment horizon associated with only a moderate deformation. The folds have axes perpendicular to the transport direction in contrast to the observations in detachments which have undergone great amounts of shear and where the fold trains that are perpendicular to the transport direction coexist with sheath folds and folds that have axes oblique or parallel to the transport direction. From comparisons between the detachment studied, more strongly deformed detachments (Marcoux *et al.* 1987; Malavielle & Ritz 1989) and highly deformed salt units (Talbot 1979), it may be inferred that in the evaporitic detachment levels, the first structures to form are folds with axes perpendicular to the transport direction. As deformation increases, differential shear (Hansen 1971; Carreras & Santanach 1973; Alsop & Holdsworth 1993) might trigger the rotation of these folds to form sheath folds, or together with the obliquity of the bedding with respect to the flow planes (Malavielle 1987) cause the formation of new folds with axes oblique or even parallel to the transport direction.

We would like to thank Súria K SA and Potasas del Llobregat SA for their collaboration in this study and G. I. Alsop and C. J. Banks for their comments and suggestions. This project has been partially financed by DGICYT PB 91-0805 and PB 91-0252.

## References

Alsop, G. I. & Holdsworth, R. E. 1993. The distribution, geometry and kinematic significance of Caledonian buckle folds in the western Moine Nappe, northwestern Scotland. *Geological Magazine*, **130**, 353--362.

Borchert, H. & Muir, R. O. 1964. *Salt deposits*. Princeton, London.

BUSQUETS, P., ORTÍ, F., PUEYO, J. J., RIBA, O., ROSELL, L., SÁEZ, A., SALAS, R. & TABERNER, C. 1985. Evaporite deposition and diagenesis in the saline (potash) Catalan basin, Upper Eocene. *In*: MILÀ, D. & ROSELL, J. (eds) *6th European Regional Meeting. Excursion Guidebook*. International Association of Sedimentologists. Lleida.

CARRERAS, J. & SANTANACH, P. 1973. Micropliegues y movimiento en los cizallamientos profundos del Cabo de Creus (prov. de Gerona, España). *Estudios Geológicos*, Madrid, **19**, 439–450.

DAVIS, D. M. & ENGELDER, T. 1985. The role of salt in fold-and-thrust belts. *Tectonophysics*, **119**, 6–88.

HANMER, S. & PASSCHIER, C. 1991. *Shear-Sense Indicators: a Review*. Geological Survey of Canada. Paper 90–17.

HANSEN, E. 1971. *Strain Facies*. Springer, Berlin.

JORDAN, P. & NÜESCH, R. 1989. Deformation structures in the Mushelkalk Anhidrites of the Schafirsheim (Jura Overthrust, Northern Switzerland). *Eclogae Geologicae Helveticae*, **82**, 429–454.

KERN, H. & RICHTER, A. 1985. Microstructures and textures in evaporites. *In*: WENK, H. R. (ed.) *Preferred Orientation in Deformed Metals and Rocks: An Introduction to Modern Texture Analysis*. Academic, London, 317–333.

KUPFER, D. H. 1968. Relationship of internal to external structure of salt domes. *In*: BRAUNSTEIN, J. & O'BRIEN, G. D. (eds) *Diapirism and Diapirs*. American Association of Petroleum Geologists, Memoir, **8**, 79–89.

MALAVIELLE, J. 1987. Extensional shearing deformation and kilometer-scale "a" type folds in a Cordilleran metamorphic Core Complex (Raft River mountains, Northwestern Utah). *Tectonics*, **6**, 423–448.

MALAVIELLE, J. & RITZ, J. F. 1989. Mylonitic deformation of evaporites in décollements: examples from the southern Alps, France. *Journal of Structural Geology*, **11**, 583–590.

MARCOUX, J., BRUN, J. P., BURG, J. P. & RICOU, L. E. 1987. Shear structure in anhydrite at the base of thrust sheets (Antalya, Southern Turkey), *Journal of Structural Geology*, **9**, 555–561.

MUÑOZ, J. A., SANS, M. & VERGÉS, J. 1994. Thrust-wedge structure in the frontal zone of the Southern Pyrenees. *Program, Expanded Abstracts and Biographies, Western Canadian and International Expertise*. Canadian Society of Exploration Geophysicists and the Canadian Society of Petroleum Geologists, Calgary.

ORTÍ, F., ROSELL L. & PUEYO, J. J. 1984. Cuenca evaporítica (potásica) surpirenaica del Eoceno Superior. Aportaciones para una interpretación deposicional. *In*: OBRADOR, A. (ed.) *Publicaciones de Geologia, Homenaje a L.Sánchez de la Torre*, 209–231.

PUIGDEFÀBREGAS, C., MUÑOZ, J. A. & MARZO, M. 1986. Thrust belt development in the Eastern Pyrenees and related depositional sequences in the southern foreland basin, *In*: ALLEN, P. & HOMEWOOD, P. (eds) *Foreland Basins*. Special International Association of Sedimentologists, Special Publication, **8**, 229–246.

——, —— & VERGÉS, J. 1992. Thrusting and foreland basin evolution in the Southern Pyrenees. *In*: MCCLAY, K. R. (ed.) *Thrust Tectonics*. Chapman & Hall, London, 247–254.

PUEYO, J. J. 1970. *Estudio Geológico de los Yacimientos Salinos de Sallent y Balsareny*. Tesis de Licenciatura. Universitat de Barcelona.

—— 1975. *Estudio Petrológico y Geoquímico de los Yacimientos potásicos de Cardona, Súria, sallent y Balsareny (Barcelona, España)*. PhD thesis, Universitat de Barcelona.

—— & SAN MIGUEL, A. 1974. Características petrológicas de las sales sódicas, potásicas y magnésicas de la cuenca potásica Catalana. *Inst. Inv. Geol. Dip. Prov. de Barcelona*, **29**, 23–49.

RAMSAY, J. G. 1967. *Folding and Fracturing of Rocks*. McGraw-Hill, New York.

RICHTER-BERNBURG, G. 1980. Interior structures of salt bodies. Bull. *Centres de Research Elf-Aquitaine*, **4**, 373–393.

SÁEZ, A. 1987. *Estratigrafía y Sedimentología de las Formaciones lacustres del tránsito Eoceno-Oligoceno del NE de la Cuenca del Ebro*. PhD thesis, Universitat de Barcelona.

SANS, M. & VERGÉS, J. in press. Fold Development Related to Contractional Salt Tectonics: Southeastern Pyrenean Thrust Front, Spain. *In*: JACKSON, M. P. A., ROBERTS, D. G. & SNELSON, S. (eds) *Salt Tectonics: A Global Perspective for Exploration*. American Association of Petroleum Geologists, Memoir.

TALBOT, C. J. 1979. Fold trains in a glacier of salt in southern Iran. *Journal of Structural Geology*, **1**, 5–18.

—— & JACKSON, M. P. A. 1987. Internal kinematics of salt diapirs. *AAPG Bulletin*, **71**, 1068–1093.

VERGÉS, J. & BURBANK, D. W. 1995. Eocene-Oligocene thrusting and basin configuration in the eastern and central Pyrenees (Spain). *In*: FRIEND, P. F. & DABVIO, C. J. (eds) *Tertiary Basins of Spain*.

——, MUÑOZ, J. A. & MARTÍNEZ, A. 1992. South Pyrenean fold-and-thrust belt: role of foreland evaporitic levels in thrust geometry. *In*: MCCLAY, K. R. (ed.) *Thrust Tectonics*. Chapman & Hall, London, 255–264.

WOODS, P. J. E. 1979. The geology of Boulby mine. *Economic Geology*, **74**, 409–418.

# Deformation in the late Permian Boulby Halite (EZ3Na) in Teesside, NE England

DENYS B. SMITH

*Geoperm, 79 Kenton Road, Newcastle upon Tyne, NE3 4NL, UK, and Department of Geological Sciences, University of Durham, Science Laboratories, South Road, Durham, DH1 3LE, UK*

**Abstract**: Differential (mainly lateral) movement of up to a few hundred metres is interpreted from borehole data to have taken place in the late Permian Boulby Halite in Teesside, northeast England, probably when it was buried to a depth of 2000 m or more. This interpretation is based on a marked variation in the present thickness of the formation, which is thought to have been almost uniformly thick when formed, together with evidence from more than 20 borehole cores of contortion, fracturing, flow-lineation and flow-brecciation; evidence of bedding-plane slip was noted in one core. Inferred boudinage in the lower part of the formation is interpreted as evidence of local extension. Unusually for rock-salt, faults cut halite in eight of the cores, and voids in fractures and faults were noted in halite in six cores.

The movement in the Boulby Halite is presumed to have been a response to inequalities in the confining pressures (i.e. pressure gradients) that were created when the underlying more competent formations were folded and faulted during Triassic or Jurassic earth movements; the displacement on all but the largest of these dislocations was absorbed by flow in the halite away from the crest of the dislocation to neighbouring less stressed areas, leading to the establishment of halite-depleted and halite-augmented areas. Up to half of the thinning in the halite-depleted areas was accomplished by the formation of sub-concordant sheets of flow-lineated (gneissic) halite in which individual crystals were flattened and stretched down the local pressure gradient, which was generally normal to the crest of the dislocation. Halite between these sheets may have been relatively undeformed, but nevertheless probably was transported between them from areas of high confining pressure to areas where pressure was lower. Overlying mainly siliciclastic strata appear not to have been transported, implying the presence of a decoupling layer or surface high in the formation.

Analysis of the records and/or cores of more than 130 boreholes (mainly brinewells) in the more recent part of a commercial brinefield in Teesside, northeast England (Figs 1, 3), has yielded much new information on the thickness and character of late Permian and younger strata in a marginal area of the Southern North Sea Basin, where few such data have been available previously. Most of the boreholes were located 70–100 m apart on a roughly rectilinear grid and many were fully cored through rocks of cycles 3 and 4 of the English Zechstein sequence and the intervening Rotten Marl; the general stratigraphy of these and adjoining strata is summarized in Fig. 2.

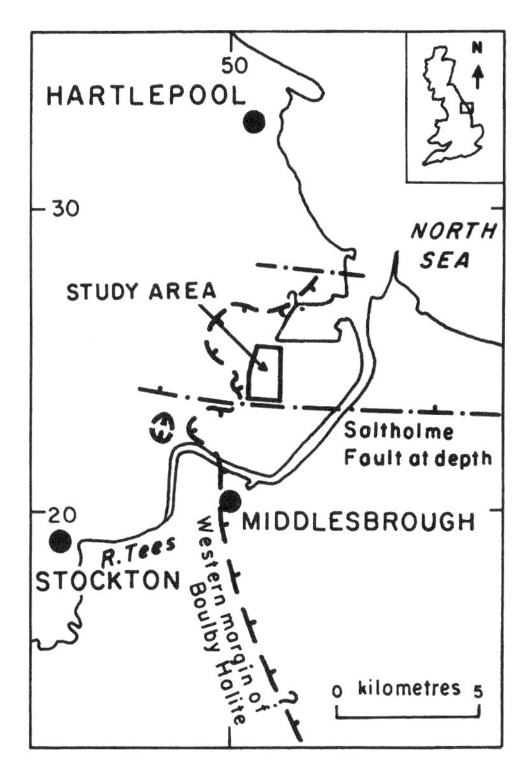


**Fig. 1.** Location of the study area.

The stratigraphy, sedimentology and palaeogeography of the late Permian evaporites and associated rocks of northeast England have been

*From* ALSOP, G. I., BLUNDELL, D. J. & DAVISON, I. (eds), 1996, *Salt Tectonics*, Geological Society Special Publication No. 100, pp. 77–87.

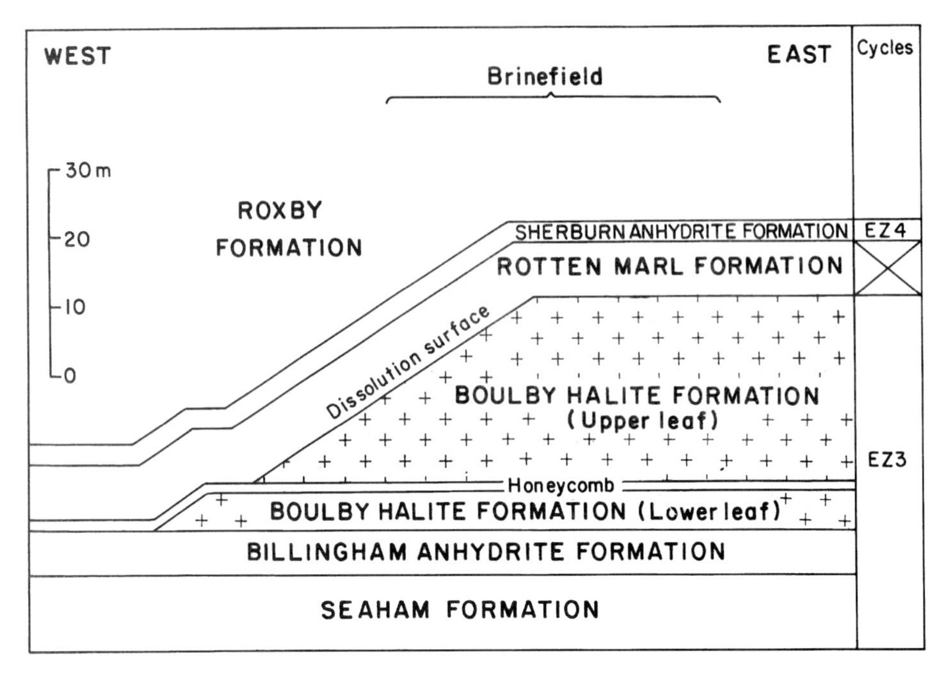


**Fig. 2.** Stratigraphy of the Boulby Halite and associated late Permian strata in the Teesside area of northeast England. The Roxby Formation is mainly of red mudstone and siltstone and the Seaham Formation is of carbonate rock.

reviewed by Hollingworth (1942), Dunham (1960) and Smith (1970, 1971, 1974, 1980*a*, *b*, 1989) and general discussion on these aspects need not be repeated. However, as the subject of this paper is deformation of the Boulby Halite and its possible causes, it is necessary first to restate the main characteristics of this formation and to note relevant aspects of its inferred depositional environments.

## Details of the Boulby Halite in Teesside

The Boulby Halite in the Teesside brinefield has a relatively uniform internal succession (Fig. 2), comprising unequal lower and upper leaves separated by a distinctive bed of halitic anhydrite locally informally known as 'the 'Honeycomb' or 'Honeycomb Rock':

| | | Typical thickness (m) |
|---|---|---|
| Upper leaf | Halite, mainly colourless to brown and red-brown, medium (5–15 mm) to coarsely (15+ mm) crystalline, with 0–5% of wisps, laminae, jagged mesh and intercrystal patches of red-brown mudstone almost throughout, but partly of grey mudstone and anhydrite near the gradational base; some beds or patches of colourless to amber pure medium-grained granular halite. | 11 |
| | Halite, mainly grey and brown, medium to coarsely crystalline, with 0–6% of very thin beds, laminae, jagged mesh and intercrystal patches of grey anhydrite; beds or patches of colourless to grey pure medium-grained granular halite; almost no mudstone. | 15 |
| 'Honeycomb' | Anhydrite, grey, finely crystalline (<1 mm), with large halite-rich upright pseudomorphs after early gypsum crystals at the top and beds of dark grey halitic dolomite and grey mudstone near the base. | 1.5 |
| Lower leaf | Halite, mainly grey, medium to coarsely crystalline, with 0–10% of laminae, jagged mesh and intercrystal patches of anhydrite and up to 11 (but commonly five to seven) thin (0.1–0.4 m) beds of halitic anhydrite (predominant), dolomite and grey mudstone that together generally form 20–30% of the unit. | 5.0 |

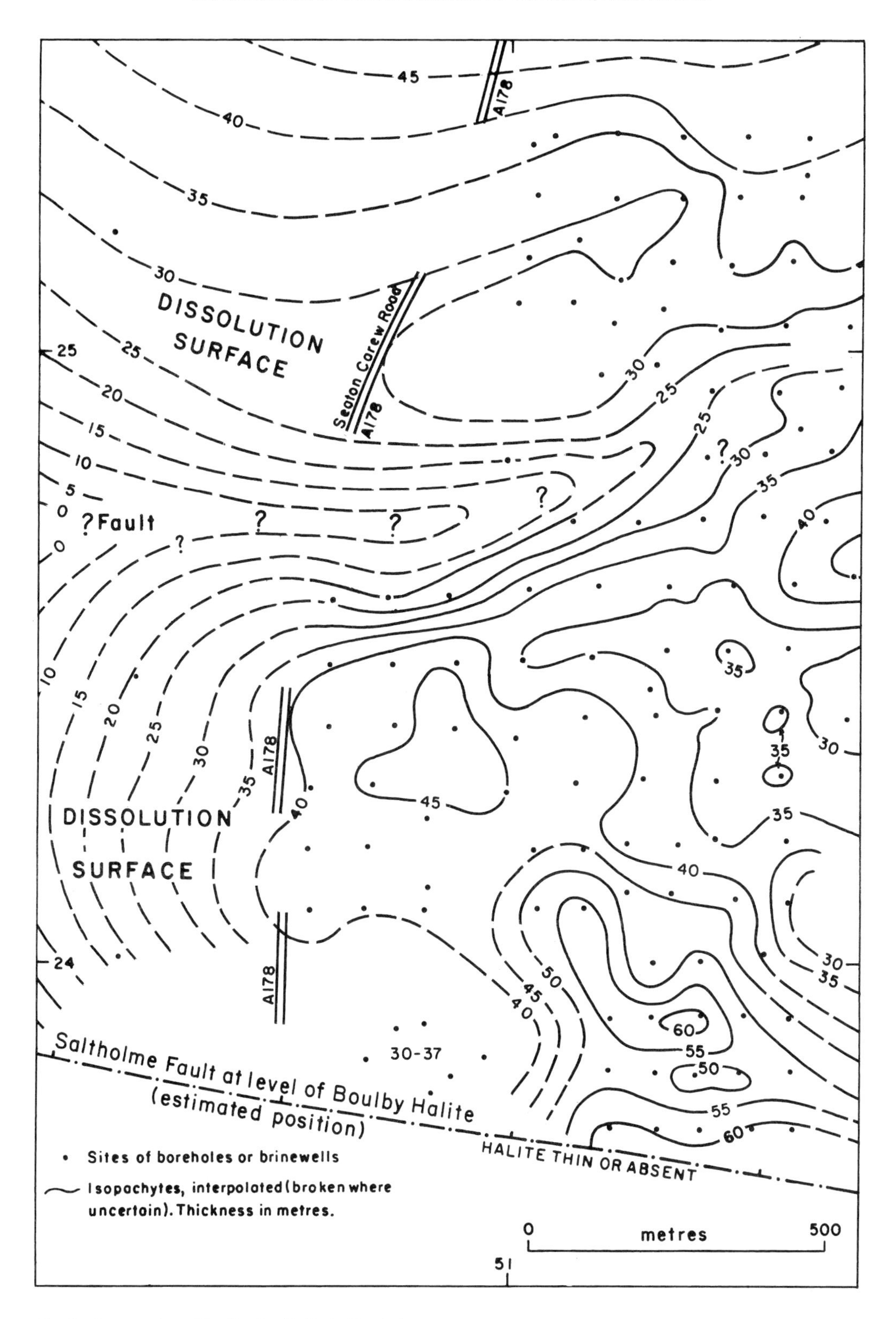


**Fig. 3.** Isopachytes of the Boulby Halite in the study area.

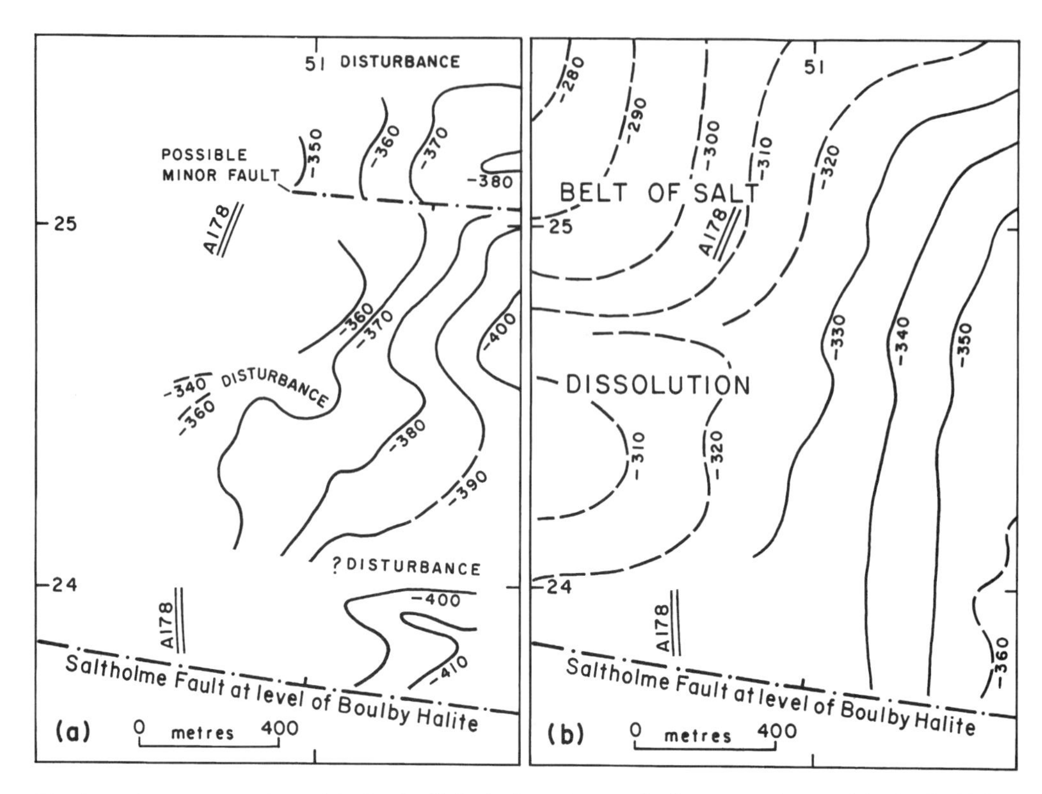


**Fig. 4.** (**a**) Contours on the base of the Boulby Halite in the study area. (**b**) Contours on the top of the Boulby Halite in the study area.

Despite marked lateral changes in thickness and lithology, and the eastward thinning of the 'Honeycomb', the broad subdivision of the Boulby Halite of Teesside into a lower unit that contains several anhydrite beds (including the 'Honeycomb') and an upper unit in which such beds are uncommon is recognizable at Boulby Mine [NZ 760 180] 25 km further east (Woods 1973), in the Whitby area of northeast Yorkshire (Stewart 1951; Smith 1971, 1973, 1974), and by wireline logs far across the Southern North Sea Basin (Smith & Crosby 1979).

Few details of the petrology of the Teesside Boulby Halite have been published, but close examination of the fabric of the halite and anhydrite has revealed hints of complex diagenetic histories comparable with those deduced by Stewart (1951) in his pioneer studies of the equivalent Middle Halite of the Aislaby (Eskdale No. 2) Borehole [NZ 843 080], near Whitby. These changes include widespread replacement of gypsum, anhydrite and subordinate dolomite by halite, the obliteration of most primary fabrics in the halite by pervasive coarse recrystallization, and the smaller-scale replacement of dolomite by anhydrite; some of these changes probably took place penecontemporaneously, but others were undoubtedly later, during burial to 2 km or more and subsequent uplift.

Although the Boulby Halite of Teesside now retains relatively few primary fabrics and sedimentary structures, enough remain to suggest that the formation accumulated on exceptionally extensive marginal flats that from time to time were flooded to depths of a few decimetres or metres by marine incursions or lagoonal expansions but were, for protracted periods, subaerially exposed (Smith 1980*a*, *b*, 1989). Acceptance of such a depositional environment, and in the absence of evidence of basal onlap against residual topography, requires that the primary depositional slope was probably negligible and the formation originally was of almost uniform thickness. It follows that any variation in the present thickness of the formation must be mainly secondary.

## Thickness variation of the Boulby Halite in Teesside

Isopachytes of the Boulby Halite of the study area (Fig. 3), interpolated from the borehole data, reveal

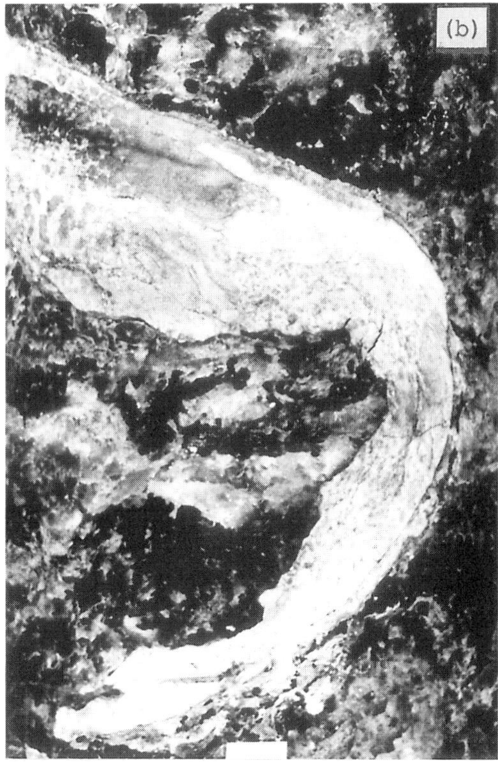


**Fig. 5.** Contortion and fractures in disarticulated beds of anhydrite (grey) low in the Boulby Halite of the study area. (**a**) E26 Borehole, depth 421.60 m; scale 40 mm. (**b**) Brinewell 167A, depth 390.50 m; scale 10 mm.

three distinct but mutually gradational patterns of thickness variation:

(1) a poorly documented pattern, mainly in the west of the study area, in which the formation thins out generally westwards in a horizontal distance of 600 to 1000 m;
(2) a semilinear, roughly W–E pattern, mainly in the south and the north-central parts of the study area, in which belts of halite are either thicker or thinner than the local average;
(3) an irregular pattern, mainly in the east of the study area, in which the halite thins and thickens apparently randomly.

*Pattern 1*

This pattern characterizes the updip (western) marginal area of the Boulby Halite and is interpreted here as a belt of halite dissolution. In it, the top and bottom of the formation converge westwards at about 2.3° to 4°, which is slightly less than the eastward dip of about 3° to 5.5° of the base of the formation; the dissolution surface is therefore sub-horizontal or slopes gently eastwards, giving rise to a secondary structural step or slight reduction of dip in overlying foundered strata.

The data are too scanty to permit a detailed interpretation of the shape of the updip dissolution surface of the Boulby Halite, which presumably owes much to the distribution of joints and fractures (i.e. fluid pathways) in underlying and overlying strata. No residual outlying halite patches have been identified along the brinefield margin to the north of the Saltholme Fault, but part of the lower leaf of the halite forms an isolated patch some 450 × 900 m across south of the fault and about two kilometres WSW of the study area (Fig. 1). This shows that dissolution of the upper leaf precedes that of the lower leaf, probably implying that undersaturated groundwaters gained access to the halite from above. Observations by L. S. Phillips (pers. comm. 1976) in the roof of Billingham Anhydrite Mine [NZ 47 22] revealed that where all the halite has been dissolved, it leaves a highly complex layered sequence, 1–3 m thick, of foundered anhydrite beds and mushy anhydritic and argillaceous residues.

### *Pattern 2*

Roughly linear W–E belts of atypically thick and somewhat thinner Boulby Halite lie on the north (downthrow) side of the Saltholme Fault, and the halite is absent or atypically thin in a narrow belt on the south side. These belts were well known to ICI geologists, especially L. S. Phillips, who attributed the thickness variations to plastic flow. Using closely spaced (i.e. 70–100 m) borehole control, Phillips (pers. comm. 1976) showed that the halite absorbed most or all of the displacement of the Saltholme Fault, which was about 50 m at the base of the formation; overlying strata appear not to have been displaced appreciably. Similar absorption of fault displacement was deduced in Boulby Mine by Woods (1979, fig. 6) in connection with a steep reverse fault that extended up from the Seaham Formation into the Boulby Halite and, farther afield, by Jenyon (1990, fig. 5.4) in connection with strike-slip faults in the southern North Sea gas fields.

The trend of the dissolution belt at the updip margin of the Boulby Halite in the study area is clearly influenced by the presence of the Saltholme Fault, presumably because the salt was thinner than average there and possibly also (though less likely) because the fault plane may have acted as a conduit for undersaturated groundwater from below. Using the configuration of the dissolution belt as a guide, it seems possible that the roughly W–E belt of thin Boulby Halite in the north-central part of the study area (Fig. 3) is also associated with a fault or monocline in the underlying strata, and there are probably others.

### *Pattern 3*

Judging from borehole data throughout the brinefield, this is the most extensive and representative pattern of thickness variation (15–63 m) in the Boulby Halite in Teesside. Within the study area the average thickness of the formation is perhaps 40 m, but for the brinefield as a whole, and excluding thicknesses in patterns 1 and 2, the average thickness is about 32–33 m.

## General structure of the late Permian strata in Teesside and its relationship to variation in the thickness of the Boulby Halite

Analysis of the geological structure of the brinefield area of Teesside, based on the borehole data, reveals a sharp disparity between the structure beneath the Boulby Halite and that above. At the base of the halite (Fig. 4a) and in underlying strata there is a substantial SE plunging anticline with complementary synclines to the north and south, whereas at the top of the halite (Fig. 4b) and in overlying strata there is a relatively simple eastward dip of about 3°. This disparity is an alternative expression of the thickness variation of the Boulby Halite, which is seen, in general, to thin over substrate eminences and to thicken into substrate hollows. Hence the Boulby Halite has moved in such a manner as to accommodate itself to many or most of the structural irregularities in the substrate and to have a generally smooth upper surface. That this upper surface is nevertheless relatively diverse in detail is indicated by local dips exceptionally up to 35° at the contact with the overlying Rotten Marl and indirectly by the abundance of minor fractures and movement planes in the Rotten Marl itself.

By comparison with the configuration of the top of the Billingham Anhydrite in the nearby Billingham Mine, as described by Wood (1950) and Raymond (1960), substrate eminences in the base of the Boulby Halite in the study area are likely to be linear and subparallel to each other, to be spaced at intervals of a few tens to a few hundreds of metres, and to take the form of minor faults and/or steep monoclines. These must have resulted in the top of the Billingham Anhydrite on the upthrow side of the dislocation being pushed up into the halite in a similar manner to that recorded in Boulby Mine by Woods (1979, fig. 6). The creation of narrow belts of anhydrite up to 15 m thick, reported in Billingham Anhydrite Mine by Raymond (1960), may also have applied differential pressure to the base of the halite, but the age of these belts is not known and they may be younger than the faults and monoclines.

## Evidence of movement of halite preserved in cores of Boulby Halite from Teesside

In addition to the thickness changes in the Boulby Halite of the study area and its surroundings,

**Fig. 6.** Strongly flow-lineated Boulby Halite on subconcordant core-break (view from above). Brinewell 155 (just outside the study area), depth 336.7 m; scale 10 mm.

compelling evidence of mass movement of the halite and associated anhydrite and mudstone was present in each of the 22 cores examined (including several from just outside the study area). This evidence comprises:

(1) folds and contortions on a decimetre to 300 m scale;
(2) faults;
(3) concordant to discordant sheets of flow lineated (gneissic) halite;
(4) flow-breccias (not in every core);
(5) variation in the number of non-halite beds present.

Of these, folds and faults are most common where the halite is of average or more than average thickness, and flow-lineation, flow-breccias and variation in the number of non-halite beds are most common in areas where the halite has been depleted (other than by dissolution) and is thinner than average.

## *Folds and contortion*

Almost total recrystallization, with consequent disarticulation, disorientation and rearrangement of bedding and reorientation of primary crystals, hinders or prevents recognition of the dip of much of the Boulby Halite; this is especially true in the high parts of the upper leaf where beds of mudstone and anhydrite are uncommon. Nevertheless, enough bedding traces remain, even in high parts of the formation, to reveal dips commonly in the range of 2–8°, and steeper dips and evidence of medium-scale contortion (Fig. 5) are readily seen in lower parts of the formation where anhydrite laminae and beds are abundant; possibly such high dips are also present in parts of the formation where dips are not apparent in cores only 100 mm in diameter, but would be visible on a larger surface. No structures that could be interpreted as the tepee-like upturned margins of primary polygons were recognized in any of the recent cores.

## *Faults*

The Boulby Halite in about half of the recent cores contains faults; these appear to be most common in the lower half of the formation where their position is made clear by the fracturing of relatively competent anhydrite and dolomite beds (Fig. 5a), but this appearance may be illusory and faults in the upper half of the formation undoubtedly cut halite in eight of the cores examined. In the relatively few examples where the fault plane was preserved, its dip ranged mainly between 30° and 60°, with a slight concentration about the middle of this range; very few fault planes had a dip exceeding 60° but a small number dip at less than 30° and two qualified as bedding-plane slip surfaces. Only one fault had a known displacement (*c.* 10 cm) and this fault was normal.

In addition to faults, the Boulby Halite in several of the cores contains pull-apart fractures up to 100 mm across in which there is no evidence of displacement between the opposing faces. These fractures, and also several of the faults, contain or are associated with ramifying veins of colourless to amber granular (predominant) and fibrous halite. Fibres in the latter are generally orientated normal to the walls of the veins, but fibres in some of the veins associated with faults are curved, indicating differential movement of the walls of the vein during crystal growth. Fibres in a number of veins are discontinuous, indicating episodic widening of the fractures and multiphase crystal growth. A small proportion of the fractures have a central void, some of which are gas-filled; some of these cavities appear to have been created by leaching, judging from the smoothness of their walls, but others appear to be unfilled parts of pre-existing larger cavities.

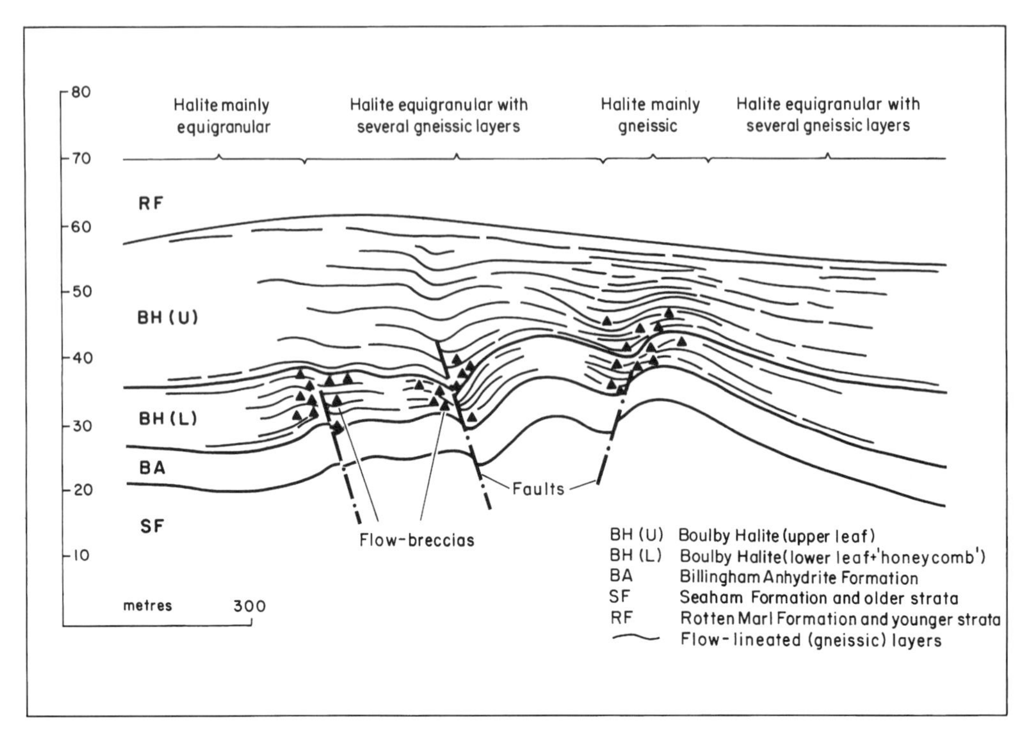


**Fig. 7.** Schematic section through the Boulby Halite of the study area, showing areas of halite depletion and augmentation and an interpretation of the distribution of sheets of flow-lineated halite. No specific orientation is intended.

## *Flow-lineated halite sheets*

Up to 30, but generally five to ten, subconcordant to concordant sheets of markedly flow-lineated (gneissic) halite are present in each of the recent cores, where they range from a few millimetres to more than 2 m in thickness. Their position in unbroken cores may generally be distinguished by parallel diffuse layering, lack of a mudstone or anhydrite mesh, generally fine grain size and colours ranging from pale grey to pale brown and to a highly distinctive pale mauve that is found only in these sheets. Closer inspection, especially of broken core surfaces, reveals that the constituent halite crystals are elongate (up to 12 mm long but generally less than 8 mm), flattened, and subparallel to each other (Fig. 6) with proportions of about 3-5:1.5–3:1. At the margins of the sheets, flow-lineated halite passes into relatively undisturbed halite in about 2–10 mm. The direction of elongation of the halite crystals is roughly constant throughout any one flow-lineated sheet and appears to be the same as that in adjoining sheets; the plane of flattening is similarly roughly constant throughout any one flow-lineated sheet and is parallel or sub-parallel with the top and bottom of the sheet. Crystal axes thus tend to be elongate parallel with the bedding, especially where the dip is less than about 10°, but a few sheets slope at more than 45°. None of the sheets of flow-lineated halite was seen to be cut by a fault and no firm conclusions may be drawn on which (if either) is the younger. Halite crystals on the faces of two faults were elongate and parallel in the direction of movement.

There is a clear pattern of vertical distribution of flow-lineated halite within the Boulby Halite of Teesside, with most sheets being concentrated at or near the top and bottom of anhydrite beds low in the formation, where sheet thicknesses of 10–50 cm are common and thicknesses exceeding 1 m are rare and mostly at or very close to the contact with the underlying Billingham Anhydrite. The less closely spaced flow-lineated halite sheets in the purer upper parts of the formation tend to be thicker than those in its lower part, but flow-lineated halite more than 1 m thick and locally more than 2 m thick (maximum thickness 2.5 m) is almost ubiquitous at or near to the top of the formation. With the exception of those at or near the bottom and top of the formation and those just below and just above the 'Honeycomb', it has not proved possible with any confidence to correlate individual sheets of flow-lineated halite even between fully cored boreholes only 80 m apart.

The proportion of flow-lineated halite relative to less strongly deformed halite in the Boulby Halite is closely related to the thickness of the formation at any one place. The proportion is lowest, less than about 5% of the whole, where the formation is 40 m or more thick, increasing to about 8–12% where it is 30 m or so thick and to about 15–30% where it is 20 m or so thick; more than 40% of the halite is flow-lineated where the formation is thinner than about 15 m, where much of the halite in the lower leaf is absent and some of the anhydrite beds are juxtaposed.

Many sheets of flow-lineated halite in the Boulby Halite Formation contain scattered rotated and abraded small clasts of anhydrite (predominant), dolomite and mudstone. By a gradual increase in the proportion of these clasts, flow-lineated halite grades into flow-breccia, perhaps implying a genetic link by which some flow-breccias evolved into impure flow-lineated halite as movement proceeded and debris was progressively ground down.

### *Flow breccias*

Mixed rocks, consisting of rotated disarticulated fragments up to a few centimetres across of anhydrite, mudstone and dolomite in a matrix of flow-lineated halite, are present in the Boulby Halite in most of the recent boreholes, but are uncommon in those where the formation is more than about 35 m thick. The longer axes of halite crystals in the matrix of these flow-breccias are deflected around rock fragments and so are less uniformly orientated than those in the flow-lineated sheets; many crystals are curved. The flow-breccias are concentrated in the lower leaf of the formation, near the inferred substrate dislocations (Fig. 7), and are interpreted here as the product of the advanced movement-induced disintegration of the primary fabric of the halite and non-halite interbeds together with the incorporation of fragments torn from adjoining beds by a combination of wedging and grinding.

### *Variation in the number of non-halite beds present*

Although some of the variation in the number of thin non-halite beds (mainly anhydrite, but with some of dolomite and mudstone) between boreholes may result from natural thinning-out, the postulated salt-flat depositional environment would be more consistent with relatively extensive individual non-halite beds and only a few natural failures between boreholes only 70–100 m apart. To account for most of the observed sharp variation in the number of such non-halite beds between boreholes on Teesside, the author infers the presence there of secondary boudinage, in which lateral extension of the formation has resulted in individual relatively competent beds being broken up into discontinuous layers of rectilinear slabs separated by halite that has filled the newly formed spaces; such boudinage is common in deformed Zechstein halite in mines in Germany and is present (Woods 1979) in distended parts of the Boulby Halite in Boulby Mine (Fig. 8), where more than 20 such beds are present. If this inference is correct, it follows that, since the gaps in one bed do not generally coincide with those in beds below and above, it is most unlikely that the full number of non-halite beds would be cut in any one borehole; this implies that there are probably more than 11 such beds in the study area. The odds against cutting such beds in any one borehole diminish with increasing extension and thinning of the formation; thus only two non-halite beds were present in each of two boreholes just north of the study area; in both of these boreholes, the formation was thin and almost all the halite in the lower leaf was flow-lineated.

## Causes of movement in the Boulby Halite in Teesside

The change in the shape of the Teesside Boulby Halite body from tabular to irregular is interpreted here as a response to changes in its confining pressure during deep burial. Judging from the areal coincidence of belts of flow-lineated halite with linear substrate eminences, these changes in confining pressures took place when the creation of

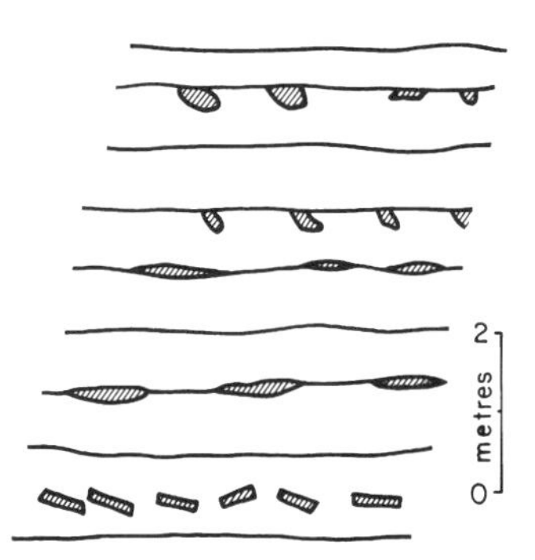


**Fig. 8.** Boudinage of anhydrite beds (shaded) in flow-lineated halite near the middle of the Boulby Halite Formation, Boulby Mine, in wall of incline near the shafts, from a sketch made by the author in 1975. Note (i) that many disarticulated anhydrite bed fragments have been rotated, and (ii) that parts of fragments at two levels high in the section have been sheared or ground off at nearly horizontal shear planes and the upper parts have been removed from the area depicted (presumably as part of flow-breccias). Note also that a vertical borehole would not pass through all the anhydrite beds.

substantial faults and steep monoclines disrupted the relatively more competent underlying Billingham Anhydrite and Seaham formations; the evidence suggests that the upward displacement of the upthrow side of these dislocations was absorbed by the Boulby Halite, lower parts of which were stretched and thinned over the newly formed eminences, and that higher parts were redistributed by flow into intervening, less highly stressed (mainly downthrown) areas. The redistribution was probably driven mainly by differential confining pressures, in an attempt to re-establish a relatively even pressure distribution such as, presumably, existed before the dislocation. A pressure gradient-governed style of movement in mobile halite was invoked by Talbot *et al.* (1982) to account for the last phases of movement of the Boulby Halite at Boulby Mine in response to Laramide (Cretaceous/Tertiary) reactivation of transcurrent faults in the Billingham Anhydrite and Seaham formations.

## Mechanism and time of movement in the Boulby Halite in Teesside

Movement of the Boulby Halite in the study area appears to have been accomplished on a range of scales by several different contributory processes. Of the processes outlined here, the creation of sheets of flow-lineated halite and flow-breccias appears to have been the main mode by which mass transfer of halite was effected, either directly or indirectly, with mainly accommodatory roles being performed by pervasive non-linear recrystallization, contortion and fracturing of the halite. That mass transfer was mainly lateral and subconcordant is indicated by the low angle of the long axes of most of the crystals in the flow-lineated sheets and of most of the sheets themselves, but the inclination of 30° or more in a few flow-lineated sheets may indicate that movement on these steeper sheets (and of apparently undeformed halite between them), was either discordant or multiphase. If it were discordant, it raises the possibility of the existence of a three-dimensional complex of flow-lineated sheets that divide at least some of the Boulby Halite in the study area into lenticular or trapezoidal masses, each a few metres thick and bounded on all sides by flow-lineated sheets. Alternative possible explanations of the relatively steep inclination of some flow-lineated sheets are (a) that they mark the position of the draped zone on the downthrow side of a fault or monocline (excluded as an explanation in most places by a lack of parallelism with adjoining sheets) and (b) (preferred) that originally gently inclined sheets formed early in the deformation of the halite body were folded during later deformation phases.

Observations of the Boulby Halite in Boulby Mine by Talbot *et al* (1982) show that the flow-lineated (gneissic) halite near the top of the formation forms a belt 50–200 m wide on the downthrow side of the reverse fault recorded by Woods (1979); beyond this it passes sharply into apparently normal undeformed halite. Striking evidence of differential extension of the halite is present in the form of metre-scale polygons that are increasingly stretched as the fault is approached, in a direction normal to the trend of the fault and parallel with the longest axis of the deformed crystals. Similar relationships probably exist in the study area, with flow-lineation and associated halite movement being approximately normal to the trend of fault-related substrate eminences. In neither area is there evidence of halite transfer on a regional scale and it is probable that the halite moved only as far as was necessary to approach equalization of its confining pressures. Presumably, this movement would have been halted when halite moving from a substrate eminence into a less highly stressed area encountered either halite under equal pressure moving in the opposite direction away from an adjoining substrate eminence or, alternatively, encountered a more rigid mass of rock such as less-deformed anhydritic mosaic halite that it could not displace; phases of mutual adjustment, perhaps involving contortion and fracturing, may have ensued.

The concentration of sheets of flow-lineated halite at or near to the base and top of the formation and at or near the base and top of the thicker of the anhydrite/dolomite interbeds is probably a response to the greater rigidity of these adjoining and interbedded rocks and its effect in focussing and transmitting pressure to the halite (see also Stewart 1951, p. 457).

Most of the processes contributing to movement in the Boulby Halite of the study area are difficult to quantify from borehole data alone, but some idea of the amount of vertical movement caused by a change from halite composed of roughly equant crystals to flow-lineated halite may be gained from a rough calculation; this calculation assumes (a) that for any given unit of halite the movement was laterally restricted by confining pressure and therefore was essentially two-dimensional, and (b) that (for example) a two-thirds reduction in height accompanied a three-fold increase in length. Applying these assumptions to Brinewell 167A, where the 28 m thick Boulby Halite includes 3.8 m of flow-lineated halite, a thinning of about 2.5 m is suggested to have resulted from the change of crystal shape; this is about half of the estimated total thinning of the formation at this site, assuming a primary thickness of 32–33 m. Such thinning would only have been possible, where pressure differentials created by the formation of the substrate eminences allowed the halite to spread or, perhaps more likely, where the eminences were large enough to cause substantial stretching of the Billingham Anhydrite–Boulby Halite interface and the immediately overlying halite.

The data do not show whether the halite deformed at the same time as the faults and monoclines were created, or whether, as seems possible, initial rapid movement of the substrate and adjoining halite was followed by protracted and progressively slower halite movement as confining pressures moved towards equilibration. There is no sign of the semi-liquefaction envizaged by Talbot *et al.* (1982) to account for the grotesque folding of parts of the Boulby Potash and uppermost Boulby Halite in Boulby Mine. The precise mechanism by which crystals in the flow-lineated sheets changed shape and move is not fully understood, though complex translation along intracrystal shear planes, and pressure dissolution and reprecipitation in the presence of fluid films may have been involved. These and other aspects of the movement and recrystallization of flow-lineated halite from the study area are being investigated by Professor C. J. Spiers of the University of Utrecht.

Determination of the details of crystallographic changes in the sheets of flow-lineated halite should go some way towards understanding movement of the Boulby Halite as a whole. Perhaps three main possibilities, not necessarily mutually exclusive, should be considered. (a) Did only the halite within the flow-lineated sheets move? (b) Did moving halite in the flow-lineated sheets function as shear planes, carrying along overlying halite that was itself not strongly deformed, and also younger, mainly clastic, strata? (c) Did moving halite in the flow-lineated sheets carry along overlying relatively undeformed halite (as in (b)) but not younger, mainly clastic, strata? Of these main possibilities, (a) seems to be least likely to satisfy all the related problems of rigidity and space, and (b) may fail because of the apparent lack of movement other than accommodation in the younger, mainly clastic strata. Possibility (c) thus appears to be the most likely of these three mechanisms of mass movement in the Boulby Halite to meet most of the observed facts; to be applicable it probably requires a degree of decoupling at or near the top of the formation that would allow the halite below to move without carrying younger clastic strata and it could be argued that the almost ubiquitous thick sheet of flow-lineated halite at or near the top of the formation may meet (or help to meet) this critical requirement.

Information on the time of deformation of the Boulby Halite is scanty and inconclusive. In the 253 Ma since it was formed, it has been buried to a depth of 2 km or more, heated for much of the time to more than 100°C, undergone protracted (probably episodic) diagenesis and been subjected to Hardegsen (early Triassic), Cimmerian (Jurassic) and Laramide (Cretaceous/Tertiary) earth movements and probably others too. Each of the three main phases of earth movements is believed to have either initiated or stimulated halite diapirism in the main North Sea Basin, but it is not clear how these movements affected relatively thin salt bodies that, like the Boulby Halite of Teesside, lay on a broad and relatively stable marginal carbonate/sulphate shelf.

In an attempt to reduce this uncertainty in their analysis of the time of deformation of the Boulby Halite and Boulby Potash at Boulby Mine, Talbot *et al.* (1982) pointed to the style of dislocation of the Billingham Anhydrite and Seaham formations at Boulby as being consistent with the effects of Laramide movement in other areas; they tentatively accepted that movement on the large reverse fault at Boulby resulted from Laramide transcurrent reactivation of faults that were originally formed or reactivated during Hardegsen and/or Cimmerian movements. If this is correct, and the faults affecting the Billingham Anhydrite and Seaham formations at Boulby and in the study area were contemporaneous, then the major substrate dislocation at Teesside was pre-Laramide. Laramide movements were envizaged by Talbot *et al.* (1982) as the trigger to the last episodes of vigorous flow in the Boulby Halite and Boulby Potash at Boulby Mine, but evidence of coeval movement in the Boulby Halite of the study area has not been recognized.

## Synthesis

Borehole data from a closely drilled brinefield in Teesside (northeast England) show that the thickness of the late Permian Boulby Halite there is sharply varied. In the western part of the brinefield, this variation results from continuing dissolution of the halite by circulating groundwaters but the remainder (and all of it to the east of the brinefield) probably resulted mainly from flow driven by inequalities of confining pressure. The effect of this flow was to move the halite laterally from areas where the underlying Billingham Anhydrite and Seaham formations had been affected by faults and steep monoclines, to areas where the configuration of the substrate was less diverse and the confining pressures were lower; substrate faults with a displacement of less than the thickness of the halite do not penetrate through it and the top of the halite is stucturally much simpler than its base. The plastic flow in the Boulby Halite thus tended to cancel out differences in tectonic pressure applied from beneath.

Examination of more than 20 complete cores

through the Boulby Halite of the brinefield has suggested that the pressure-equalizing lateral movement was not uniform throughout the formation but was concentrated in relatively thin subconcordant layers of halite with a gneissic (flow-lineated) fabric and additionally in its lower part, by flow-breccias of anhydrite and other debris in a matrix of gneissic halite. Relatively thicker sheets or lenses of apparently undeformed halite separate the layers of gneissic or flow-brecciated halite except where the formation is especially thin, where gneissic halite predominates. These relationships are tentatively interpreted to indicate that bulk movement of the halite may have been accomplished by the lateral movement of large sheets and lenses of mainly undeformed halite that slid past each other at different rates between subparallel to gently anastomosing thin layers of gneissic halite that acted as glide-planes. A widespread thick sheet of gneissic halite at or near the top of the formation may have acted in a decoupling role, thus explaining the apparent lack of extension in overlying mainly clastic strata.

I acknowledge with thanks the release by ICI plc of stratigraphical and lithological data from many of its more recent boreholes in Teesside and its permission to publish this account. Thanks are also extended to M. K. Jenyon, D. J. Shearman and C. J. Talbot for valuable discussion during the preparation of the text, and to I. Davison and R. Holmes for much helpful guidance as referees.

## References

DUNHAM, K. C. 1960. Syngenetic and diagenetic mineralization in Yorkshire. *Proceedings of the Yorkshire Geological Society*, **32**, 229–284.

HOLLINGWORTH, S. E. 1942. Correlation of gypsum–anhydrite deposits and associated strata in the north of England. *Proceedings of the Geologists' Association*, **53**, 141–151

JENYON, M. K.1990. *Oil and Gas Traps*. Wiley, Chichester.

RAYMOND, L. R, 1960. The pre-Permian floor beneath Billingham, County Durham, and structures in overlying Permian sediments. *Quarterly Journal of the Geological Society of London*, **106**, 297–315.

SMITH, D. B. 1970. The Permian and Trias. *In*: HICKING, G. (ed.) *The Geology of Durham County*. Transactions of the Natural History Society of Northumberland, Durham and Newcastle upon Tyne, **41**, 61–91.

——1971. Possible displacive halite in the Permian Upper Evaporite Group of north-east Yorkshire. *Sedimentology*, **17**, 221–232.

——1973. The origin of the Permian Middle and Upper Potash deposits of Yorkshire: an alternative hypothesis. *Proceedings of the Yorkshire Geological Society*, **39**, 327–346.

——1974. The Permian. *In*: RAYNER, D. H. & HEMINGWAY, J. E. The geology and mineral resources of Yorkshire. *Yorkshire Geological Society, Special Publication*, **2**, 115–144.

——1980*a*. Permian and Triassic rocks. *In*: ROBSON, D. A. (ed.) *The Geology of North East England*. Natural History Society of Northumbria, Special Publication, 36–48.

——1980*b*. The evolution of the English Zechstein. *Contributions to Sedimentology*, **9**, 7–34.

——1989. The late Permian palaeogeography of north-east England. *Proceedings of the Yorkshire Geological Society*, **47**, 285–312.

—— & CROSBY, A. 1979. The regional and stratigraphical context of Zechstein 3 and 4 potash deposits in the British sector of the southern North Sea and adjoining land areas. *Economic Geology*, **74**, 397–408.

STEWART, F. H. 1951. The petrology of the evaporites of Eskdale no. 2 boring, east Yorkshire. Part II The Middle Evaporite Bed. *Mineralogical Magazine*, **39**, 445–475.

TALBOT, C. J., TULLY, C. P. & WOODS, P. J. E. 1982. The structural geology of Boulby (Potash) Mine, Cleveland, United Kingdom. *Tectonophysics*, **85**, 167–204.

WOOD, F. W. 1950. Recent information concerning the evaporites and the pre-Permian floor in south-east Durham. *Quarterly Journal of the Geological Society of London*, **105**, 327–346.

WOODS, P. J. E. 1973. Potash exploration in Yorkshire: Boulby Mine pilot borehole. *Transactions of the Institution of Mining and Metallurgy*, **82**, 99–106.

——1979. The geology of Boulby Mine. *Economic Geology*, **74**, 409–418.

# The past of a future syntaxis across the Zagros

C. J. TALBOT[1] & M. ALAVI[2]

[1]*Hans Ramberg Tectonic Laboratory, Institute of Earth Sciences, Uppsala University, S-752 36 Uppsala, Sweden*

[2]*Research Institute for Earth Sciences, Geological Survey of Iran, PO Box 13185-1494 Tehran, Iran*

**Abstract**: Longitudinal components of the Zagros mountain chain change in character and width across N–S trending zones of strike-slip transfer faults lying between 51° and 54° E. To the northwest, a fold-thrust belt with consistent SW vergence has a width of *c.* 220 km in front of an imbricate belt *c.* 160 km wide. To the southeast, an imbricate zone *c.* 80 km wide is fronted by a gently tapering festoon of upright periclines that is *c.* 350 km wide and punctured by over a hundred emergent salt diapirs.

Pre-Zagros stages of the transfer zones in the Zagros are preserved on the Arabian platform and the two most obvious bound what we call the incipient Qatar syntaxis. This is at an early stage of one of the many syntaxes that compartmentalize the Alpine–Himalayan mountain chain. We use structures in the Hormuz salt to map and gauge the tectonic pulse of basement blocks that jostled as ocean basins opened and closed diachronously like zip fasteners along the Tethyan margin of Gondwana. This incipient syntaxis was a lithospheric key that went up while others went down during the rifting and riffling, but not drifting, of the still-born Hormuz basin we call Proto-Tethys.

Like many older orogens, the length of the Alpine–Himalayan orogen is punctuated by syntaxes (Sarwar & De Jong 1979). These bends and tucks, *c.* 350–400 km apart, divide the orogen into compartments with different characters and histories (Fig. 1).

Recent studies of orogenic segmentation in Pakistan (e.g. Davis & Lillie 1994) have focussed on the mechanical differences between the thin-skinned contractional wedges between the syntaxes. Understanding how orogenic wedges differ in active orogens obviously has useful lessons for

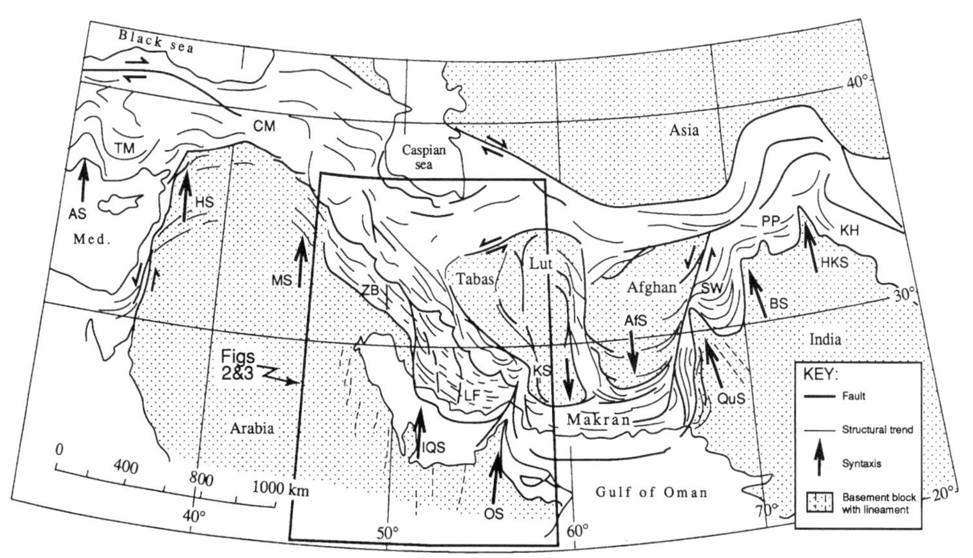


**Fig. 1.** Outline map of Alpine–Himalayan mountain chain indicating syntaxes and wedges and outlining the area considered here (after Sarwar & De Jong 1979; Alavi 1991; Dykstra & Birnie 1979). Syntaxes alternate with orogenic compartments from west to east: AS, Antalyo syntaxis; TM, Taurus mountains; HS, Halab syntaxis; CM, Caucausus mountains; MS, Mosul syntaxis; ZB, Zagros fold-thrust belt; IQS, Incipient Qatar syntaxis; LF, Zagros fold-diapir festoon in Laristan; OS, Oman syntaxis, KS, Kerman syntaxis, Makran wedge; AfS, Afghan Syntaxis; QuS, Quetta syntaxis; SW, Sulaiman wedge; BS, Bannu syntaxis; PP, Potwar plateau; HKS, Hazara-Kashmir syntaxis; KH, Kashmir Himalaya.

*From* ALSOP, G. I., BLUNDELL, D. J. & DAVISON, I. (eds), 1996, *Salt Tectonics*, Geological Society Special Publication No. 100, pp. 89–109.

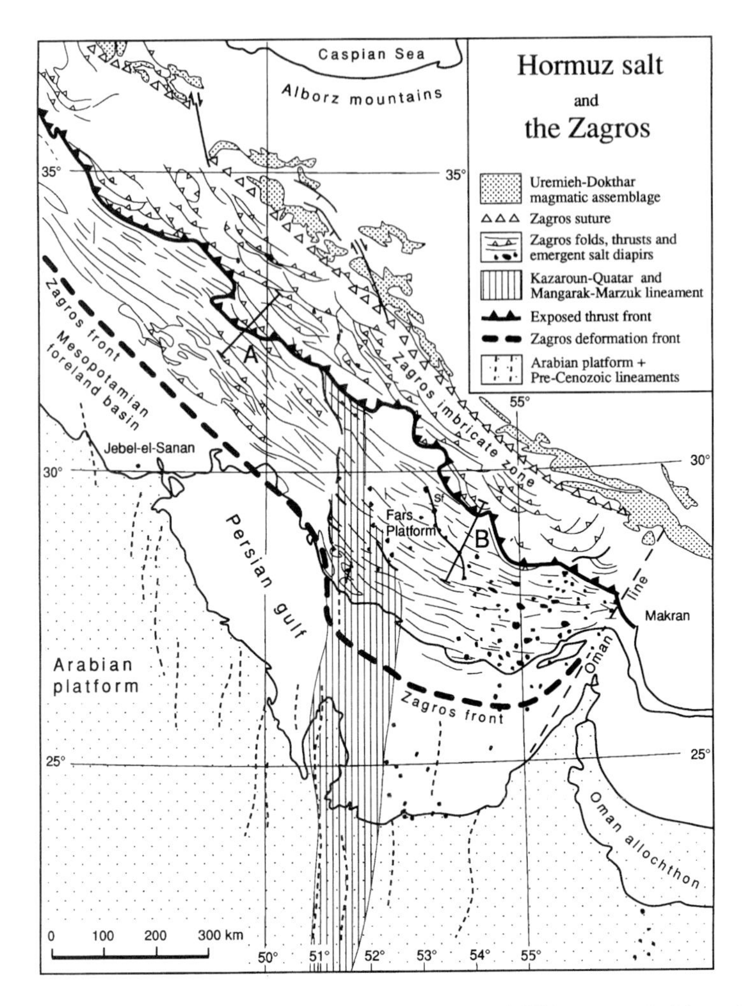


**Fig. 2.** The N–S Kazerun–Qatar and Mangarak–Marzuk lineaments extend 450 km across the Zagros between 51° and 53°E and extend 800 km further south onto the Arabian platform. Together they bound the incipient Qatar syntaxis. Notice that the topographic Zagros deformation front in the NW Zagros fold-thrust belt is behind the front defined using the shapes of oil fields on Fig. 3.

older orogens. The four segments discussed by Davis & Lillie (1994) represent different stages in the closure of an ocean basin and the subsequent contraction of sutured continental crust. The Neo-Tethyan ocean is still open to the south of the Makran wedge but has closed in both Pakistan to the east and in Iran to the west (Fig. 1). In the Makran, flysch-type sediments are accreting in a 300 km wide prism with a moderate 4° taper (Davis & Lillie 1994). To the east, beyond the Quetta syntaxis (Fig. 1), a thick continental margin sequence in the festoon-shaped Sulaiman wedge is at the stage where ocean–continent transitional crust is underthrusting Asia. Davis & Lillie (1994) speculated that the moderate 3° taper seen in profiles of the Sulaiman fold-thrust lobe might be due to ductile creep of warm carbonates as deep as 20 km. This novel scenario contrasts with the classic picture to the east, where a thin cover sequence detaches over a thin layer of salt. In northern Pakistan (Fig. 1), continental collision has doubled the thickness of the continental crust in two very different wedges on either side of the main Himalayan (Hazara–Kashmir) syntaxis. The Kashmir Himalaya is the most mature collisional wedge. It is a narrow (60 km wide) and strongly deformed fold-thrust belt with larger (7°) taper and constant SW vergence. This geometry is consistent with high traction along the base of the orogenic cover (Davis & Lillie 1994). By contrast, the Potwar Plateau to the west has an extremely gentle taper (≤1°) with remarkably little internal deformation in a thin

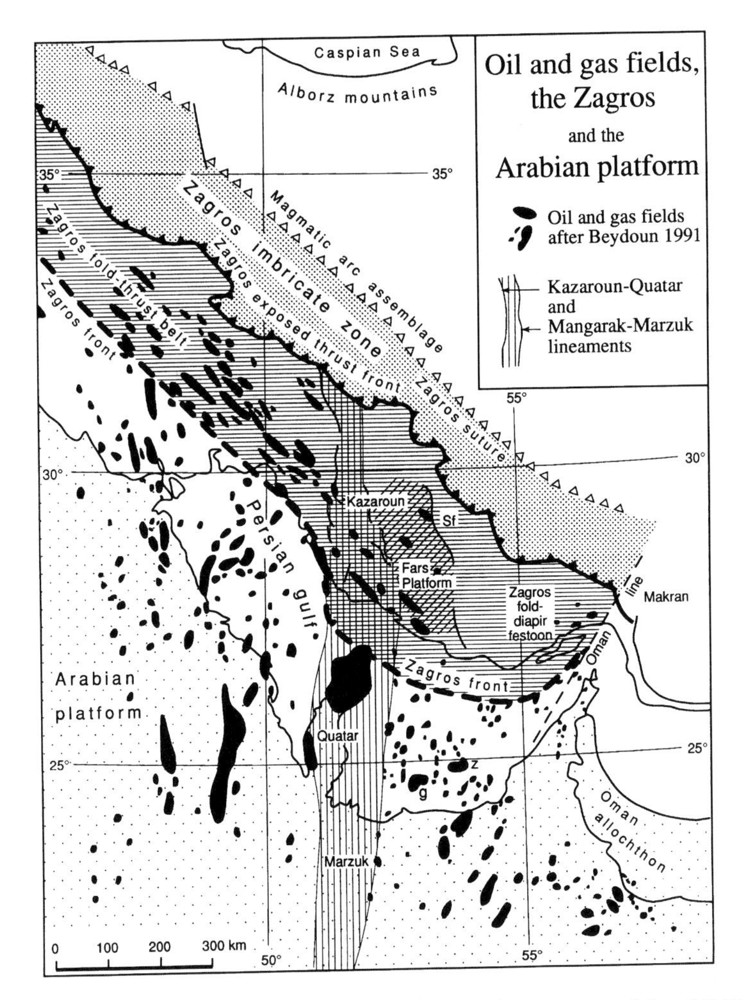


**Fig. 3.** The incipient Qatar syntax and the Zagros deformation front stand out on a map of the oil fields of the Middle East (adapted from Beydoun 1991). Sf is the Sarvestan fault. The Ghasha and Zakum fields are labelled g and z, respectively.

cover sequence taken to have transmitted compressive stresses over a salt decollement to the salt range along its leading edge (Butler *et al.* 1987).

This paper focusses on the tectonic development of the Zagros mountains west of the Makran. We distinguish two new wedges: the Zagros fold-thrust belt to the northwest of what we will refer to as the incipient Qatar syntaxis, and the Zagros fold-diapir festoon to the southeast (Fig. 2). These two wedges are additional to those listed by Davis & Engelder (1987) and Davis & Lillie (1994) and have exhibited different styles of deformation. We attribute these differences in style to the presence of Hormuz salt which accumulated unconformably on the Precambrian basement under one wedge but not under the other. We concentrate more on the history of the incipient Qatar syntaxis than the mechanical differences in the adjoining orogenic compartments. However, just as salt adds taste to a good meal, so salt can amplify the regional tectonics (Talbot 1992). We use the structures of Hormuz salt to feel the pulse of the northern margin of Gondwana as it was segmented by Tethyan basins; we then go forward in time and use Hormuz lithofacies to reconstruct the Neoproterozoic to mid-Cambrian tectonostratigraphy. Stamfli *et al.* (1991) considered the simple separation of Tethyan margins into Palaeo-and Neo-Tethys as inadequate and so inserted the gulf of Permo-Tethys into the story. Here we add earlier chapters (and details to Husseini (1989)) by distinguishing the lithofacies of an aborted palaeo-ocean that we call Proto-Tethys.

## The Zagros Orogen

Previous subdivisions of the Zagros mountains (e.g. Stöcklin 1968) have been into four orogen-parallel zones (reviewed in Alavi (1994)): the Uremieh–Dokthar magmatic arc assemblage, the Sanandaj–Sirjan imbricate zone, the Simply Folded Belt, and the Mesopotamian foreland basin with the Persian Gulf (Figs 2 and 3).

The paleo-ocean of Neo-Tethys opened to the NE of the Sanandaj-Sirjan imbricate zone in the Permian (in Turkey) and Trias (in Iran). In the Late Cretaceous; Neo-Tethys closed along the Zagros suture lying between the magmatic arc and the imbricate zone. The complex Sanandaj-Sirjan imbricate zone (with some low-grade metamorphism) used to be separated from the unmetamorphosed Simply Folded Belt by the Main Zagros Thrust (Stöcklin 1968). However, this structure has recently been demoted from a single steep reverse fault or crush zone to a series of low-angle thrusts indistinguishable from others throughout the imbricate zone (Alavi 1994). The boundary to the imbricate zone shown on Figs 2 and 3 is merely the southwesternmost Zagros thrust that is exposed locally.

We here divide what has previously been known as the Simply Folded Belt of the Zagros into two new but distinct along-strike segments. These segments lie NW and SE of a zone that strikes N–S across the NW–SE tectonic grain of the Zagros between 51° and 53° E (Figs 2 and 3). This zone lies between what we call the Kazerun–Qatar lineament (to the west) and the Mangarak–Marzuk lineament (to the east). We interpret the intervening block as a future syntaxis that preserves the earlier stages of processes now obscured by continental contraction in Pakistan. This incipient syntaxis consists of the 460 km long N–S zone bounded by the Kazerun and Mangarak fault zones across the Zagros mountains, together with the Qatar and Marzuk lineaments that extend for a further 800 km southward across the Arabian platform (Figs 2 and 3).

The deformation front of the Zagros Simply Folded Belt has been driving the foreland basin in front of it as it has propagated episodically to the SW since the Late Cretaceous. The cover sequence is still undeformed on the Arabian platform but has been approximately doubled in thickness in the high plateau of the Zagros as a result of NE–SW shortening across the Simply Folded Belt. Hydrocarbon fields that are elongate along N–S or NNE–SSW trends on the Arabian platform are deformed to NW–SE Zagros trends behind the Zagros front (Fig. 3). The Zagros deformation front defined by the shape of oil and gas fields (Fig. 3) is intermediate in depth between Zagros fronts defined by seismicity and surface topography. Northwest of the Kazerun fault zone, the seismic front follows the south-western limit of earthquakes with kinematic solutions of NW–SE trending thrusts at depths of ≤15 to 20 km (Jackson & McKenzie 1988, Ekström & England 1989). This seismic front is offset approximately 160 km along the Kazerun fault zone. Six moderate-sized earthquakes along the Kazerun fault zone with NS strike-slip solutions have been attributed to right-lateral transtension along old faults in the basement (Baker *et al.* 1993).

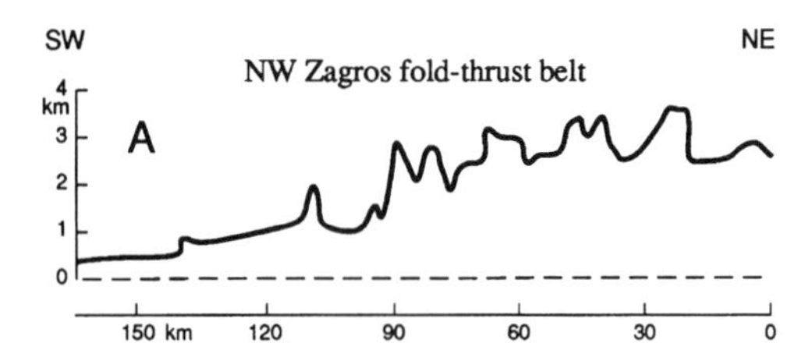


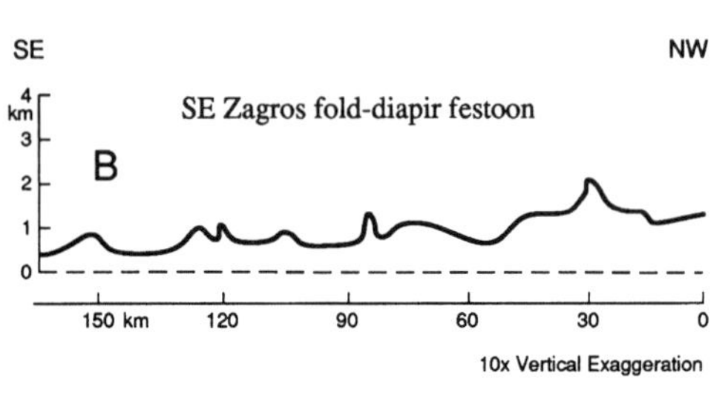


**Fig. 4.** Topographic profiles across the Zagros mountains along lines A and B indicated on Fig. 2.

At the surface, the topographic front of the Zagros mountains is currently about 50 km behind the deep seismic front to the NW of the Kazerun fault zone, but more nearly coincident with the 50 fathom water-depth contour we take to bound the topographic front SE of the Mangarak fault zone. The topographic front of the Zagros is offset by *c.* 200 km along the incipient syntaxis.

### *Three transfer zones and the incipient Qatar syntaxis*

Kent (1979) described three right-lateral transcurrent fault zones decorated by salt extrusions trending obliquely across the Zagros: the Kazerun, Mangarak and Sarverstan fault zones (Fig. 2). These zones display complex anomalous relationships between strike-slip faults that trend NS, NNW or NE, and Zagros folds and thrusts that trend NW–SE. All three are narrow zones of low topography bordered by broad blocks with higher topography. All terminate in strike-parallel thrusts with NW–SE Zagros trend (Figs 5 and 6). The northern

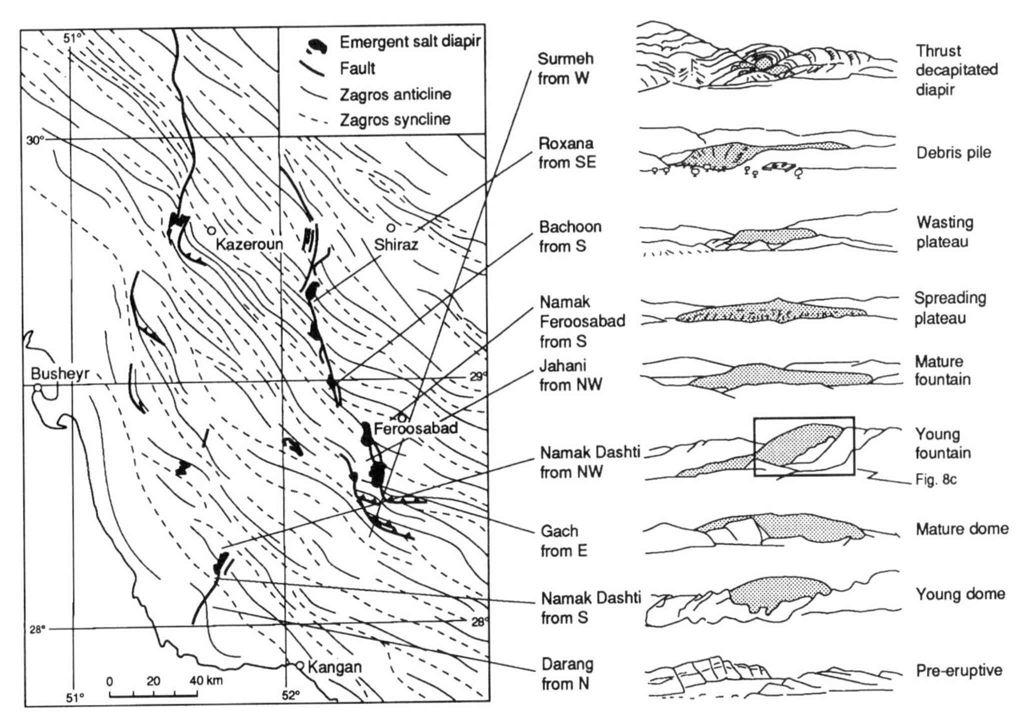


**Fig. 5.** Map of the Kazerun and Mangarak fault zones (to left and right, respectively) and sketches (traced from photographs) of some of the (stippled) Hormuz extrusions along them.

end of each fault zone swings to the NW; it is unclear whether this curvature is primary or is due to an increase in Zagros shortening towards the suture, or to left-lateral strike-slip shear along the suture.

The Kazerun fault zone is 450 km long, and the Mangarak fault zone is 150 km long some 60 km to the east (Fig. 5). About 100 km still further NE, the 100 km long Sarvestan fault zone trends NNW near Fasa in the inner Zagros (Kent 1979). We take the Sarvestan fault zone, the least well known transcurrent fault zone across the Zagros and one that we have not worked along, to bound the north-eastern corner of the Fars platform. The Fars platform is an area with a comparatively thin Permian to Eocene cover sequence, not pierced by salt structures, that lies between the Mangarak fault zone and 54°E (Fig. 2).

The Kazerun and Mangarak fault zones are a pair of anastomosing fault zones that trend generally N–S across the Zagros Simply Folded Belt between 51° and 53° E (Fig. 5). They act as oblique footwall ramps that transfer NE–SW shortening across the steep and narrow Zagros fold-thrust belt in the NW from the wider Zagros fold-diapir festoon with its lower taper in Laristan in the SE (Kent 1958). The fault blocks to the east of both the Kazerun and the Mangarak fault zones are structurally higher and more deeply eroded than blocks to the west (Kent 1979). This difference in structural level can reach 5 km in the north but varies with location and depth; at the south end of the Kazerun fault zone, the offset in the top of basement was thought to be *c.* 1–1.5 km (by Templeton, personal communication 1977) while Jurassic beds are offset only *c.* 100 m.

Both the Kazerun and Mangarak fault zones consist of en-echelon segments of steep right-lateral strike-slip faults with local throws that are both reverse and normal. However, whereas anticlines die out towards older faults along the Kazerun fault zone (Baker *et al.* 1993), anticlines culminate and are offset by younger faults along the Mangarak fault zone (Figs 5 and 6). With NNW trends oblique to the main NW Zagros trend, one might expect right-lateral strike-slip transfer zones to be generally transpressive. However, both zones are decorated by diapirs of Hormuz Formation that rose into rhombic cavities indicated by offsets of the traces of anticlines (Fig. 6). Such cavities play the role of (upside down) pull-apart basins at intervals of *c.* 22 ±6 km. Later, when discussing Zagros kinematics, we will argue that these salt bodies are reactive synshortening diapirs that have risen, or are still rising, wherever Zagros anticlines are offset along

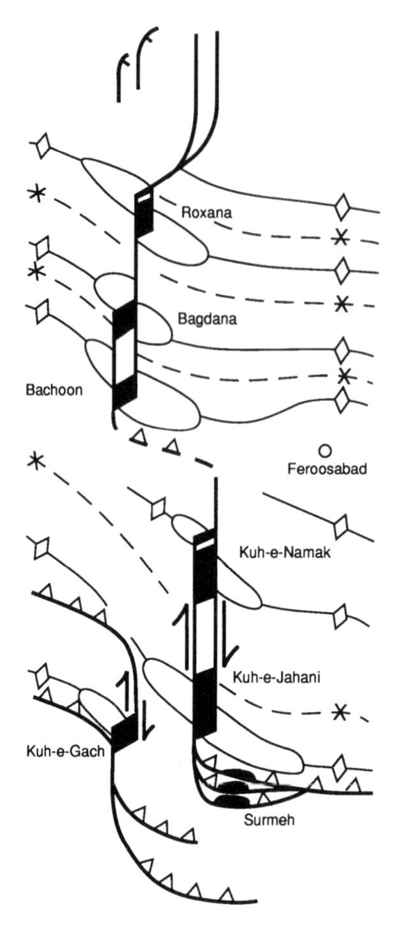


**Fig. 6.** Current kinematics of the Mangarak fault zone stylized to an almost continuous pattern of faults. Salt extrusions are in black; elliptical outlines are anticlinal culminations.

individual strike-slip faults (as diapirs have been interpreted to have risen along jogs in strike-slip faults in the Sverdrup basin in Arctic Canada (Schwerdtner & Osadez 1983)). Along the Mangarak fault zone, five diapirs have risen into rhombic openings pulled apart at releasing right-lateral en-echelon offsets along four or five parallel strike-slip faults (Fig. 6). Thrusts splay southeastward from the southern end of each segment of the Mangarak zone; one such group of thrusts dices a beheaded pre-Zagros diapir now devoid of salt at Surmeh (top right on Fig. 5)).

Changes in tectonic style both along and across the Kazerun and Mangarak fault zones are due to more than merely the level of erosion of thin-skinned fault zones (Kent 1979). This is because deformation styles relate to changes in thickness and facies at several levels in the regional stratigraphy between the Present and Precambrian. We will describe some of these changes later and show how the alignment of salt extrusions along the faults, together with their current seismicity (Baker *et al.* 1993), suggest that these zones are due to repeated reactivation of Precambrian faults in the underlying basement. The Kazerun and Mangarak fault zones represent fundamental along-strike divisions of the Zagros Simply Folded Belt into the compartments described in the next two sections.

## *NW Zagros fold-thrust belt*

NW of the Kazerun fault zone, the Simply Folded Belt of the Zagros is a long (1400 km), straight and narrow (220 km) seismically active wedge with a profile taper of *c.* 2° (Fig. 4). The NW Zagros fold-thrust belt stretches *c.* 1400 km between the Kazerun fault zone and the Mosul syntaxis in front of a 160 km wide Zagros imbricate zone (Fig. 1). Thrust-cored anticlines are 30–50 km long, 8–16 km wide and 2–3 km in amplitude, and verge consistently to the SW in a 6–10 km thick cover sequence consisting largely of platform carbonates.

No diapirs of Hormuz salt reach the surface in the NW Zagros fold-thrust belt (Fig. 2) and we suppose that any underlying Hormuz sediments are anhydrite or carbonates with only subsidiary salt. As well as a lack of emergent diapirs, there are also comparatively few emergent thrusts in the NW Zagros fold-thrust belt (Fig. 2). We follow O'Brien (1957) and attribute the structural style (and the 2° taper) of the fold-thrust belt (Fig. 4) to control by a shallow Miocene salt rather than a deep Hormuz salt. The disharmony and different wavelengths of folds above and below the Miocene salt demonstrate décollement along it (Talbot *et al.* 1987). Letouzey *et al.* (in press) show how a shallow décollement results in blind rather than emergent thrusts. Their models explain how anticlines with fishtail mapforms and triangular zones in profiles are so common in the NW Zagros fold-thrust belt (O'Brien 1957).

## *SE Zagros fold-diapir festoon in Laristan*

The long straight and closely-spaced thrust-cored anticlines of the Zagros fold-thrust belt abruptly broaden SE of the Mangarak and Sarvestan fault zones into an arcuate festoon of short wriggly and relatively gentle folds punctured by *c.* 120–130 emergent diapirs of Hormuz salt in Laristan. The SE Zagros fold-diapir festoon has a profile taper of <1° (Fig. 4) and reaches a width of *c.* 360 km between boundaries tucked 500-600 km apart. The tight eastern tuck in the festoon is along the NE-trending western boundary of the Oman syntaxis which separates the Zagros continent–continent collision zone from the Makran subduction zone offshore

(Figs 1–3). The SE Zagros fold-diapir festoon is characterised by upright and symmetric periclines with doubly plunging sinusoidal axes only a few tens of kilometres long and *c.* 15 ± 3 km apart on either side of tight synclines. Whereas diapirs are localized in fault zones along the incipient syntaxis, they are spatially dispersed in Laristan. The festoon was originally underlain by a thickness of Hormuz salt (estimated at 1–1.5 km (Kent 1970)) sufficient for it to rise diapirically through a cover sequence 6–12 km thick.

The Salt Range in Pakistan (Fig. 1) defines a broad arc around the gentle (<1°) taper of the Potwar Plateau that is also detached above a (Hormuz-equivalent) salt (Butler *et al.* 1987). In contrast to the many anticlines and intrusions in the SE Zagros fold-diapir festoon, the Potwar Plateau is almost unfolded and devoid of diapirs. We attribute this difference to the salt being significantly thicker beneath the SE Zagros fold-diapir festoon than beneath the Potwar Plateau.

The pre-orogenic histories of the Zagros and its future syntaxes could be laboriously reconstructed within the Zagros mountains themselves. However, we take a simpler approach and look southward, beyond the Persian Gulf in the foreland basin, to the Arabian platform. Here, in platform sediments still untouched by the Zagros orogeny, precursors of the structures in the syntaxis are represented by pre-orogenic diapirs rising from salt pillows (Kent 1987). All three transfer zones across the Zagros are in line with clear N–S lineaments on the Arabian platform (Figs 2, 3 & 9). We will refer to the two lineaments extrapolating the Kazerun and Mangarak fault zones *c.* 800 km southward as the Kazerun–Qatar and Mangarak–Marzuk lineaments because they bound local structures with those names.

## Salt pillows

N–S or NNE–SSW lineaments in the Zagros foreland basin and on the Arabian platform are constrained by the locations and shapes of the many giant and few supergiant oil and gas fields (Fig. 3). Most of these fields occur in gentle and essentially unbroken periclines in Tertiary sediments with flank dips of <5° (Kent 1979, Alsharhan 1989). The periclines are thought to be draped over deep pillows of Hormuz salt that were triggered by early to mid-Cretaceous differential sedimentation onto overburden weakened by movements along old-established basement faults (Alsharhan 1989).

On the Arabian platform to the NW of the Kazaroun–Qatar lineament, these pillows and their drape folds are elongate with a roughly N–S orientation known as the Arabian trend (Edgell 1992). To the SE of the Mangarak–Marzuk lineament, some of the drape folds on the Arabian platform trend NNE (Fig. 3). Many of the emergent diapirs onshore in Laristan may also have been aligned in lines that rose from pillows elongate along a NE trend before their pattern was distorted by Zagros shortening (Kent 1979, Fig. 7).

A large N–S trending pillow indicated by thinning of Jurassic cover at the southern end of the Kazerun fault zone is both the northernmost and earliest pillow reported in the literature (Kent 1958, p. 259). Pillowing of salt generally began in Jurassic times in the west of the region (Beydoun 1991, p. 58) and in the Early Cretaceous in the east (Alsharhan 1989, p. 272). The same tendency to young eastward is seen in the oil and gas fields. Another structural pattern is that salt pillows on the platform tend to change northward to diapirs-rooted-to-pillows under the foreland basin, and then to diapirs-emergent-from-deflated-pillows in the Zagros orogen.

We will return to considering tectonic timing after explaining how the diapirs register the tectonic pulse (stresses) and sample the stratigraphy of the deep décollement that controls the regional tectonics and hydrocarbon accumulations. Here we interpret the salt structures on the Arabian plate as having been initiated by a wave of instability that travelled eastward along the northern edge of Gondwana long before the closure of Neo-Tethys led to the subsequent Zagros deformation. We take most pre-, syn- and post-Zagros salt structures to have been elongate or aligned in N–S or NNE trending rows along narrow fault zones separating large stable blocks that were jostled by a succession of tectonic movements which we shall discuss later. Two possible exceptions are the Ghasha and Zakum fields (g and z on Fig. 3), which trend W–E in the southern gulf and may record the first distant effects of Zagros shortening (Alsharhan 1989, p. 271). This shortening may have been transmitted through a stiff cover acting as a stress guide as described in the Potwar Plateau (Butler *et al.* 1987).

## Diapirs of Hormuz sequence

There is a vast literature on the extrusions of Hormuz salt in the Zagros mountains and Oman (see references in Kent (1958, 1979, 1987), Gansser (1960), Stöcklin (1968) and Talbot & Jarvis (1984). These spectacular fountains and flows of Precambrian salt spreading over Recent sediments offer subaerial examples of submarine sheets of allochthonous salt currently spreading off the shores of the US Gulf coast (Talbot 1993) and Yemen (Heaton *et al.* in press). Gravity spreading of salt extrusions has been common in other sedimentary basins in the past (e.g. northern Germany in the Campanian, Nordkapp basin in the Jurassic).

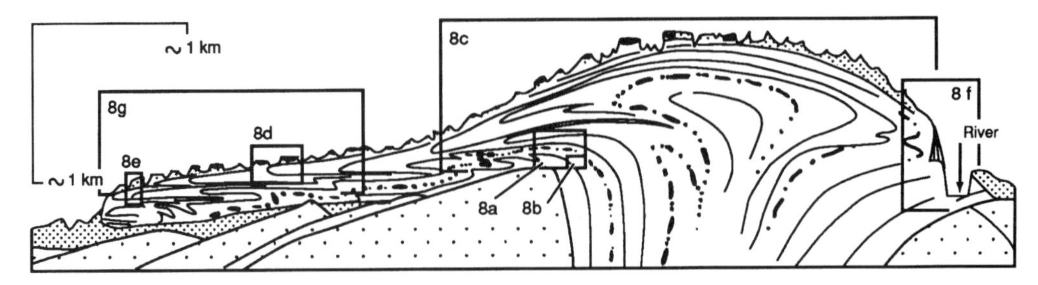


**Fig. 7.** Cartoon of internal structures based on salt fountains along the Kazerun and Mangarak fault zones. Boxes indicate locations of photographs in Fig. 8. Bedding is picked out by colour banding or boudinaged limstone beds and subparallels both non-slip and free salt boundaries. Steep layering in the orifice spreads in extrusive salt to parallel the parabolic top surface which is fretted to pinnacles by dissolution. The namakier (salt sheet) advances like a tank track, with recumbent flow folds indicating where it has surmounted obstructions.

In front of the high Zagros, diapirs are typically 5–10 km across and extrude up to 1.5 km above sea-level before spreading several kilometres beyond their vents as sheets of allochthonous salt known as 'namakiers', or glaciers of salt (from *namak*, the Farsi for salt). Each of the extrusions along the Mangarak fault zone shows a different stage of evolution (Fig. 5). Topographic domes of salt (Kuh-e-Gach, Fig. 5) later develop the distinctive parabolic profiles of viscous fountains with a summit dome (Kuh-e-Jahani and Namak, Dashti) that collapses (Talbot 1993) when the deep source is exhausted or disconnected (Kuh-e-Namak, Feroosabad). After that, the emergent salt is dissolved to leave a debris-filled crater (as at Roxana, Fig. 5).

Although bedding survives in extruded salt as compositional layering, all depositional salt fabrics have recrystallized. Layering in the salt (or in residual soils after salt dissolution) generally parallels the rigid bottom and side boundaries of all salt bodies, wherever they are exposed (Talbot 1979, 1993). This is so whether the country rock contacts are subvertical in exposed vents, have irregular dips away from the vent, or are subhorizontal over bedrock or recent alluvium ± debris shed from the advancing salt flow (Fig. 7). Traced upward, the axial surfaces of upright folds exposed in deeply eroded vents (Figs 8a, b) turn to either parallel gentle (<20°) slopes controlled by flow (Fig. 8d), or steeper slopes truncated by erosion (Fig. 8f).

Two thin namakiers spread from the salt fountain at Kuh-e-Namak (Dashti) and flow down the flanks of an anticlinal culmination disrupted by the southern end of the Kazerun fault zone (Talbot 1979). They are sufficiently steep not to be mantled by thick soils (Fig. 8c) and their internal deformation structures and fabrics are well exposed (Fig. 8d). Rather than erode their channels, namakiers flow over pre-existing topographies. Trains of internal flow folds and ductile shears thin and repeat the layers, and separate, mill and disperse inclusions wherever the flowing salt thickens to pass channel obstructions (Talbot 1979). A 'menacing maze of salt pinnacles . . . often protected by perched erratic blocks' (Kent 1970, p. 80) characterizes vigorous young extrusions. Older salt extrusions (Fig. 5) are progressively cloaked by insoluble debris in which conical collapsed craters eventually coalesce to form badland topographies (Fig. 8g).

The basic structure of each namakier is that of a recumbent sheath fold with a down-slope profile like a tank-track (Figs 6 and 8d). Isoclinal, similar-type recumbent folds (Fig. 8d and e) characterize the distal portions of salt sheets around all of the salt extrusions along the Mangarak fault zone (Fig. 7). The basal levels of extrusive salt are typically dirty mélanges smeared to conformity with the irregular tops of water-sorted deposits shed from the advancing salt front. Namakiers in the Zagros flow over their own debris like other nappes that have advanced over emergent thrusts. Whereas some namakiers have the steep snouts overriding terminal moraines typical of advancing ice flows (Fig. 8g), others have the distal feather edges characteristic of retreating glaciers.

## Current salt flow

A matter of great debate during optimistic discussions in the 1970s about storing radioactive wastes in salt diapirs (Gera 1972) was whether the namakiers of Iran are still flowing (Kent 1970) or only flowed temporarily in the past like hot lavas (Gussow 1966).

The first field study devoted entirely to a Zagros salt extrusion established that Kuh-e-namak (Dashti) expands and contracts as an elastic solid

**Fig. 8.** Photographs of (**a**) steep layering (emphasized by black lines) in the vent exposed at Bachoon (see Figs 5 or 6 for locations); (**b**) steep pegmatite of halite to bottom right of area in photo (**a**). Scale bar is 10 cm long. (**c**) The salt fountain at Kuh-e-Namak (Dashti) seen from the NW. The northern namakier flows out of the picture to the left. (**d**) The tank-track recumbent fold just behind the snout in the northern namakier at Kuh-e-Namak (Dashti). (**e**) An individual pegmatite of halite repeated by recumbent folds in the SE snout of Kuh-e-Namak (Feroosabad). (**f**) River cliffs (*c.* 100 m high) of perhaps *c.* 80 vol% of salt partially obscured by accumulations of grey impurities along the northern edge of Kuh-e-Jahani. (**g**) Since the clean younger Hormuz salt in the southern snout of Kuh-e-Jahani dissolved, this lobe has consisted of grey remnant soils rich in anhydrite or gypsum.

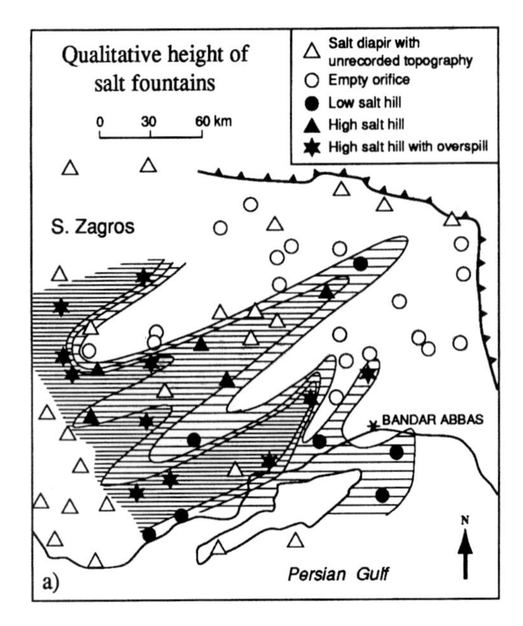


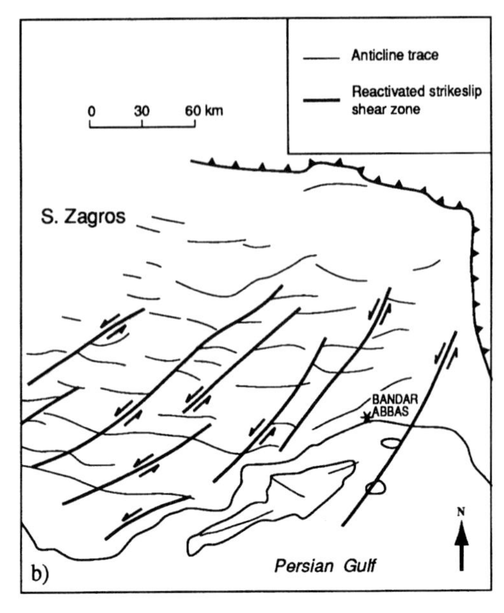


**Fig. 9.** Relative heights of salt extrusions (**a**) vary much more than the local rainfall, which is *c.* 100–200 mm $a^{-1}$ near Bandar Abbas in Laristan. Contours of qualitative extrusion height (data mainly from Harrisson (1930) and Talbot (1992)) follow neither contours of rainfall nor the trends of Zagros folds (**b**); instead they follow a pattern of strike-slip faults oblique to the Zagros trend (Fürst 1979).

when it is dry, but flows as a ductile fluid after a fractured brittle lid, about 10 m thick, is dampened by rain (Talbot & Rogers 1980). Daily avalanches from a cliff high above the southern namakier during the rainy season signal ductile flow in the summit dome. It is not only direct rainfall that softens extrusive salt; brine springs flow from near the summit of many salt extrusions even in the dry season, and drilling into salt in the Iranian oil fields has produced large volumes of brine at very high pressures (Gansser 1992, p. 839).

Field studies inspired a series of laboratory experiments that explain how <0.1 wt% of enclosed brine can weaken confined rock salt to an essentially viscous but crystalline fluid (Urai *et al.* 1986). Dispersed cubic inclusions of brine can be smeared by very low shear stresses into continuous films that act as dramatically mobile boundaries to halite grains. The rock mass flows by ions dissolving from damaged grains of halite, diffusing across the brine-filled grain boundaries, and precipitating new undamaged grains of a different shape (Urai *et al.* 1986). This process becomes cryptic after stress-drop annealing (Urai *et al.* 1986), when surface tension disperses brine-films back through rounded brine fingers to isolated brine inclusions that start as spheres but mature to cubes. Whereas rocks have long been known to flow by thermally assisted creep, field studies of the surface flow of salt in the Zagros helped towards the recognition that other metamorphic rocks flow by water-assisted dynamic recrystallization.

Other theoretical problems raised by the extrusions of the Hormuz sequence in Iran and Oman were the space problem (Kent 1979), and how it was possible for the salt to lift large dense inclusions at least 5 km up to the surface (Gansser 1992). The space problem can be solved by the salt having risen reactively along the spaces opened along faults. Jackson & Vendeville (1994) considered the space to be created by normal faults, we consider the space to be created at releasing bends along strike-slip faults (Fig. 6). The inclusion problem was solved by numerical models (Weinberg 1993) showing that salt rising at documented velocities of a few millimetres per year, with experimentally derived power-law exponents of stress of $n = 4$ to 5, can indeed lift the largest salt-entrained blocks known in Iran (5000 m × 2000 m × 125 m).

Flow of rock salt at the surface emphasizes the effectiveness of salt décollements at the higher temperatures and pressures expected at depth.

## *Extrusion topographies gauge current tectonic pressure*

Past workers considered the topographies of extrusions of the Hormuz sequence to vary more or less systematically across the Zagros (Ala 1974, Kent

1979). They reported vigorous salt fountains in the lower rainfall (100–400 mm $a^{-1}$) of the coastal ranges but thick piles of insoluble Hormuz debris burying dissolved salt in regions of higher precipitation (400–800 mm/a) inland. However, whether an extrusion of Hormuz sequence is a vigorous salt fountain, a crater clogged with insoluble debris, or a narrow (e.g. 25 km × 3 km) streak of coloured soils after a diapir has been smeared along a steep fault in the Sanandaj–Sirjan imbricate zone (Figs 2 and 8) depends on more than just the ratio of rates of salt rise and dissolution (Talbot & Jarvis 1984; Talbot 1993). Extrusive vigour is also likely to depend upon the thickness of the salt source (vigorous extrusions from thick sources) and its extrusion history. This is clear in the region around Bandar Abbas (Fig. 9) where salt extrusions still connected with their source layer act as gauges of tectonic pressure along the deep décollement. Topographies of salt extrusions emphasize a set of strike-slip faults that continue activity after the local arrival of the Zagros front but these are not considered further here (see Fürst 1976; Talbot 1992). Even where diapir topographies do vary systematically from south to north, as along the Kazerun and Mangarak fault zones (Fig. 5), the gradation is more likely to reflect tectonic history than rainfall. The closer a diapir is to the Zagros suture, the longer it has been reactivated or initiated by the arrival of the advancing Zagros thrusts and/or folds and the longer it has had to exhaust its deep source layer.

## *Influence of salt on Zagros folding in the Tertiary*

The apparently haphazard relationship between salt diapirs and the Zagros folds is because most of the salt structures predate Zagros shortening in one form or another. Kent (1979) related emergent diapirs to Zagros folds in a designated part of Laristan and counted 13 diapirs as emergent on anticlinal culminations, 14 on anticlinal noses, eight on the flanks, two in synclines and left three unclassified.

What happens when the Zagros front eventually reaches any particular area depends on the nature of the salt structures already there. One of two things happens to already emergent diapirs. Where salt withdrawal has emptied the deep decollement, the resultant weld between the overburden and the basement pins the end of a future Zagros antiform (Kent 1979) or synform (Koyi 1988). All ten of the diapirs emergent in the southern gulf in front of the Zagros front are in synclines due to salt withdrawal (Kent 1979, fig. 9). Diapirs firmly pinned by welds become smeared along the thrusts that behead them (see Surmeh, top right on Fig. 5). These are unlikely ever to fountain and instead probably pass through a stage of being sub-circular breccia pipes before they are sheared to listric profiles and their planforms are squeezed to the ellipses elongate along the Zagros trend seen in the high Zagros (Koyi 1988).

Where sufficient autochthonous salt remains along the deep decollement, the advancing Zagros front rucks up deep, ridge-like salt mullions that may spawn a second suite of syn- or post-shortening diapirs (Koyi 1988).

In developing their thesis that most salt structures are triggered by (thin- or thick-skinned) extensional faulting, Jackson & Vendeville (1994) underplayed syn-shortening diapirism in the Zagros. Shortening was indeed irrelevant to the initiation of the pre-Zagros salt pillows, but not to the numerous cases where the Zagros front migrated along pre-Zagros salt pillows already aligned along the older transfer faults that became oblique footwall ramps during the Zagros orogeny. Deep pillows that had previously remained conformable, like those still on the platform, suddenly pierced and surfaced and contributed distinctive dark Hormuz clasts to the surrounding stratigraphy, which was characterized by pale colours (Kent 1979). Linear depotroughs deepened along or across the faults above the deflating pillows. In the Miocene, the Mangarak fault zone was the boundary between the Dezful gypsiferous facies to the NW and Razak (red-bed) facies of the Fars Group to the SE. This group thickens into a depotrough shared by both Kuh-e-Namak (Feroosabad) and Kuh-e-Jahani (Fig. 5) along the Mangarak fault zone (Kent 1979). Similarly, part of the Fars group also thickens into depocentres where NW trending Zagros thrusts swing into oblique footwall ramps along the Kazerun fault zone. Strong mutual feed-back between sedimentation and salt flow was increasingly delayed behind the Zagros front as it advanced southwestward. Pliocene siliciclastic deposits thicken to several times their usual thickness in the syncline south of a diapir along the Kazerun fault zone where a local 20° unconformity excising the Mishan Group records the deflation of the underlying pillow (Kent 1979). Similarly localized deposition resulting from the sudden deflation of a deep pillow is recorded by tilted Bahktyari conglomerates onlapping the crest of a fishtail anticline near the SW corner of Bachoon salt diapir (Fig. 5). Vast volumes of Hormuz salt that were extruded along the Kazerun and Mangarak fault zones in Miocene times were probably eroded and recycled as the salt in the Fars Group, which provided the shallow décollement responsible for such fishtail anticlines along the NW Zagros fold-thrust belt (O'Brien 1957, fig. 1).

## Discussion

### *Tectonic pulse from formation and deformation of Hormuz salt*

Rocks older than Jurassic are rarely exposed in the Zagros Simply Folded Belt and are deeply buried beneath the foreland. Consequently, the little we know of what happened near the Persian Gulf between the end of deposition of Hormuz salt in mid-Cambrian and the first documented salt movement in Jurassic comes from information in and near the diapirs.

Hormuz salt appears to have been restrained by the strength of its stiff carbonate overburden for *c.* 300 Ma (Kent 1958, Jackson & Vendeville 1994). Despite its positive buoyancy since early Palaeozoic burial, thickness and facies changes around the salt structures throughout the gulf region indicate that none started to rise until Jurassic or Early Cretaceous times. This pre-Zagros triggering of salt structures has been attributed to (Palaeo) Tethys beginning to close in the Jurassic (Kashfi 1976), to opening or closure of the NE trending Oman basin with its foredeep parallel to the Oman line (Alsharhan 1989), and to rift and drift of Neo-Tethys at some time between the Permian and earliest Cretaceous times (Jackson & Vendeville 1994). However, while attributing the earliest salt pillowing, piercement and most vigorous salt venting to regional extension of the north-dipping shelf during the opening of Neo-Tethys, Jackson & Vendeville (1994, p. 64) did not explain how salt pillows with N–S trends relate to N–S extension. Kent (1987, p. 31) considered the alignment of the salt structures at high angles to what later became the Zagros suture to demand an origin invoking tear rather than tension faults, and we agree with this view.

We consider both the closure of Paleo-Tethys in the Alborz mountains and the Early to Late Triassic opening of Neo-Tethys along what is now the Sanandaj–Sirjan imbricate zone to have been diachronous. However, we recognize no sign of either event in Zagros salt tectonics. Instead, we consider the most likely tectonic candidate for starting salt movement to have been the reactivation of basement blocks along the margin of Gondwana, which accompanied probable Jurassic subduction of ocean floor northward before the diachronous closure of Neo-Tethys along the Zagros suture began in the early Cretaceous.

Simultaneous opening or closure of a basin requires separation or rotation of large rigid blocks. Diachronous opening or closure of basins requires flexibility in the opposing sides. Basement blocks defined by old faults provide such flexibility. Zip fasteners engage or disengage rows of teeth that interlock along flexible borders; basins can open or close by strike-slip shuffling of teeth-like blocks that may not interlock so well when they close. Before disengagement or after opening, marginal basins of Tethys were like keyboards along which disturbances travelled as slow waves of vertical flexure (riffling). We will refer to basement blocks defined by transfer faults at high angles to the incipient marginal basins of Tethys as lithospheric keys. These 'lithospheric keys can be rifted and riffled, or drifted and riffled.

In Jurassic and Cretaceous times, we visualize diachronous zip-fastener-like subduction of Neo-Tethys as riffling the Gondwana keyboard along the subducting Arabian plate both before and after closure. We attribute initiation of the salt structures to local manifestations of subduction of Neo-Tethys before Zagros shortening of the sutured crust. Open-ocean subduction could account for the N-S alignment of pre-Zagros salt pillows and rows of pre-Zagros diapirs above the N–S edges of keys in an already segmented basement.

The way in which salt pillows have subsequently developed has depended on their distance from the Tethyan margin. Salt far to the south has remained conformable in deep salt pillows that grew slowly at nearly constant rates until the end of the lower Cretaceous (Edgell 1992, fig. 7). Diapirs closer to the unstable margin were more sensitive gauges of regional tectonic pressure. Spreads of distinctive dark Hormuz inclusions in the pale cover rocks record their extrusion in three main episodes: Lower Cretaceous, Eocene–Miocene and Recent. These episodes occurred when riffling of basement keys controlled depositional thicknesses and facies in the shallow marine sediments accumulating above them.

We now attempt to reconstruct the history of the region mainly using evidence brought up from depth with emergent salt. We start by using the inclusions in the salt to map its lithofacies.

## Lithofacies of Hormuz sequence

Extrusive diapirs of the Hormuz sequence throughout the region rise through an indurated overburden consisting predominantly of carbonates. Only along the major transfer faults does a syn-Zagros cover sequence relate to local depotroughs, and only around some of the offshore islands were Tertiary shales sufficiently ductile to turn up in peripheral collars (Kent 1979, p. 121). Consequently, rather than inclusions in the Hormuz salt sampling the surrounding ductile overburden (as happens in diapirs downbuilt near the US Gulf coast: (Kupfer 1979)), the vast majority of inclusions in diapirs near the Persian Gulf were part of the original

Hormuz sequence (Kent 1958, 1970, 1979; Gansser 1960, 1992).

It has long been recognized that stromatolites and trilobites in different parts of the Hormuz sequence indicate ages ranging from late Neoproterozoic to mid-Cambrian. Until now there has been little attempt to discern details in the Hormuz story. However, our recent work on Zagros salt extrusions suggests that future mapping could distinguish not only Hormuz lithofacies but also divide an older from a younger Hormuz sequence. Here we sketch how such a division is possible in one area before tentatively extrapolating the distinction more broadly.

Kuh-e-Jahani (28°37′ N, 52° 25′ E) is the southernmost and currently the most voluminous salt extrusion along the Mangarak fault zone (Figs 5 & 6). The summit dome of this vigorous 11 km × 7 km oval salt fountain rises 1,485 m above sea-level. Five rings of different coloured soils trace funnel-shaped red layers concentric with both the distal flow front and the flanks of the summit dome (Kent 1958). The outer red layer consists of beds of red limestone and marl with a conformable breccia of black dolostone veined by sphalerite. We do not yet know the ages of the red layers, nor whether there is structural repetition among them. Nonetheless, they divide Kuh-e-Jahani into two portions. An outer periphery consists of distinctive grey and buff anhydritic soils (Fig. 8g). These are residual after halite that was cleaner and more uniform (Fig. 8f) than the multicoloured salt encircled by the red bands. Among the pale soils are numerous blocks of chloritic gabbro and vesicular, fine-grained diabase that are missing from other diapirs nearby.

Traction between salt diapirs and their country rocks turns the stratigraphies of their bedded source layers inside out within the diapir (Jackson & Talbot 1989). Stratigraphic inversion is usual beneath the axial surfaces of the 'tank-tracks' in salt diapirs spread near the surface (Fig. 7). The uniform soil residual after clean peripheral salt at Kuh-e-Jahani (Fig. 8g) is so different from elsewhere that Kent (1958) referred to it as a flow of mud-breccia. However, we interpret it as younger, purer allochthonous Hormuz succession encircling a core of multicoloured Hormuz sequence that is older.

Small volumes of late Hormuz sequence also occur around the peripheries of some of the other extrusions along the Mangarak fault zone. However, the significant volume of clean salt with igneous inclusions extruded at Kuh-e-Jahani is conspicuous by its absence, not only in the southernmost extrusion along the Kazerun fault zone 80 km to the SW, but also at the southernmost extrusion along the Mangarak zone only 6 km to the SW of Kuh-e-Jahani (Kent 1958). A strong facies boundary is implied between Jahani and Gach (Figs 5 and 6) in late Hormuz time, when both clean salt and mafic igneous activity appear to have been confined to a narrow trough now trending nearly N–S.

## *Early cyclic Hormuz sequence*

What we interpret as the older (mainly Proterozoic?) Hormuz sequence at Kuh-e-Jahani and elsewhere along the Mangarak fault zone consists of thick beds of multicoloured salt and anhydrite (or their derivative soils) interbedded with dark foetid dolomites and thin red, purple or greenish sandstones, siltstones or marls (and some local yellow-brown orthoquartzites). These rock types generally match the autochthonous equivalent succession (without significant salt) on the Iranian plateau (Stöcklin 1968). The early Hormuz sequence is therefore typical of the Hormuz that contains huge rafts of bedded sediment exposed in most emergent diapirs in and near the Persian gulf, including Oman (Kent 1970). Abundant haematite, sulphur and local concentrations of base-metals indicate that the early Hormuz sequence accumulated in reducing conditions. The algal mats common in the dolostones indicate sabkha-type facies. These lithologies indicate that in early Hormuz times, an unknown number of cycles of salt, anhydrite and carbonate were deposited over large areas of shallow water with only local and temporary emergence. The early Hormuz was therefore cyclic like the Zechstein sequence of northern Europe (Kent 1979). Indeed, the Palaeozoic sequence throughout the region stayed shallow until Cretaceous turbidites indicate Stöcklin´s (1968) seaway deepening along the current Sanandaj–Sirjan imbricate zone.

## *Late Hormuz sequence*

The large volume of uniform grey or buff anhydritic salt at Kuh-e-Jahani is reminiscent of comparatively pure salt sequences like the Louann salt in the Gulf of Mexico. However, whereas the Louann salt covered a large region and is devoid of igneous inclusions, the uniform younger Hormuz salt near the Persian Gulf contains undated inclusions of chloritic gabbro and vesicular diabase and is noticeably more localized, like the Sedom salt of the Dead sea.

The Mangarak trough, which we infer to have deepened in late Hormuz times may have been confined to a half-graben as long (*c.* 200 km), narrow (12 km) and deep (1.5 km) as the Dead Sea basin, where the *c.* 1 km thick Sedom salt of Early Miocene age now rises in a line of three diapirs. Like the salt in the Dead Sea (Ben-Avraham *et al.* 1990), the clean late Hormuz salt in the Mangarak trough also accumulated in an isolated basin pulled

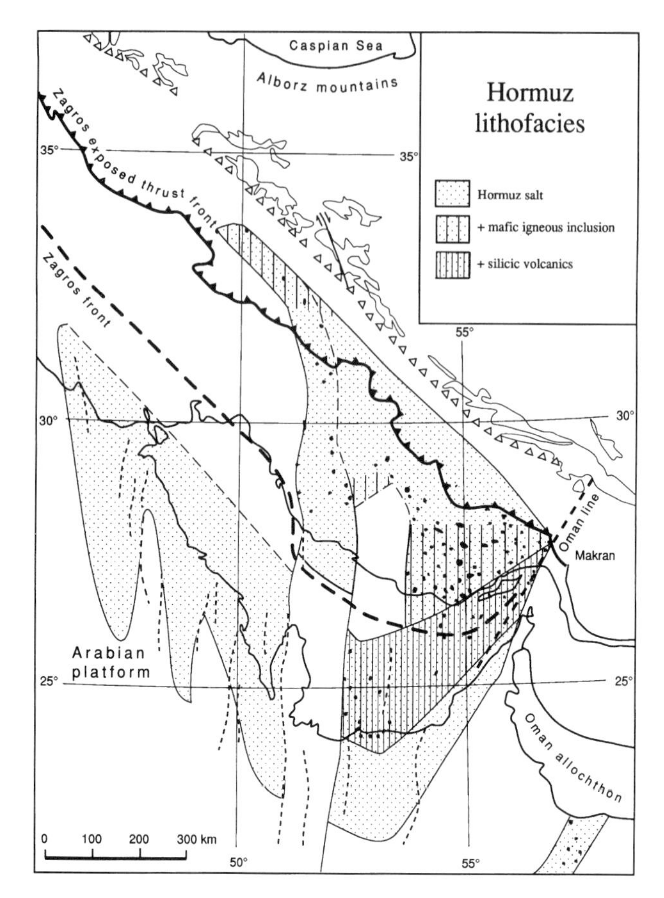


**Fig. 10.** Distinguishing Hormuz evaporite facies on the same basis as Figs 2 and 3 maps the basement keys under Proto-Tethys (based on Harrisson (1930), Kent (1958, 1970, 1979), Gansser (1960, 1993), Beydoun (1991) and Edgell (1992)). The basic assumption behind this map is that diapirs or pillows signal Hormuz salt of significant thickness while their absence implies insignificant salt thickness. Halite facies give way northward to dolomites of Hormuz-equivalent Soltanieh facies.

apart at a releasing bend along a plate-boundary fault inherited from the tectonic grain of its Pan-African basement. The main difference between these two rhombic basins was that strike-slip displacement was left-lateral along the Miocene Dead Sea and right-lateral along the late Hormuz Mangarak trough (Fig. 6).

## *Hormuz geography*

The same subdivision of the Hormuz succession into cyclic early Hormuz and uniform later Hormuz that is possible at Kuh-e-Jahani may also be applicable to the Hormuz sequence elsewhere in the Zagros. Thus Kent (1987, p. 31) mentions a distinctive evaporitic and dolomitic unit overlying terrestrial siliciclastics in Chah Benu diapir (100 km north of the coast at 55.0° E). Gansser (1992, p. 837) emphasized that Kuh-e-Anguru (90 km ESE of Chah Benu) is devoid of significant inclusions. Indeed, many previous workers have alluded to two Hormuz sequences separated by a few hundred metres of carbonates and red beds (O'Brien 1957, Alsharhan 1989, Husseini & Husseini 1990; Beydoun 1991) but nobody has distinguished their palaeogeographies.

Because of dissolution, we cannot rely on previous descriptions of the salt itself to constrain early and late Hormuz geographies. Until the extrusions are remapped, only published reports of the igneous inclusions help limit the late Hormuz basins. This is because there are no igneous rocks in the regional cover above mid-Cambrian so that all the igneous inclusions were intrusive and extrusive into the Hormuz sequence (Kent 1979; Gansser 1992). The link we infer between igneous activity and uniform

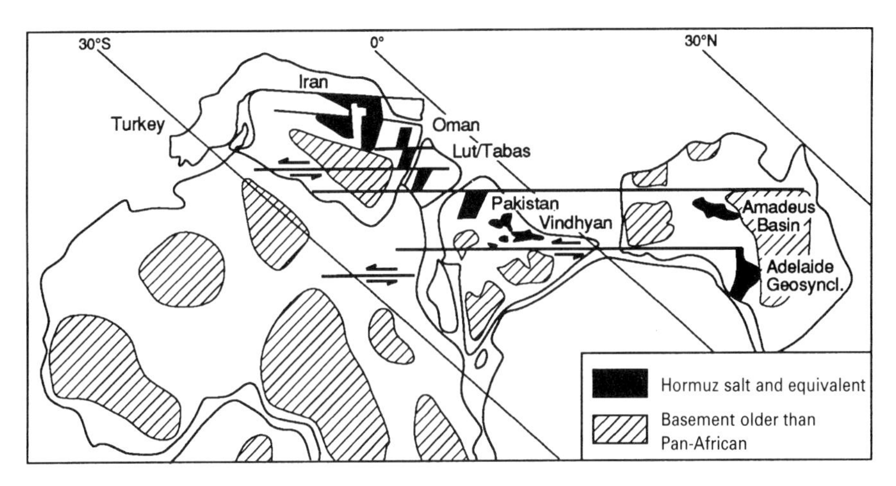


**Fig. 11.** Basins of Hormuz salt in Iran, Oman (Kent 1987), Lut-Tabas block (Stöcklin 1968), and equivalent sequences in Pakistan (Butler *et al.* 1987), the Vindhyan basin of India (Prasad 1984; Dey 1991) and Australia (Lindsay 1987; Kennedy 1993; Van der Bord, pers. comm. 1992). Neoproterozoic palaeolatitudes are from Scotese & McKerrow (1990).

clean salt is inevitably obscured, not only by dissolution but also by magma erupting in the pure younger Hormuz after having passed through the older cyclic Hormuz sequence. Descriptions of most individual diapirs merely list spilitic diabase as among 'the usual suite of Hormuz inclusions' (Harrisson 1930); they seldom mention whether the igneous and sedimentary inclusions are mutually exclusive as they certainly are at Kuh-e-Jahani. On the contrary, vulcanicity was not everywhere confined to otherwise clean salt, for Kent (1979, p. 125) mentioned mafic lavas or tuffs as being conformable in rafts of cyclic sediments at Chah Benu.

Despite all these qualifications, a surprisingly clear picture emerges (Fig. 10) from a careful re-reading of Harrisson (1930), Kent (1958, 1970, 1979, 1987) and Gansser (1992) leavened by our own limited field observations. Inclusions of intrusive diabase in diapirs in the Sanandaj–Sirjan imbricate zone and inland Laristan (Fig. 10) give way southward to extrusive mafic pillows and agglomerates along the coast (Kent 1958). By contrast, rhyolites, trachytes and porphyries are ubiquitous in the emergent diapirs forming the Gulf islands (Fig. 10). Silicic and mafic rocks occur together in only one diapir along the Iranian coast.

Igneous inclusions which are common at Kuh-e-Jahani are conspicuous by their positive absence to the north and west until they reappear 500 km away in the solitary salt diapir of Jebel-el-Sanan (Fig. 2) near the Iraq–Kuwait border (Kent 1970 p. 83). Igneous inclusions are also missing from the cyclic salt sequence in Oman except for some altered porphyries and rhyolitic tuffs in the southernmost emergent diapir (Kent 1987, p. 29).

The only exotic components in the Hormuz diapirs of Iran are very rare blocks of metamorphic rocks reported in extrusions near the southern coast. These are assumed to have fallen from fault scarps of basement during deposition (Kent 1979) or soon afterward (Alavi 1991a, b). Individual blocks of garnetiferous limestone, metamorphosed mudstone, schist, mafic mylonites, tonalite and a 3000 m$^3$ block of gneissic granite are listed by Kent (1979 ) and Gansser (1992). These are taken to be blocks of the Pan-African basement exposed in the Arabian–Nubian craton *c.* 800 km to the SW. Rocks with basement characteristics exposed in central Iran, west of the Tabas block (Fig. 1) and hidden deep beneath the Zagros, are also interpreted as Pan-African (Alavi 1991).

On a larger scale (Fig. 11), mafic intrusions occur in the Hormuz type sequence near Kerman (Stöcklin 1968). Butler *et al.* (1987, p. 354) do not mention igneous rocks among the red marls, gypsum and dolomites passing upward to gypsum and thick (630 m) salt with some potash in the Hormuz-equivalent Salt Range Formation in Pakistan. Basic tuffs and gypsum are reported in the cyclic shallow-marine and continental Neoproterozoic to Cambrian sequence (with pseudomorphs after halite at one level) forming rugged hills in the Vindhyan basin that inverted across northern India (Prasad 1984; Dey 1991). Igneous rocks appear to be missing from the shallow-marine cyclic Hormuz-equivalent successions containing salt in the Amadeus basin of central Australia

(Lindsay 1987, Kennedy 1993). However, igneous inclusions are present in the breccias which might indicate spreads of insoluble debris after extrusions of Hormuz-equivalent salt in the Adelaide geosyncline of south Australia (C. J. Van der Bord, pers. comm. 1992). Hormuz salt and equivalent sequences may have floored a broad tropical basin stretching from Australia through India, Pakistan and the Lut-Tabas block to Oman, with Iran being closest to a pole of spreading somewhere further to what is now the WNW (Fig. 11).

### *Proto-Tethys*

Hormuz lithofacies boundaries in the Persian Gulf area (Fig. 10) have three main orientations: NW–SE, N–S and NE–SW. The NW trending boundaries define a basin that was parallel to later Tethyan basins along the margin of Gondwana and was sufficiently long, broad and shallow to be the likely repository of the early cyclic Hormuz sequence. In the absence of palaeomagnetic evidence for any significant ocean basin having opened near the Persian gulf in Hormuz times (Scotese & McKerrow 1990) we will adapt Kinsman's concept of a proto-ocean that failed to spread and refer to the Hormuz basin that rifted and riffled but failed to drift (Figs 10–12) as Proto-Tethys.

NW-trending boundaries of Proto-Tethys near the Persian Gulf are noticeably asymmetric (Fig. 10). Those to the SW are irregular (primary feather edges offset by faults?). Those to the NE (Fig. 12c) have smooth (fault?) traces and were probably controlled by the NW trending Najd strike-slip faults (Fig. 12b) in the Pan-African basement (Alsharhan 1989, Husseini & Husseini 1990). Proto-Tethys was segmented by cross-marginal troughs during the vulcanicity that accompanied accumulation of the younger Hormuz sequence (Fig. 12d). Abrupt edges of Hormuz lithofacies, which now trend N–S and NE–SW bound blocks that parallel and were almost certainly inherited from, the *c.* 1000–600 Ma old terrain-accretion fabric (Fig. 12a) that preceded the Najd faults in the Pan-African basement (Alsharhan 1989). The lithospheric keyboard riffled by the mantle beneath the younger Hormuz basins also appears on regional maps of isopachs of Permo-Carboniferous strata above the Hercynian unconformity (Eyles 1993, fig. 16.19) and free air gravity anomalies and aeromagnetic maps described by Edgell (1992).

### *Proto-Tethys rifting and riffling but not drifting*

Instead of Proto-Tethys opening to an ocean, the basement keyboard ceased to play at about the stage where the lithosphere had rifted and riffled but not drifted). The keys along the edge of Gondwana were stilled and buried beneath Cambrian to Lower Ordovician carbonates and clastic sediments with no sign of any local volcanics (Figs 12d and 12e). Proto-Tethys was still-born when igneous activity migrated to open Palaeo-Tethys further to the north (Figs. 12d and 12e). Huge volumes of mid-Ordovician to mid-Devonian mafic to intermediate intrusive and extrusive igneous rocks occur in the Zagros imbricate zone, central Iran and the Alborz mountains: these are taken to record the opening of Palaeo-Tethys because they are neither metamorphosed nor deformed by either subduction or orogeny (Sengor & Burke 1978).

Husseini & Husseini (1990, fig. 5) appear to have attributed the Hormuz basins in the Persian Gulf region to aborted spreading ridges trending NE, being offset along Najd transform faults now trending NW. However, impressed by the Mangarak trough and Fig. 9, we agree with Stamfli *et al.* (1991) and reverse these kinematics, thinking of the aborted rifts as now trending NW with potential transforms faults now trending NE.

The width and complexity of the crust stretched beneath the present Persian Gulf suggested to Hempton (1987) that the whole of the north Arabian continental margin was on the lower plate or footwall when Palaeo-Tethys eventually opened in Ordovician time by simple extensional lithospheric shear to the N or NE. This may be so, but the lithospheric keys along the zip fastener that failed to open Proto-Tethys (Fig. 10) add earlier details to the saga that began with Proto-Tethys and continues even now with the Red Sea (Fig. 12).

We do not think of Proto-Tethys as an aulacogen, a rift inland of a continental margin where one branch of a triple rift junction failed to drift to an ocean. Instead we rationalise the still-birth of the half-graben of Proto-Tethys (Fig. 12d) in terms of a model of lithospheric extension by simple shear (Wernicke 1985; Stamfli *et al.* 1991; Hempton 1987; Brun *et al.* 1994). The order in which rifting was followed by early Hormuz salt deposition, and then uplift and volcanism along late Hormuz insert basins, implies passive rather than active rifting (Sengör & Burke 1978).

The thick sedimentary cover on the broad gentle shelf on the SW side of Proto-Tethys in Iran is symptomatic of a footwall of a basin in lithosphere extended by simple shear (Figs 10–12). Hormuz salt has insignificant thickness on the narrow shelves on the opposing hanging wall. The incipient Qatar syntaxis and Fars platform were defined by the only transfer faults that reversed the general asymmetry. To prime the future syntaxis, two particular lithospheric keys went up rather than down, perhaps because two strips of the Pan-African basement had distinctive characters.

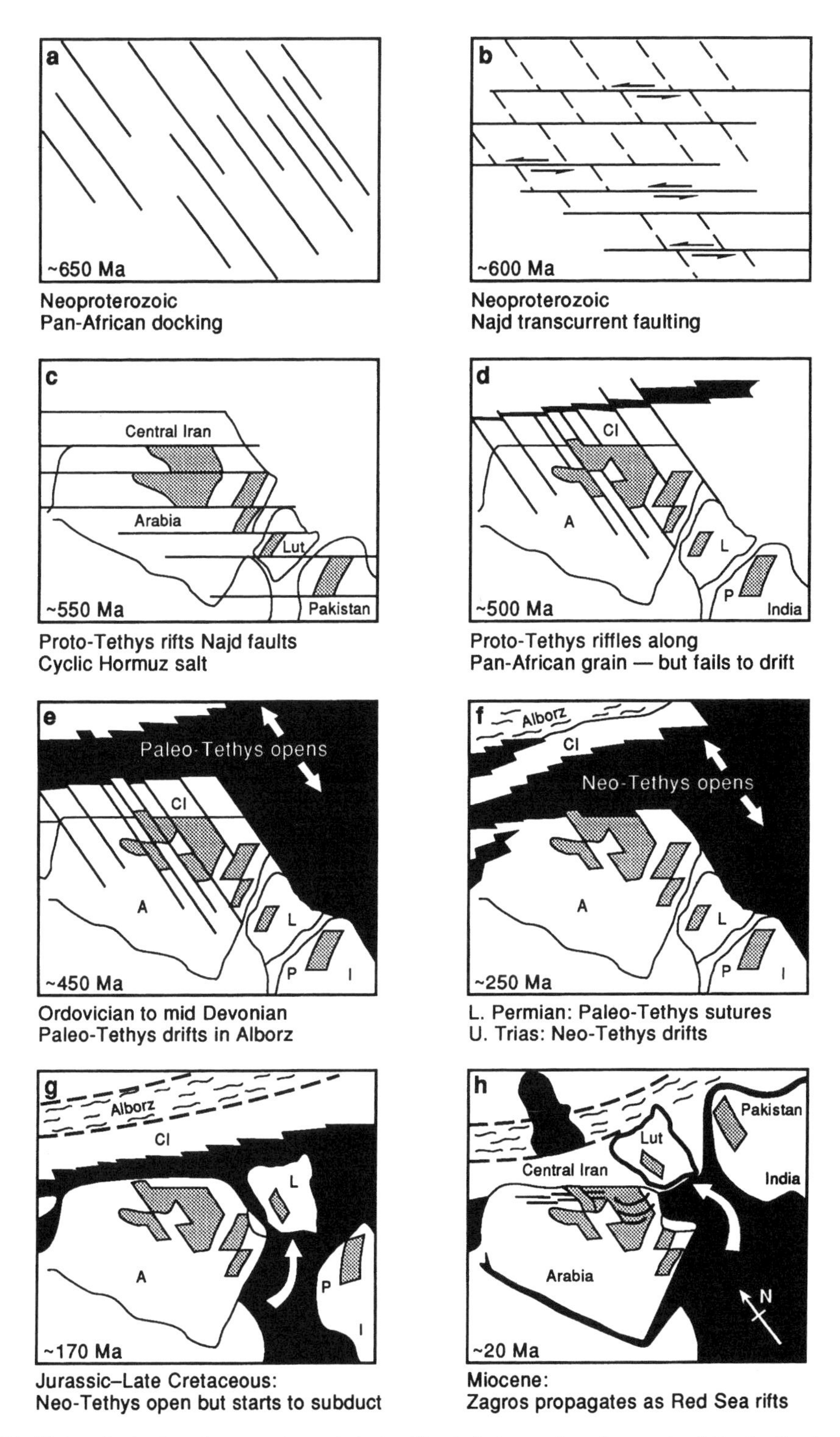


**Fig. 12.** Tethyan basins have been opening and closing like zip fasteners along the margin of Greater Gondwana since the Pan-African orogeny brought east and west Gondwana together. The Red Sea is just the latest in a long list of Tethyan basins.

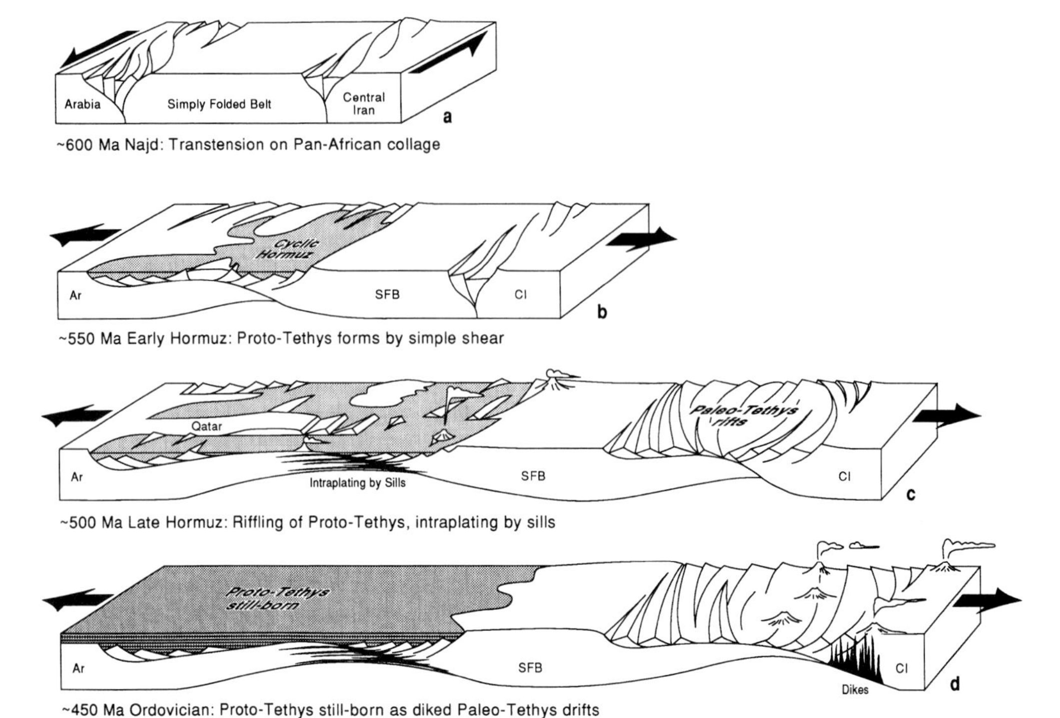


**Fig. 13.** Proto-Tethys may have rifted when a flower structure along a Najd strike-slip fault matured to an initial breakaway. The Arabian plate riffled and most of it subsided as slow extension pulled it off the passively bulging asthenosphere. Qatar became an incipient syntaxis because it rose on a hangingwall between the Kazerun–Qatar and Mangarak–Marzuk transfer faults. Proto-Tethys may have failed to drift because lower Cambrian intraplating involved sills. Palaeo-Tethys may have rifted and drifted further north because it involved dykes (adapted from Wernicke (1985); Hempton (1987), Stamfli *et al.* (1991) and Brun *et al.* (1993)).

We follow Husseini & Husseini (1990) and interpret the smooth NW trending boundary along the NE side of Proto-Tethys to have initiated as a steep strike-slip Najd fault (Fig. 13a). We assume that this steep fault rotated to become the master fault along which most of the slow ductile N–S extension of Proto-Tethys occurred (Fig. 13b). We infer that this master fault dipped northward under the NW Zagros fold-thrust belt and the SE Zagros fold-diapir festoon, but southward under the Qatar arch and the Fars platform (Fig. 12). Slow lithospheric extension pulled the Iranian hangingwall off the (passive or reactive) isostatically bulging asthenosphere. The active main detachment fault remained steep to the NE but was progressively deactivated to the SW as it was rotated to a listric accommodation fault (Brun *et al.* 1994) with gentle dips on the shoulder of the Arabian footwall (Fig. 13c). The many basin-parallel faults anticipated by this model are assumed to be buried beneath the post-rift early Hormuz cyclic salt in which the lithofacies map two depotroughs, a hanging wall half-graben bound to the north by the main detachment fault, and an older footwall basin with a southern edge bound by the inactivated listric accommodation fault (Fig. 13c).

We infer from the pattern of igneous activity between the Mangarak–Marzuk fault zone and Oman syntaxis that mafic intraplating melted the deepest N-dipping Arabian footwall and led to silicic melts erupting to the south (Fig. 13c). Mafic sills are more efficient than dykes at heating and softening continental crust, but less efficient at disrupting the lithosphere. We infer that the slow extensional rifting of Proto-Tethys failed in mid-Cambrian because the passively bulging asthenosphere intruded sills rather than dykes (Fig. 13c). Palaeo-Tethys was opened by Ordovician to mid-Devonian dykes further north (Fig. 13d).

Because Proto-Tethys rifted and riffled but failed to drift to an ocean, it became a remarkably successful hydrocarbon kitchen where the extraordinary hydrocarbon riches in and beside the current Persian Gulf were brewed and trapped. This is because the still-born ocean settled to a long and extraordinarily wide, almost flat-bottomed, shallow epeiric shelf-sea on the passive margin of

Gondwana until the late Palaeozoic (Beydoun 1991, p. 19). The NE margin of this 2000 km wide shelf was marked by continental magmatism and the accumulation of siliciclastic red-beds in the Sanandaj–Sirjan zone during Silurian, Devonian and Carboniferous times (Alavi 1991).

Late Permian rifting and Upper Triassic drifting (Beydoun 1991, p. 19) further shredded the margin of Gondwana when Neo-Tethys started to open westward, like a zip fastener, along what is now the Zagros imbricate zone (Fig. 12f). The closure of Palaeo-Tethys soon afterwards (Fig. 12f) is recorded by a magmatic arc (or ophiolites) along the suture in the Alborz mountains (Alavi, in review). The diachronous zipping open (Fig. 12f) and closure (Fig. 12g) of NeoTethys is recognisable in the age of the oil and gas reservoirs which young from Jurassic in the west to middle Cretaceous in the east. Later, as the Zagros front arrived in each area, oil either escaped or remigrated into Asmari reservoirs that were sealed by the Miocene evaporitic sequence and folded by Zagros shortening (Dunnington 1967).

The facies boundaries defining the keys in the Tethyan margin of Gondwana beneath the Gulf extrapolate southward along small circles across the whole of the Arabian platform to transform faults in the spreading ridges of the Gulf of Aden and Red Sea, which rifted 25 Ma BP in the Miocene (Fig. 12h) and began drifting at 5 Ma BP. The most significant lithospheric key along the Zagros, the incipient Qatar syntaxis, is on the same small circle as the dog-leg between the Red Sea and the Gulf of Aden. It is also in line with, and much the same width as, the Ethiopian rift valley. Such relationships raise the possibility that transform margins in the current spreading ridge were guided by the same old lines of weakness built into the basement by the Pan-African orogeny. This would make the Red Sea merely the present installment of the saga that began with Proto-Tethys riffling the keys along Najd faults in the Pan-African basement in late Proterozoic times.

## Conclusions

The mantle is still playing the same keyboard in the Tethyan margin of the African and Arabian plates that was constructed by Pan-African docking and diced by Najd faults. Some keys played more important roles than others during each glissand or riffle along the lithospheric keyboard. The incipient Qatar syntaxis assumed particular importance after it went up rather than down along the margin of Proto-Tethys in about mid-Cambrian times. The key destined to become a syntaxis perhaps 20 Ma in the future was chosen *c.* 500 Ma in the past, when it rose too high for Hormuz salt to survive on it.

Salt deposited on the depressed keys did not influence the distant opening or closure of Palaeo-Tethys, but strongly influenced subsequent suturing and shortening in the Zagros. A deep décollement of salt amplified the style of rifting, drifting and closure of Tethyan ocean basins and strongly influenced the style of riffling and shortening of sutured continental crust. The hydrocarbon fields of the region are huge because Proto-Tethys aborted to a flat shallow shelf that jostled only gentle when various Tethyan basins opened further north. This gentle riffling was sufficient to collect impressive quantities of hydrocarbons in periclines draped over the deep salt pillows that map the edges of the lithospheric keys, but not enough to drift Proto-Tethys to an ocean.

The Kazerun–Qatar and Mangarak–Marzuk lineaments were first introduced as part of a keynote talk at the Salt Tectonics meeting of the combined Petroleum and Tectonics Groups of the Geological Society of London on 14–15 September 1994.

We thank Hemin Koyi, Hassan Shahrivar, and Esmael Haidari for their invaluable help in the field. Christina Whelström and Jeanette Bergman-Weihed processed our diagrams with great skill and patience. Hemin Koyi suggested improvements to an early draft and reviews by Peter Szatmari, Ian Davison and Robert Hooper helped improve the presentation. Robert Hooper suggested using the term oblique footwall ramps for the Kazerun and Mangarak fault zones. Field work was financed by the Swedish Foundation for Natural Science and the Geological Survey of Iran.

## References

ALA, M. A. 1974. Salt diapirism in southern Iran. *AAPG Bulletin*, **58**, 1758–1770.

ALAVI, M. 1991a. Sedimentary and structural characteristics of the Paleo-Tethys remnant in northeastern Iran. *Bulletin of the Geological Society of America*, **103**, 983–992.

—— 1991b. Tectonic map of the Middle East (at scale 1:5,000,000). *Geological Survey of Iran, P.O. Box 13185-1494, Tehran, Iran*,

—— 1992. Thrust tectonics in the Bidalood region: NE Iran. *Tectonics*, **11**, 360–370.

—— 1994. Tectonics of the Zagros orogenic belt of Iran: new data and interpretations. *Tectonophysics*, **229**, 211–238.

ALSHARHAN, A. S. 1989. The petroleum geology of the United Arab Emirates. *Journal of Petroleum Geology*, **12**, 253–288.

BAKER, C, JACKSON, J. A. & PRIESTLY, K. 1993. Earthquakes on the Kazerun line in the Zagros mountains of Iran: strike-slip faulting within a fold-thrust belt. *Geophysical Journal International*, **115**, 41–61.

BEN-AVRAHAM, Z., TEN BRINK, U. & CHARRACH, J. 1990. Transverse faults at the northern end of the southern basin of the Dead sea graben. *Tectonophysics*, **180**, 37–47.

BEYDOUN, Z. R. 1991, *Arabian Plate Hydrocarbon*

*Geology and Potential – a Plate Tectonic Approach.* American Association of Petroleum Geologists Studies in Geology, **33**.

BRUN, J-P., SOKOUTIS, D. & VAN DEN DREISCHE, J. 1994. Analogue modelling of detachment fault systems and core complexes. *Geology*, **22**, 319–322.

BUTLER, R. W. H. COWARD, M. P. HARWOOD, G. M. & KNIPE, R. J. 1987. Salt control on thrust geometry structural style, and gravitational collapse along the Himalayan front in the Salt Range of northern Pakistan *In*: LERCHE, I. & O'BRIEN, J. J. (eds) *Dynamical Geology of Salt and Related Structures.* Academic, New York, 301–337.

DAVIS, D. M. & ENGLEDER, R. 1987. Thin-skinned deformation over salt. *In*: LERCHE, I. & O'BRIEN, J. J. (eds) *Dynamical Geology of salt and related structures*. Academic, New York, 301–337.

—— & LILLIE, R. 1994. Changing mechanical response during continental collision: active examples from the foreland thrust belts of Pakistan. *Journal of Structural Geology*, **16**, 21–34.

DEY, R. C. 1991. Trans-Aravalli Vindhyan evaporites under the semi-desertic plains of western India; significance of depositional features. *Quarterly Journal of the Geological, Mining and Metallurgical Society of India*, **37**, 136–150.

DUNNINGTON, H. V. 1967. Stratigraphic distribution of oilfields in the Iraq–Iran–Arabia basin. *Journal of the Institute of Petroleum*, **53**, 129–161.

DYKSTRA, J. D. & BIRNIE, R. W. 1979. Segmentation of the Quarternary subduction zone under the Baluchistan region of Pakistan and Iran, *In*: FARAH, A. & DE JONG, K. A. (eds) *Geodynamics of Pakistan.* Geological Survey of Pakistan, Quetta, 319–323

EDGELL, H. S. 1992. Basement tectonics of Saudi Arabia as related to oil field structures. In: RICKARD, M. J. *et al.* (eds) *Basement Tectonics.* Kluwer, Dordrecht, 169–193.

EKSTRÖM, G. & ENGLAND, P. 1989. Seismic strain rates in regions of distributed continental deformation. *Journal of Geophysical Research*, **94**, 10231–10257.

EYLES, N. 1993. Earth glacial record and its tectonic setting. *Earth Science Reviews*, **35**, 3–150.

FÜRST, M. 1976. Tektonic und diapirismus der östeerlichen Zagrosketten. *Zeitschrift der Deutschen geologischen Gesellschaft*, **127**, 183–225.

GANSSER, A. 1960. Über Schlammvulkane und Saklzdome. *Viertejahrsschrift der Naturforschenden Gersellschaft i Zürich*, **105**, 1–46.

—— 1992. The enigma of of the Persian dome inclusions. *Eclogae Geologica Helveticae*, **85**, 825–846.

GERA, F. 1972. Review of salt tectonics in relation to disposal of radioactive wastes in salt formation. *AAPG Bulletin*, **83**, 3551–3574.

GUSSOW, W. C. 1966, Salt temperature: a fundamental factor in salt dome intrusion. *Nature*, **210**, 518–519 (and reply to discussion: *Nature,* **211**, 1387–1388).

HARRISON, J. V. 1930. The geology of some salt-plugs in Lariistan (southern Persia). *Quarterly Journal of the Geological Society, London*, **86**, 463–522.

HEATON, R., JACKSON, M. P. A., BAMAHOUD, M. & NANI, A. S. O. (in press). Superposed Neogene extension, contraction and salt canopy emplacement in the Yemeni Red Sea. *In*: JACKSON, M. P. A., ROBERTS, D. G. & SNELSON, S. (eds) *Salt Tectonics: A Global Perspective for Exploration.* American Association of Petroleum Geologists, Memoir.

HEMPTON, M. R. 1987. Constraints on Arabian plate motion and extensional history of the Red Sea. *Tectonics*, **6**, 687–705.

HUSSEINI, M. I. 1989. Tectonic and depositional model of late Precambrian–Cambrian Arabian & adjoining plates. *AAPG Bulletin*, **73**, 1117–1131.

—— & HUSSEINI, S. I. 1990. Origin of the Infracambrian salt basins of the middle East *In*: BROOKS, J. (ed.) *Classic Petroleum Provinces.* Geological Society, London, Special Publication, **50**, 279–292.

JACKSON, J. A. & MCKENZIE, D. P. 1988. The relationship between plate motions and seismic moment tensors and the rates of active deformations in the Mediterranean and the Middle East. *Geophysical Journal of the Royal Astronomical Society*, **93**, 45–73.

—— & TALBOT, C. J. 1989. Anatomy of mushroom-shaped diapirs. *Journal of Structural Geology*, **11**, 211–230.

—— & VENDEVILLE, B. C. 1994. Regional extension as a geologic trigger for diapirism. *Bulletin of the Geological Society of America*, **106**, 57–73.

KASHEI, M. S. 1976. Plate tectonics and structural evolution of the Zagros geosynclines, southwestern Iran. *Bulletin of the Geological Society of America*, **87**, 1486–1490.

KENNEDY, M. J. 1993. The Undoolya Sequence; a new stratigraphic unit syn-depositional adjacent to a late Proterozoic salt structure; Amadeus Basin, Central Australia. *Australian Journal of Earth Science*, **106**, 57–73.

KENT, P. E. 1958. Recent studies of south Persian salt plugs. *AAPG Bulletin*, **422**, 2951–2972.

—— 1970. The salt plugs of the Persian Gulf region. *Transactions of the Leicester Literary & Philosophical Society*, **64**, 56–88.

—— 1979. The emergent Hormuz salt plugs of southern Iran. *Journal of Petroleum Geology*, **2**, 117–144.

—— 1987. Island salt plugs in the middle east and their tectonic implications. *In*: LERCHE, I. & O'BRIEN, J. J. (eds) *Dynamical Geology of salt and related structures*. Academic, New York, 3–37.

KOYI, H. 1988. Experimental modeling of role of gravity and lateral shortening in Zagros mountain belt. *AAPG Bulletin*, **72**, 1381–1394.

KUPFER, D. H. 1979. Internal kinematics of salt diapirs: Discussion. *AAPG Bulletin*, **73**, 939–942.

LETOUZEY, J. COLLETA, B., VIALLY, R. & CHERNETTE, J. C. (in press). Evolution of salt-related structures in compressional settings. *In*: JACKSON, M. P. A., ROBERTS, D. G. & SNELSON, S. (eds) *Salt Tectonics: A Global Perspective for Exploration.* American Association of Petroleum Geologists, Memoir.

LINDSAY, J. F. 1987. Upper Proterozoic evaporites in the Amadeus basin, central Australia and their role in basin tectonics. *Bulletin of the Geological Society of America*, **99**, 852–865.

O'BRIEN, C. A. E. 1957. Salt diapirism in south Persia. *Geologie en Mijnbouw, Nieue Serie*, **19**, 357–376.

PRASAD, B. 1984. *Review of Vindhyan Supergroup of SE Rajasthan.* Geological Survey of India Miscellaneous publications, **50**, 31–40.

SARWAR, G. & DE JONG, K. A. 1979. Arcs, oroclines, syntaxes: The curvature of mountain belts in Pakistan. *In*: FARAH, A. & DE JONG, K. A. (eds) *Geodynamics of Pakistan*. Geological Survey of Pakistan, Quetta, 341–349.

SCHWERDTNER, W. N. & OSADEZ, K. 1983. Evaporite diapirism in the Sverdrup basin: New insights and unsolved problems. *Bulletin of Canadian Petroleum Geology*, **31**, 27–36.

SCOTESE, C. R. & MCKERROW, W. S. 1990. Revised world maps and introduction. *In*: MCKERROW, W. S. & SCOTESE, C. R. (eds) *Palaeogeography and Biogeography*. Geological Society, London, Memoir **12**, 1–21.

SENGÖR, A. M. C. & BURKE, K. 1978. Relative timing and volcanism on Earth and its tectonic implications. *Geophysical Research Letters*, **5**, 419–421.

STAMFLI, G., MARCOUX, J. & BAUD, A. 1991. Tethyan margins in space and time. *Paleogeography, Palaeoclimatology & Palaoecology*, **87**, 373–409.

STÖCKLIN, J. 1968. *Salt Deposits of the Middle East*. Geological Society of America, Special Paper, **88**, 158–181.

TALBOT, C. J. 1979. Fold trains in a glacier of salt in southern Iran. *Journal of Structural Geology*, **1**, 5–18.

—— 1992: *Quo Vadis* Tectonophysics? with a pinch of salt! *Journal of Geodynamics*, **16**, 1–20.

—— 1993. Spreading of salt structures in the Gulf of Mexico. *Tectonophysics*, **228**, 151–166.

—— & JARVIS R. J. 1984. Age, budget and dynamics of an active salt extrusion in Iran, *Journal of Structural Geology*, **6**, 521–533.

—— & ROGERS, E. A. 1980. Seasonal movements in a salt glacier in Iran. *Science*, **208**, 395–397.

——, KOYI, H., SOKOUTIS, D. & MULUGETA, G. 1987. Identification of diapirs formed in horizontal compression. *Bulletin of Canadian Petroleum Geologists*, **36**, 91–94.

URAI, J. L., SPIERS, C. J. ZWART, H. J. & LISTER, G. S. 1986. Weakening of rock salt by water during long term creep. *Nature*, **324**, 554–557.

WEINBERG, R. F. 1993. The upward transport of inclusions in Newtonian and power-law salt diapirs. *Tectonophysics*, **228**, 141–150.

WERNICKE, B. 1985. Uniform-sense normal simple shear of the continental lithosphere. *Canadian Journal of Earth Sciences*, **22**, 108–125.

# Development of salt-related structures in the Central North Sea: results from section balancing

PETER G. BUCHANAN[1], DANIEL J. BISHOP[2] & DAVID N. HOOD[3]

[1]*Oil Search Ltd., NIC Haus, PO Box 1031, Champion Parade, Port Moresby, Papua New Guinea*

[2]*Department of Geology and Geophysics, The University of Edinburgh, Grant Institute, West Mains Road, Edinburgh, UK (Present Address: Department of Geology, Victoria University of Wellington, PO Box 600, Wellington, New Zealand)*

[3]*Esso Exploration and Production UK Ltd, Mailpoint 508, Esso House, Victoria Street, London SW1E 5JW, UK*

**Abstract**: The geometric and kinematic evolution of salt-related structures in the Central North Sea has been studied using structural restoration and balancing techniques. Strain associated with fault displacements, compaction and isostasy has been systematically removed, thus producing a graphical representation of the temporal and spatial distribution of salt structures through time. The results clearly demonstrate that the Central North Sea can be sub-divided into three structural provinces based on the type of salt structure present: (1) the Central Graben which is characterized by large offset extensional faults with broad salt swells and tall salt diapirs, (2) the Eastern platform comprising the Norwegian–Danish Basin which is dominated by symmetrical salt swells that are typically welded in intervening lows and deeply buried, and (3) the Western platform which is characterized by salt swells and gravity-driven, thin-skinned extension. By restoring the deformation at both a local and a regional scale, it becomes apparent that the Permian salt has controlled the development of structures throughout the Mesozoic and Tertiary in the Central Graben region and the Western Platform, whereas on the Eastern Platform the salt was primarily active during the Triassic only. This has led to a variation in the evolution of structural styles across the Central North Sea. The restorations also suggest that salt dissolution has not been a significant feature during the development of diapirs in this area.

Section restoration techniques were initially developed and applied to regions of contractional tectonics (Dahlstrom 1968) and later adapted for areas of extensional tectonics (Gibbs 1983). Only recently has the method been successfully applied to areas which have undergone salt tectonics (e.g. Hossack & McGuinness 1990; Rowan 1993; Bishop *et al.* 1995). However, deformed salt provinces are increasingly becoming important in hydrocarbon exploration (e.g. North Sea and Gulf Coast) and an understanding of the often complex structural history is critical in delineating the relative timing of deformation, the geometric evolution of structures and the possible complex hydrocarbon migration paths which might have existed throughout the evolution of a basin.

This paper presents some initial results from an on-going study in which several sections through the Central North Sea (Fig. 1) have been retro-deformed, systematically incorporating the removal of deformation, compaction and isostatic effects. The results of three specifically chosen sections, all parallel to the major tectonic transport direction, are here presented as they are taken to be representative of the database of eight such sections. Two of the sections are located on either margin of the Central Graben and one is a regional profile across the entire Central North Sea. The results clearly illustrate the detailed evolution of the Central Graben and in particular, they highlight distinct differences between the salt structures which evolved on the Eastern and Western platforms respectively. The sequential removal of deformation and its effects, enable several analyses to be undertaken on the sections and a quantification to be made on the rate of extension and possible salt loss through time.

The results presented in this paper are by no means an exhaustive study and the conclusions are therefore somewhat limited. However, the restorations do highlight some important considerations and demonstrate the power of computer-aided structural balancing and modelling techniques to illustrate graphically the sequential evolution of complex structures.

## Methodology

The method applied in the restoration of the Central North Sea sections was initially described by Rowan (1993) and further developed by Bishop *et al.* (1995) and is therefore only briefly described

*From* ALSOP, G. I., BLUNDELL, D. J. & DAVISON, I. (eds), 1996, *Salt Tectonics*, Geological Society Special Publication No. 100, pp. 111–128.

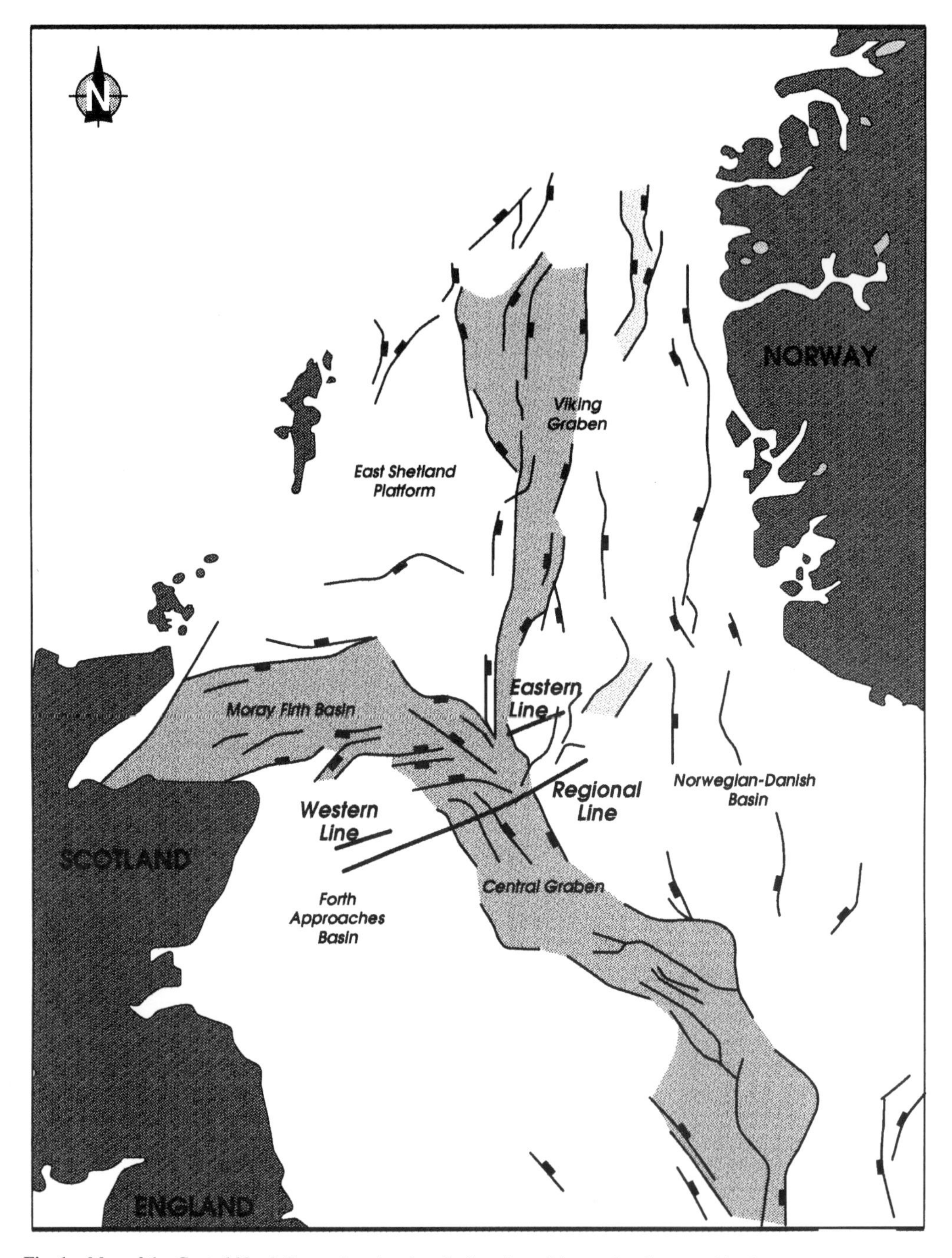


**Fig. 1.** Map of the Central North Sea region showing the location of the section lines used in the restorations.

here. The restoration methodology involves the removal of major deformation effects related to fault movements, compaction and isostacy. Such modelling at the accuracy presented here would not be possible without the use of modern computer software applications. In this study, the commercially available Geosec programme, developed by CogniSeis Development, was used in the reconstructions.

Starting with a depth converted section (Fig. 2a),

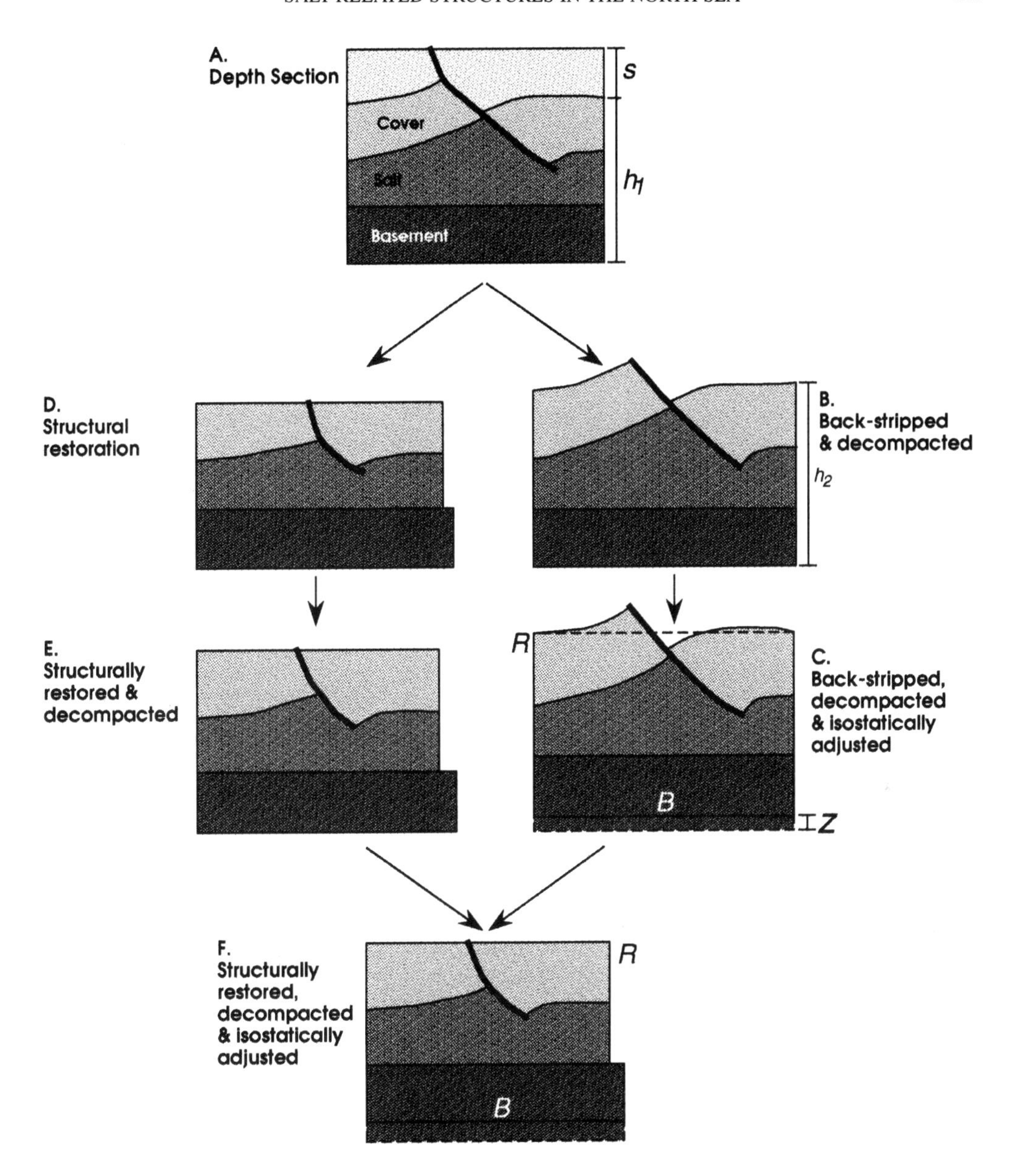


**Fig. 2.** Methodology involved in restoring a cross-section involving salt (adapted from Rowan, 1993). (a) Original depth-converted section. (b) Back-stripped and decompacted section using base-line as a reference which remains horizontal. (c) Isostatically adjusted section with assumed regional showing relative paleo-bathymetry. (d) Restored flattened section. (e) Decompacted section. (f) Juxtaposition of restored and decompacted allochthonous cover onto isostatically adjusted basement. ($h_1$, original thickness of supra-salt section; $h_2$, decompacted thickness; $s$, removed overburden; $Z$, isostatic response; $R$, regional; $B$, baseline).

backstripping is performed down to the horizon which is to be the reference for subsequent restoration (Fig. 2b). Once the overburden has been removed, the remaining section can be decompacted such that expansion is from the baseline upwards. The decompaction function utilises the empirical compaction curves derived by Sclater & Christie (1980) and Schmoker & Halley (1982). In the sections described in this paper, the basement and salt are designated solid and thus undergo no expansion or geometrical alteration during decompaction. Typical 'average' lithologies were used to represent each interpreted seismostratigraphic unit (see Bishop *et al.* 1995). Lithological inhomogeneities within units were not modelled.

It is assumed that the addition of load to a section in the form of sedimentary overburden will produce an isostatic subsidence in order to balance forces in the lithosphere. Therefore, the removal of that load during backstripping should be accompanied by a commensurate amount of isostatic rebound of the entire section. It is possible to model isostatic compensation in the Central North Sea using simple Airy (one-dimensional) isostasy which ignores the flexural rigidity of the crust. It may be argued that a flexural isostatic model is more appropriate. However, the flexural rigidity of the Central North Sea is relatively low, and the effects of flexural rigidity are also less apparent on short baseline sections. Therefore, it is assumed that errors caused by ignoring flexural isostatic effects are not significant compared to other sources of error in the restorations, such as deformation mechanisms and time-depth conversion. The pragmatic approach is to use Airy isostatic compensation, which is not only easy to apply, but provides very satisfactory results, as already demonstrated by Rowan (1993) and Bishop *et al.* (1995). Furthermore, the integration and smoothing of multiple 1D isostatic calculations produces a profile resembling that of flexural bending (of a beam with a low elastic thickness).

As shown by Rowan (1993), the calculation of Airy isostasy ($Z$) requires the following parameters to be known: the overburden thickness (s), the vertical thickness of the compacted section at each point across the section ($h_1$) and the decompacted vertical thickness of the section at the same points across the section ($h_2$). These values can then be input into the following equation:

$$Z = [s - (h_1 - h_2)]\,(\rho_c - \rho_w)/(\rho_m - \rho_w)$$

where $\rho_c$ is the average density of crust (2.8g/cm$^3$), $\rho_w$ is the density of water (1.0g/cm$^3$) and $\rho_m$ is the average density of the mantle (3.3g/cm$^3$).

The output from such a calculation is an uplift profile across the entire section which reflects the Airy isostatic compensation. The section is then uplifted from its horizontal base using this profile and can be considered isostatically adjusted (Fig. 2c). Following the method described by Rowan (1993), a smooth curve can be drawn along the top of the section which connects blocks which appear to have been relatively stable during the period covered by the restoration and as such, have been unaffected by salt movements (Fig. 2c). This curve may be a uniform gradient, or even a horizontal line. Such a curve is thought by Rowan (1993) to represent a palaeobathymetric profile. It is of practical use, since the section can be structurally restored to this profile, or 'regional' line, rather than to an arbitrary horizontal line, as has been more common in many previous restoration studies. In the sections presented here, the palaeobathymetric profiles are almost straight lines, which by coincidence connect the flattened crests of tilted suprasalt fault blocks. This suggests that the crests have been flattened by erosion, close to the palaeobathymetric curve, either at the sea bed, or close to sea level.

True decompaction is at least a two-dimensional process since material is moving within the plane of the section and is not simply being displaced in a vertical manner. Accurate decompaction can only be achieved if material points are tracked during the restoration process within the plane of the section such that the overburden in the deformed and restored state is known for each point (Geiser *et al.* 1991). In order for this to be calculated, both the deformed state section and the restored state section need to be available. Therefore, despite the section having already been decompacted, a secondary and separate procedure is undertaken first to restore the structural section (Fig. 2d) and then to decompact it (Fig. 2e). This truly decompacted section can then be swapped into the isostatically compensated section using the calculated regional profile for reference (Fig. 2f).

This methodology effectively restores the suprasalt cover and basement as separate entities thereby enabling the salt to change geometry and cross-sectional area through time. It is widely accepted that it is legitimate to allow the salt cross-sectional area to vary during restoration (e.g. Rowan & Kligfield 1989; Hossack & McGuinness 1990; Schultz-Ela 1992; Seni 1992) despite the cross sections being located parallel to the major transport direction and orthogonal to the salt structures and intervening lows.

The final product arising from the above procedure will be a series of sections which are geologically, geometrically and kinematically balanced and graphically illustrate the structural evolution through time.

### *Thermal subsidence*

The methodology employed in restoring the cross sections presented here ignores the effect of thermal subsidence. Thermal effects are difficult to quantify or differentiate from overburden or tectonically induced components in the total deformation field. Both thermal and overburden-induced subsidence are isostatic processes and are likely to be coeval, making differentiation between the two equivocal. In the North Sea, the thickness of the post-rift deposits is such that a significantly larger amount of extension than is observed would be required if thermal collapse of the lithosphere was the sole tectonic mechanism. Furthermore, there has been considerable thermal uplift in northern Britain during the time of proposed thermal subsidence in the North Sea (Brodie & White 1995). The uplift has demonstrably produced large amounts of erosion and this has likely fed the overburden-driven subsidence modelled in the Central North Sea. Therefore, modelling of thermal subsidence is not deemed necessary at the present time, given the limited control on the parameters required to obtain meaningful results, and is as such beyond the scope of this paper. Furthermore, thermal subsidence will have only introduced a modification of the section after the development of the salt structures. With such low deviatoric stresses induced by thermal subsidence, an alteration of the spatial distribution and geometry of salt is unlikely to be significant. If thermal collapse were a major driving force during the post-rift phase, then it would probably produce regional variations in subsidence and would have no discernible effect on the relationship between individual structures or between the position of basement relative to cover.

## Results

The results are presented here in the order in which the restorations were produced, that is from present day to Permian. However, it is beneficial to view the sections in a forward sense, starting with the Permian and working towards the present.

### *Regional Section*

Figure 3 shows the results from the modelling performed on the regional section. The original depth interpretation (Fig. 3a) shows that the Central Graben is the deepest part of the Central North Sea region and consists of two sub-basins separated by a significant horst block (the Montrose–Forties High). The Western Platform is markedly different to the Eastern Platform in that it dips around 5–7 degrees towards the east in contrast to the Eastern Margin which has only a moderate dip of approximately 1–2 degrees in an easterly direction. The difference in basement dip and elevation is reflected in the variation in thickness of the upper Tertiary cover, which thickens dramatically from west to east, but most rapidly across the Western Platform. The average thickness of salt is also variable across the section, being thickest in the Central Graben and thinnest over the Eastern margin. The salt has been completely withdrawn in places on the Eastern Platform such that welding has occurred locally. The salt forms diapirs and pillows throughout the Central North Sea, with more symmetrical and pronounced structures in the east. However, in the west, large listric growth faults occur which detach into the salt and are associated with gentle roll-over structures in their hanging walls and salt diapirs in their footwalls.

In Fig. 3b, the section is shown restored to the top Palaeocene. A significant amount of decompaction is seen to have occurred in the Palaeocene unit (around 45%). The slope of the Rotliegend basement has been reduced on the Western Platform to approximately 2–3°, whilst on the Eastern Platform, the slope of basement has increased to about the same amount. The palaeo-slope of basement across the Central North Sea during the Palaeocene was relatively consistent between the two major tilt blocks separated by the Central Graben. The average thickness of salt remains similar across the section to the present-day configuration.

In the early Cretaceous (Fig. 3c), subsidence was more pronounced over the Western Platform with the greatest thickness of sediment being deposited in the Central Graben. A Jurassic high separates the Western Platform subsidence from a more limited subsidence over the Eastern Platform. Again, there is little change in the geometry or thickness of salt across the Central North Sea.

The restoration to top Jurassic (Fig. 3d) shows that subsidence and sedimentation was restricted to the Central Graben during that time, with the development of two sub-basins of roughly equal proportions. This is clearly the time of maximum localized extension and subsidence in the Central Graben.

Restoration to Late Triassic (Fig. 3e) indicates the basement slope to be in the order of 1° towards the east across the entire Central North Sea region with locally a more pronounced easterly dip above the Eastern Platform. The Triassic is locally welded to basement at this time. In the east, the basement becomes virtually horizontal when the section is restored to the early Triassic (Fig. 3f). This restoration demonstrates that the geometry of salt structures across the Central North Sea region was consistent, with salt pillows and diapirs separating irregular Triassic basins and sub-basins. The

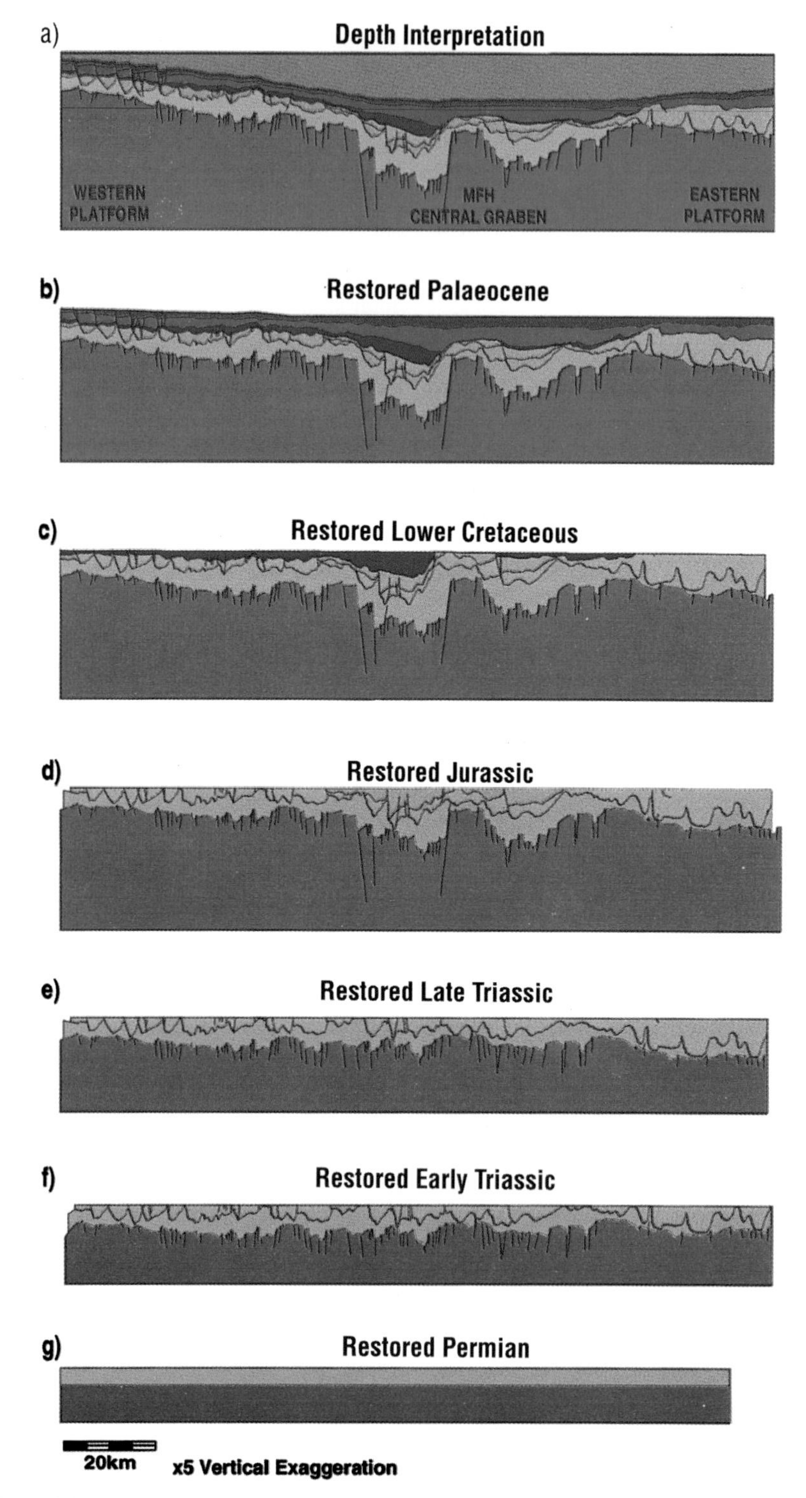


**Fig. 3.** Sequential restoration of regional line. (a) Depth section. (b) Restored top Palaeocene. (c) Restored lower Cretaceous. (d) Restored Jurassic. (e) Restored late Triassic. (f) Restored early Triassic. (g) Restored top Permian. (MFH, Montrose–Forties High).

welding of the Triassic in the east is still shown, although the isostatic adjustment of the basement is probably inaccurate because of the unquantified effect of basement faulting. It is interesting to note that the average height of the tallest diapirs across the section is approximately the same.

The section has been restored to the Permian (Fig. 3g). The derived thickness for the Permian is explained in the discussion and is here represented by a uniform thickness salt layer overlying an undeformed basement. The change from a uniform thickness of salt, to relatively extreme variations in thickness has implications for the flow of salt into, and around the Central Graben region.

### *Western Platform Section*

The sequential restoration of the Western Platform section is depicted in Fig. 4. A detailed analysis of this section (Fig. 4a) has been given previously in Bishop *et al.* (1995). The depth interpretation (Fig. 4b) shows a strongly asymmetric structural architecture with down to the basin listric growth faults evenly spaced and detaching onto salt. These faults are offsetting Palaeocene reflectors indicating Tertiary activity and often are associated with minor antithetic faults. The salt is thinnest in the hanging wall to the main faults and thickest in the footwall. The Tertiary section steps down towards the east in a series of terraces of equal proportions. The basement is relatively strongly tilted towards the east at approximately 6°.

Restoration to the top Palaeocene (Fig. 4c) shows a reduction in the tilt of the basement to around 3–4°. There is only minor off-set of the Palaeocene unit across the main basin faults suggesting they were relatively inactive during this time. There is a minor increase in salt thickness across the section from west to east although there is very little change in the geometry of the top salt horizon.

Figure 4d illustrates the section restored to top Chalk. The thickness of the Chalk changes in the western part of the section where it is off-set by the major growth faults. It is worth noting that these faults appear relatively inactive in the up-dip part of the section where changes in the thickness of the chalk appear to relate more to changes in the shape of the underlying diapirs. The dip in the basement is now partitioned into a western, semi-horizontal portion and an eastern, easterly dipping portion. In the lower Cretaceous (Fig. 4e), it is mainly minor graben bounding faults which show evidence of being active and these are located in the eastern part of the section. However, there is one salt-detached fault which is active in the centre of the section and juxtaposes lower Cretaceous in the hanging wall onto Triassic and Zechstein in the footwall.

In the Triassic (Fig. 4f), the basement maintains a gentle dip towards the east above which the Zechstein salt forms a number of pillows. In places on the western part of the section, the salt has almost reached the surface. Intra-Triassic reflectors imaged on the seismic (Fig. 4a) indicate that some of the Triassic basins are fault bounded as the reflectors diverge towards the next salt swell, terminating against it and suggesting normal movement along the salt-Triassic interface. However, in adjacent basins, the intra-Triassic reflectors onlap both sides of the lower bounding salt horizon and have a synclinal morphology. The main control on sedimentation in such basins may be a combination of salt withdrawal and possibly dissolution (mainly soluble salts such as halite) plus differential sedimentary loading by Triassic sediments. Therefore, Triassic sedimentation is localised by (a) growth faults, (b) fluvial systems and (c) a positive feedback relationship between salt withdrawal/dissolution and sedimentary loading (e.g. Hodgson *et al.* 1992; Penge *et al.* 1993).

Restoration to Permian has been achieved by removing the extension in the basement and maintaining a constant area for salt (Fig. 4g). The thickness of the salt horizon is assumed to be that of the stable margin at the extreme western end of the section. This section as well as the restored Triassic section implies there is a discrepancy between the amount of extension in the supra-salt cover and the basement of approximately 15–20%.

### *Eastern Platform section*

Figure 5 illustrates the sequential structural evolution of the Eastern Platform section (Fig. 5a). The depth interpretation shows a constant thickness of Tertiary across the section (Fig. 5b) with a large salt diapir dominating the structure. Although there is a strong irregularity at the top basement resulting from numerous faults, the average dip of the top basement surface is sub-horizontal. The diapir is virtually isolated from neighbouring salt swells as a result of the base of the supra-salt section being very close to top basement. The top Balder (near top Palaeocene) shows a gentle anticlinal form centred upon the underlying salt diapir. Restoration of the section to near top Palaeocene (Fig. 5c) illustrates very little active extension occurred during the Tertiary with the sole exception of a single, easterly-dipping extensional fault which detaches onto the flank of the underlying diapir. Apart from some minor thinning in the footwall of the extensional fault, the Palaeocene is of nearly uniform thickness. The top Chalk restoration (Fig. 5d) shows minor extensional faulting above the crestal region of the main diapir and a gentle warping of the Chalk sequence above the diapir, such that the Chalk is thinnest above the crestal region despite the

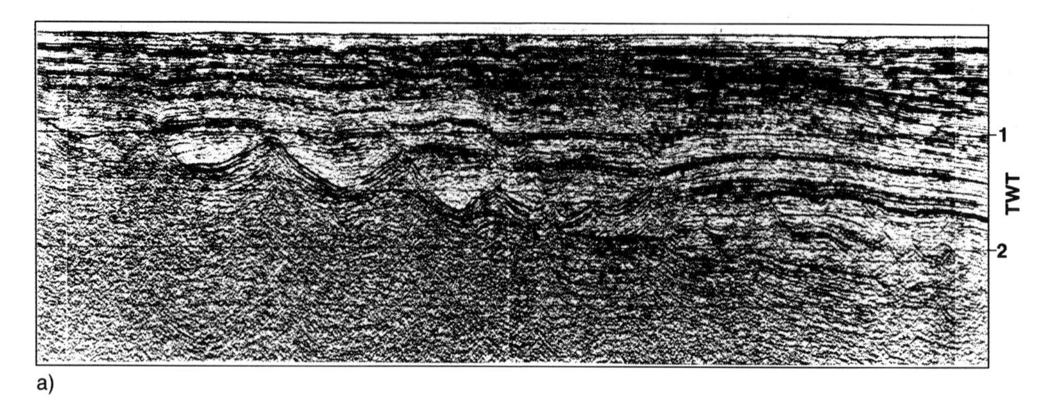


**Fig. 4.** Sequential restoration of Western Platform section. (a) Seismic section. (b) Depth interpretation. (c) Restored top Palaeocene. (d) Restored top Chalk. (e) Restored lower Cretaceous. (f) Restored top Triassic. (g) Restored top Permian.

localization of faulting in the same region. There is a minor increase in the thickness of salt in the section.

Restoring the section to top Hidra (upper Cretaceous) shows there was greater subsidence in the west and that the graben structure located in the crestal region was active (Fig. 5c). During Cromer Knoll (lower Cretaceous) and Kimmeridgian times, the section underwent a constant subsidence with little extensional faulting (Fig. 5f,g). Subsidence during upper Jurassic Oxfordian times was located immediately above the crestal region of the main underlying salt diapir (Fig. 5h). The restricted Jurassic basin was generally symmetrical with significant extensional faulting producing stepped terraces on both margins of the basin. Despite there being apparent subsidence above the salt diapir, there is a greater salt area and thickness in comparison to the Cretaceous.

Restoration to the Triassic highlighted some major displacement problems at top Zechstein level (Fig. 5i) which showed reverse offset in places. This can be explained by a loss of salt out of the plane of section or, conversely, movement of the Triassic sequence oblique to the line of section. The control on salt movement during the early phase of extension, when the basement slope was negligible, would likely result in dome-shaped salt swells. This would imply that during the early phase of diapirism, displacement vectors of either the salt or the cover can be complex and three-dimensional. However, during subsequent extension when basement slopes become more pronounced, plane strain deformation is likely to occur as a particular structural mechanism and orientation predominate. Nevertheless, the general form during the Triassic shows a less pronounced diapir with approximately the same structural configuration as in the present-day section. In Fig. 5i, the basement has been restored and arbitrarily inserted into the section to highlight the difference in net section lengths between the cover and the underlying basement where there is a discrepancy of about 10%.

## Discussion

### *Salt area and thickness*

The methodology employed in the restoration of the cross sections presented here assumes that rock volumes above and below the salt horizon behave in a dominantly plane strain manner and that substantial movement into or out of the plane of the section has not occurred. This is likely to be the case where salt mobility and cover extension are induced by regional slope, as is the case in the Central North Sea as well as in places such as the Gulf Coast. As the sub-salt basement has been restored isostatically, whilst the supra-salt cover has undergone both isostatic uplift, decompaction and physical restoration of fault displacements, the intervening salt mass is free to change shape and area. In this way, it is possible to obtain a quantification through time of the change in salt area. In the sections presented here there are relatively few places where salt actually reaches the surface (except in the Triassic along the Western Platform) and, therefore, salt dissolution is not considered to be a major factor affecting salt area through time.

In Fig. 6, the salt in the regional section has been flattened on its base to an arbitrary horizontal datum. A most striking observation of this datumed

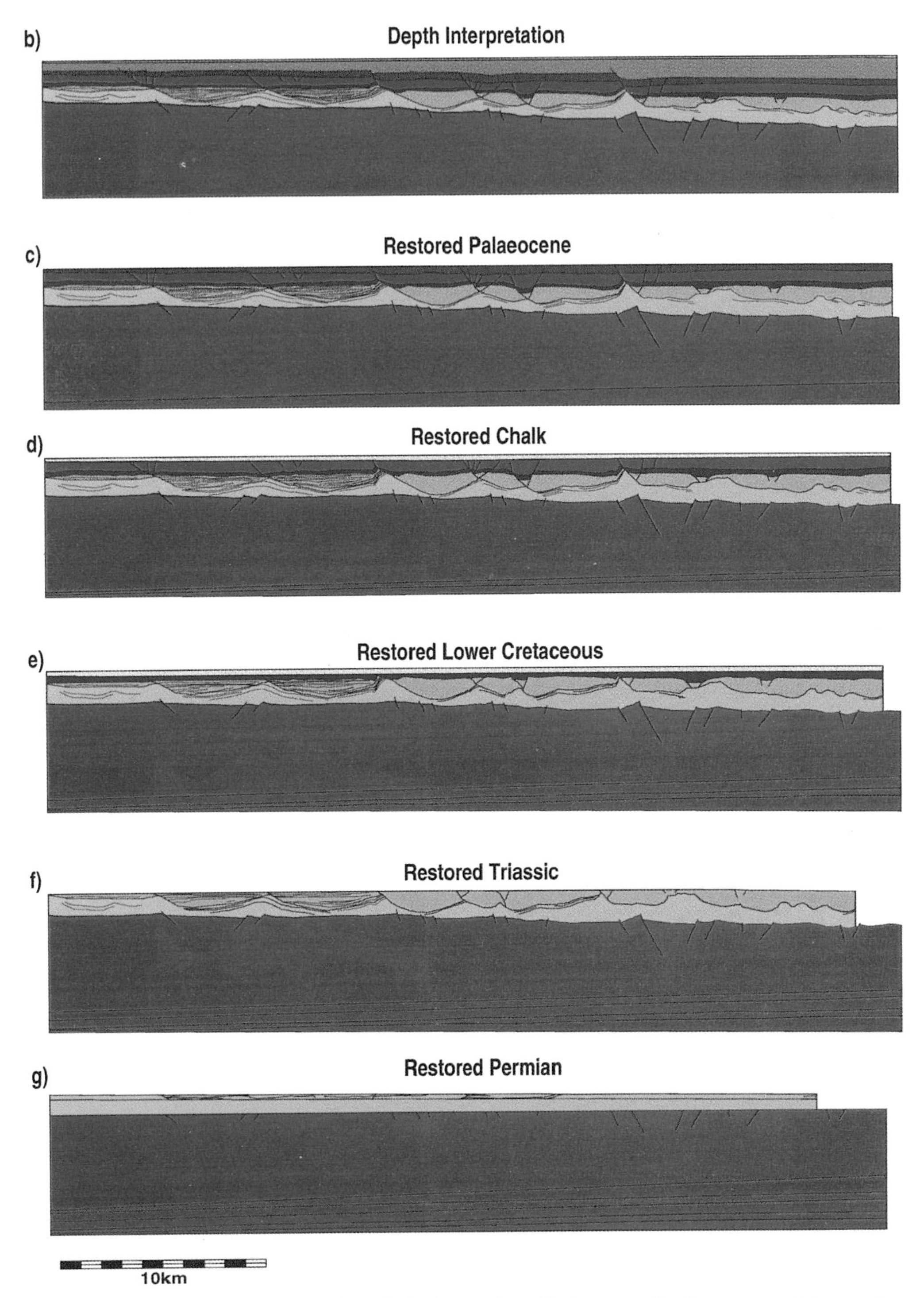


salt body is that the major 'diapirs' on both the western and eastern ends of the section have almost exactly the same height of approx 945 m. A line drawn at this elevation across the section shows other diapirs towards the centre of the section having the same height as well, although there also exist both smaller and larger diapirs across the section.

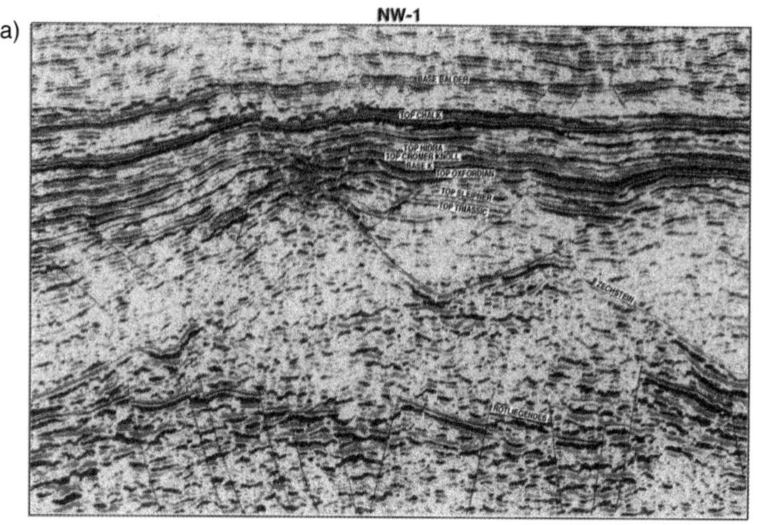
a)
NW-1
BASE BALDER
TOP CHALK
TOP HIDRA
TOP CROMER KNOLL
BASE K
TOP OXFORDIAN
TOP SLEIPNER
TOP TRIASSIC
ZECHSTEIN
ROTLIEGENDES

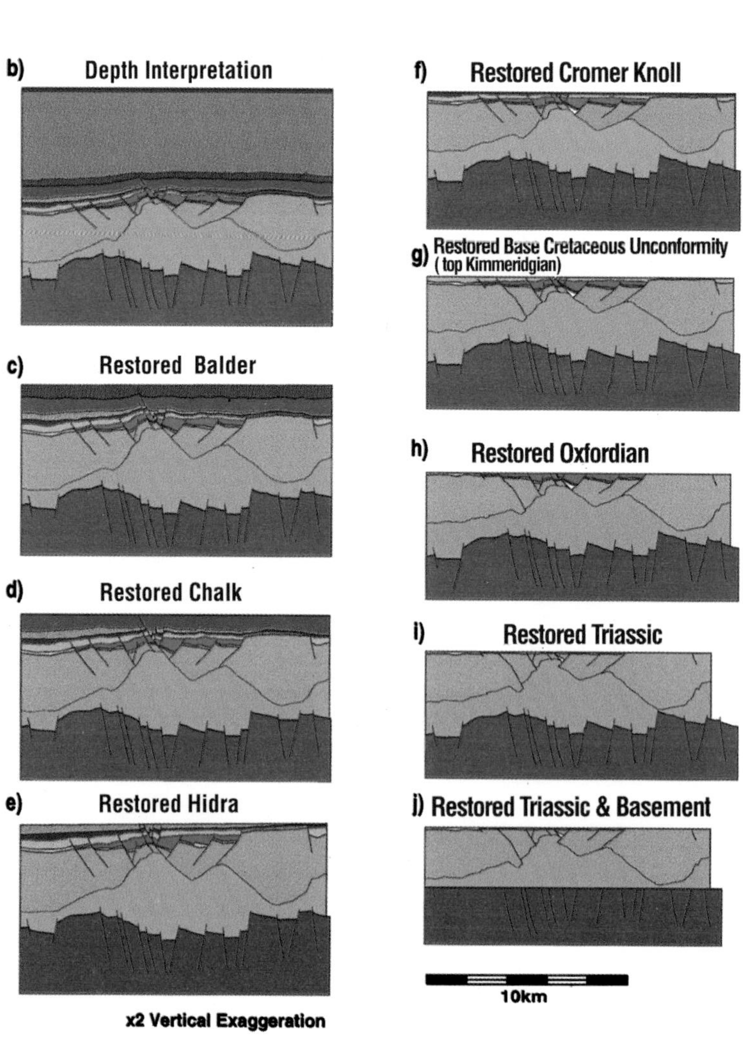
b) Depth Interpretation
c) Restored Balder
d) Restored Chalk
e) Restored Hidra
x2 Vertical Exaggeration
f) Restored Cromer Knoll
g) Restored Base Cretaceous Unconformity ( top Kimmeridgian)
h) Restored Oxfordian
i) Restored Triassic
j) Restored Triassic & Basement
10km

Regional Section

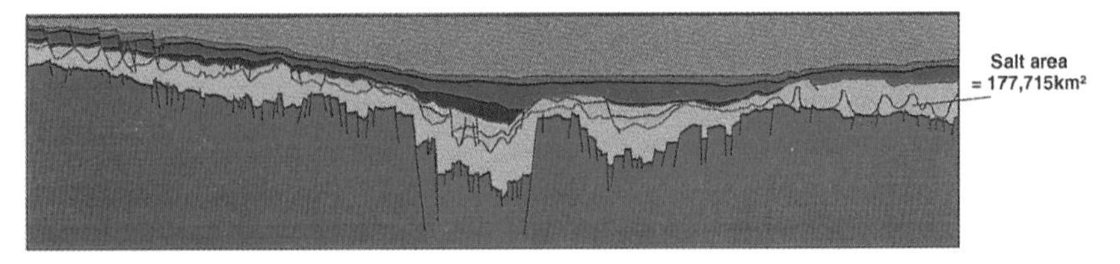


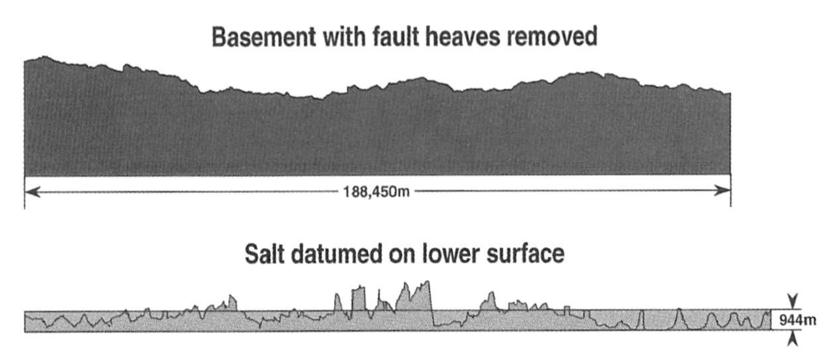


**Fig. 6.** Diagram showing the salt mass from the Regional Section flattened onto a lower horizontal datum. The average height of major diapirs across the basin is 945 m which when divided into the area of salt from the present day section (177 700 km$^2$), gives an estimated original length of 188.0 km. This equates to the 188.5 km calculated from restoring displacement along basement faults.

Restoration of the numerous seismically imaged faults in the basement produced a figure for the length of the Permian section of 188.5 km. The salt area in the present-day regional section is calculated to be 177 700 km$^2$. Assuming that there has been negligible loss of salt from the section, the average thickness of the original salt layer can be calculated by dividing the total salt area by the length of the restored section, giving a figure of approx 945 m for the thickness of the salt layer. This figure is the same as the height of the major diapirs.

It is possible that the salt diapirs were formed in response to sedimentary loading from above, during the Triassic. This would have caused downbuilding of the sediment into the salt layer, and withdrawal of salt beneath areas of downbuilding. Salt withdrawal may have been balanced by movement of salt upwards, forming diapirs between the areas of downbuilding. Salt may also have been lost by dissolution of diapirs near the surface.

If this model is correct and assuming the original salt layer had a relatively consistent thickness across the basin, then the diapirs must presently be taller than the thickness of the original salt layer, implying that the salt layer was originally much less than 1 km thick. This discrepancy could possibly be explained if the cross section restoration was incorrect, and the original thickness of the salt layer was much less than 1 km. Another explanation might be that the measured diapir heights are not typical, and that in reality most diapirs are greater than 1 km in height. It is also possible that the assumption of constant cross-sectional area for the salt body is incorrect, and that there have been large amounts of salt dissolution and movement of salt in or out of the plane of section. Furthermore, the coincidence in diapir height and calculated Permian salt thickness could simply be fortuitous.

However, there are some problems with the downbuilding model as an explanation for what is observed. The reasonably consistent height of the major diapirs could be explained if sedimentary loading was uniform across the region, but the alluvial systems which existed during the Triassic in the North Sea would have been unlikely to have spread their deposits equally over the area. Also, it is

**Fig. 5.** Sequential restoration of Eastern Platform section. (a) Seismic section. (b) Depth interpretation. (c) Restored near top Palaeocene. (d) Restored top Chalk. (e) Restored top Hidra (upper Cretaceous). (f) Restored top Cromer Knoll (lower Cretaceous). (g) Restored base Cretaceous unconformity (top Kimmeridgian). (h) Restored top Oxfordian (upper Jurassic). (i) Restored top Triassic. (j) Restored top Triassic & Basement.

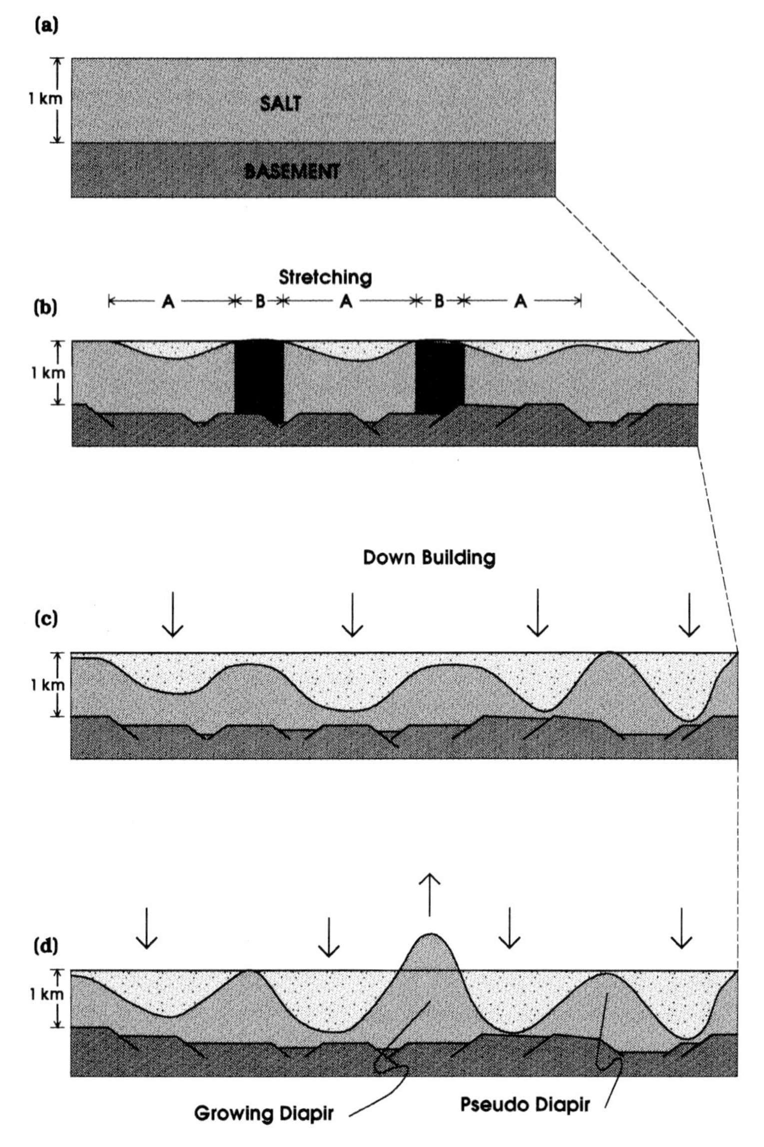


**Fig. 7.** Schematic diagram illustrating the attenuation model proposed here to describe the development of pseudo-diapirs. Stretching of the initial 1 km thick salt layer (a) results in localised subsidence and sedimentary downbuilding (b). The process continues until welding occurs at which time the diapir shaped salt bodies become effectively frozen (c). Stretching of the salt layer due to regional extension causes the creation of accommodation space. Reduced extension with continued downbuilding subsequently produces diapirism with some diapirs growing in height above their original thickness of 1 km (d). (A, zones of stretching; B, zones of undisturbed salt).

evident in many salt bodies on the Norwegian side of the Central graben, that there is a lack of disturbance of internal reflectors (e.g. Fig. 4a), which is not what would be expected if there had been major upward diapiric movements of salt.

In answer to the problems outlined above, we propose a new model that can explain all the observations made in this paper. In particular, our model explains why the calculated original salt thickness matches the height of the major diapirs. The new model is illustrated schematically in Fig. 7. The observed geometry of the salt body may be explained by stretching and thinning of the original salt layer due to the crustal extension measured in the basement. This model suggests that accommodation space is generated by lateral extension which, in order to maintain compatability between cover and basement, results in a thinning process in the salt layer. Such a thinning process would subdue the diapiric rise of salt induced by downbuilding which requires no lateral extension to develop. Faults within the basement provide points of instability at the base of the salt layer, causing the salt layer to thin more in some places than in others, and this would be exacerbated by sediment loading from above. Thus the primary diapirs represent remnants of a salt layer that originally had a nearly uniform thickness, and are therefore as high as the original layer was thick, i.e. about 1 km. Whilst sediment downbuilding almost certainly did occur during the Triassic, it is suggested that it was not the main driving mechanism, but only the result of lateral salt movements. It is postulated that the main driving mechanism was crustal extension.

The model proposed explains the features of the Triassic restorations, but not all the post-Triassic salt structures. Downbuilding and salt buoyancy are thought to have been important driving mechanisms in the Central Graben since the Triassic, whilst gravity-driven processes are thought to have become important on the platform west of the Central Graben. At any given time and place, it is likely that there was more than one halokinetic driving mechanism, and it is usually hard, or even impossible, to discover which mechanism was most important. We suggest that although other mechanisms may have been important also, Triassic salt movements were at least in part driven by crustal extension, resulting in localized thinning of the salt layer. Therefore, the model proposed here is an end member, but in reality, a hybrid salt thinning – downbuilding mechanism is probably the most appropriate model for the Central North Sea.

The importance of salt dissolution as a phenomenon remains uncertain. It is not possible to estimate the amount of salt dissolution in the Triassic–Jurassic. However, the greater the assumed salt loss by dissolution, the greater the calculated thickness of the original salt layer. 1 km represents a minimum thickness, assuming negligible salt loss. Using the independent evidence of diapir heights, 1 km seems to be a good estimate for the original salt layer thickness, which implies that salt dissolution was indeed minimal.

If our model is correct, it has important implications for the use of the term 'diapir' to describe the salt structures. The term is generally used to describe salt columns that have undergone vertical movement, but in our model, no vertical movement is implied, only lateral movement due to stretching and localized thinning of the salt layer. Thus, many of the diapirs observed in the North Sea may not be diapirs *sensu stricto*, but rather what we term 'pseudo-diapirs' (Fig. 7).

The salt budget for the section is readily apparent (Fig. 6) with a net increase in the Central Graben and a corresponding decrease on the margins, with a greater net decrease on the Eastern Platform (where there exists a greater overburden). Therefore, active diapirism and associated structures would be expected in the central part of the Central North Sea whilst passive down-building would be the dominant process on the margins. Indeed, large and tall diapirs are very common in the deepest parts of the East, West and Norwegian sectors of the Central Graben, although none are intersected by the regional line presented in this paper. These have been the object of much exploration activity over the years, particularly in the Norwegian 'Chalk' fields. They have already been widely reported in the literature and are probably driven by buoyancy effects (Jenyon 1990). They may also be partly driven by compression from the 'sides' of the basin due to down-slope movement of salt and/or supra-salt allochthonous cover, i.e. squeezed up and out in a similar manner to that described by Vendeville *et al.* (1994) for the Nordkapp Basin.

If this model is appropriate and it still needs to be tested on other sections in the area, then salt is flowing downslope into the deeper parts of the basin. Restorations of both marginal sections are in agreement with this model as they both show a net decrease in salt area through time (Figs 4 & 5). The process of salt flow on the Western Platform is further stimulated by the tectonic slope developed since the Jurassic. In this region, gravity-driven, thin-skinned tectonics has significantly altered the earlier Triassic pinch and swell type geometries (Bishop *et al.* 1995).

## *Amount of extension*

Figure 8 shows a plot of measured extension in the cover versus time for the three sections.

Despite the poor correlation in the absolute timing of the main extensional pulses, all sections show continuous extension through time with a relatively sharp decay in the rate of extension in the Jurassic. As a generalization, it can be stated that supra-salt extension was dominant in the Permian through to the early Jurassic, after which, subsidence and downbuilding became the major process affecting the development of salt structures. Total amounts of extension calculated for the Central North Sea region fall in the range 6–8%.

The 10–20% discrepancy between the line lengths of the restored cover and basement sequences for both marginal sections suggests that decoupling in the salt horizon has occurred and that supra-salt material has moved into the central

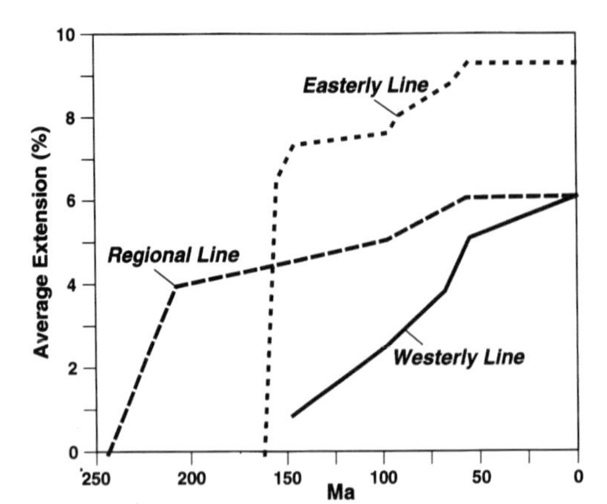


**Fig. 8.** Plot of extension against time for the supra-salt sequence for each of the three sections.

portion of the basin. Movement of material toward the centre of the basin means that we might expect shortening in the centre of the basin and possibly uplift and erosion as well. However, the crustal extension observed in the Central Graben probably exceeds any shortening due to movement of the supra-salt material, effectively masking the expected contraction.

There are likely to be significant amounts of sub-seismic, extensional faulting in the basement which remain unresolved and therefore, caution should be applied to the accuracy of the results (see Pickering, *et al.* 1995). Of interest is the greater differential line lengths between cover and basement seen on the western section, where gravity has had a greater affect on translating the cover down-dip over the basement.

### *Salt geometry*

The restorations suggest that salt withdrawal rather than salt injection is the dominant process affecting diapir geometry, excluding the tall diapirs in the vicinity of the Central Graben. Extension of the basement creates regional accommodation space and localized salt withdrawal cells develop in response. The withdrawal cells serve to focus fluvial sedimentation which in turn, exacerbates the withdrawal process and results in downbuilding. The resulting geometry of the salt is one of smooth, sinusoidal swells and basins, where the swells are relatively inactive through time and simply separate regions of active withdrawal.

In the Central Graben, where it is postulated that large volumes of salt have accumulated in post-Jurassic times, there is a net increase in salt and active diapirism occurs by buoyancy instability and lateral compressional stresses. Here, diapirs grow and, depending on relative density contrasts and rates of growth, can form mushroom-shaped profiles, although none are intersected by the sections presented here. In contrast, the platform diapirs are passive in that they develop due to extension and local attenuation in the early stages. The crude but perceivable layering within the Zechstein salt on the Eastern Platform (Fig. 5a) is sub-horizontal which attests to the relatively stable nature of the 'diapir' through time.

Future work will concentrate on restoring the intra-Triassic reflectors within the individual basins between the salt diapirs. This will hopefully demonstrate the detailed nature of the withdrawal process, which by examination of the onlap patterns, appears to be relatively erratic with basin depocentres migrating back and forth within the overall basin system.

## Model for the evolution of the Central North Sea

From the restorations, it is possible to propose a model for the development of the Central North Sea region highlighting and summarizing the various differences between the two margins with regards to the evolution of salt-related structures.

### *Permian–Triassic*

Figure 9a shows the Rotliegend basement to be undeformed and overlain by a maximum 1 km of Zechstein salt. Extension during the ?Late Permian through Triassic was pervasive throughout the region with widespread, relatively small scale faults inducing salt mobility in the overlying Zechstein (Fig. 9b). It appears that, in places, there was actual extensional faulting occurring in the supra-salt sequence immediately above the basement faults and can thus be considered a form of soft-linked system as it is unlikely the two faults were actually linked through the salt layer (Fig. 9c). However, physical extension along detachments in the salt layer resulted in a disturbance of the upper horizontal surface producing areas of subsidence juxtaposed above the faults. In addition, there was mass movement of the salt body towards the centre of the incipient basin producing non-fault related depressions in the upper salt surface. The two mechanisms appear to have combined in places to produce an associated roll-over-syncline pair which is clearly demonstrable in the intra-Triassic markers and the halite layers in the top salt (Fig. 4a). This structural association is also recognised on parallel sections (see Bishop *et al.* 1995) and in the southern North Sea (see Stewart & Coward 1995).

The basement driven, extensional deformation in

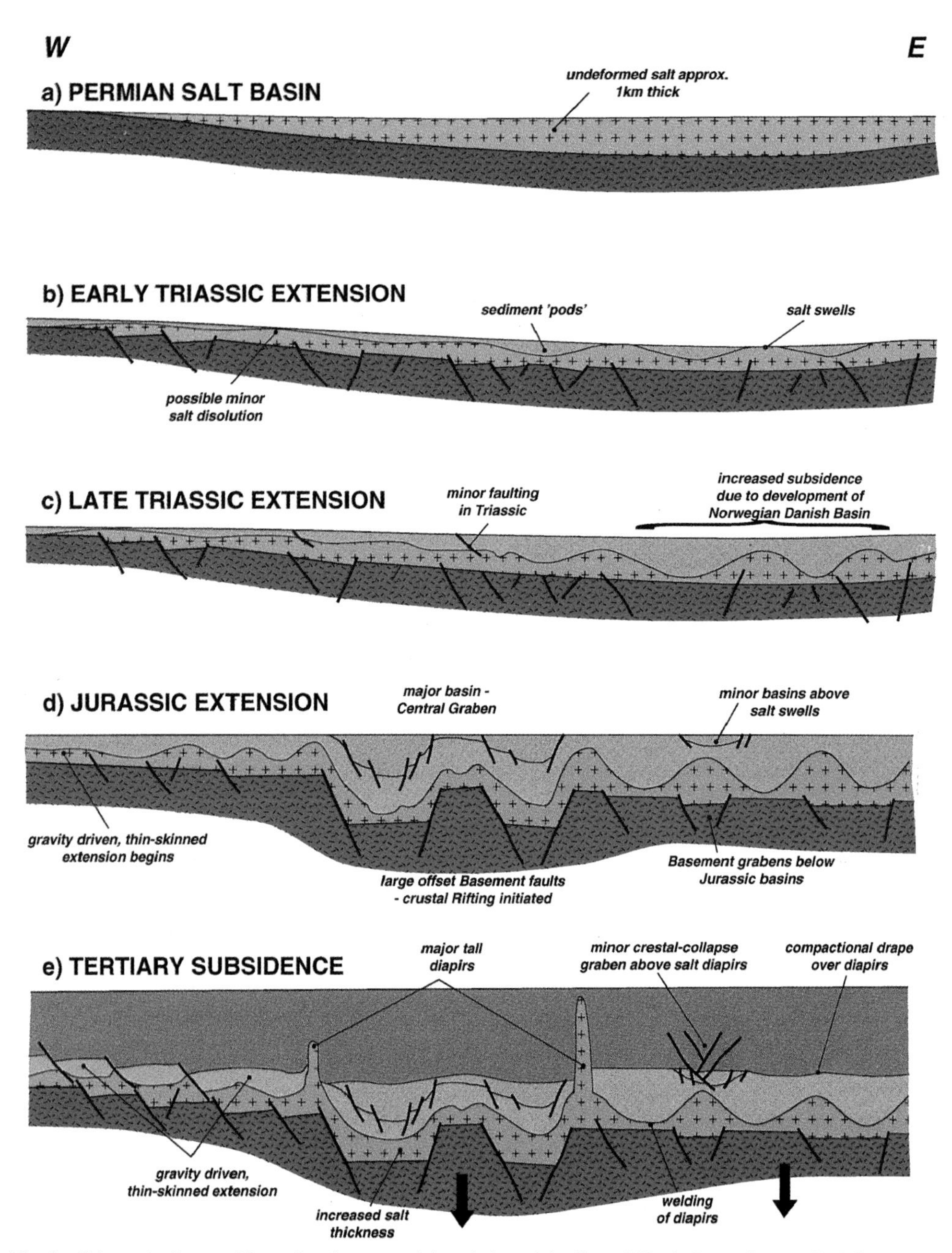


**Fig. 9.** Schematic diagram illustrating the sequential evolution of the Central North Sea region. (a) Undeformed Permian sequence. (b) Pervasive deformation across the region during the Permian-Triassic. (c) Development of basins during Triassic. (d) Development of late Triassic Norwegian–Danish Basin in the east. (e) Concentrated extension in the Central Graben during the Jurassic. (f) Passive subsidence during the Cretaceous. (g) Relative uplift in the west, significant subsidence in the east with minor amounts of extension.

the salt was enhanced by sedimentary down-building in the developing depressions. This proceeded until in places, the Triassic basins became welded to the basement. The intervening salt between the neighbouring basins is therefore considered to be at its original thickness, and the fact that these salt walls are of almost exactly the same height across both margins supports this model. It is possible that there was local dissolution of salt where some diapirs reached the surface, but the majority appear to taper upwards and terminate before being exposed at the surface. The irregular nature of the pod-like basins reflects the lateral shift in the depocentres which would be expected in a fluvial system with strongly meandering river systems.

Current investigations show complex relationships between intra-Triassic reflectors where it appears that there are several phases of sub-basin development within the main Triassic basins and complex onlapping configurations exist. After the onset of extension, where salt and basement appear to have had a degree of linkage, the salt facilitated decoupling of the cover from the basement. This is possibly the result of the localisation of sedimentation above the faults, which induced down-building and salt withdrawal at a rate greater than the accommodation space provided by extension. This is seen in the restorations by a lack of communication during later stages between salt and basement faults.

In the Late Triassic, there was an increase in subsidence on the Eastern Margin which was associated with the opening of the Norwegian–Danish Basin (Fig. 9c). As the sedimentary overburden increased in thickness, so the process of down-building continued with the result that there was more widespread welding and sealing of the basins. This would have served to reduce dramatically the continued movement of salt on the Eastern Margin by cutting off the supply.

### *Jurassic*

During the Jurassic, the southward propagation of the South Viking Graben localized extension and the Central Graben proper was initiated (Fig. 9d). The restoration of the regional line and the net salt distribution at present (Figs 3 & 6) suggest that the downward displacement of the two major keystone blocks in the Central Graben, had an important effect in facilitating the displacement of salt into the Central Graben from the neighbouring margins. This was probably the result of a combination of forces, in particular (a) sediment down-building along the margins of the Central North Sea, (b) gravity, (c) frictional drag induced by down-slope displacement of the overburden and (d) the upward evacuation of salt in the Central Graben into tall diapirs responding to buoyancy effects (e.g. Nalpas & Brun 1993; Vendeville & Jackson 1992). However, there was also major subsidence with Jurassic sedimentation in the Central Graben depocentres and associated growth faulting along the margins.

The Jurassic sedimentation was concentrated in the Central Graben region, although there were smaller areas of subsidence on the platformal areas. The restoration to top Jurassic of the Eastern Margin line shows a basin juxtaposed immediately above the salt diapir or wall (Fig. 9d). This could be interpreted in several ways; that salt has moved down into the Central Graben producing crestal collapse due to salt withdrawal, or that there has been Jurassic extension below the salt body which is reflected in the crestal subsidence, or that there has been extension of the cover where it is thinnest (i.e. above the salt body).

A measure of the area formed by the basement graben shows that it approximates to the area in the decompacted Jurassic basin, suggesting that basement extension is being directly reflected in the supra-salt sequence. A reason for this might be that the salt body is effectively isolated by lateral welding and thus, can no longer act independently or serve to mask basement deformation. Similarly, when the section was decompacted, the apparent drape over the salt body disappeared suggesting differential compaction occurred across the diapir, which behaved as a solid mass. Often, such warping of strata over deeper salt bodies is considered evidence of diapirism as the two processes will produce the same geometry. It is therefore important to decompact such sections in order to differentiate compactional folding from active diapirism.

### *Cretaceous to Present*

During the Cretaceous very little change occurred in the salt structures with widespread and uniform subsidence across the basin. Some salt withdrawal from the margins continued to feed the major active, buoyancy driven diapirs developing on the flanks of the Central Graben.

In the Tertiary, a major uplift event took place centered around Scotland producing a significant tilt of the Western Platform of up to 6° towards the east (Fig. 9e). This induced gravity driven, thin-skinned extensional tectonics remarkably similar to that seen in the present Gulf Coast (Bishop *et al.* 1995). Previous down-to-the basin extensional faults were reactivated to form large listric growth faults detaching on top salt with map patterns displaying typical configurations of *en echelon,* curvilinear fault traces (Fig. 10). The movement down slope of the supra-salt allochthonous cover caused the lateral squeezing upwards of flank diapirs in the Central Graben. This forced them to grow and expand upwards under an increased differential stress regime and in so doing, they possibly sourced their salt supply from the platformal margins. Hence, there is a significant apparent salt loss from the restored Western Platform section during Tertiary times. These tall salt diapirs are in effect contractional structures which balance the

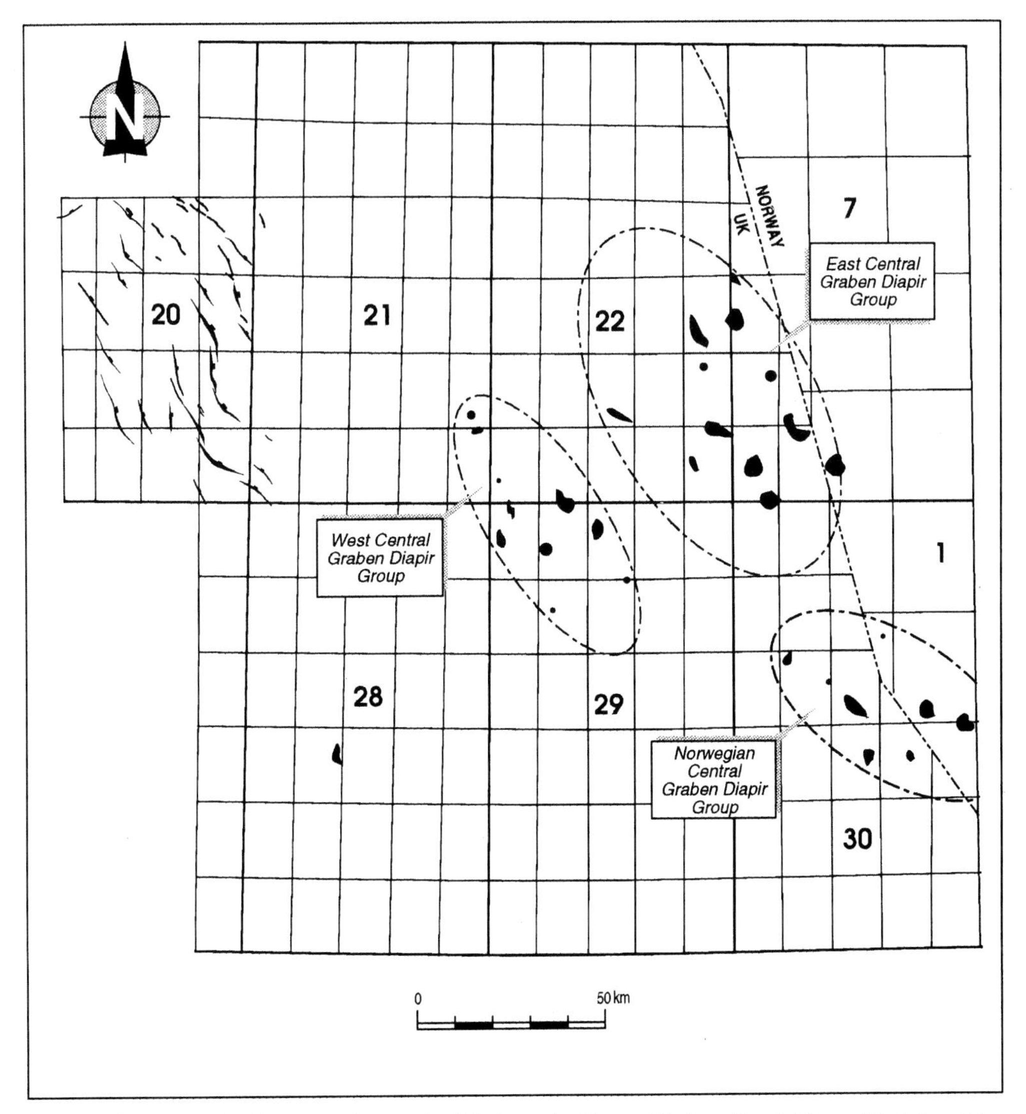


**Fig. 10.** Map pattern showing supra-salt extensional faults on the Western Platform (Quad. 20) possibly balanced by tall salt diapirs in the deeper parts of the Central Graben.

up-dip extension on the Western Platform (Bishop *et al.* 1995). The map shows their distribution is limited to the deeper parts of the basin, downslope from the large listric faults to which they are kinematically linked (Fig. 10). Similar dynamics have been described in the Southern North Sea by Stewart & Coward (1995) where extension up-dip balances contractional folds down-dip.

On the Eastern Platform during the Tertiary, there was relatively little activity with continued subsidence leading to an extremely thick Tertiary sequence. The overburden at this time in conjunction with the increased welding and lack of basement slope resulted in virtually no further salt movement as the system became mechanically locked. This resulted in relatively strong differential compaction across the salt body, with associated small-scale extensional faulting in the crestal regions of the diapirs. These extensional faults often reactivate earlier Jurassic faults related to salt withdrawal.

The restoration work was undertaken using the Geosec structural modelling software developed by CogniSeis Development. Conoco U.K. and PGS Nopec are thanked for permission to publish the western and regional seismic lines. Saga Petroleum a.s. and Esso Norge a.s. are thanked for permission to publish the seismic data and results of the modelling work on the Eastern margin section. Oil Search Ltd and Esso Exploration and Production U.K. Limited are thanked for financial support in the publication of the colour figures. Simon Skirrow is thanked for drafting the figures. Finally the authors would like to thank Simon Stewart, Ian Davison and Mike Hudec for thorough reviews of the manuscript and for making many helpful suggestions.

## References

BISHOP, D. J., BUCHANAN, P. G. & BISHOP, C. J. 1995. Gravity-driven thin-skinned extension above Zechstein Group evaporites in the Western Central North Sea: an application of computer-aided section restoration techniques. *Marine and Petroleum Geology*, **12**, 115–135.

BRODIE, J. & WHITE, N. 1995. The link between sedimentary basin inversion and igneous underplating. *In*: BUCHANAN, J. G. & BUCHANAN, P .G. (eds) *Basin Inversion*, Geological Society, London, Special Publication **88**, 21–38.

DAHLSTROM, C. D. A. 1968. Balanced cross-sections. *Canadian Journal of Earth Sciences,* **6**, 743–757.

GEISER, J., RATLIFF, R., KLIGFIELD, R. & MORRIS, A. 1991. Simultaneous decompaction and restoration: an improved method for basin modelling. *AAPG Bulletin,* **75**, 579.

GIBBS, A. D. 1983. Balanced cross-section construction from seismic sections in areas of extensional tectonics. *Journal of Structural Geology,* **5**, 153–160.

HODGSON, N. A., FARNSWORTH, J. & FRASER, A. J. 1992. Salt-related tectonics, sedimentation and hydrocarbon plays in the Central Graben, North Sea, UKCS. *In*: HARDMAN, R. F. P. (ed.) *Exploration Britain*. Geological Society, London, Special Publication, **67**, 31–63.

HOSSACK, J. R. & MCGUINNESS, D. B. 1990. Balanced sections and the development of fault and salt related structures in the Gulf of Mexico. *Geological Society of America, Abstracts and Programme*, **22**, A48.

JENYON, M. K. 1990. *Oil and Gas Traps*. Wiley, Chichester.

NALPAS, T. & BRUN, J. P. 1993. Salt flow and diapirism related to extension at crustal scale. *Tectonophysics*, **228**, 349–362.

PENGE, J., TAYLOR, B., HUCKERBY, J. A. & MUNNS, J. W. 1993. Extension and salt tectonics in the East Central Graben. *In*: PARKER, J. R. (ed.), *Petroleum Geology of Northwest Europe: Proceedings of the 4th conference.* Geological Society, London, 1197–1210.

PICKERING, G., BULL, J. & SANDERSON, D. J. 1996. Scaling of fault displacements and implications for the estimation of sub-seismic strain. *In*: BUCHANAN, P. G. & NIEUWLAND, R. (eds), *Modern Developments in Section Validation Techniques*. Geological Society, London, Special Publication, **99**, 11–26.

ROWAN, M. G. 1993. A systematic technique for the sequential restoration of salt structures. *Tectonophysics,* **228**, 331–348.

—— & KLIGFIELD, R. 1989. Cross section restoration and balancing as an aid to seismic interpretation in extensional terranes. *AAPG Bulletin,* **73**, 955–966.

SCHMOKER, J. & HALLEY, R. 1982. Carbonate porosity versus depth: a predictable relation for south Florida. *AAPG Bulletin*, **66**, 2561–2570.

SCHULZ-ELA, D. D. 1992. Restoration of cross sections to constrain deformation processes of extensional terranes. *Marine and Petroleum Geology,* **9**, 372–388.

SCLATER, J. G. & CHRISTIE, P. A. F. 1980. Continental stretching: an explanation of the post-mid-Cretaceous subsidence of the Central North Sea Basin. *Journal of Geophysical Research,* **85**, 3711–3739.

SENI, S. J. 1992. Evolution of salt structures during burial of salt sheets on the slope, Northern Gulf of Mexico. *Marine and Petroleum Geology,* **12**, 34–37.

STEWART, S. A. & COWARD, M. P. 1995. A synthesis of salt tectonics in the southern North Sea, U.K. *Marine and Petroleum Geology,* **12**, 457–476.

VENDEVILLE, B. C. & JACKSON, M. P. A. 1992. The rise of diapirs during thin-skinned extension. *Marine and Petroleum Geology,* **9**, 354–371.

——, NILSEN, K. T. & JOHANSEN, J. T. 1994. Using concepts derived from tectonic experiments to improve the structural interpretation of salt structures: the Nordkapp Basin as an example. *In*: BUCHANAN, P. G. (ed.) *Modern Developments in Structural Interpretation, Validation and Modelling* . Petroleum Group Meeting, Geological Society, London, 21–23 February (abstracts).

# Salt tectonism in the Persian Gulf Basin

H. S. EDGELL

*8 Barkly Crescent, Canberra, ACT2603, Australia*

**Abstract:** The Persian Gulf Basin is an elongate, margin sag-interior sag, sedimentary basin spanning the last 650 Ma along the northeastern subducting margin of the Arabian Plate and is the largest basin with active salt tectonism in the world. This basin is asymmetrical in NE–SW cross section with sediments thickening from 4,500 m near the Arabian Shield to 18,000 m beside the Main Zagros Reverse Fault. Halokinesis in the Persian Gulf Basin originates where major intersecting basement faults cut the buoyant salt beds of the Neoproterozoic Hormuz Series (1.5 to 2.5 km thick) and the equivalent Ara Formation of Oman where salt thickness reaches 4 km. Another source of salt is the Oligo-Miocene Gachsaran Formation, forming Lali salt plug. Major basement fault trends control salt tectonism in the basin, namely the extensional N–S Arabian Trend, the sinistral NE–SW Aulitic Trend, the dextral Erythaean Trend and the extensional E–W Tethyan Trend. The main types of salt structures in the Persian Gulf Basin are salt domes, salt pillows, salt walls, salt piercements, rim anticlines, turtleback structures, disharmonic folds, orogenic fold fillings and dissolution drapes. Diapiric oil fields, account for 60% of the 600 billion barrels of recoverable oil reserves of the Persian Gulf Basin and have grown continuously since Late Jurassic, or since Permian in some large structures, such as Bahrain.

The Persian Gulf Basin is an elongate margin-sag–interior sag basin developed during the last 650 Ma along the northeastern subducting margin of the Arabian Plate. This sedimentary basin is some 2600 km long (i.e. from northern Syria to easternmost Oman) and is from 900 to 1800 km wide, forming the richest petroleum province in the world (Fig. 1).

In NE–SW cross-section, the basin is asymmetrical (Figs 2 and 3), with sediments thickening from only a few hundred metres just northeast of the Arabian Shield to 18000 m immediately south of the Main Zagros Reverse Fault. The sedimentary sequence in the Persian Gulf Basin extends from Upper Proterozoic to Recent with only an occasional hiatus, although Recent sediments are only being deposited in the Persian Gulf itself,

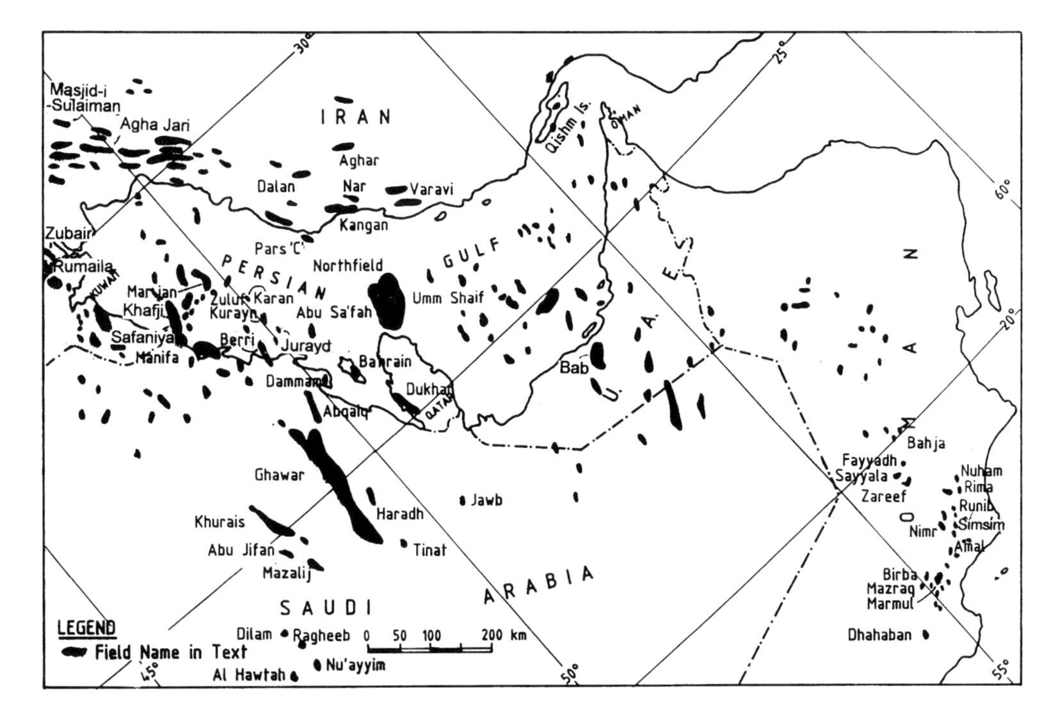


**Fig. 1.** The main salt diapiric areas of the Persian Gulf Basin with principal oil and gas fields.

*From* Alsop, G. I., Blundell, D. J. & Davison, I. (eds), 1996, *Salt Tectonics*, Geological Society Special Publication No. 100, pp. 129–151.

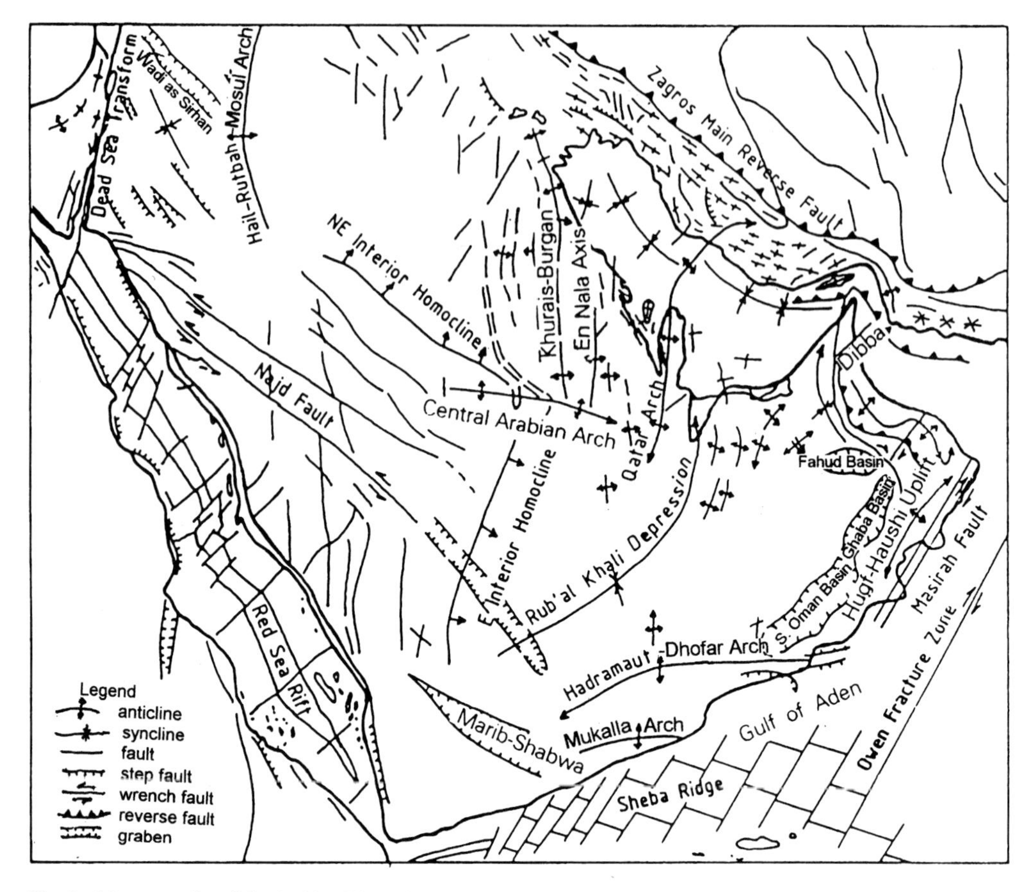


**Fig. 2.** Megatectonics of the Arabian Plate, showing major structural trends in the Persian Gulf Basin.

which is the present-day basinal remnant of the much more extensive Persian Gulf Basin, and after which the basin is conventionally named. Following Late Proterozoic extensional rifting of the Arabian Plate along the N–S Arabian Trend, dextral faulting occurred along the line of the present-day Main Zagros Reverse Fault. This led to the development of a number of evaporite basins in the latest Proterozoic (Neoproterozoic), extending through southern Iraq, southwestern Iran, Kuwait, northeastern Saudi Arabia and all the Persion Gulf, except for the Qatar Arch. In addition, Neoproterozoic evaporite basins were developed in southern and southeastern Iran, the United Arab Emirates, (UAE) parts of northeastern Saudi Arabia and Oman (Figs 4 & 5). Thick evaporite beds were deposited throughout most of the Persian Gulf Basin forming the Neoproterozoic Hormuz Series, which is up to 2.5 km thick, and the equivalent Ara Formation of Oman, where thickness of salt reaches 4 km. In northeastern Saudi Arabia, the deposition of salt took place in the latest Proterozoic between uplifted, N–S trending, basement horsts (Fig. 4) and, in some cases, underlies oil fields as much as 150 km inland, as at Jawb Field. Salt in the evaporitic Oligo-Miocene Lower Fars Series, or Gachsaran Formation, also formed the Lali Ambal salt plug and important gravity glide and disharmonic structures in Iraq and southwestern Iran. However, almost all the major salt diapirism in the Persian Gulf Basin originates from the very thick and deep evaporites of the Neoporoterozoic Hormuz Series and the equivalent Ara Formation of Oman.

Diapiric movement of this deep-seated Neoproterozoic salt has been initiated by repeated extensional and strike-slip movements in the crystalline Precambrian basement complex. This diapirism usually occurs where intersecting basement fault trends cut part of the overlying sedimentary sequence, providing localized areas of pressure relief for the upward movement of buoyant salt. These salt diapirs either reach the surface as piercements and salt fissures in areas of active tectonism, or dome up the overlying strata. Four

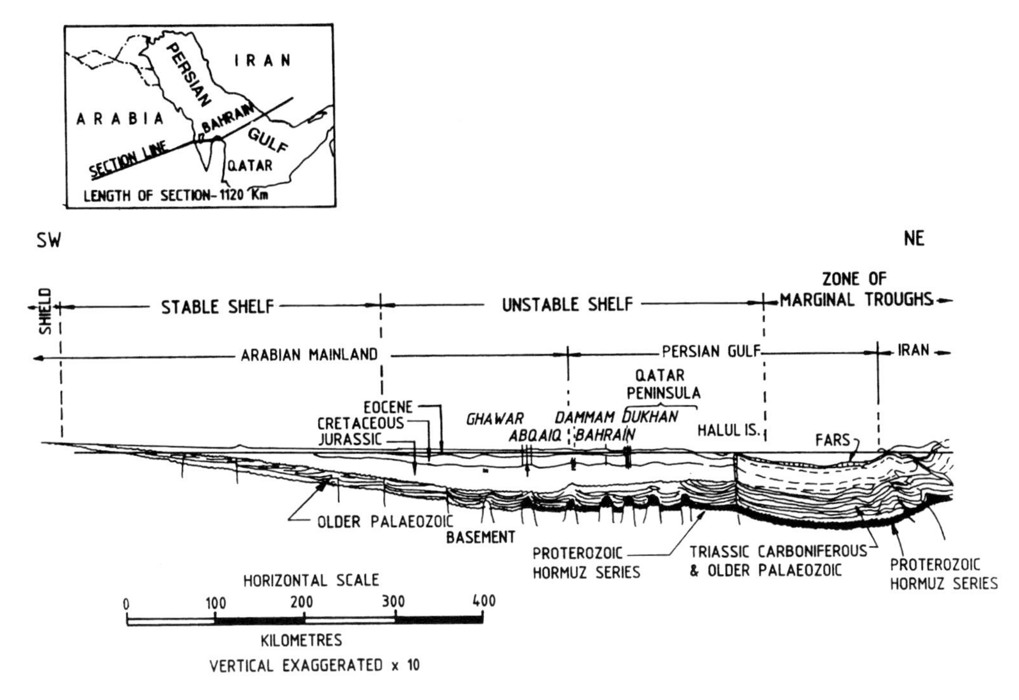


**Fig. 3.** Geological cross-section of the Persian Gulf Basin showing the Neoproterozoic Hormuz Series salt (black) and basement block faulting.

major fault trends control salt tectonism in the Persian Gulf, namely the N–S Arabian Trend (extensional), the NE–SW Aulitic Trend (left-lateral strike-slip), the NW-SE Erythraean Trend (right-lateral strike-slip, and the E–W Tethyan Trend (extensional) (Wissman *et al.* 1942). Many diapirs in southern Iran are aligned along the Arabian Trend as in Fars Province and the southern Persian Gulf (Edgell 1973; Ala 1974; Talbot & Alavi 1994; Koyi 1988) but the majority have a NE–SW fault alignment (Kent 1979; Fürst 1970, 1976).

## Megastructure of the Persian Gulf Basin

The megastructure of the Persian Gulf Basin is controlled by the edge of the Arabian Shield on the southeastern side and by the South Anatolian Fault and the Main Zagros Reverse Fault on the northern and northeastern side. It forms an elongate, asymmetrical sedimentary basin, whose present axis is sigmoidal (Schroeder 1944), being convex northward against the Taurus Ranges in Syria and northern Iraq and concave northward against the Zagros Ranges, with its axis close to the Iranian coast of the Persian Gulf. On its western end, the basin is terminated by the Dead Sea Transform Fault, and its eastern end is determined by the Dibba and Masirah Transform Faults and the Gulf of Aden Rift (Fig. 2).

Basically, the Persian Gulf Basin has developed along the northern edge of the Arabian Plate where it is subducting beneath the Turkish and Iranian plates. Its development has taken place since Late Proterozoic when a number of evaporitic basins developed (Fig. 5), interfingering between the N–S Arabian Trend extensional uplifts, and bordered to the northeast by the line of Main Zagros Reverse Fault. The Zagros Fault Line seems to have acted as an ancient uplifted border to the Neoproterozoic Hormuz Salt Series evaporitic basin (Stöcklin 1968*b*), or at least as a major, right-lateral strike-slip fault line (Husseini & Husseini 1990). From their inception, about 650 Ma BP, these Neoproterozoic evaporite basins have been followed by almost continuous Phanerozoic sedimentation. Recent sedimentation is still taking place throughout the more than 900 km long Persian Gulf, which is the present-day geosynclinal expression of the 2600 km long Persian Gulf Basin.

The strongly asymmetrical nature of the Persian Gulf Basin is due to its history as a margin sag–interior sag basin (Kingston *et al.* 1983), influenced by maximum subsidence close to the Iranian plate, so that sediment thicknesses range from 18000 m just south of the Main Zagros Reverse Fault (Morris 1977) to just a few hundred metres near the crystalline Arabian Shield. The Persian Gulf Basin is basically a northeasterly thickening

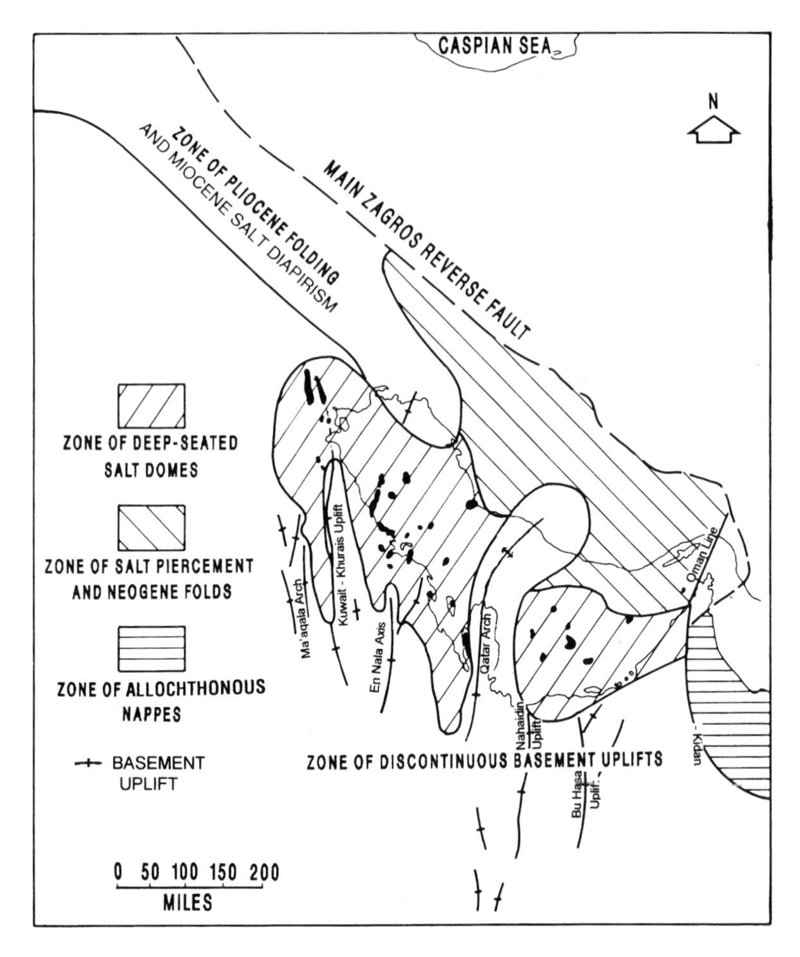


**Fig. 4.** Tectonostratigraphic provinces of the Persian Gulf area.

sedimentary wedge (Fig. 3) or a one-sided compressional basin with youngest reservoirs towards the deformed belt (North 1985). This basin has been periodically affected by extensional block faulting along the N–S 'old grain of Arabia' and the E–W Tethyan Trend (Henson 1951), by episodic wrench faulting along NE–SW and NW–SE trends, as well as by compressional faulting and folding, particularly throughout southern and southwestern Iran and adjacent areas of Iraq. It is, basically, rejuvenated jostling of basement blocks produced by these four trends and compressional folding in Iraq and Iran that has produced the major oil field structures of the Persian Gulf Basin and, especially, the salt diapirs which are so important in updoming petroliferous structures.

The Persian Gulf Basin consists of a number of NW–SE trending, geotectonic units (Henson 1951), such as the Interior Homocline, Arabian Platform (divided into a Stable Shelf and an outer Unstable Shelf), and a Zone of Marginal Troughs, including the Zagros Fold Belt, limited on the northeast by the Main Zagros Reverse Fault (Fig. 3). Lesser, NW–SE geotectonic units include the graben-like Wadi as Sirhan Depression and the Marib–Shabwa Basin (Fig. 2). There are also many marked N–S, extensional, Arabian Trend structures crossing the Persian Gulf Basin (Mina *et al.* 1967), not just one major N–S Qatar–Kazerun lineament as indicated by Talbot and Alavi (1994). They are the Hail–Rutbah–Mosul Arch, Khufaisah–Mubayhis Uplift, Jaham–Ma'aqala–Wariah Uplift, Rumaila–Minagish Uplift, Burgan–Wafraa–Juraybi'at–Khurais Uplift, En Nala Axis Uplift (Aramco 1959), Qatar Arch, Matti Uplift, Nahaidin Anticlinal Uplift, Mushash Anticlinal Uplift, Bu Hasa–Kidan Anticlinal Uplift, Ras Musandam Uplift, Butabul–Zauliyat Rise, and the Hugf–Haushi Uplift. These extensional uplifts are major elements of the megatectonic framework of the Persian Gulf Basin and

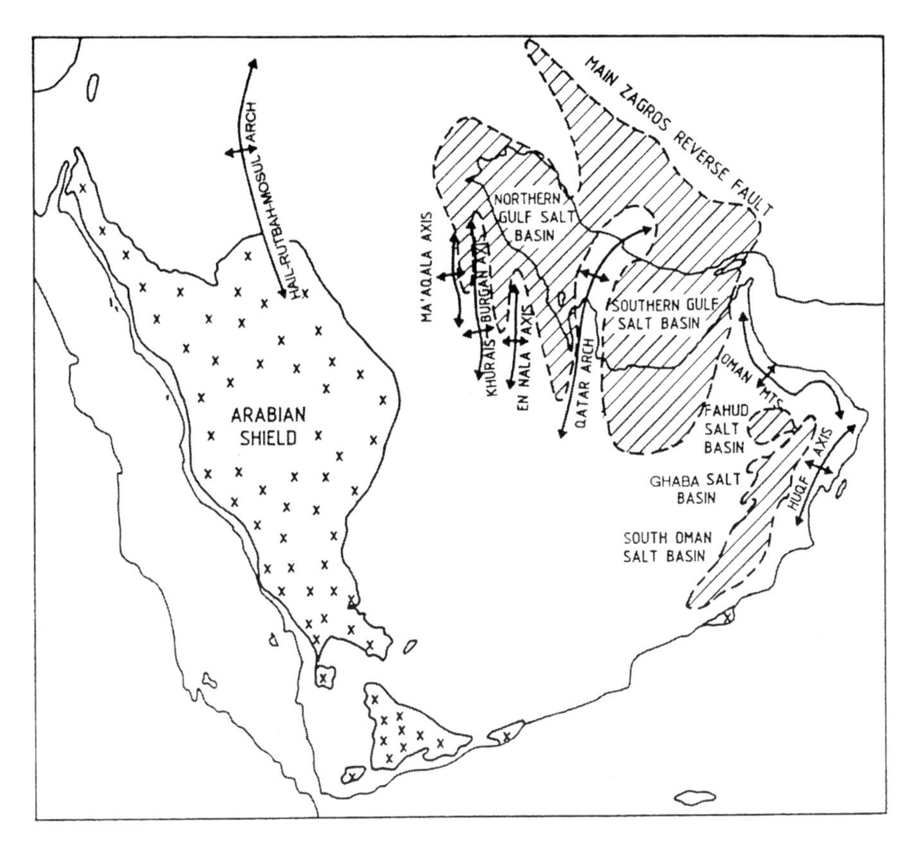


**Fig. 5.** Neoproterozoic salt basins of the Persian Gulf Basin and major basement uplifts.

extend many hundreds of kilometres: 350 km in the case of the En Nala Axis, 500 km for the Burgan–Khurais Uplift (really extending from Mazalij to Raudhatain), nearly 500 km for the Qatar Arch, while the Hail–Rutbah–Mosul Arch is some 800 km long (Fig. 2) The geotectonic influence of these uplifts and their intervening depressions is seen not only on the Arabian side of the Persian Gulf Basin and in the Persian Gulf itself, but also on the Iranian side. This is seen in the Central Basin (between the Burgan–Khurais and En Nala Uplifts), which extends onto the Iranian side as the Dizful Embayment. Similarly, the Qatar Arch extends into southern Iran, where its west-bordering associated Kazerun–Mangarak Fault Zone forms a N–S lineament some 450 km long. The influence of the N–S Ras Masandam Uplift is also seen to deflect the Zagros Ranges and embay the Iranian coast in the Strait of Hormuz in association with the northerly trending Oman Fault Line.

These repeatedly rejuvenated uplifts have existed for 650 Ma and the Qatar Arch nearly divides the Neoproterozoic Northern Gulf Salt Basin from the Southern Gulf Salt Basin (Figs 4 and 5). The coeval South Oman-Ghaba Salt Basin is similarly bordered by uplifts and the Fahud Salt Basin is now separated by a carbonate platform from the nearby South Oman-Ghaba Salt Basin (Murris 1980). These four Neoproterozoic salt basins (Fig. 5) are of immense significance to salt diapirism in the Persian Gulf Basin and were probably interconnected, together with the Salt Range of Pakistan, before plate displacement (Gorin *et al.* 1982), while Zharkov (1984) refers to them as the Iran–Pakistan Salt Basin.

As the Arabian Plate has subducted (Jackson *et al.* 1981), the axis of deposition has progressively migrated southwestward from just south of the Main Zagros Reverse Fault in Late Cretaceous time to the middle of the Zagros Ranges in Eocene time and to its Recent axial position just off the Iranian coast. Similarly, the Qatar Arch has moved southwestward from a position in the middle of the present-day Persian Gulf in the Cretaceous, to just north of the Qatar Peninsula in the Eocene, and to its present structural high about 50 km northwest of Doha.

Major NE–SW trending elements in the megatectonic framework of the Persian Gulf Basin (Fig. 2) include the Hadramaut–Dhofar Arch, the South

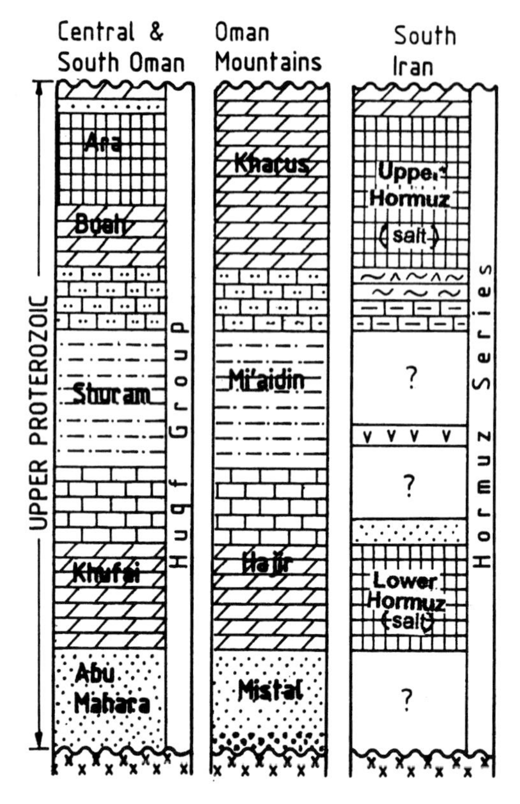


**Fig. 6.** Comparative stratigraphy of Neoproterozoic strata in the Persian Gulf Basin.

Oman-Ghaba Salt Basins (probably connected), the Rub'al Khali Depression and the Palmyra Fold Belt. The conspicuous alignment of salt piercements in southeastern Iran (Fürst 1970, 1976; Kent 1970, 1979; NIOC 1978) is due to reactivated sinistral basement faulting, which is of megatectonic significance, since this faulting appears to have successively deflected the general NW–SE trend of the Zagros Ranges to E–W in Laristan and even NE–SW to the north of Bandar Abbas, as seen in Kushk Kuh (de Böckh *et al.* 1929, plate 5; British Petroleum 1956; 1964; Heim 1958, fig. 4).

Megastructures along E–W trends are relatively few, but include the Central Arabian Arch and associated Central Arabian Graben System, as well as the Taurus Fold Belt of southern Turkey.

Another megatectonic element of the Persian Gulf Basin is the Oman Mountains ophiolite-radiolarite complex, emplaced as NW–SE trending, obducted allochthonous nappes (Fig. 4) in Late Cretaceous time (Glennie *et al.* 1974).

## Salt-related stratigraphy of the Persian Gulf Basin

The principal salt stratigraphy of the Persian Gulf Basin (Fig. 6) exists just above the Precambrian crystalline basement complex at the base of the sedimentary sequence. It consists of the Hormuz Series (Pilgrim 1908), which is composed of massive salt, anhydrite, foetid limestone, dark dolomite and some red sandstone and shale (Kent 1970), being too thick and lithologically diverse to be considered a formation as originally named by Blanford (1872). In thickness, the Hormuz Series is more than 1000 m in the Zagros Range, and 525 m to an estimated 2285 m beneath certain anticlines in South Iran, the latter figure estimated from salt infilling anticlines by Player (1969). It is up to 2500 m thick in parts of the Persian Gulf (Edgell 1989, 1991). The age of the Hormuz Series is considered to be Late Proterozoic, since it is found beneath thick Lower Cambrian red sandstones of the Lalun Formation. Although generally strongly disturbed, some sequence can be recognized in the Hormuz Series. Schroeder (1946) and Huber (1977) have recognized a Lower Hormuz, mainly evaporitic deposit, consisting of 'white and light red, coarse crystalline rock salt with, probably condesary, flow-induced banding of impure salt and anhydrite with crushed dolomite, basic igneous rocks and red siltstone'. Huber (1977) notes that brine analysed from associated salt springs shows the presence of a certain amount of potash salts and sulphates. The Upper Hormuz crops out on the rim of the Hormuz Island salt piercement (Wolf 1959). According to Huber (1977) it consists there of evaporites alternating wth fetid, gray, thin-banded dolomite, rhyolites, diabase flows, purple-gray tuffaceous sandstones and mudstones. Some 200 m of non-saline rocks, mostly red marl (Kent 1958) are exposed in the Jehani salt piercement of Fars Province (Huber 1977). An equivalent evaporitic sequence (Ara Formation), with massive halite beds, minor anhydrite and sandstone and some stromatolitic dolomite stringers, occurs in the salt basins of south Oman, where it is 1775 m in type thickness (Hughes Clarke 1988) and up to 4000 m thick in parts of the South Oman and Ghaba salt basins. The Ara Formation is laterally equivalent to the Hormuz Series and was probably its genetic continuation before major strike-slip faulting reshaped the western margin of the South Oman and Ghaba Salt Basins (Visser 1991). In age, the Ara Formation is dated as between 570 and 620 Ma (Hughes-Clarke 1988), which also supports the Infracambrian (Riphaean) age of the more extensive Hormuz Series. Fossils from the Ara Formation (Mattes & Conway Morris 1990) include typical Neoproterozoic (Late Riphaean to Vendian) acritarchs, such as *Protosphaerium* spp. and *Stictosphaeridium* spp., as well as the cyanobacterial genus *Angulocellaria* Vologdin and the primitive shelly fossil *Cloudina*, seen elsewhere in the latest Precambrian (Ediacaran). The Ara Formation is the uppermost formational unit of five formations, which comprise the thick Upper Proterozoic Hugf Group in southern and central Oman (Fig. 6). This group is up to 7000 m thick in the Ghaba Salt Basin (Visser 1991), consisting of basal sands and shales followed by interbedded carbonates and sandstones, capped by the evaporites of the Ara Formation. The significance of the Neoproterozoic Hugf Group as a source rock (Fig. 12), has been shown by Grantham *et al* (1987), who demonstrated that oil in all reservoir rocks of Oman shows the chemical fingerprint of Proterozoic sourced oil.

The overlying Palaeozoic strata of the Persian Gulf Basin are predominatly clastic, except for Devonian and Permian carbonates, but contain no salt deposits

Salt layers appear in the Middle Triassic Jilh Formation of northeastern Saudi Arabia but are of no significance for salt tectonism. However, Upper Jurassic, Riyadh Group evaporites, including anhydrite and salt, are especially important in the oil-bearing structures of Saudi Arabia where they provide the caprock, such as the Hith Anhydrite cap to Arab Formation and separate anhydrite caprocks to the Arab B, Arab C and, especially, Arab D carbonate reservoirs (Steinecke *et al.* 1958).

In Kuwait and its offshore, most of the Riyadh Group anhydrites and limestones are replaced by rock salt having a cumulative thickness of 200 m (Owen & Nasr 1958; Stöcklin 1968a). These rock salt layers are seen in the Greater Burgan Field (Brennan 1990), but do not appear to have played any significant role in salt diapirism in Kuwait, which is primarily caused by the movement of deep-seated Neoproterozoic salt. Upper Jurassic salt, (unrelated to the evaporites of Saudi Arabia), plays a major role in the formation of oil fields in the Marib–Shabwa Basin of Yemen. This is due to the Sab'atayn Formation,which causes salt piercements in the Shabwa area (i.e. the Shabwa, Milh Khirwa, 'Ayad and Milh M'qah salt plugs), the salt plugs of Beihan, as well as the Safir salt plug in the Ramlat Sab'atayn and three salt plugs in the Mintaq area (Beydoun 1964; Paul 1990). Some 280 m of saliferous deposits are exposed in the 'Ayad salt piercement. Salt diapirism emanating from Upper Jurassic salt has domed up the Shabwa and Alif oil fields.

The Cretaceous System in the Persian Gulf Basin has no significant salt deposits and Lower Tertiary salt is only found as thin interbeds within the anhydrites of the Lower Eocene Rus Formation. An Oligocene Asmari salt basin occurs in the eastern Persian Gulf between Bandar-e Lengeh and Dubai, known as the Asmari–Chel Salt Basin (Huber 1977), although without known surface diapirism.

Upper Tertiary evaporites of the Lower Fars Series, or Gachsaran Formation, occur in southwestern Iran and adjacent northeastern Iraq, as well as in northern Syria. These evaporites contain up to several hundred metres of salt, associated with anhydrite, grey marl and thin limestones. They form part of a NW–SE trending salt basin, some 1200 km long and averaging 150 km in width (Lees 1931; Stöcklin 1961). A small salt piercement formed by this Neogene salt is the Lali Ambal salt plug in southwestern Iran. Salt diapirism in the Lower Fars is mostly due to flow of this mobile series over competent Asmari Limestone anticlines and below the competent Middle Fars, or Mishan Formation. Both of these competent formations act as control units and buckle disharmonically since they are separated by incompetent material (salt and anhydrite) of the Lower Fars, or Gachsaran Formation (Price & Cosgrove 1990). This leads to frontal thrusts and folds, as well as frequent disharmonic folding (Figs. 13 and 14). The mechanisms of this diapirism are attributed to rotation of intervening synclines (Dunnington 1968), to salt bulges (O'Brien 1957), to decollement (Falcon 1969), and to disharmonic buckle folding (Price & Cosgrove 1990). Apart from the importance of these saliferous evaporites as cap-rocks in Iran and Iraq, they form a very distinctive type of diapirism along many of the Zagros anticlines, due to the mobility of the Lower Fars, which is squeezed out over the southwestern fronts of Asmari anticlines by gravity and compression

The entire sedimentary sequence in southwestern Iran and Iraq has been divided into five tectonostratigraphic groups (O'Brien 1957). In descending sequence these are:

(1) The Passive Group – comprising the thick Neogene Bakhtiari Conglomerate, Agha Jari Formation sandstones, Mishan Formation marls and limestones and the uppermost anhydrites and thin limestones of the Lower Fars Series, or Gachsaran Formation. This is sometimes referred to as the Semi-Competent Group (Ion *et al.* 1951), and has also been renamed the Incompetent Group (Colman-Sadd 1978; Koyi 1988).
(2) The Upper Mobile Group – consisting of the thick salt, anhydrite and marls of the Lower Fars Series, or Gachsaran Formation.
(3) The Competent Group – including all strata from the top of the Oligo-Miocene Asmari Limestone to the base of the Lower Cambrian Lalun Sandstone.
(4) the Lower Mobile Group composed of the thick salt and evaporites of the Neoproterozoic Hormuz Series.

(5) The Basement Group – consisting of underlying Precambrian crystalline rocks.

This tectonostratigraphic subdivision of the sedimentary sequence applies in whole, or in part, to most of the Persian Gulf Basin and is useful in emphasizing the salt-related tectonism of the Upper and Lower Mobile groups.

## Halokinesis in the Persian Gulf Basin

The main factors to be considered in halokinesis in the Persian Gulf Basin are origin of salt tectonism, mechanisms of salt movement, rate of growth of salt structures, and the regularity of spacing of salt structures.

### *Origin of salt tectonism*

Precipitation of thick salt beds under extremely arid conditions is an obvious prerequisite before salt tectonism can begin, and salt layers less than 500 metres thick have little chance of forming diapirs even under thick overburden (Price & Cosgrove 1990). The very thick Neoproterozoic Ara Formation salt of central and southern Oman also contains thin interbeds of aeolian sands, which are clear evidence of arid conditions. Halite precipitation requires restricted basin conditions with intermittent inflow of seawater, little or no sediment influx, and evaporation to about one-tenth of the volume of the original restricted marine waters. The fact that thick intervals of potassium salts are interbedded in the massive halite deposits of the Ara Formation are a further indication that the restricted basin had reached the final phases of evaporation (Mattes & Conway Morris 1990).

Continued basinal subsidence is also required for salt to accumulate. The NW–SE trending Hormuz Series salt is considered to be synrift (Husseini & Husseini 1990), while the NE–SW trending South Oman Salt Basin is a rift, bounded by extensional faults (Mattes & Conway Morris 1990). Much of the Hormuz Salt Series in coastal areas of Saudi Arabia has been shown to have formed in N–S trending extensional rifts in the older cystalline Precambrian basement (Edgell 1987, 1991), but the main NW–SE control is by early dextral faulting along the Zagros Line, as has been suggested by Stöcklin (1968*a*) and Husseini (1988), with the Zagros Fault Line forming a northeastern border to the Hormuz Salt Series. Exceptions are the evaporites of the Proterozoic Desu Formation near Kerman (Huckreide *et al.* 1962) and the salt of the Ravar Formation in eastern Iran (Stöcklin 1961).

In addition, a sufficiently thick salt sequence needs to accumulate to allow for salt tectonism, possibly of the order of 100 m, as well as an adequate overburden to pressure the underlying, more buoyant salt (density 2.16 $g/cm^{-3}$) to rise into overlying sediments (density 2.4–2.8 $g/cm^{-3}$). The Persian Gulf Basin provides good examples, where salt beds of the Middle Triassic Jilh Formation appear to have been insufficiently thick and continuous to have caused salt diapirism. However, the thick salt deposits of the Hormuz Series, and its equivalent Ara Formation, although in excess of 1000 m thick, on geological evidence did not undergo salt diapirism until Permian times, or in the Late Jurassic (Player 1969), by which time from 2500 to 3000 m of Palaeozoic sedimentary strata had accumulated as overburden.

Similarly the Neogene Lower Fars Series (Gachsaran Formation), where salt is estimated at 1200–1830 m thick, has had a 4000 m thick overburden of clastic Pliocene to Miocene Bakhtiari, Agha Jari and Mishun Formations before diapirism.

### *Mechanisms of halokinesis in the Persian Gulf Basin.*

Repeated extensional and wrench faulting in the basement has cut the overlying Neoproterozoic salt sequence of the Persian Gulf Basin and has provided lines of weakness along which the salt has slowly escaped upward, generally doming up overlying strata. The significant role of extensional faulting in triggering salt diapirism has been stressed by Jenyon (1985), Price & Cosgrove (1990) and Jackson & Vendeville (1994), and earlier, under the term basement block folding, by Henson (1951). Seismic profiles over oil field anticlines in coastal Saudi Arabia show the increasing presence with depth of reactivated faults in the deep subsurface. This is considered to be a major factor in providing pathways for the upward movement of Neoproterozoic salt. It also explains the remarkable increase in structural closure with depth, characteristic of both continuous and discontinuous growth anticlines and domes, typically found in the Persian Gulf Basin. In many cases, active intersecting basement faults have allowed the bouyant salt to escape to the surface as salt piercements (Fürst 1970, 1976; Edgell 1987, 1991, 1992*a*). It is probable that intersecting basement wrench and Zagros trend faults have created local pull-aparts, allowing for rapid salt piercement, as along the Kazerun–Mangarak Fault (Talbot & Alavi 1994). The presence of enclosed pieces of Precambrian granites and schists in the salt piercements of southern Iran indicates the plucking of these rocks from basement fault scarp faces (Kent 1970), or from basement fault wedges (Huber 1977) which act to trigger diapiric intrusion.

Where basement faults intersect, as is shown by the second derivative of the potential field of gravity in Dammam Dome and Abu Sa'fah Field (Edgell

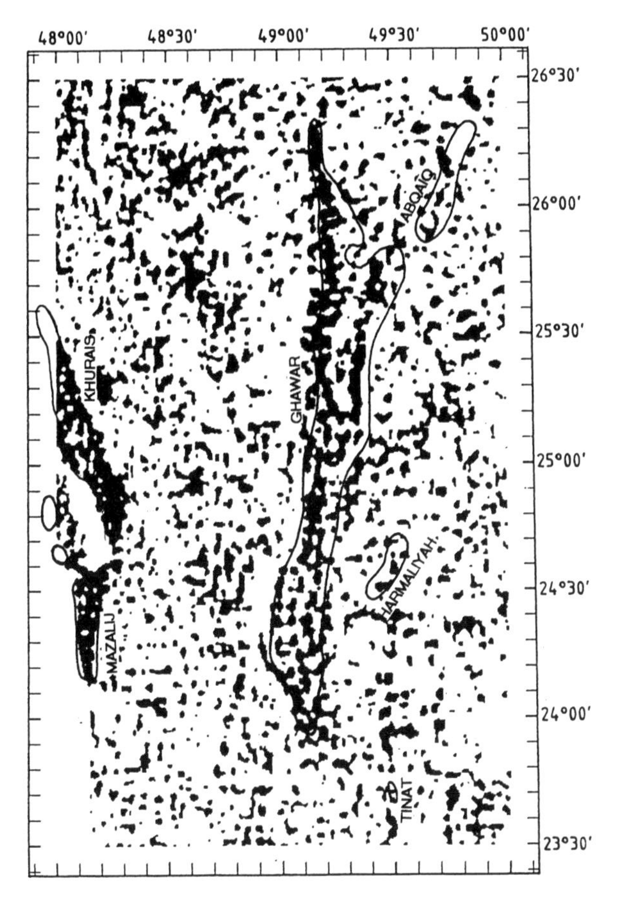


**Fig. 7.** Gravity map of onshore Saudi Arabian oil fields showing Arabian Trend basement uplifts. Salt tectonic oil field structures in Kuwait, southern Iraq, coastal Saudi Arabia, Qatar, UAE and the Persian Gulf shown an opposite negative anomaly but are still dominantly along the Arabian Trend (digitized running difference gravity modified from Barnes (1987); solid lines outline oil fields).

1992*a*), and inferred for Greater Burgan Field (Stöcklin 1961; Murris 1980), the resulting salt dome structures are subcircular. Gravity maps, especially using running difference gravity, show the dominant role of Arabian Trend basement uplifts and associated faults (Fig. 7). They differentiate between the Zone of Discontinuous Basement Uplifts with positive gravity anomalies, as seen in Ghawar Field (Barnes 1987), and the Zone of Deep-seated Salt Domes in most of the Persian Gulf (Fig. 4), where distinct negative gravity anomalies are found over major oil and gas field structures, although these are still mainly aligned along the Arabian Trend. Some examples are Dukhan, Abu Safah, Karan, Rumaila, Zubair, Abu Hadriyah and Khursaniyah fields with negative anomalies of 2 mgal or more, and Dammam Dome with a negative anomaly of some 5 mgal, while the Bahrain anticline has an even larger negative gravity anomaly.

In cases where one main basement fault has allowed the upward diapiric flow of salt, typical salt wall structures have developed, as is seen in the Dukhan anticline. This major oil field is a salt tectonic structure, as shown by a strong gravity low on its larger northern end, by multiple culminations with time, by its position on the easternmost updip of the Northern Gulf Neoproterozoic Salt Basin, and by its continuous growth, (QGPC & AQPC 1991). A characteristic feature of oil and gas field structures cored by deep-seated salt in the Persian Gulf Basin is their continuous growth (Fig. 8).

The mobility of salt when buried at depth under a normal geothermal gradient has been studied by Gussow (1966), who showed that salt buried at a depth of over 7000 metres can have a temperature of 270°C, when it behaves as a true plastic and flows indefinitely. In addition, the rheological behaviour of rock salt is greatly weakened at shallower depths in the presence of even small amounts

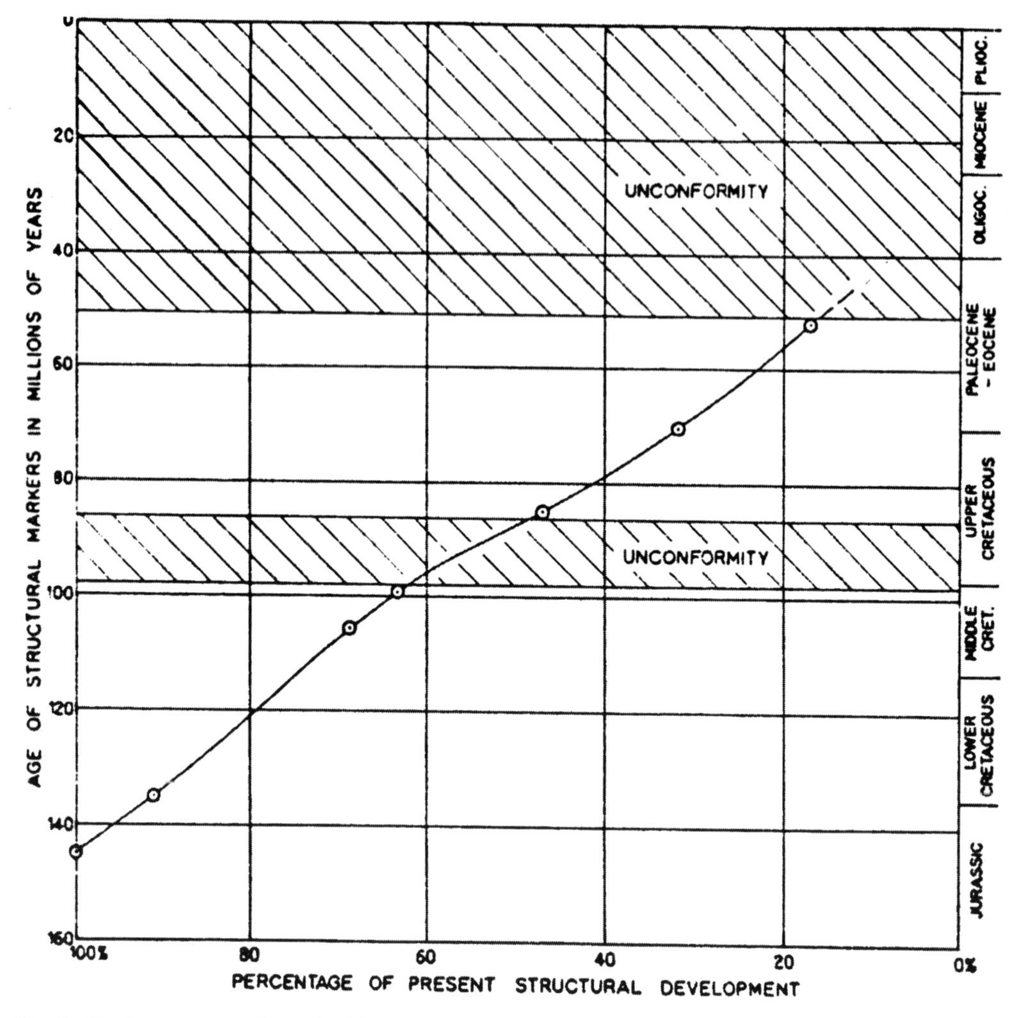


**Fig. 8.** Continuous structural growth with time, starting from Late Jurassic in Dukhan anticline (Sugden 1962); typical of Persian Gulf Basin salt-cored oil and gas fields.

of free water (brine), usually 0.1–1.0 wt% in natural rock salt, as shown by Urai *et al.* (1986) and Spiers (1994). Degree of salt mobility is influenced by four factors: (1) heat of burial; (2) salt thickness and purity; (3) water present with salt due to dehydration of associated gypsum; (4) presence of free water (brine) in polycrystalline halite, which greatly increases the rate of creep (Jenyon 1986). The Hormuz Series salt has an overburden of at least 12000 m in most of the Persian Gulf and adjacent areas. It can be expected to have a temperature of well above 270°C, and consequently to have great fluidity as compared to overlying clastics and carbonates. Free water is unlikely at this heat of burial.

Apart from the mobility and buoyancy of salt as compared with surrounding strata (Carey 1962), some probable mechanisms for the rise of deep-seated salt are pathways provided by: intersecting rejuvenated basement faults; pull-aparts due to wrench fault deflections; compressional wrench fault deflections; reactivation of extensional grabens with thicker salt fills; transpression of salt-filled grabens; reactivated normal dip-slip faults in the basement; instability of thicker salt at the foot of tilt blocks; and drape-over positive basement blocks, which have been repeatedly uplifted.

Salt piercements form where frequently rejuvenated major basement faults intersect (i.e. conjugate basement wrench faults of the rhegmatic shear pattern of the underlying crystalline Arabian Shield). They are seen especially, where predominant wrench faulting occurs in areas which are still seismically active, as in southeastern Iran, north of

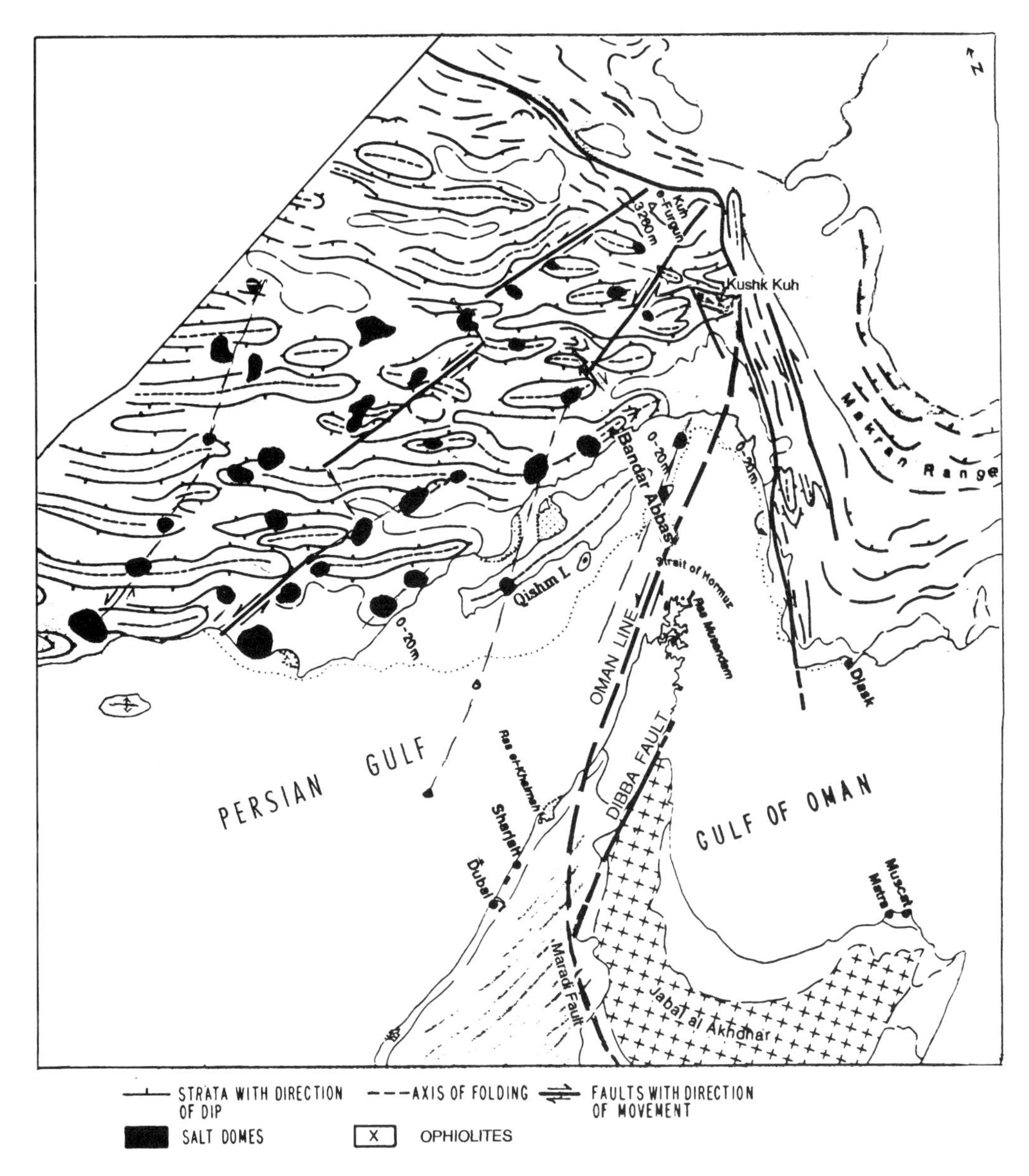


**Fig. 9.** Salt piercements from the Neoproterozoic Hormuz Series in southeastern Iran and the eastern Persian Gulf. Note their relation to deep, sinistral basement wrench faults, particularly where these intersect Zagros trend axes and faults.

Bandar Abbas, and in much of the eastern Persian Gulf and Strait of Hormuz (Fig. 9). The mechanism of relatively rapid salt piercement is repeated fault release of the lower specific gravity salt weakened by associated brine, so that it often extends as mountains of mobile salt from which salt glaciers emanate, as noted by Richardson (1926, 1928), Lees (1929), Harrison (1931), O'Brien (1955), Heim (1958), Walther (1972), Trusheim (1974), and Talbot & Jarvis (1984). Examples are Kuh-e Anguru (Fig. 10), and Kuh-e Bam, where extrusive salt has broken through the axial crest of the anticline and flowed southward. There are over 100 of these surface salt piercements in southeast Iran (Harrison 1968), generally with diameters of 5–12 km (1–8 km Hirschi 1944). They are generally suboval in plan, although mushroom-shaped in vertical section, some with relatively low sloping overhangs (e.g. Talbot & Jarvis 1984, fig. 4b), but most are steeper tongues of salt partly covering

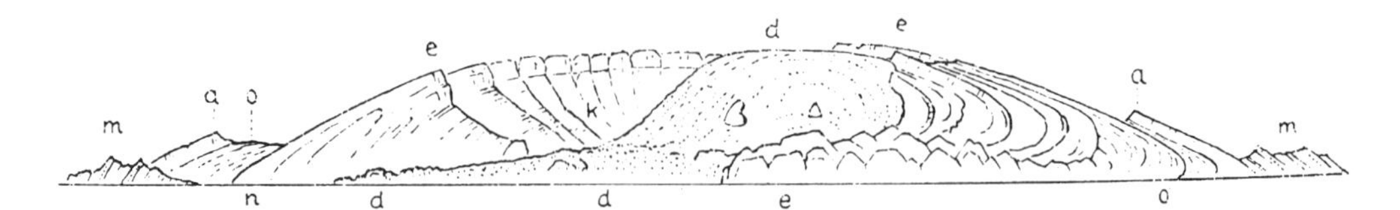


**Fig. 10.** Salt glacier (namakier) flowing from the Hormuz salt piercement in Kuh-e Anguru brachyanticline, southeastern Iran as sketched by Heim (1958), looking northwest. Key: d = salt glacier with included carbonate blocks; k = Senonian–Eocene marl; e = Eocene limestone; o = Oligocene green marl; a = Oligo-Miocene Asmari Fm.; m = Miocene marl with sandstone interbeds.

eroded 'flat-irons' on the flanks of anticlines (Harrison 1968). Many lie in synclines and most have no direct assocation with the weaker apices of anticlines. Other salt piercements are fault-associated salt fissures, only 100 m wide. The conspicuous NE–SW alignment of the Hormuz Series salt piercements in southeastern Iran has been noted by Bruderer (1960), Fürst (1970), NIOC (1978) and Edgell (1987, 1991), who attributed this alignment to left-lateral strike-slip faulting, while Kent (1970, 1979) remarked that the NE–SW distribution of piercements in southeastern Iran suggested basement control oblique to Zagros trends. Koyi *et al* (1993) note that diapirs in the North Sea are spatially related to basement faults, which trigger diapirism, although diapirs are not necessarily directly above the basement faults. Hospers *et al* (1988), also for the North Sea, attempt to show that diapir relation to basement faults is accidental, with diapir trends due to palaeodepth contours. Fürst (1976) classified salt piercements of southeast Iran as: (1) active salt stocks; (2) inactive salt stocks; and (3) incipient salt stocks. The evolution of salt plugs (salt extrusions) as given by Ala (1974) is: (1) salt bulge (incipient stage); (2). salt breaks through the surface disruptively; (3) salt glaciers dissipate the salt mass; and (4) salt removal leaves an amphitheatre with collapsed material and exotic blocks.

In the western U.A.E. and its offshore there is a distinct N–S alignment of salt diapirs observed by Bruderer (1960, Fig. 5), and Abu Dhabi Dept of Petroleum *et al.* (1981, Fig. 1.4). This has been interpreted by Edgell (1987, 1991) as N–S Arabian Trend basement faulting which has triggered the growth of diapirs along this alignment. In the eastern U.A.E. and its offshore most diapirs trend NE–SW parallel to the alignment of fault-related diapirs in nearby southeastern Iran and are considered to be triggered by the same direction of basement faulting (Abu Dhabi Dept of Petroleum *et al.* 1981, Fig. 1.4).

The mechanism of rim anticline formation is the rapid intrusion of very large salt piercements, causing salt withdrawal and rim synclines around a primary diapir followed by a ring of secondary halokinetic diapirs, or rim anticlines (Jenyon 1986, fig. 4.11). Similar peripheral folds have been produced in more sophisticated polymer and glass-bead modelling by Alsop (1994). The Persian Gulf Island of Abu Musa is entirely a Hormuz salt piercement with a diameter of 4 km. Rim synclines immediately surround this large surface piercement, and further out peripheral rim anticlines fold strata down to base Jurassic in oil-producing structures, such as the Mubarak Field (Edgell 1987), which produces 15000 BPD from five separate stacked reservoirs (Fig. 11).

Turtleback structures, such as the Marmul Field, have been formed in the South Oman Salt Basin partly by initial salt withdrawal, but primarily by shallow subsurface dissolution of salt pillows and ridges of Neoproterozoic Ara Formation between which Cambro-Ordovician Haima Group continental sandstones were deposited. Shallow dissolution of salt diapirs by near-surface and subsurface meteoric water, as evidenced by unconformities, has resulted in structural inversion (Heward 1990, fig. 4). A schematic cross-section of this type of turtleback structure from South Oman (is shown in Fig. 12), from Aley & Nash (1985).

Dissolution drapes occur over residual piles of relatively insoluble dolomite, shale and thin sandstone once interbedded in the dominantly salt sequence of the Neoproterozoic Ara Formation, which also contains interbedded dolomite, shale and thin sandstone after subsurface salt dissolution, as in the Amal South Field and Simsim Field (Al-Marjeby & Nash 1986; Heward 1990).

Disharmonic structures are common throughout southwestern Iran and adjacent areas of Iraq. They are formed by the southwest flow of the Upper Mobile Group Gachsaran Formation over competent Asmari Limestone anticlines and beneath competent limestones of the Mishan Formation (Middle Fars). This occurred as the Zagros Ranges were gradually uplifted during the Pliocene (Falcon 1961, 1969) and has led to spectacular disharmonic folding. Individual, NW–SE trending Asmari Limestone anticlines are 30–50 km long and have amplitudes of 3000–4500 m (O'Brien 1957). The mobile Lower Fars Series, or Gachsaran Formation, up to 2000 metres thick, has filled synclines between

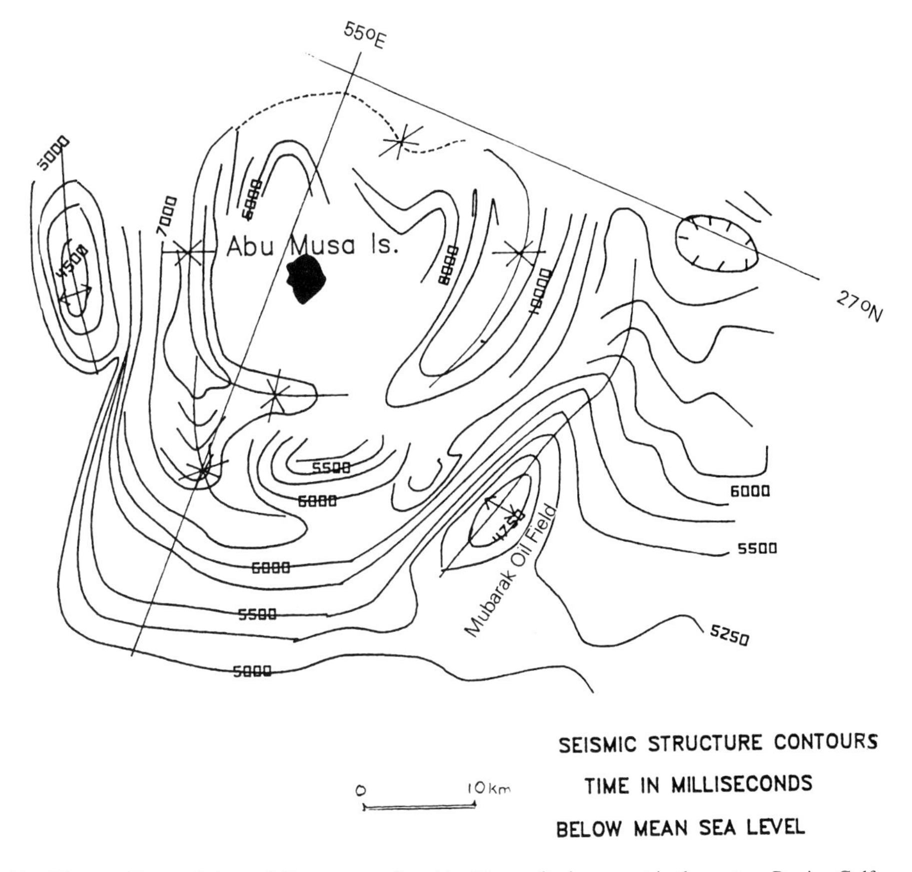


**Fig. 11.** Rim synclines and rim anticlines surrounding Abu Musa salt piercement in the eastern Persian Gulf.

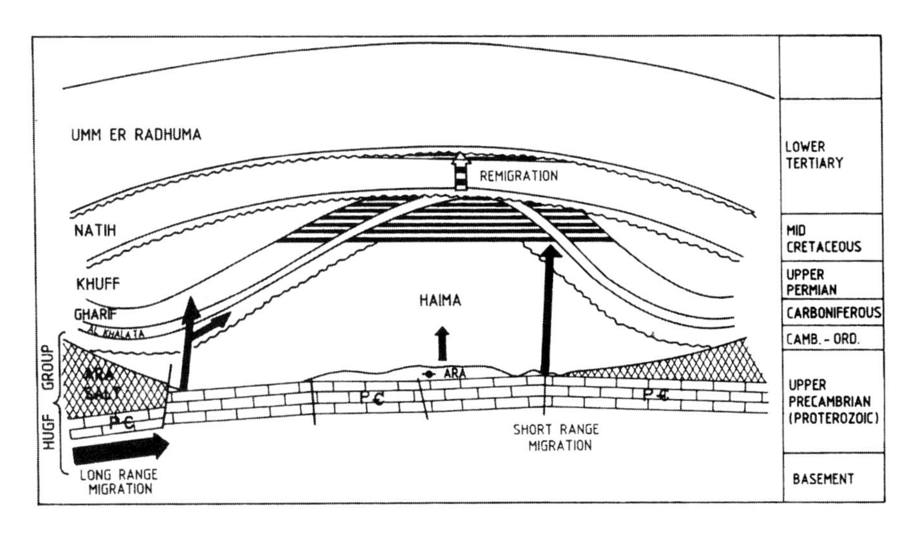


**Fig. 12.** Turtleback structure on the east flank, South Oman Salt Basin due to structural inversion caused by Neoproterozoic (Ara Formation) salt dissolution. Note migration and remigration paths of Hugf Group, Proterozoic sourced oil. (Schematic geological cross-section modified from Aley & Nash 1985).

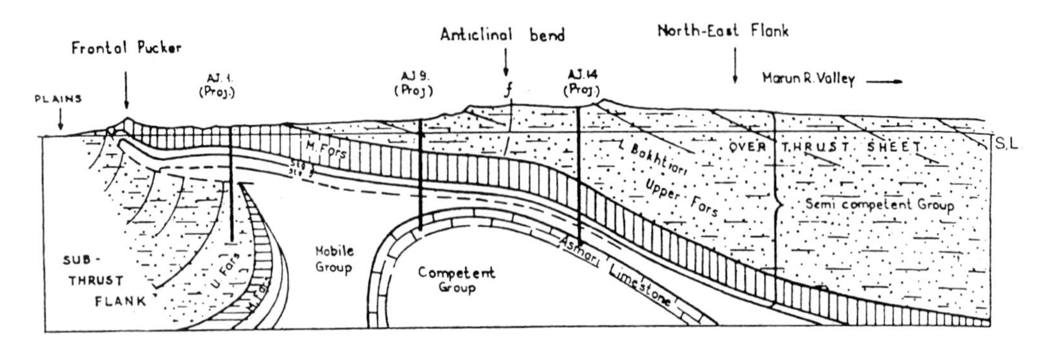


**Fig. 13.** Disharmonic folding of the Agha Jari Oil Field, southwestern Iran, where uplift of the Zagros Ranges has caused the thick, mobile Lower Fars Series salt to flow over the competent, underlying Asmari Limestone and beneath the relatively competent Middle Fars limestones and marls. A SW displaced frontal fold with a thrust plane and a frontal bulge has formed in the Lower Fars Mobile Group (after Ion *et al.* 1951).

Asmari folds. These synclines have then been successively rotated along the decollement planes provided by the mobile Gachsaran Formation as the Zagros Range were uplifted, resulting in the eventual squeezing out of Gachsaran Formation salt and anhydrite to the southwest of Asmari fold axes with frontal folds and thrusts, as seen in Agha Jari (Fig. 13) and Kirkuk oil fields (Dunnington 1968). This is referred to as incompetent folding by Heim (1958), but is commonly known as disharmonic folding, largely caused by a combination of flow in confined incompetent strata and compression in the Zagros Ranges. The Lower Fars, or Gachsaran Formation, salt is squeezed out from the crests of Asmari anticlines, leaving shallow synclines overlying anticlinal axes (de Bockh 1929; Lees 1931, 1952), as seen in the Masjid-i Sulaiman Oil Field (Fig. 14). The Lower Fars salt is also squeezed into deep synclines bordering the anticlinal areas and the thickness of the mobile salt is so great that it breaks through the thick synclinal overburden forming active diapiric salt ridges. These are the so-called Omega structures of Busk (1929), seen in Tul-e Bazun on the northeast flank of Masjid-i Sulaiman Field where Lower Fars salt is thick (Fig. 14), and are also shown by O'Brien (1957, fig. 12b). These disharmonic folds in Iran are termed injection folds by Beloussov (1959) who comments that 'the injection phenomena are associated not so much with gravitational "surfacing" as with dynamic sideward squeezing of a plastic mass from the crests of dome-like uplifts (reflected block folds) during the growth of these uplifts, and under conditions where the overburden offers resistance'. Rönnlund & Koyi (1988) comment on a 41% lateral shortening of the Zagros Ranges, caused by decollement and compression.

Orogenic fold filling is interpreted as taking

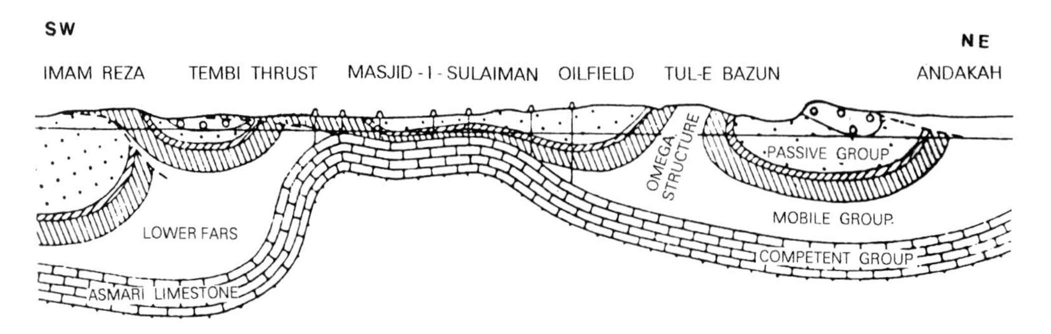


**Fig. 14.** Disharmonic folding and diapirism in Masjid-i Sulaiman Oil Field, where plastic Lower Fars Series salt and anhydrite have been squeezed between the competent Asmari anticline and the overlying Passive Group clastics. A diapiric Omega-type salt–anhydrite ridge is seen at Tul-e Bazun where thick Fars salt has broken through the Passive Group (modified from Lees 1953).

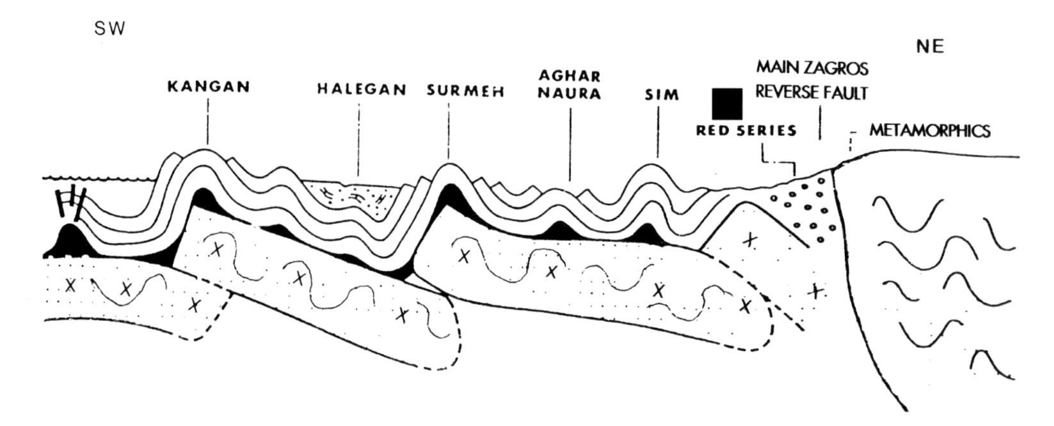


**Fig. 15.** Orogenic fold-filling structures in southern Iran, where Neoproterozoic Hormuz Series salt (black) is interpreted as forming the cores of large anticlines (schematic NE–SW geological cross-section adapted from Comby *et al.* 1977).

place beneath large box-folds of the southern Zagros Range, where Neoproterozoic Hormuz salt is thought to fill the overlying sedimentary folds by flowage along a plane of decollement between the crystalline basement and the overlying folded Phanerozoic sequence. Falcon (1969) gives Kuh-e Kainu as an example, while Comby *et al.* (1977) explain the large gas-bearing structures of southern Iran, such as Kangan and Aghar, in a similar manner; however they also interpret the basement as a series of underlying southwest thrust blocks (Fig. 15) and indicate detachment folding due to decollement along the mobile Hormuz Series salt above the crystalline basement. Fürst (1976) also considers that the anticlines, mostly brachyanticlines, of southeastern Iran have a salt cushion. On the basis of experimental modelling, Koyi (1988) suggests a three-fold division of salt piercements in the same region into prebuckling and synbuckling, postbuckling diapirs, and states that 'one-half to two-thirds of the Hormuz Salt diapirs predate the Zagros orogeny. This concept was initiated by Colman-Sadd (1978) and omits the fact that many Zagros anticlines can be shown to have begun substantial growth in the Cretaceous as the proto-Zagros range (Falcon 1974), and probably in Jurassic times (Huber 1977). Geological cross-sections of southeastern Iran by National Iranian Oil Company (1978) also show the anticlines as underlain by Hormuz Series salt swells. Vita-Finzi (1986) has given an uplift rate of 7 mm $a^{-1}$ for the Zagros shore of southeastern Iran by growth of anticlines underlain by Hormuz Series salt and by Holocene horizontal shortening of 26 mm $a^{-1}$.

## *Rate of growth of salt structures*

The rate of growth of salt structures in the Persian Gulf Basin is clearly not uniform as can be seen by some dormant salt piercements, such as Jabal San'am, while others are very active indeed, such as Kuh-e Namak (Bushehr Province). In addition, most non-piercement salt structures, where overlying strata are updomed, have not shown any Neoproterozoic salt in deep drilling, even to 7600 m in the Dammam Deep Test, although distinct negative gravity anomalies indicate the presence of salt. The underlying diapiric Hormuz Series salt is thus deep-seated, and there is quite sufficient space between Precambrian basement at 13400 m in the Dammam area and the deepest well drilled to 7600 m. Most of the Persian Gulf oil field structures, mainly domes and brachyanticlines, are growth structures which show increased structural closure with depth and also deep underlying faults. For example, the Dukhan Anticline shows very little structural closure at the surface, but has a structural closure of 488 m at the Jurassic Arab-D level and a closure of 641 m in the Permian Khuff reservoir. A maximum structural closure at the base of the Lower Palaeozoic Lalun Sandstone is, therefore, not more than 1000 m. Sugden (1962) has shown that the Dukhan anticline grew continually from Late Jurassic to at least Miocene (Fig. 8). This would constitute a rate of salt growth of 1000 m in the last 102 Ma, or 0.01 mm $a^{-1}$. The more rapidly growing salt piercement, forming Kuh-e Namak in Bushehr Province, has grown through an overlying sequence of nearly 15240 m during the last 156 Ma since Late Jurassic. Upward salt growth in this large active piercement is therefore at a rate of about 0.10 mm $a^{-1}$, although a rate of 170 mm $a^{-1}$ is deduced by Jackson & Talbot (1986), who refer to it as 'a fountain'. Most domes and anticlines cored by Neoproterozoic Hormuz salt do not seem to have moved during the long time interval from Late Proterozoic to Early Mesozoic, presumably due to insufficient overburden. The Zagros orogeny certainly induced the more rapid rise of salt, and rates of growth of Hormuz Series salt plugs have been interpreted by Player (1969) at 0.3–2 mm $a^{-1}$. An exceptional case of rapid salt growth is noted by Player (1969) for the dyke-like Parag salt piercement just northwest of Lar, where a sliver of Hormuz salt only 100 m wide has pierced through the Bakhtiari Conglomerate. He concludes that a thickness of 12200 m of sedimentary strata overlying the Neoproterozoic Hormuz Series has been penetrated in the last 3 Ma, giving a rate of growth of 4.06 mm $a^{-1}$. Estimates of the growth rate of salt diapirs elsewhere are 0.3 mm $a^{-1}$ in northern Germany (Trusheim 1957, 1960; Sannemann 1968), 0.01–0.55 mm $a^{-1}$ in East Texas (Seni & Jackson 1984*a,b*), and 0.01, 0.046 and 0.05 mm $a^{-1}$ for three salt diapirs in the North Sea (Jenyon 1986). These figures for the rate of growth of salt structures in other areas are very similar to those calculated by the author for the Persian Gulf Basin. Braitsch (1971) states that the presence of large haematite and pyrite crystals (up to 2 cm long), in addition to equally large crystals of authigenic dolomite and apatite, in Persian Gulf salt domes indicates a long period of growth, so that circulation of only small amounts of pore solution, or higher temperatures must be assumed. Large haematite crystals up to 40 mm long, as well as banded haematite, are found commonly on the salt piercements of Latzk and Hormuz Islands (Schroeder 1946; Wolf 1959; Walther 1972).

With the younger salt of the Lower Fars, or Gachsaran Formation, which once had an overburden of 4000 m of Upper Tertiary clastics, the rate of growth for the Lali Ambal salt plug would be approximately 0.27 mm $a^{-1}$. This is a generally faster rate than the growth of most Hormuz Series salt structures and may explain why there are so few piercements from the Gachsaran Formation salt, with Lali Ambal salt plug being almost a unique example. The presumption is that this Miocene salt has not yet had time to form salt piercements, except in a few isolated instances.

### *Spacing of salt structures*

There is a very well defined spacing of salt piercements in southern and southeastern Iran (Fig. 9), mainly as a result of repeatedly rejuvenated, intersecting basement faults and active seismotectonics to the present day, with many micro-earthquakes taking place on a daily basis. As a result, an observed spacing of 28 km has been given for the Zagros diapir province of southern Iran (Koyi 1988; Jackson *et al.* 1990), where diapirism is related to the Neoproterozoic Hormuz Series salt. Trusheim (1974) has noted an average 20 km spacing of Hormuz salt diapirs in southeastern Iran.

In northeastern Saudi Arabia, an attempt has been made to relate oil fields to a presumed 40 km spacing of N–S lineaments of the Arabian Trend (Al-Khatieb & Norman 1982). In fact, many coastal and offshore salt structures are closer, with Abu Hadriya and Khursaniyah being about 30 km apart, while only about 25 km separates the Jana and Jurayd salt structures. The salt-cored oil field anticlines of southern Kuwait are even more closely spaced, with the Minagish, Abduliya, Dharif, Khashman, Magwa and Ahmadi structures each being separated by about 10 km.

As can be seen from the northerly structural trends of most of the salt-cored oil field domes and brachyanticlines in the offshore UAE (Fig. 1), the spacing is variable but they are generally not more than 50 km apart. On the east flank of the South Oman Salt Basin, there are about 26 salt-related oil field structures in a distance of 400 km, giving an average spacing along that flank of about 15 km.

In general, the spacing of Neoproterozoic salt diapirs in the Persian Gulf Basin is between 10 and 50 km, largely dependent on the spacing of basement fault structures, on salt layer thickness, on roof subsidence, and on lateral variations in loading and/or facies. Trusheim (1960) noted that salt walls in northern Germany are 8–10 km apart, while Sannemann (1968; fig. 4) shows a salt-stock family from the same area with a spacing of about 20 km between salt diapirs. Calculations by Turcotte & Schubert (1982) gave a diapir separation of 10–15 km for northern Germany and stated that the separation ($D$) between salt intrusions should be $D = 2.568z$, where $z$ is the depth to the salt layer. Rönnlund & Koyi (1988) used Fry's centre-point spacing strain analysis on models and for 150 salt diapirs in the eastern Persian Gulf area to deduce an isotropic, minimum average wavelength of 13–20 km. Price & Cosgrove (1990) reviewed diapir spacing and found that a a salt layer needed to be 500 m thick before significant lateral migration is achieved, even at a salt temperature of 250°C. They found that roof subsidence of 1 km was required for a spacing of 15 km, and for a separation of 10 km the roof subsidence needed to be 2.2 km, assuming a diapir radius of 2 km and depth to the mother salt of 5 km. Since roof subsidence is related to salt layer thickness, this may explain the close spacing of diapirs in the South Oman Gheba Basins where thickness of the Ara Salt is up to 4000 km.

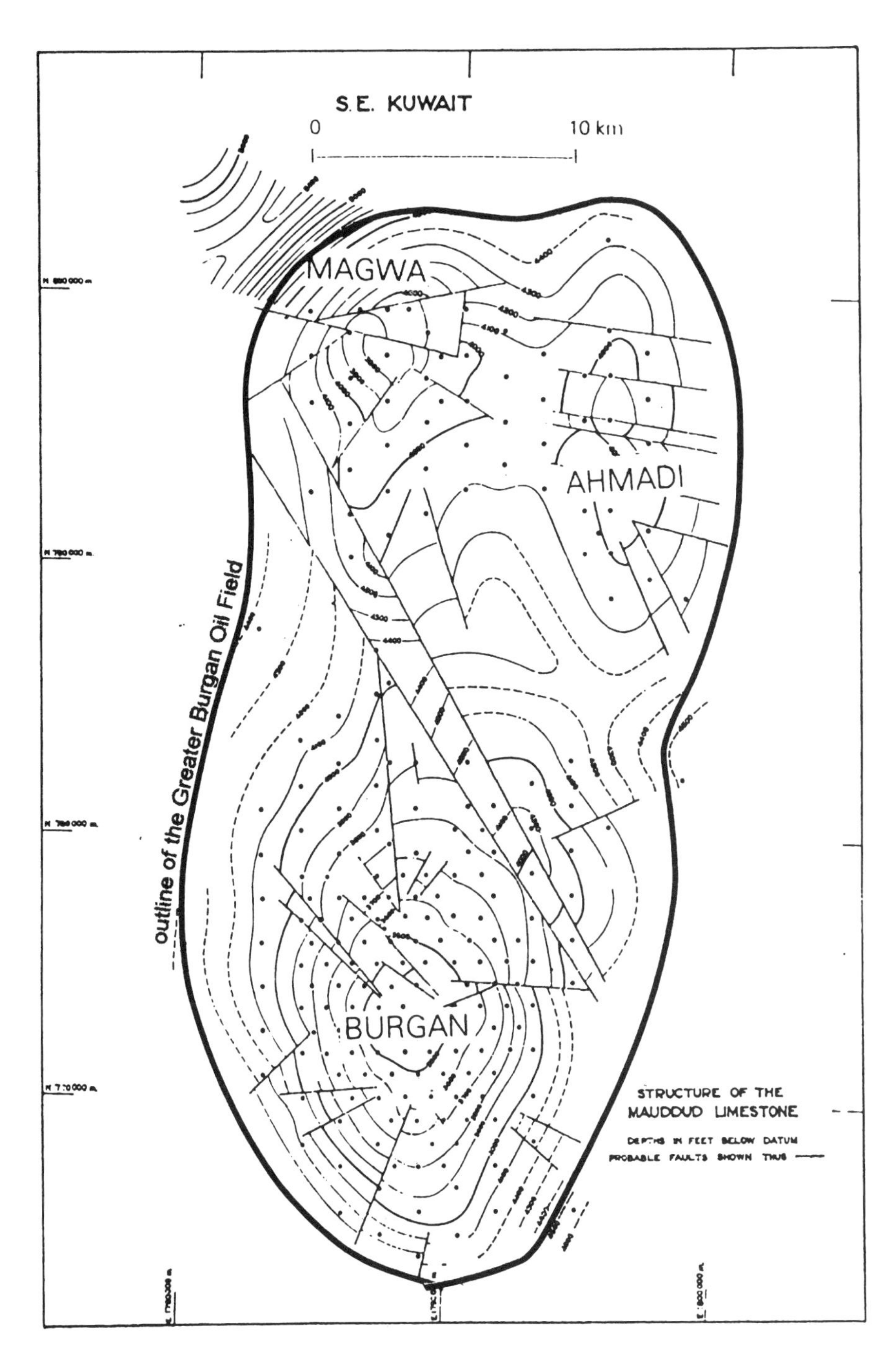


**Fig. 16.** Salt dome structure of the Greater Burgan Oil Field showing typical radial faulting (modified from Fox 1956).

## *Types of Persian Gulf Halokinetic Structures*

Principal types of geological structures due to salt tectonism in the Persian Gulf Basin are listed below with examples of each.

1. *Salt Domes* are subcircular, often with radial faulting, due to slow upwarping of overlying strata by deep-seated salt, e.g. Dammam Dome, Burgan (Fig. 16), Raudhatain, Umm Shaif and Yibal fields. The terms salt stocks, salt plugs,

and salt domes are synonymous – they are all diapiric in that they cut overlying strata while salt pillows are not discordant. (Trusheim 1960; Price & Cosgrove 1990).

2. *Salt pillow structures* are basically brachyanticlines underlain at great depth by salt pillows e.g. Khafji, Jana, Bahrain, Bu Hasa and Bab fields may fall in this category, as well as the salt-cushioned anticlines of S. Iran. North Sea seismic examples are given by Jenyon (1986) and Owen & Taylor (1983).
3. *Salt wall structures* are very elongated anticlines cored at depth by salt walls along single basement fault trends with forced folding of overlying strata, together with their diapirism (QGPG & AQPC 1991), e.g. Dukhan Field.
4. *Piercement salt structures* occur where salt is actively extruding at the surface e.g. Kuh-e Anguru (Fig. 10), Abu Musa (Fig. 11), Hormuz and Larak islands, and Kuh-e Namak in Bushehr Province, or where the salt plug is at a passive stage with a lower relief and covered by Hormuz Series debris.
5. *Rim anticlines* occur around large active salt piercements where the rapid extrusion of a major salt column has caused multiple peripheral folds, especially peripheral synclines due to salt withdrawal with intervening rim anticlines over secondary halokinetic diapirs. e.g. Mubarak Field (Fig. 11).
6. *Turtleback structures* in Southern Oman (Heward 1990) are quite different from secondary halokinetic diapirs and form by initial salt withdrawal, but mainly by near surface salt dissolution, creating depocentres for Cambro-Ordovician clastics. These later underwent structural inversion in S. Oman so that Permian to Tertiary strata became draped over the hard clastic cores, e.g. Marmul Field (Fig. 12).
7. *Disharmonic folds* are caused by the mobile salt and evaporites of the Oligo-Miocene Lower Fars Series, or Gachsaran Formation, where they are squeezed between the underlying competent Asmari Limestone and the overlying Passive Group of Agha Jari and Bakhtiari formations (O'Brien 1957). This caused disharmonic buckle folding over the Agha Jari (Fig. 13), and Masjid-i Sulaiman (Fig. 14) fields, with Omega diapirs (Busk 1929) of extruded evaporite ridges adjacent to the latter.
8. *Gravitational spreading structures* are represented in southern Iran by the many surface salt piercements with active salt extrusion. The mobile salt rises to considerable heights (1485 metres in Kuh-e Namak, Bushehr Province) and the mobile salt mountain spreads laterally in the form of salt glaciers (namakiers) under the influence of gravity as Lees (1929) and Talbot & Jarvis (1984) have so well described.
9. *Orogenic fold filling* apparently occurs in the compressional structures of the Simply Folded Belt (Falcon 1969) of southern and southwestern Iran, giving rise to large box anticlines and the large gas-bearing anticlines of the Kangan area, where Neoproterozoic salt is believed to be interposed between basement and the overlying Phanerozoic sequence, filling the cores of anticlines (Comby *et al.* 1977). e.g. Kuh-e Kainu, and the Nar, Varavi, Kangan and Aghar gas fields (Fig. 15).
10. *Dissolution drapes* result from the localized dissolution of Neoproterozoic salt forming small anticlines draped over dolomite and shale residue (Heward 1990), which were once interbedded with Ara Formation salt of southern Oman. e.g. Amal South Field and Simsim Field (Heward 1990).

## *Salt Diapirism and Oil Abundance in the Persian Gulf Basin.*

The abundance of oil in the Persian Gulf Basin, which contains 65% of the ultimate recoverable reserves in the world, has often been remarked on (Law 1957; Halbouty *et al.* 1970, Kamen-Kaye 1970, Kent & Warman 1972, Edgell 1991, 1993). Some of the main reason for prolific hydrocarbons in the Persian Gulf Basin are as follows.

1. Its early geological basinal history, wherby thick Neoproterozoic evaporite basins with interbedded carbonates and shales were formed as a Prototethyan precursor bounded by the NE–SE Zagros Line and interdigitated between extensional uplifts of the northeastern edge of the Afro-Arabian Plate. These thick Neoproterozoic beds were rich in hydrocarbons and still provide the main source rocks for all oil in Oman.
2. The uplift of the Zagros Ranges in the Pliocene, so that the marine sedimentary wedge on the northeast edge of the subducting Arabian Plate was folded in Iran, Iraq and Syria and did not continue to be consumed by subduction beneath the Iranian and Turkish microplates, despite Holocene folding and uplift at 1.9–7.4 mm $a^{-1}$ in southeastern Iran (Vita-Finzi 1979).
3. The thick, mainly marine sedimentary sequence of the Persian Gulf Basin, up to 18000 m thick, and the immense volume of shallow water sediments with occasional anaerobic intervals, as well as the large size of the basin (2600 km long and more than 1000 km wide).
4. The presence of rich source rocks at several levels ranging from the Neoproterozic Hugf

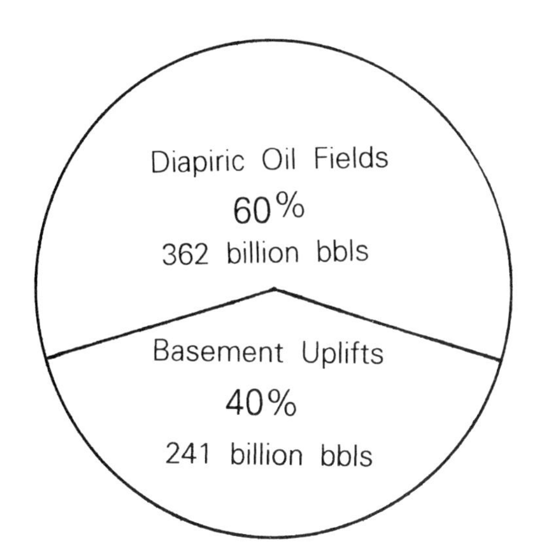


**Fig. 17.** Diagram showing that Persian Gulf Basin oil fields are predominantly of salt diapiric origin, accounting for 60% of the basin's ultimate recoverable oil reserves or about 40% of the world's oil.

Group and Hormuz Series, through Lower Palaeozoic organic shales, such as the Qusaiba Shale, to dark hydrocarbon-rich Jurassic rocks, Lower Cretaceous shales, and Upper Cretaceous to Lower Tertiary marls.

5. Excellent carbonate and sandstone reservoir rocks with high porosity and permeability, such as the Burgan Sands, the Arab Formation carbonates, the Khuff Formation carbonates and the fractured Asmari Limestone.
6. Caprocks of salt, anhydrite and shale sealing the reservoirs at many different levels from Upper Proterozoic to Miocene, providing multiple stacked reservoirs.
7. A sedimentary sequence with deposits of all ages from Late Proterozoic to Holocene formed in warm, shallow, organic-rich, marine conditions with no lengthy depositional interruptions and few unconformities
8. An abundance of very large, domal and anticlinal structures, with many simple folds.
9. Continuous, or intermittent, structural growth of major folds, due to salt diapirism or repeated basement block uplift.
10. The rarity of extrusive or intrusive rocks and their absence from the drilled sequence in the majority of the basin.
11. The presence of deep-seated salt diapirism, which accounts for some 60% of all the oil field structures of the Persian Gulf Basin (Fig. 17).
12. The Neoproterozoic Hormuz Salt, which has provided surfaces of decollement for folding in the Zagros Range of Iran and Iraq.

## Conclusions

Active salt tectonism in the Persian Gulf Basin has produced some of the largest and most prolific oil field structures in the world.

Sequences of massive salt are present in Upper Proterozoic (Riphaean) strata of much of the Persian Gulf Basin, primarily in the Hormuz Salt Series and the equivalent Ara Formation. They are themselves underlain by, and intercalated with, hydrocarbon-rich sediments, so that Neoproterozoic strata are both proven source rocks for all reservoirs in Oman and provide productive Precambrian dolomite reservoirs.

In Yemen, thick salt of the Upper Jurassic Sab'atayn Formation underlie the Marib-Shabwa Basin and is the cause of diapiric oil field structures in the Alif and Shabwa fields.

The Miocene salt basin formed by the mobile Lower Fars, or Gacharan Formation, plays an important part in sealing the long, whaleback Asmari anticlines of the Zagros Range, as well as in their structural development.

Salt structures of the Persian Gulf Basin include salt domes (e.g. Dammam Dome), salt pillows (e.g. Khafji Field), salt walls (e.g. Dukhan Field) salt piercements (e.g. Hormuz Island), namakiers or salt glaciers (e.g. Kuh-e Anguru), rim anticlines (e.g. Mubarak Field), turtleback structures (e.g. Marmul Field), gravity glide folds (e.g. Kangan Gas Field) and dissolution drapes (e.g. Simsim Field).

The rate of growth of salt is between 0.01 mm $a^{-1}$ in deep-seated salt diapirs and 0.3 mm $a^{-1}$ in active piercements. The fastest rate of salt movement is probably less than 1 mm $a^{-1}$ in the young Miocene Gachsaran Formation salt.

Basement extensional and wrench faulting, which has often been rejuvenated, has played an important role in the mechanism of salt tectonics. The intersecting normal faults of the Arabian Trend and the lateral strike-slip faults of the Aulitic Trend have provided avenues for the upward, buoyant movement of deep-seated Neoproterozoic salt. This has often produced salt piercements, but in many instances the salt has gradually updomed overlying strata, producing almost continuous growth structures.

Large salt-cored domes and brachyanticlines predominate amongst the prolific oil fields of the Persian Gulf Basin, although the giant Ghawar oil field is an exception, being produced by underlying block uplift of basement (Aramco 1959). Most salt diapirs in the Persian Gulf Basin did not begin growth until Late Jurassic, although some have grown since Permian times.

Some 60% of the ultimate recoverable oil reserves of the Persian Gulf Basin originate from salt tectonism, and 40% of the known world oil reserves are, thus, due to salt diapirism in this basin.

## References


ABU DHABI DEPT. OF PETROLEUM (ADDP), ABU DHABI MARINE AREAS & ABU DHABI OIL CO. 1981. Geology of Abu Dhabi, *In*: PINNINGTON, D. J. (ed.) *Schumberger Well Evaluation Conference, United Arab Emirates/Qatar.* Éditions Galilée, France, 4–16.

ALA, M. A. 1974. Salt diapirism in southern Iran. *AAPG Bulletin,* **58**, 758–1770.

ALEY, A. A. & NASH, D. F. 1985. A summary of the geology and oil habitat of the eastern flank hydrocarbon province of South Oman. *Proceedings, Seminar on Source and Habitat of Petroleum in Arab Countries*, OAPEC, Kuwait, 521–541.

AL-KHATIEB, S. O. & NORMAN, J. W. 1982. A possibly extensive crustal fracture system of economic interest. *Journal of Petroleum Geology*, **4**, 319–327.

AL-MARJEBY, A. & NASH, S. F. 1986. A summary of the eastern flank hydrocarbon province of South Oman. *Marine and Petroleum Geology*, **3**, 306–314.

ALSOP, I. 1994. Physical modelling of overburden deformation adjacent to salt diapirs. *Salt Tectonics Conference, Programme and Abstracts*, Petroleum and Tectonic Groups of the Geological Society, London, 2.

ARABIAN AMERICAN OIL COMPANY (ARAMCO) 1959. Ghawar Oil Field, Saudi Arabia. *AAPG Bulletin,* **43**, 434–454.

BARNES, B. B. 1987. Recent developments in geophysical exploration for hydrocarbons. *Short Course Handbook on Hydrocarbon Exploration*, King Fahd University of Petroleum and Minerals Press, Dhahran, 1–12.

BELOUSSOV, V. V. 1959. Types of folding and their origin. *International Geology Review,* **1**, 1–21.

BEYDOUN, Z. R. 1964. *The Stratigraphy and Structure of the Eastern Aden Protectorate*. Overseas Geology and Mineral Resources, Supplement Series, Bulletin Supplement, **5**, 107.

BLANFORD, W. T. 1872. Note on the geological formations seen along the coasts of Biluchistan and Persia from Karachi to the head of the Persian Gulf, and on some of the Gulf islands. *Records of the Geological Survey of India*, **5**, 41–45.

BRAITSCH, O. 1971. *Salt Deposits, Their Origin and Composition*. Springer, Berlin.

BRENNAN, P. 1990. Greater Burgan Field., *In*: BEAUMONT, E. A. and FOSTER, N. H. (eds) *Structural Traps 1, Tectonic Fold Traps*. Treatise of Petroleum Geology, Atlas of Oil and Gas Fields, American Association of Petroleum Geologists, Tulsa, 103–128.

BRITISH PETROLEUM COMPANY, LIMITED. 1956. Geological Maps and Sections of south-west Persia. *20th International Geological Congress, Proceedings, Mexico.* Six geological maps (Scale 1:1,000,000), four sheets of geological cross sections, seven sheets of stratigraphical columns. London, Edward Stanford Ltd.

—— 1964. Geological maps, columns and sections of the High Zagros of south-west Iran. *22nd International Geological Congress, Proceedings, Delhi*, 16 sheets (Scale 1:250,000). London, Edward Stanford Ltd.

BRUDERER, W. 1960. Le Bassin pétrolier du Golfe Persique. *4e Congrès National du Pétrole*, Association Française des Techniciens du Pétrole, Deauville, **2**, 7–28.

BUSK, H. G. 1929. *Earth Flexures*. Cambridge University Press.

CAREY, S. W. 1962. Folding. *Journal of the Alberta Society of Petroleum Geologists*, **10**, 95–144.

COLMAN-SADD, S. P. 1978. Fold devlopment in Zagros Simply Folded Belt, Southwest Iran. *AAPG Bulletin,* **62**, 984–1003.

COMBY, O., LAMBERT, C. & COAJOU, A. 1977. An approach to the structural studies of the Zagros Fold Belt in the EGOCO Agreement Area. *Proceedings of the Second Geological Symposium of Iran*, Tehran, 103–159.

DE BÖCKH, H., LEES, G. M. & RICHARDSON, F. D. S. 1929. Contribution to the stratigraphy and tectonics of the Iranian ranges. *In*: GREGORY, J. W. (ed.) *The Structure of Asia*. Methuen & Co., London, 58–176.

DUNNINGTON, H. V. 1968. Salt-tectonic features of Northern Iraq. *In*: MATTOX, R. B. *et al.* (eds) *Saline Deposits*. Geological Society of America, Special Paper, **88**, 183–227.

EDGELL, H. S. 1973. Stratigraphic variations in Interior Fars, South Iran and their tectonic implications. *The First Iranian Geological Symposium*, Tehran Abstracts, 30.

—— 1987. Structural analysis of hydrocarbon accumulations in Saudi Arabia, *Short Course Handbook on Hydrocarbon Exploration*, King Fahd University of Petroleum and Minerals Press, Dhahran, Saudi Arabia.

—— 1989. Infracambrian salt basins of the Persian Gulf and their role in prolific hydrocarbon generation. *28th International Geological Congress, Washington, D.C.*, Abstracts, 433–434.

—— 1991. Proterozoic salt basins of the Persian Gulf area and their role in hydrocarbon generation. *Precambrian Research*, **54**, 1–14.

—— 1992*a*. Basement Tectonics of Saudi Arabia as related to oil field structures. *Proceedings of the Ninth International Conference on Basement Tectonics*, Canberra **3**, 169–193.

—— 1992*b*. Halokinetic origin of prolific Persian Gulf oil fields. *American Association of Petroleum Geologists International Conference and Exhibition* Sydney, Abstracts, 48.

—— 1993. Extensional basin tectonics in Saudi Arabia and their role in giant hydrocarbon accumulations. *International Conference on Basin Tectonics and Hydrocarbon Accumulation*, Nanjing, Abstracts, 44–45.

FALCON, N. L. 1961. Major earth-flexuring in the Zagros Mountains of South-West Iran. *Quarterly Journal of the Geological Society*, London, **117**, 367–376.

—— 1969. Problems of the relationship between surface structure and deep displacements illustrated by the Zagros Range. *In*: KENT, P. E. *et al.* (eds), *Time and Place in Orogeny*. Geological Society, London, Special Publication, **3**, 9–22.

—— 1974. Southern Iran: Zagros Mountains. *In*: SPENCER, A. M. (ed.) *Mesozoic–Cenozoic Orogenic Belts – Data for Orogenic Studies*. Geological Society, London, Special Publication, **4**, 199–211.

FOX, A. F. 1956. Oil Occurrences in Kuwait. *20th*

*International Geological Congress, Mexico*, Symposium sobre jacimientos de petroleo y gas, 131–148.

FÜRST, M. 1970. Stratigraphie und Werdegang der östlichen Zagrosketten (Iran). *Erlanger geologische Abhandlungen*, **80**, 1–51.

—— 1976. Tektonik und Diapirismus der östlichen Zagrosketten. *Zeitschrift der deutschen geologischen Gesellschaft*, **127**, 183–225.

—— 1990. Strike-slip faults and diapirism of the S.E. Zagros Ranges. In: *Proceedings of the Symposium on Diapirism with Special Reference to Iran*. Geological Survey of Iran, **2**, 149–181.

GLENNIE, K. W., BOEF, M. G. A., HUGHES-CLARKE, M. W., MOODY-STUART, M., PILAAR, W. F. H. & REINHART, B. M. 1974. Geology of the Oman Mountains. *Verhandelingen Koninklijk, Nederlands geologisch mijnbouwkundig Genootschap*, **31**, 423.

GORIN, G. E., RACZ, L. G. & WALTER, M. R. 1982. Late Precambrian–Cambrian sediments of the Hugf Group, Sultanate of Oman. *AAPG Bulletin*, **66**, 2609–2627.

GRANTHAM, P. J., LIJMBACH, G. W. M., POSTHUMA, J., HUGHES-CLARKE, M. W. & WILLINK, R. J. 1987. Origin of crude oils in Oman. *Journal of Petroleum Geology*, **11**, 61–80.

GUSSOW, W. C. 1966. Salt temperature: A fundamental factor in salt dome intrusion. *Nature*, **210**, 518–519.

—— 1968. Salt diapirism: importance of temperature, and energy source of emplacement, *In*: BRAUNSTEIN, J. and O'BREIN, G. D. (eds) *Diapirism and Diapirs*. American Association of Petroleum Geologists, Memoir, **8**, 16–52.

HALBOUTY, M. T., MEYERHOFF, A. A., KING, R. E., DOTT, R. H., KLEMME, H. D. & SHABAD, T. 1970. World's giant oil and gas fields, geologic factors affecting their formation, and basin classification. *In*: HALBOUTY, M. T. *Geology of Giant Petroleum Fields*. American Association of Petroleum Geologists, Memoir, **14**, 502–555.

HARRISON, J. V. 1931. Salt domes in Persia. *Journal of the Institute of Petroleum Technology (London)*, **17**(91), 300–320.

—— 1968. Geology, *In*: FISHER, W. B. (ed.) *The Land of Iran*, The Cambridge History of Iran, **1**(2), Cambridge University Press, 111–185.

HEIM, A. 1958. Beobachtungen über Diapirismus. *Eclogae Geologicae Helvetiae*, **51**, 1–32.

HENSON, F. R. S. 1951. Observations on the geology and petroleum occurrences of the Middle East. *Proceeding of the Third World Petroleum Congress*, The Hague, 118–140.

HEWARD, A. P. 1990. Salt removal and sedimentation in Southern Oman. *In*: ROBERTSON, A. H. F. *et al.* (eds) *Geology and Tectonics of the Oman Region*. Geological Society London, Special Publication, **49**, 637–651.

HIRSCHI, H. 1944. Über Persiens Saltzstocke. *Schweizerische Mineralogische und Petrographische Mitteilungen*, **24**, 30–57.

HOSPERS, J., RATHORE, J. S., JIANHUA, FENG, FINNSTROM, E. G. & HOLTHE, J. 1988. Salt tectonics in the Norwegian–Danish Basin. *Tectonophysics*, **149**, 35–60.

HUBER , H. 1977. *Explanatory notes to geological map of Iran*. National Iranian Oil Company, Exploration and Production Group, **GR 340**, 1–170, (unpublished).

HUCKREIDE, R., KÜRSTEN, M. & VENZLAFF, H. 1962. Zur Geologie des Gebeites zwischen Kerman und Sagand. *Beihefte zum Geologischen Jahrbuch*, Hannover, **51**, 1–197.

HUGHES-CLARKE, M. W. 1988. Stratigraphy and rock unit nomenclature of the oil-producing area of Interior Oman. *Journal of Petroleum Geology*, **11**, 5–60.

HUSSEINI, M. I. 1988. The Arabian Infracambrian extensional system. *Tectonophysics*, **148**, 93–103.

HUSSEINI, M. I. & HUSSEINI, S. I. 1990. Origin of the Infracambrian Salt Basins of the Middle East. *In*: BROOKS, J. (ed.) *Classic Petroleum Provinces*. Geological Society, London, Special Publication, **50**, 279–292.

ION, D. C., ELDER, S. & PEDDER, A. E. 1951. The Agha Jari oilfield, south-west Persia, *3rd World Petroleum Congress, The Hague, Proceedings*, Section 1, 162–186.

JACKSON, J. A., FITCH, T. J. & MCKENZIE, D. P. 1981. Active thrusting and evolution of the Zagros fold belt. *In*: MCCLAY, K. R. & PRICE, N. J. (eds) *Thrust and Nappe Tectonics*. Geological Society, London, Special Publication, **9**, 371–379.

JACKSON, M. P. A. & TALBOT, C. J. 1986. External shapes, strain rates, and dynamics of salt structures. *Bulletin of the Geological Society of America*, **97**, 305–323.

—— & VENDEVILLE, B. C. 1994. Regional extension as a geologic trigger for diapirism. *Bulletin of the Geological Society Society of America*, **106**, 57–73.

——, CORNELIUS, R. R., CRAIG, C. H., GANSSER, A., STÖCKLIN, J. & Talbot, C. J. 1990. *Salt Diapirs of the Great Kavir, Central Iran*. Geological Society of America, Memoir, **177**.

JENYON, M. K. 1985. Fault-associated salt flow and mass movement. *Journal of the Geological Society, London*, **142**, 547–553.

—— 1986. *Salt Tectonics*. Elsevier, Barking.

KAMEN-KAYE, M. 1990. Geology and Productivity of Persian Gulf Synclinorium. *AAPG Bulletin*, **54**, 2371–2394.

KENT, P. E. 1958. Recent studies of south Persian salt plugs. *AAPG Bulletin*, **42**, 2951–2972.

—— 1970. The salt plugs of the Persian Gulf Region. *Leicester Literary and Philosophical Society, Transactions*, **64**, 55–58.

—— 1979. The emergent Hormuz salt plugs of southern Iran. *Journal of Petroleum Geology*, **2**, 117–144.

—— & WARMAN, H. R. 1972. An environmental review of the world's richest oil-bearing region – the Middle East. *Proceedings of the 24th International Geological Congress, Montreal*, Section 5, 142–152.

KINGSTON, D. R., DISHROON, C. P. & WILLIAMS, P. A. 1983. Global basin classification system. *AAPG Bulletin*, **67**, 2175–2193.

KOYI, H. 1988. Experimental modeling of role of gravity and lateral shortening in Zagros Mountain Belt. *AAPG Bulletin*, **72**, 1381–1394.

——, JENYON, M. K. & PETERSEN, K. 1993. The effect of basement faulting on diapirism. *Journal of Petroleum Geology*, **16**, 285–312.

LAW, J. 1957. Reasons for Persian Gulf oil abundance. *AAPG Bulletin*, **41**, 51–69.

LEES, G. M. 1929. Salzgletscher in Persien. *Mitteilungen der Geologischen Gesellschaft in Wien*, **24**, 29–34.

—— 1931. Salt – some depositional and deformational problems. *Journal of the Institute of Petroleum Technology (London)*, **17**, 259–280.

—— 1952. Foreland folding. *Quarterly Journal of the Geological Society, London*, **108**, 1–34.

—— 1953. Persia. *In*: ILLING, V. C. (ed.) *The World's Oilfields, The Eastern Hemisphere, The Science of Petroleum*, **6**, 73–82.

MATTES, B. W. & CONWAY MORRIS, S. 1990. Carbonate/evaporite deposition in the Late Precambrian–Early Cambrian Ara Formation of Southern Oman. *In*: ROBERTSON, A. H. F. *et al.* (eds) *The Geology and Tectonics of the Oman Region*. Geological Society, London, Special Publication, **49**, 617–636.

MINA, P., PARAN, Y. & RAZAGHNIA, M. 1967. Geological and geophysical studies and exploratory drillings of the Iranian continental shelf, Persian Gulf. *7th World Petroleum Congress*, **2**, 871–904.

MORRIS, P. 1977. Basement structure as suggested by aeromagnetic surveys in south-west Iran. *Proceedings of the Second Geological Symposium of Iran*, Tehran, 294–307.

MURRIS, R. J. 1980. Middle East stratigraphic evolution and oil habitat. *AAPG Bulletin*, **64**, 597–618.

NATIONAL IRANIAN OIL COMPANY (NIOC) 1978. *Geological Maps and Sections of Iran*. Six geological map sheets, explanatory notes, accompanying cross sections and tectonic maps, Tehran.

NORTH, F. K. 1985. *Petroleum Geology*. Allen and Unwin, Boston.

O'BRIEN, C. A. E. 1955. Salztektonik in Südpersien. *Zeitschrift der deutschen geologischen Gesellschaft*, Hanover, **104**, 803–813.

—— 1957. Salt diapirism in South Persia. *Geologie en Mijnbouw*, 337–376.

OWEN, P. F. & TAYLOR, N. G. 1983. A salt pillow structure in the southern North Sea. *In*: BALLY, A. W. (ed.) *Seismic Expression of Structural Styles*, AAPG Studies in Geology Series, **2**, 2.3.

OWEN, R. M. S. & NASR, S. N. 1958. Stratigraphy of the Kuwait-Basra area. *In*: WEEKS, L. G. (ed.). *Habitat of Oil, a Symposium*. American Association of Petroleum Geologists, Tulsa, 1252–1278.

PAUL, S. K. 1990. People's Democratic Republic of Yemen: a future oil province. *In*: BROOKS, J. (ed.) *Classic Petroleum Provinces*. Geological Society, London, Special Publication, **50**, 329–339.

PILGRIM, G. E. 1908. *The Geology of the Persian Gulf and adjoining portions of Persia and Arabia*. Memoir of the Geological Survey of India, **34**, 1–177.

PLAYER, R. A. 1969. *Salt Plug Study*. Iranian Oil Operating Companies, Geological and Exploration Division, Report No. **1146**, (unpublished).

PONIKAROV, V. P., KAZMIN, V. G., MIKHAILOV, I. A. *et al.* 1967. *The Geology of Syria, Explanatory Notes on the Geological Map of Syria. Scale 1:500,000, Part 1, Stratigraphy, Igneous rocks and Tectonics*. Ministry of Industry, Damascus.

PRICE, N. J. & COSGROVE, J. W. 1990. *Analysis of Geological Structures*. Cambridge University Press.

QATAR GENERAL PETROLEUM CORPORATION (QGPC) & AMOCO QATAR PETROLEUM COMPANY (AQPC). 1991. Dukhan Field-Qatar, Arabian Platform. *In*: FOSTER, N. H. & BEAUMONT, E. A. *Structural Traps V, Treatise of Petroleum Geology, Atlas of Oil and Gas Fields*. American Association of Petroleum Geologists, Tulsa, 103–120.

RICHARDSON, R. K. 1926. Die Geologie und die Salzdome im südwestlichen Teil des Persischen Golfes. *Verhandlungen des Naturhist.medizinischen Vereinigung. Heidelberg, N.F.*, **15**, 375–424.

—— 1928. *Weitere Bemerkungen zu der Geologie und den Salzaufbrüchen am Persischen Golf*. Centralblatt fur Mineralogie, Geologie und Palaontologie, Abt. B, Stuttgart, 43–49.

RÖNNLUND, P. & KOYI, H. 1988. Fry spacing of deformed and undeformed modeled and natural salt domes. *Geology*, **16**, 465–468.

SANNEMANN, D. 1968. Salt-stock families in northwestern Germany, *In*: BRAUNSTEIN, J. & O'BREIN, G. D. *Diapirism and Diapirs – A Symposium*. American Association of Petroleum Geologists, Tulsa, Memoir, **8**, Tulsa, 261–270.

SCHROEDER, J. W. 1944. Essai sur la structure de l'Iran. *Eclogae Geologicae Helvetiae*, **37**, 37–81.

—— 1946. Géologie de l'Île de Larak. Contribution a l'étude des dômes de sel du Golfe Persique. Comparaison avec la Salt Range. *Archives de Sciences phys, natur.*, Geneva, **28**), 1–18.

SENI, S. J. & JACKSON, M. P. A. 1983*a*. Evolution of salt structures, East Texas Diapir Province, Part 1: Sedimentary record of halokinesis. *AAPG Bulletin*, **67**, 1219–1244.

—— & —— 1983*b*. Evolution of Salt Structures, East Texas Diapir Province, Part 2: Patterns and rates of halokinesis. *AAPG Bulletin*, **67**(8), 1245–1274.

SPIERS, C. J. 1994. Microphysical aspects of the flow of rocksalt. *Salt Tectonics Conference, Programme and Abstracts*, Petroleum and Tectonics Groups of the Geological Society, London, 32.

STEINECKE, M., BRAMKAMP, R. A. & SANDER, N. J. 1958. Stratigraphic relations of Arabian Jurassic oil, *In*: WEEKS, L. G. (ed.) *Habitat of Oil, a Symposium*. American Association of Petroleum Geologists, Tulsa, 1294–1329.

STÖCKLIN, J. 1961. Lagoonal formations and salt domes in East Iran. *Bulletin of the Iranian Petroleum Institute*, **3**, 29–46.

—— 1968*a*. Salt deposits of the Middle East. *In*: MATTOX, R. B. *et al.* (eds.) *Saline Deposits*. Geological Society of America, Special Paper, **88**, 157–181.

—— 1968*b*. Structural history and tectonics of Iran: a review. *AAPG Bulletin*, **52**, 1229–1258.

SUGDEN, W. 1962. Structural analysis and geometrical prediction for change of form with depth of some Arabian plains-type folds. *AAPG Bulletin*, **46**, 2213–2228.

TALBOT, C. J. 1981. Sliding and other deformation mechanisms in a glacier of salt, S Iran, *In*: MCCLAY, K. R. & PRICE, N. J. (eds) *Thrust and Nappe Tectonics*. Geological Society, London, Special Publication, **9**, 173–183.

—— & ALAVI, M. 1994. Salt extrusions along the Kazaroun–Qatar and Mangarak–Marzuk lineaments across the Arabian Platform and Zagros Fold-Thrust Belt. *Salt Tectonics Conference, Programme and Abstracts*, Petroleum and Tectonic Groups of the Geological Society, London, 37.

—— & JARVIS, R. J. 1984. Age, budget and dynamics of an active salt extrusion in Iran. *Journal of Structural Geology*, **6**, 521–533.

TRUSHEIM, F. 1957. Über Halokinese und ihre Bedeutung fur die strukturelle Entwicklung Norddeutschlands. *Zeitschrift der Deutschen Geologischen Gesellschaft*, Hanover, **109**, 11–158.

—— 1960. Mechanism of salt migration in northern Germany. *AAPG Bulletin*, **44**, 1519–1520.

—— 1974. Zur Tektogenese der Zagros-Ketten Süd Irans. *Zeitschrift der deutschen geologischen Gesellschaft Hanover*, **125**, 119–150.

TURCOTTE, D. L. & SCHUBERT, G. 1982. *Geodynamics, Application of Continuum Physics to Geological Problems*. Wiley, New York.

URAI, J. L., SPIERS, C. J., ZWART, H. J. & LISTER, G. S. 1986. Weakening of rock salt by water during long-term creep. *Nature*, **324**, 554–557.

VISSER, W. 1991. Burial and thermal history of Proterozoic source rocks in Oman. *Precambrian Research,* **54**, 15–36.

VITA-FINZI, C. 1979. Rates of Holocene folding in the coastal Zagros near Bandar Abbas, Iran. *Nature*, **278**, 632–634.

—— 1986. *Recent Earth Movements, an Introduction to Neotectonics*. Academic, London.

WALTHER, H. W. 1972. Über Salzdiapire in Sudöst-Iran. *Geologisches Jahrbuch*, **90**, 359–388.

WISSMANN, H. VON, RATHJENS, C. & KOSSMAT, F. 1942. Beiträge zur tektonik Arabiens. *Geologisches Rundschau*, **33**, 221–353.

WOLF, J. R. 1959. The geology of Hormuz Island at the entrance of the Persian Gulf. *Geologie en Mijnbouw*, **21**, 390–395.

ZHARKOV, M. A. 1984. *Paleozoic Salt Bearing Formations of the World*. Springer, Berlin.

# An insight on salt tectonics in the Angola Basin, South Atlantic

FIVOS SPATHOPOULOS

*Amerada Hess Limited, 33 Grosvenor Place, London SW1X 7HY, UK*

**Abstract:** Salt movements have dominated the post-Aptian tectonic process of the West Africa Salt Basin, which extends from northern Namibia to southern Cameroon. The thick salt layer was deposited during the Aptian within the shallow rift basins of the proto-South Atlantic and was overlain by an extensive carbonate platform. As a result of the basinward tilt of the margin, the carbonate platform and the overlying Upper Cretaceous sequence slid on the salt layer and broke up into smaller blocks. This process caused a mobilization of the salt and the formation of salt rollers, salt diapirs and a whole suite of raft tectonics structures such as turtlebacks, carbonate rafts and severe folding at the toe of the salt basin. Although this process is seen throughout the Angola Salt Basin, the areal distribution of the salt movement-induced structures shows a significant differentiation. Burial history curves also show important differences along the margin in the post-Aptian subsidence. Analysis of recent seismic data suggests that the thin-skinned extension needed for salt and raft tectonics is accommodated basinward by folding of the overburden up to a point where all the extension has been taken up. From there to the toe of the salt layer, the overburden is largely undisturbed. The salt and raft tectonics in the Angola Basin was the main factor that influenced the distribution of oil reserves in the region and its understanding is very important for hydrocarbon exploration there.

The part of the West African margin which includes the oil-producing Lower Congo and Kwanza Basins of Angola and Zaire (Fig. 1) was formed predominantly by E–W extension associated with the opening of the South Atlantic, and is dissected by a large number of approximately N–S trending normal faults which mostly dip to the west (Brice *et al.* 1982; Schlumberger 1991). The basins are bounded to the east and are underlain, in their continental part, by the Precambrian Mayombe metamorphic system (Brice *et al.* 1982). The Lunda Ridge forms the southern boundary of the Kwanza Basin, and the Ambrizette Spur divides the Kwanza from the Lower Congo Basin (Fig. 1). Because of their large oil reserves, both areas have been the focus of intense oil exploration and production since 1910. The data collected became the basis of numerous publications on the geology and tectonics of these basins (Brognon & Verrier 1966; Brice *et al.* 1982; Schlumberger 1991). As it became obvious that salt-induced structures were responsible for all post-rift oil accumulations, new tectonic models for the salt basins of West Africa (Brown 1993; Gaullier *et al.* 1993), as well as of Brazil (Mohriak *et al.* 1990; Cobbold & Szatmari 1991; Demercian *et al.* 1993; Szatmari *et al.* 1996) and the Gulf of Mexico (Hongxing *et al.* 1994), have been published in the last few years. These make use of modern seismic data and salt kinematic laboratory experiments conducted in the USA and France. These models explain the post-rift evolution of the basins by referring to raft tectonics, a thin-skinned tectonic process responsible for the creation of a series of structures such as rafts, grabens and salt rollers.

Most of the work published to date about the South Atlantic salt basins has been related to specific areas of Brazil, Angola, Congo and Gabon, where detailed analyses of local salt and raft tectonics were presented. In this paper, basin evolution and raft tectonics are examined more regionally, both in a latitudinal (from Congo to southern Angola) and a longitudinal (from landward basin edge to seaward edge of salt layer) sense. This gives us the opportunity to study the large-scale processes of raft tectonics as well as the variations that occur not only between basins but also within the same basin.

The database on which this study is based consists of seismic lines from the GECO-PRAKLA GWA88 seismic survey plus stratigraphic tops from 115 wells all along the Angola–Zaire margin (Fig. 2).

## Tectonics and geology

### *1. Tectonic concepts*

Before discussing the tectonic history of the Angola–Zaire basins, it is necessary to introduce certain concepts and theories relating to the formation of the observed structures. ‘Raft

*From* ALSOP, G. I., BLUNDELL, D. J. & DAVISON, I. (eds), 1996, *Salt Tectonics*, Geological Society Special Publication No. 100, pp. 153–174. 

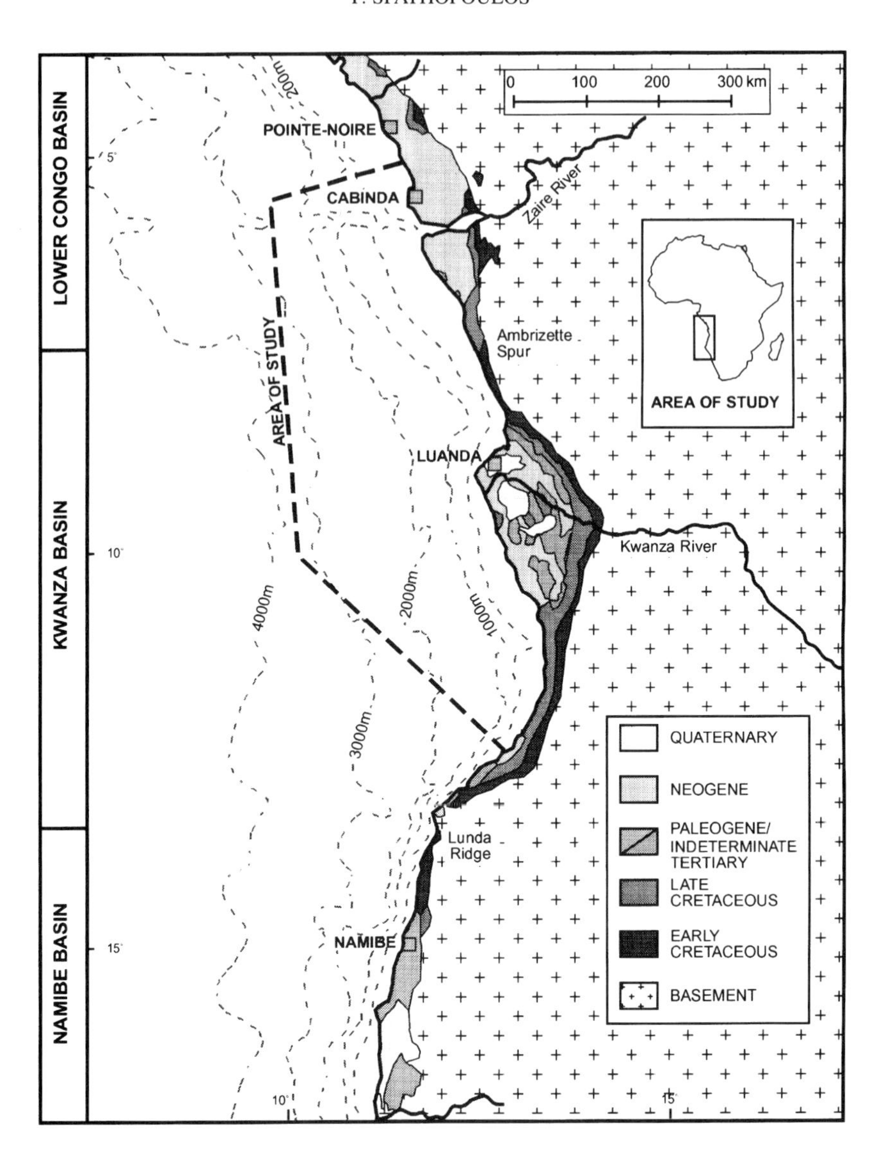


**Fig. 1.** Study area location map. The onshore geology and the main rivers are also shown.

tectonics' is the downslope movement of rigid blocks of sediment due to gravity-driven sliding on a décollement layer (usually evaporites or shales). It is a thin-skinned tectonic process whereby the footwalls and hangingwalls of fault blocks become completely separated; the fault blocks are then called 'rafts' (Duval *et al.* 1992). The downslope movement produces listric faults and grabens. These grabens are filled with syndeformational strata (Lundin 1992) and can be large; for example, the Quenguela Graben in the Kwanza Basin is 20 km wide and 90 km long (Verrier & Castello Branco 1972) and the Gaivota Graben is 15–20 km wide and 225 km long (Duval *et al.* 1992). The mechanism by which sediment infilling occurs within a graben formed by salt rise and fall has been discussed by Vendeville & Jackson (1992*a,b*) and Lundin (1992). With continued extension between two blocks, the salt rises passively to fill the increasing gap until the supply of salt to the diapir cannot meet the demand and the diapir starts to fall (Vendeville & Jackson 1992*a,b*). As the diapir sags, its crest becomes a subsiding trough which fills rapidly with sediments. If the downslope movement does not cause the complete separation of blocks, but simply the dissection of the sedimentary layer by listric faults, the fault blocks remain in contact and are called 'pre-rafts'.

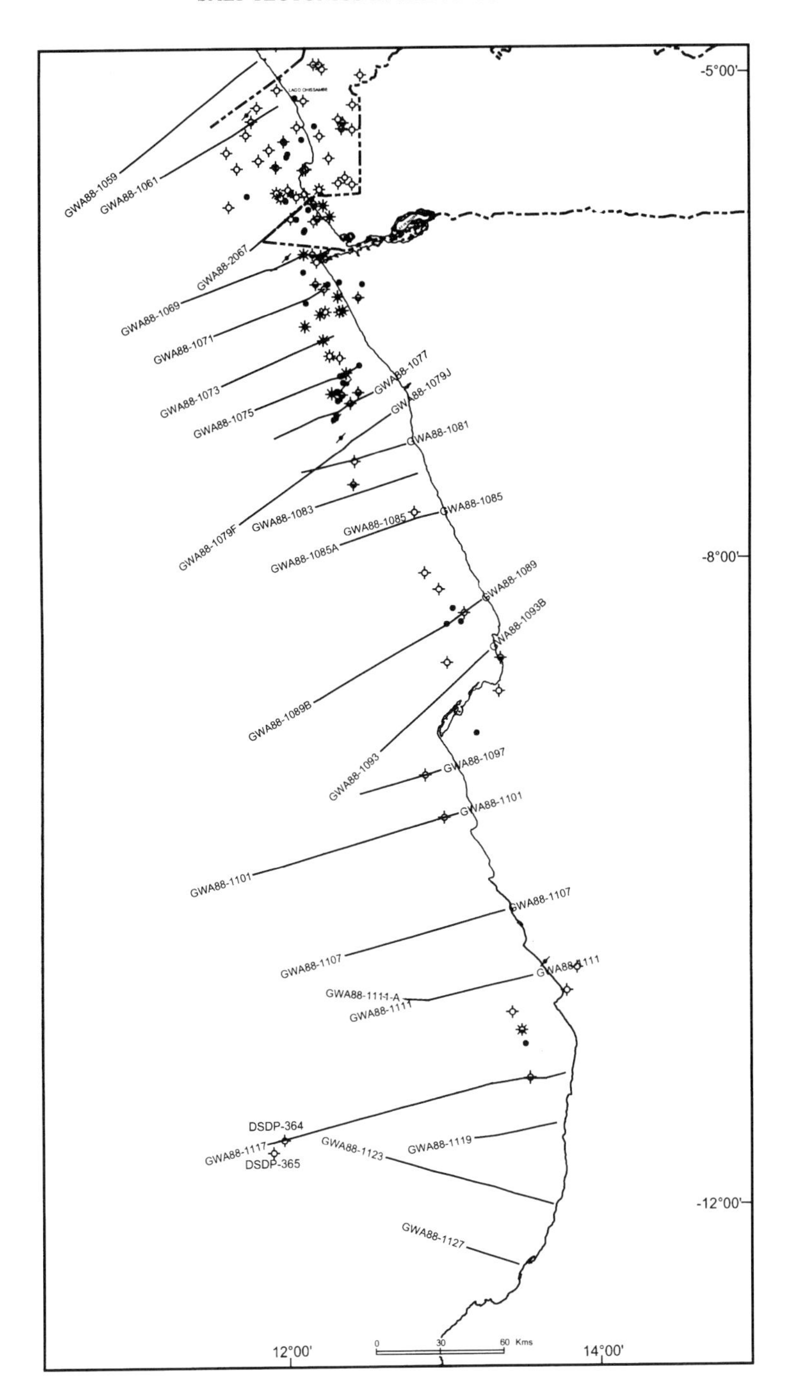


**Fig. 2.** Database map with the seismic lines used in this study. Wells shown on the map are those for which the burial history curve was calculated.

When the décollement layer is evaporitic, its thickness is progressively reduced throughout the whole process of formation of rafts and pre-rafts due to downslope salt flow and dissolution. When most of the salt is removed under the faulted blocks, their downslope movement will cease and they will be welded (or grounded) on the pre-salt layers (Lundin 1992).

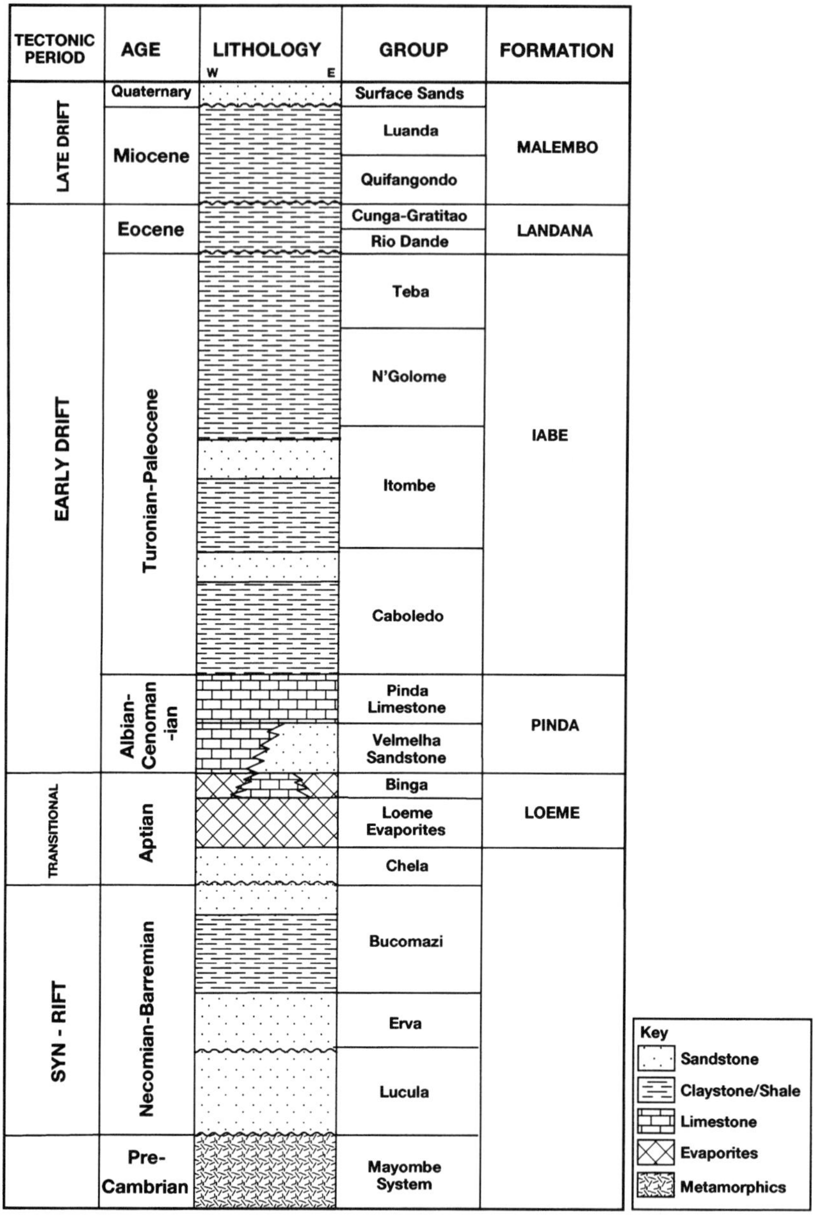

| TECTONIC PERIOD | AGE | LITHOLOGY (W–E) | GROUP | FORMATION |
|---|---|---|---|---|
| LATE DRIFT | Quaternary | | Surface Sands | MALEMBO |
| | Miocene | | Luanda | |
| | | | Quifangondo | |
| EARLY DRIFT | Eocene | | Cunga-Gratitao | LANDANA |
| | | | Rio Dande | |
| | Turonian-Paleocene | | Teba | IABE |
| | | | N'Golome | |
| | | | Itombe | |
| | | | Caboledo | |
| | Albian-Cenoman-ian | | Pinda Limestone | PINDA |
| | | | Velmelha Sandstone | |
| TRANSITIONAL | Aptian | | Binga | LOEME |
| | | | Loeme Evaporites | |
| | | | Chela | |
| SYN - RIFT | Necomian-Barremian | | Bucomazi | |
| | | | Erva | |
| | | | Lucula | |
| | Pre-Cambrian | | Mayombe System | |

Key: Sandstone; Claystone/Shale; Limestone; Evaporites; Metamorphics

**Fig. 3.** Stratigraphy of the Angola–Zaire basins.

## *Tectonic history*

The tectonic history of the Angola and Zaire Basins can be broadly divided into the following periods.

(i) The *syn-rift period* extended from Early Neocomian to Aptian. Normal faulting produced rift grabens and basement-tilted fault blocks which are the dominant structural features in the pre-salt section. The sediments that accumulated in the topographic lows during this period were Neocomian fluvial sandstones and conglomerates overlain by organic-rich shales (Fig. 3) deposited in deep, anoxic lakes (Brognon & Verrier 1966; Brice *et al.* 1982; Schlumberger 1991).

(ii) During the *transitional period* (Aptian), a change from rift to drift tectonics occurred and was marked by a southerly marine transgression into the Angolan rift basins (Brognon & Verrier 1966; Schlumberger 1991). The initial deposit was a thin layer of sand (Chela Sandstone) deposited in a coastal to shallow-water environment. Subsequent cycles of marine incursion into the restricted basins resulted in the accumulation of a thick evaporitic layer (Loeme Evaporites) composed mostly of halite followed by anhydrite (Fig. 3). The evaporites are, in places, more than 1000 m thick and in the middle of the basin attain a thickness of approximately 3000 m (Bolli *et al.* 1978).

(iii) The *early drift period* started immediately after salt deposition ceased (Early Albian) and is characterized by a major marine transgression (Tillement 1987). This event resulted in the formation of a carbonate ramp during Albian–Early Cenomanian (Fig. 3) in a sabkha to shallow marine environment. The deposition of carbonates (Pinda Formation) was interrupted intermittently by influxes of terrestrial clastics (Tillement 1987). As soon as the Albian carbonates attained a thickness of a few hundred metres, or less in some areas, they were dissected by numerous normal listric faults, most of which dip oceanward (Brice *et al.* 1982). These faults, probably caused by gravity sliding of the carbonate section on the evaporite layer (Duval *et al.* 1992), created the carbonate pre-rafts and rafts. Salt moved under these listric faults to fill the void between the separating fault blocks (Duval *et al.* 1992), creating salt rollers (Fig. 4.1).

Carbonate sedimentation ceased in the Cenomanian and was replaced by clastic deposition. Slowing down of subsidence during the Upper Cretaceous and Early Tertiary allowed transgressive shale and sand units of Cenomanian–Eocene age (Fig. 3) to yoke the carbonate pre-rafts together (Duval *et al.* 1992) (Fig. 4.1). Epeirogenic uplift of southwest Africa, which was initiated during Late Cretaceous (Sahagian 1988), caused strong basinward tilting of the Angola margin. In the Early Tertiary, the whole post-salt sedimentary sequence began to separate into rafts which started to slide downslope (Fig. 4.2) on the evaporite décollement layer (Duval *et al.* 1992). The salt rose passively to fill the gap between the rafts up to a point where the supply of salt could not keep pace with the increasing void and the salt started to fall, creating local depocentres (Lundin 1992). The movement continued during the Oligocene and Miocene, probably assisted by the increased sedimentary load as the shelf prograded during the Oligocene sea-level drop and in response to the Miocene uplift of the southern Africa region (Lunde *et al.* 1992; Walgenwitz *et al.* 1990).

(iv) The *late drift period* began in the Oligocene with a significant erosional event. A major sea-level drop led to the removal of a large part of the Cenomanian–Eocene section (Schlumberger 1991), producing a major unconformity. The overlying Miocene regressive sequence is characterized by turbiditic sands and shales infilling the grabens (Fig. 4.3). Uplift of the southern part of the African continent in the late Tertiary resulted in deep erosion onshore (Lunde *et al.* 1992).

## Seismic interpretation – structural provinces

For the mapping of the geological horizons, the GECO-PRAKLA GWA88 seismic survey was used. The seismic data were calibrated to stratigraphic tops taken from unpublished well information. While the base of the salt can be clearly traced as a strong reflector, its top can be mapped with accuracy only in the areas with dense well control. In the deep-water areas, the top of the diapirs is picked with limited accuracy. Confidence in the mapping of the top salt reflector in these areas was added by the stratigraphic tops and times provided by DSDP Site 364 (Bolli *et al.* 1978), which is crossed by line GWA88-1117 (Fig. 2). Although the mapping of the post-salt reflectors presented no problems, the pre-salt reflectors are poorly imaged on the seismic data. The interpretation was mainly focused on the post-salt structures and the salt tectonics. Based on the raft tectonics features and the nature of the salt bodies, five distinct N–S trending tectonic provinces characterized by thin-skinned extension or contraction were defined. Seismic examples from each of these provinces can be seen in Figs 5–9.

### *Extensional provinces*

The *unrafted province* exists in the onshore and the eastern part of the offshore Lower Congo Basin, as well as off the southern Kwanza Basin. It is characterized by the presence of Albian–Cenomanian pre-rafts yoked together by Upper Cretaceous–Eocene strata (Fig. 5). Very few of the listric normal faults which dissect the Albian carbonate ramp continue upwards through the Upper Cretaceous. Salt rollers underlie most of the listric faults.

The *raft province* lies to the west of the unrafted region (Fig. 6) and is separated from it by a swarm of large, basinward dipping listric normal faults. These faults sole out on the salt layer and are not directly connected with structural features within the pre-salt section. The raft province is characterized by alternating rafts and grabens, the size of which varies along the margin (5–40 km). Correlation of seismic horizons with wells, together with published information, indicate that all grabens are filled with sediments of Tertiary age (Brognon & Verrier 1966; Verrier & Castello Branco 1972;

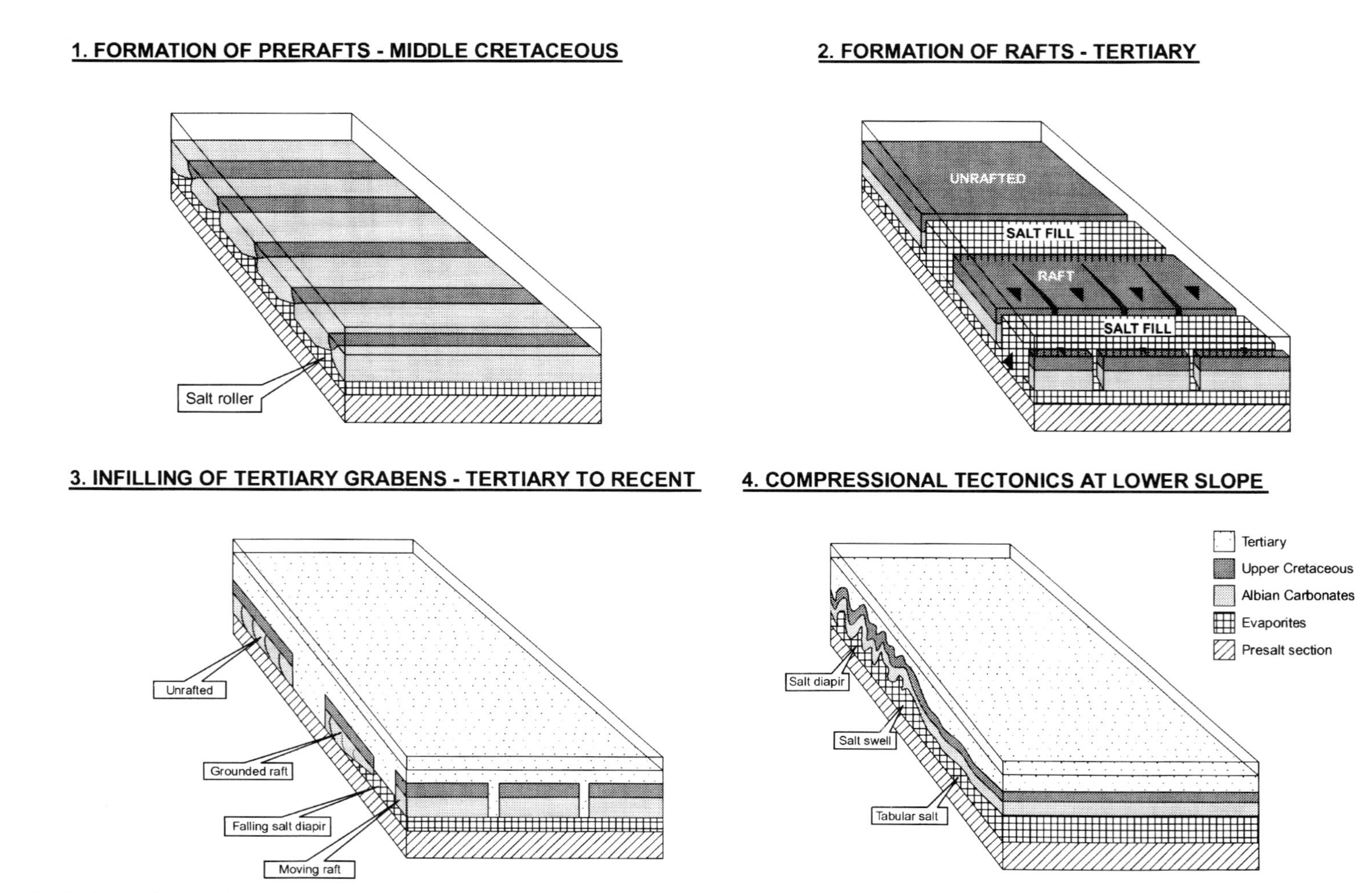


**Fig. 4.** Schematic post-salt tectonic history of the Angola–Zaire basins.

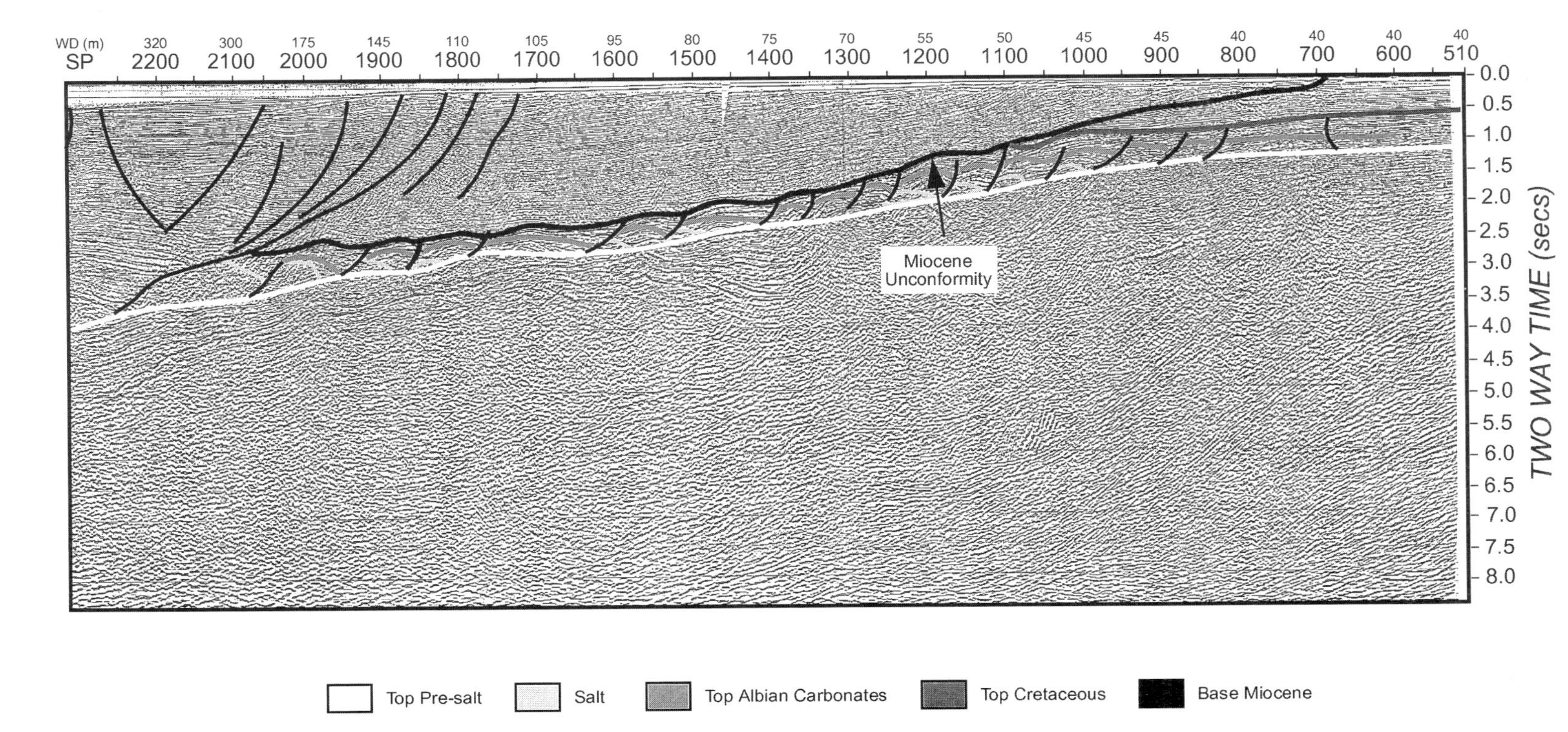


**Fig. 5.** Seismic line across the unrafted province in the Lower Congo Basin. Note the pre-rafts separated by listric faults and yoked together by Late Cretaceous sediments and the Base Miocene unconformity.

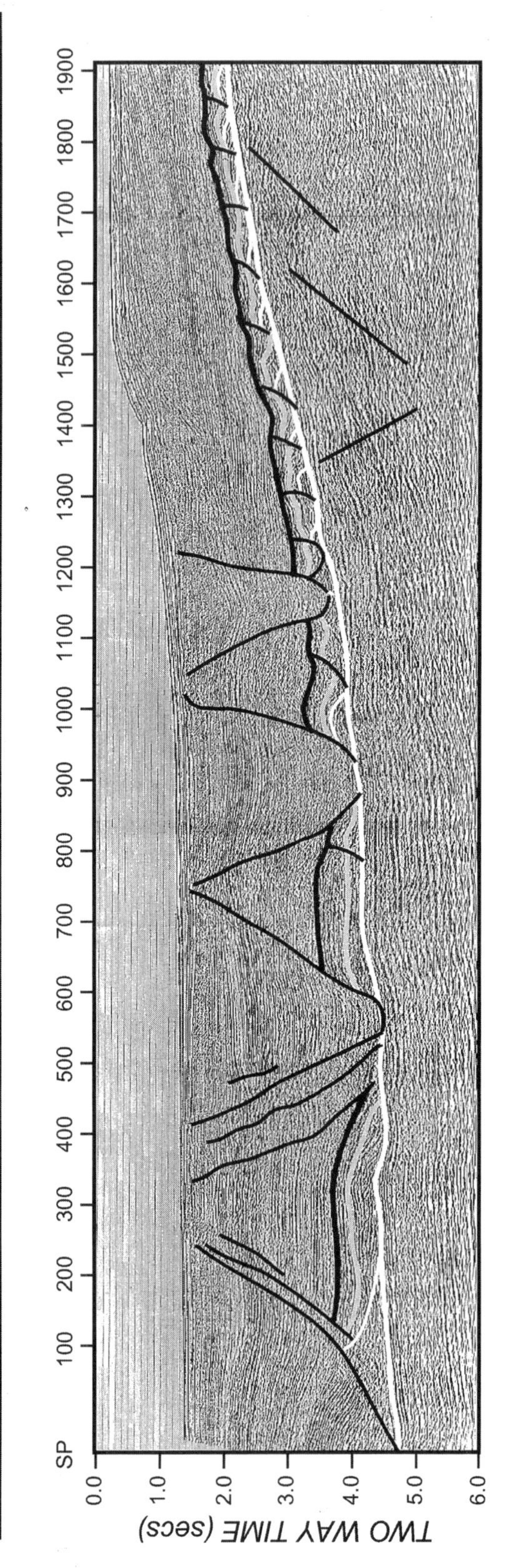


**Fig. 6.** Seismic line across the raft province in the northern Kwanza Basin, showing three rafts which have been detached from the unrafted zone on the east and the intervening Tertiary grabens.

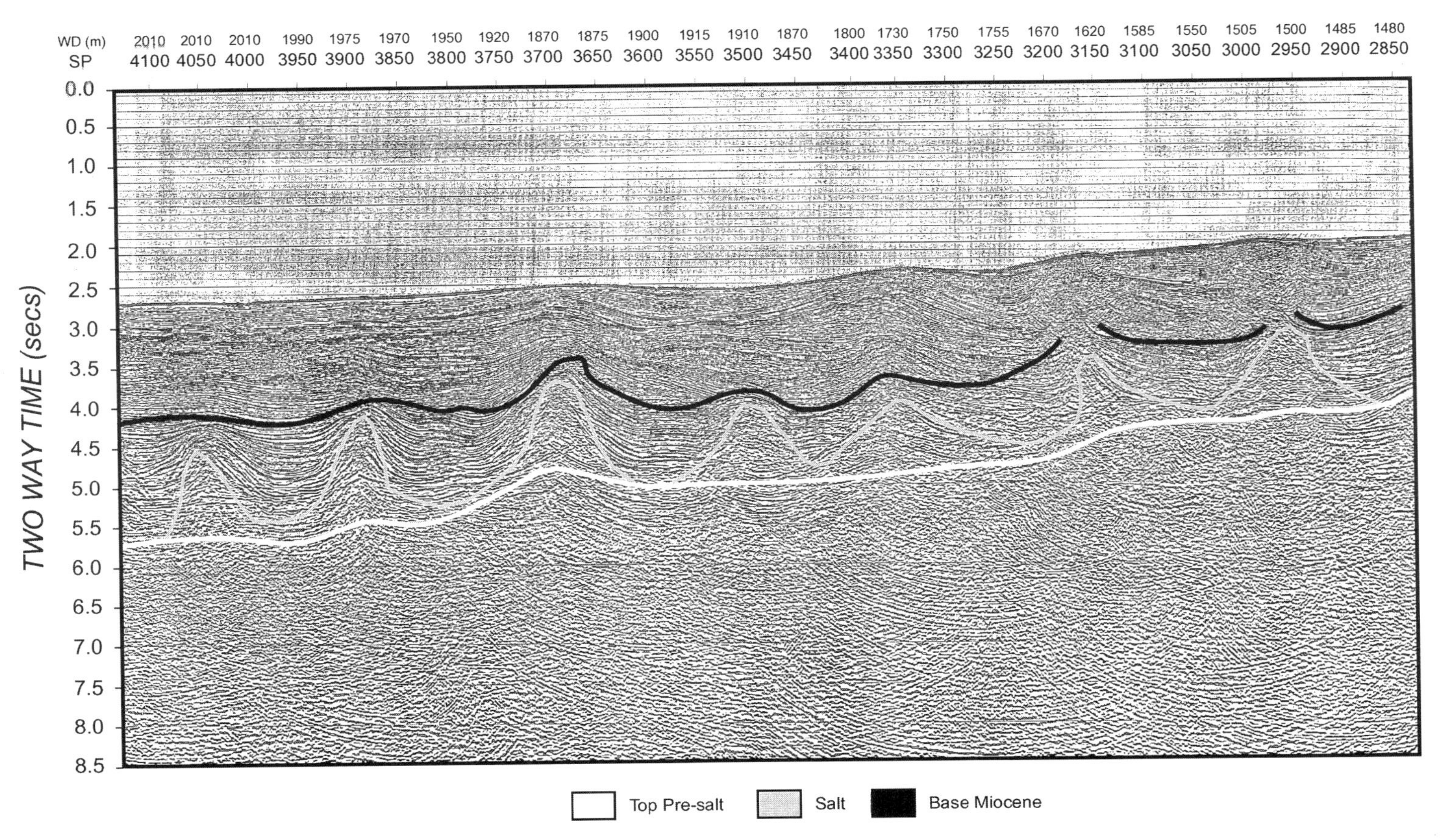


**Fig. 7.** Seismic line across the salt diapir province in southern Kwanza Basin.

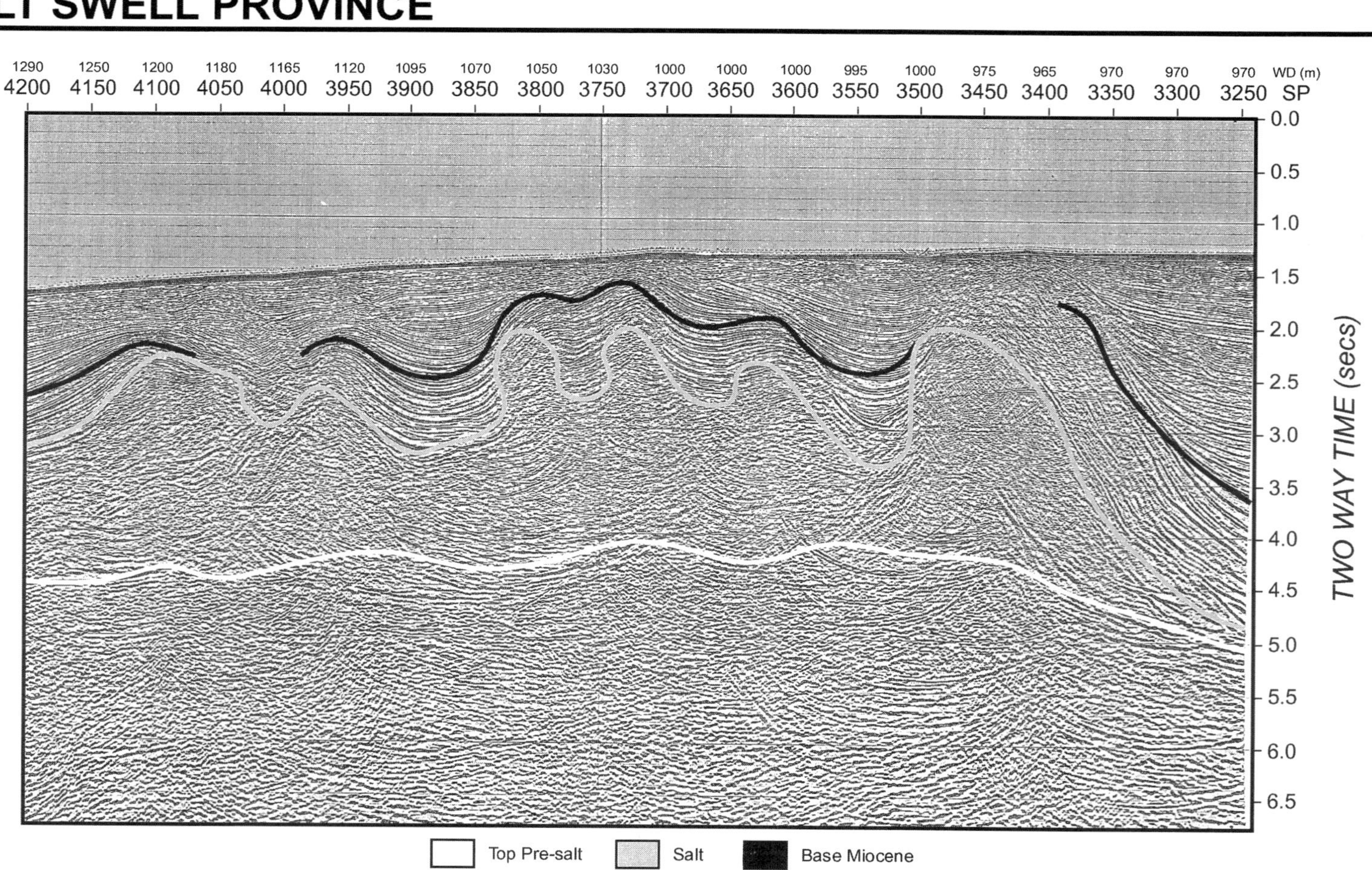


**Fig. 8.** Seismic line across the salt swell province in southern Kwanza Basin. Note the abrupt increase in the thickness of the salt layer at shot point 3300.

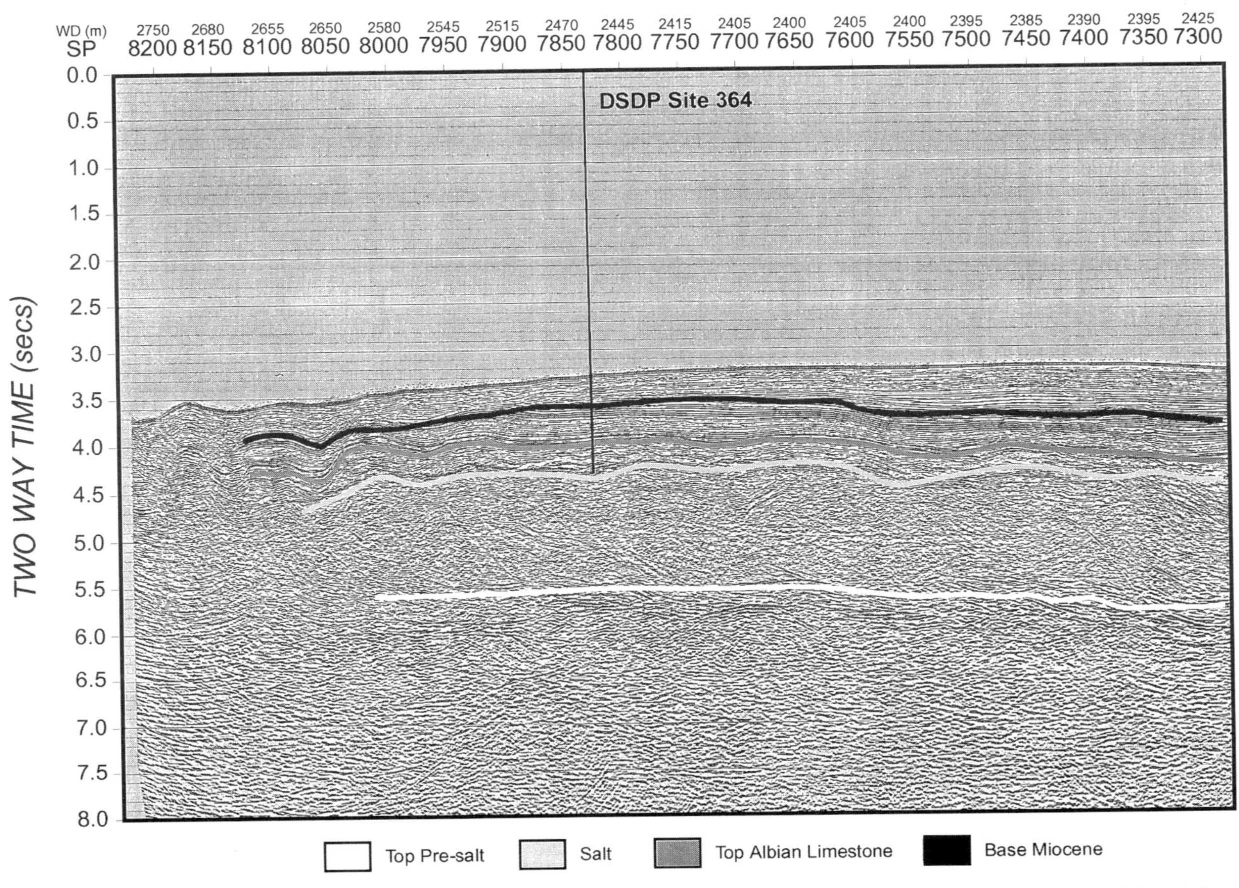


**Fig. 9.** Seismic line across the tabular salt province in southern Kwanza Basin. The basement uplift at the left-hand end of the line is the edge of the salt basin and probably constitutes the fixed point against which all upslope extension must be accommodated.

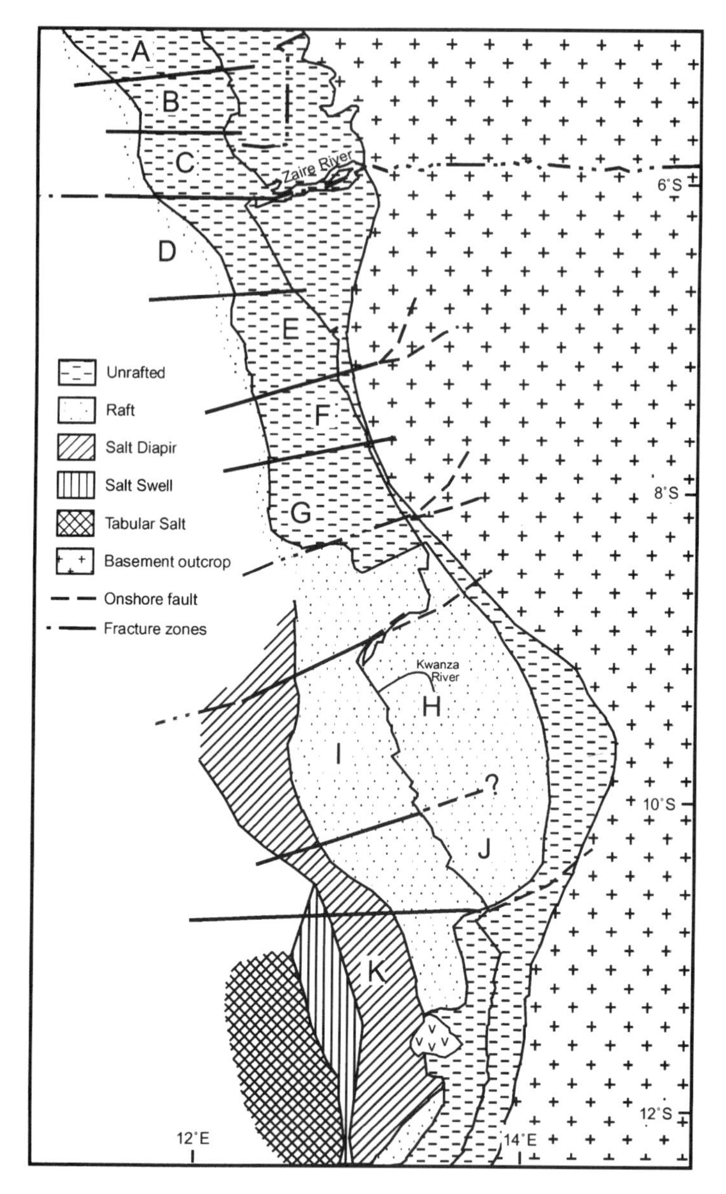


**Fig. 10.** Areal distribution of the unrafted, raft, salt diapir, salt swell and tabular salt provinces. The large extent of the raft province in the Kwanza Basin is clearly observable. Sections A to K are also shown. The onshore faults and the fracture zones are taken from the Geological Map of Angola (1988).

Lundin 1992). The salt layer in the province is very thin to absent. Large subsiding salt diapirs are found only locally, in grabens where the salt has not yet been depleted.

## *Contractional provinces*

The *salt diapir province* is located to the west of the raft province, in water depths of more than 500 m. The province is characterized by several salt diapirs (Fig. 7), each about 5 km in diameter, which have pierced the folded rafts. In some places the rafts are underlain by a layer of salt 200–300 ms in thickness. It appears that the salt diapirs are not isolated features but are connected to form N–S striking salt walls.

The *salt swell province* is the westward continuation of the salt diapir province and lies in water

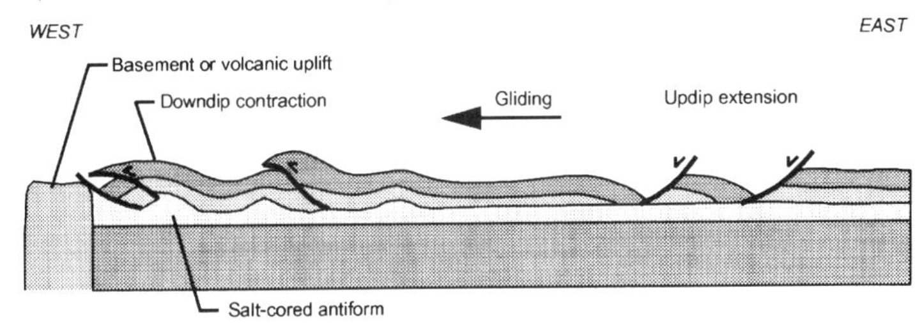


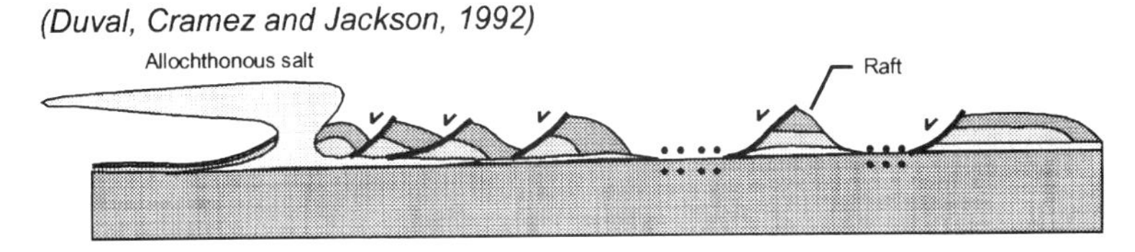


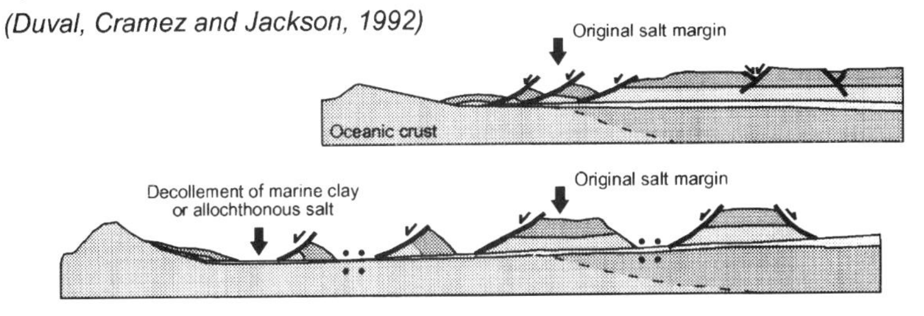


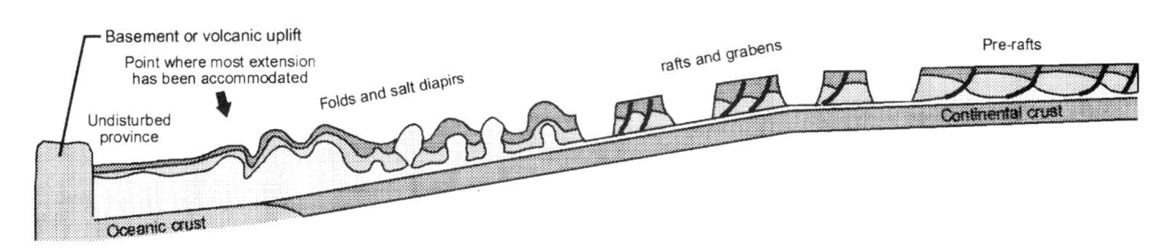


**Fig. 11.** Models for the creation of lateral space for raft tectonics. Models 1, 2, and 3 are from Duval *et al.* (1992). Model 4 is the one proposed by this study.

depths of more than 1000 m. Its main characteristics are a very thick evaporite layer (1000–2000 ms) with salt swells (*sensu* Jenyon 1986) and diapirs on top (Fig. 8), and the strongly faulted and folded post-salt sedimentary layers. The abrupt change in the thickness of the salt layer observed at the boundary between this province and the previous one can be attributed to the basinward expulsion of salt by the prograding Upper Cretaceous–Recent sediment wedges. Experimental proof of this process has been provided by Hongxing *et al.* (1994).

The *tabular salt province* lies in water depths of more than 1500 m. It constitutes the westward continuation of the salt swell province and is characterized by a thick (1500–2000 ms) evaporite layer, the top of which is almost flat (Fig. 9). The salt is overlain by a relatively thin (500–1000 ms) Albian–Recent sedimentary section.

The areal extent of the above provinces can be seen on Fig. 10. The location of the boundary between unrafted and rafted provinces varies from place to place along the margin. In the southern parts of the Lower Congo and Kwanza Basins the boundary is near (50–70 km) the landward basin edge. In the rest of the Kwanza Basin and to the north of 7°S (northern Angola, Zaire and Cabinda) the boundary lies at a considerable distance (100–120 km) from the basement outcrop.

## Accommodation of the thin-skinned extension

From the above presentation of the geological history of the Angola–Zaire Basins it is evident and well recognized that their post-rift structure has largely been determined by raft tectonics. However, in raft tectonics, as in any form of thin-skinned extension, the problem of frontal space to accommodate the downslope movement of the blocks is crucial. Published models which try to provide this space (Figs 11.1–11.3) include the formation of a fold-and-thrust belt at the downdip edge of the basin, the compressive upward displacement of the salt and the formation of new oceanic crust onto which the last downdip rafts will slide (Duval *et al.* 1992). Line GWA88-1117, included in the database of the present study (Fig. 2), crosses the whole margin from near the coastline up to the edge of the salt basin (Fig. 12) and ties with DSDP Site 364 which encountered Albian–Aptian carbonates and reached the top of salt (Bolli *et al.* 1978). This line allows us to study the accommodation of the updip extension needed for raft tectonics. Basinwards from the pre-rafts, a graben formed by thin-skinned extension exists and this is followed, further to the west, by a province dominated by salt diapirs which penetrate folded rafts. Adjacent to the latter, there is a province of thick salt with swells and a province of thick, tabular salt. The edge of the salt basin is defined by a basement or volcanic high (Bolli *et al.* 1978) that acts as a fixed point against which all compression must be accommodated. It can be suggested, therefore, that the downslope, extensional movement which creates rafts and grabens is translated on the lower slope into compressional folding of the Albian–Miocene section. Salt has moved into the void created by the folds and forms salt diapirs. After most of the extensional movement has been accommodated by downdip folding, the post-salt section and the salt layer remain little disturbed (Figs 4.4 and 11.4). Indications of some folding near the salt basin edge can be seen on line GWA88-1117 (Figs 9 & 12) and were also found in the Albian carbonates in DSDP Site 364 (Bolli *et al.* 1978).

Displacement of autochthonous salt as salt tongues into the post-salt sequence, as postulated in model 2 (Fig. 11.2), may also occur, especially at the boundary between the extensional and contractional provinces. In that sense, model 2 can be considered as part of the more general model 4 (Fig. 11.4), representing the transition between the extensional and contractional regimes.

## Variation in raft tectonics along the Angola–Zaire margin

Mapping of unrafted and rafted provinces reveals that, apart from the above longitudinal division, the margin can also be split up into latitudinal units on the basis of the width of the unrafted province. Accompanying this variation, there is also a significant change in the style of raft tectonics along the Lower Congo and Kwanza Basins, which can be seen in Fig. 13. Off Cabinda and Zaire, a narrow (*c.* 10 km) graben created by gravitational sliding separates the unrafted province from the first raft (cross-section 1, Fig. 13). To the south of the Zaire estuary, a wide (90–100 km) unrafted section is observed, followed by large grabens filled with Tertiary sediments and underlain by subsiding salt layers (cross-sections 2 and 3, Fig. 13). Further southwards, the unrafted section becomes smaller (50–60 km) and the number of rafts and grabens increases (cross-section 4, Fig. 13). In the northern and central parts of the Kwanza Basin the unrafted section is small in width (*c.* 10 km), while several large (15–20 km wide) grabens and rafts exist, bounded to the west by a compressional province of folded post-salt sediments, salt diapirs and swells (cross-sections 5 and 6, Fig. 13). Although no seismic lines in the onshore Kwanza Basin were available for this study, published maps and cross-sections indicate that large grabens exist very near the basement outcrop (Brognon & Verrier 1966). The style of raft tectonics changes further off southern Kwanza Basin, where the unrafted section has a width of 50–60 km (cross-sections 7 and 8, Fig. 13). There is only one graben in this area, immediately followed by a province of strongly folded Albian–Miocene sequence and a very thick salt layer with diapirs and swells.

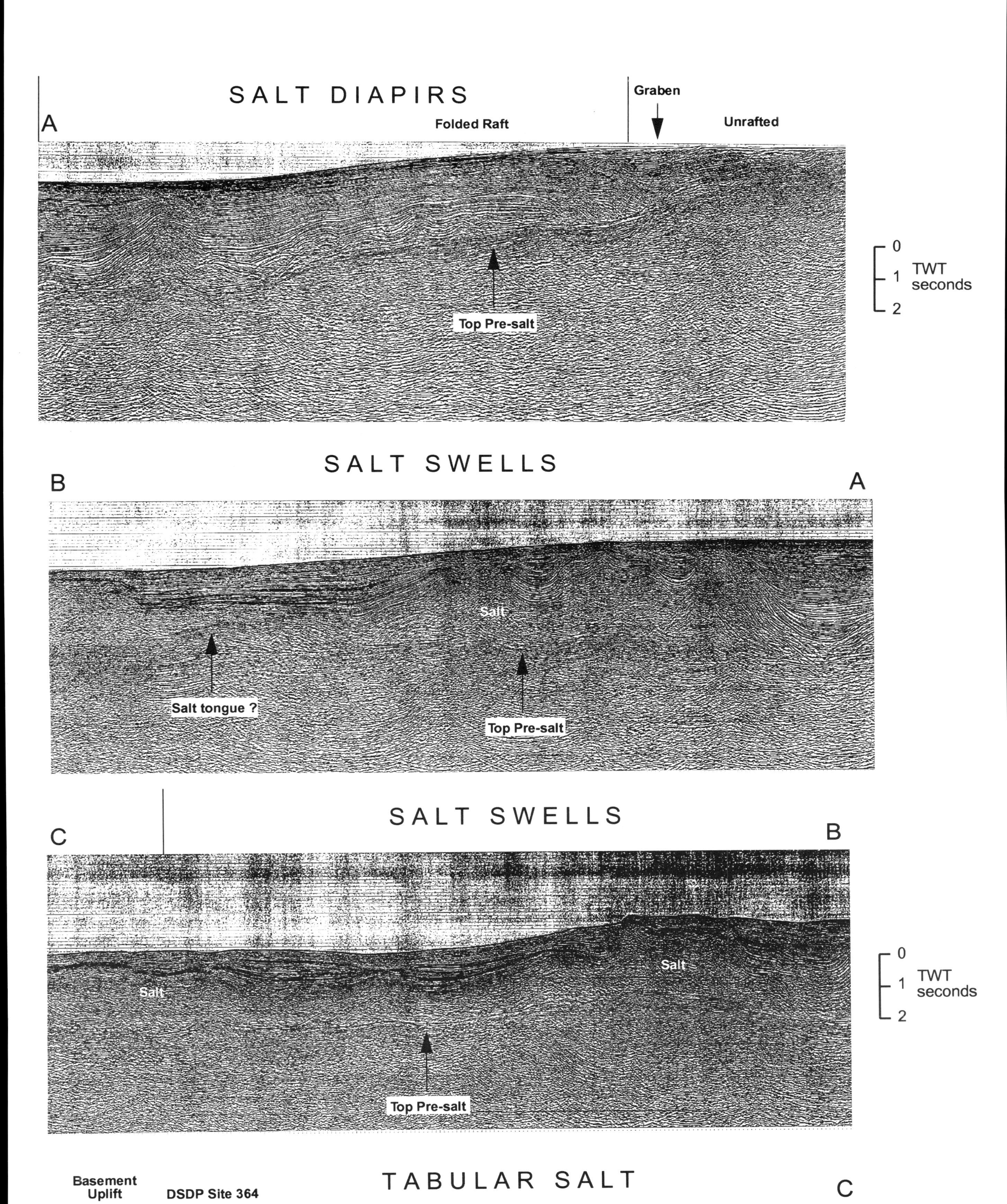


**Fig. 12.** Seismic line GWA88-1117. For location see Fig. 2. The line is presented in four parts and letters A, B and C indicate continuation points.

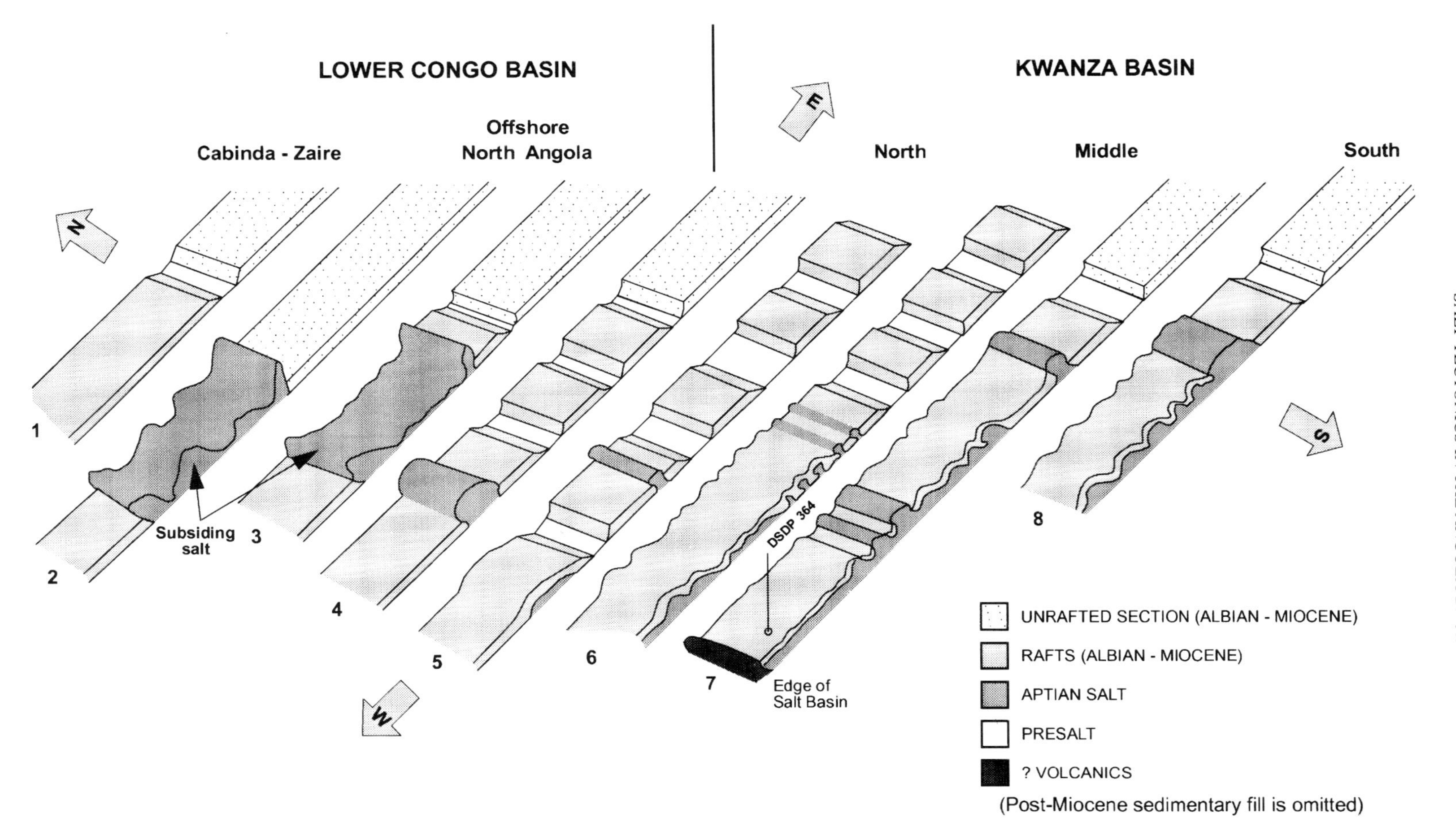


**Fig. 13.** Patterns of raft tectonics along the Angola–Zaire basins. The post-Miocene sedimentary fill is omitted for clarity. The unrafted section attains its maximum width off northern Angola and almost disappears in the Kwanza Basin. Note also the extent of the contractional regime (expressed by folding) in the same basin. The salt bodies in cross-sections 2 and 3 are subsiding and large grabens are being created.

TYPICAL BURIAL HISTORY CURVES FOR SECTIONS A - F

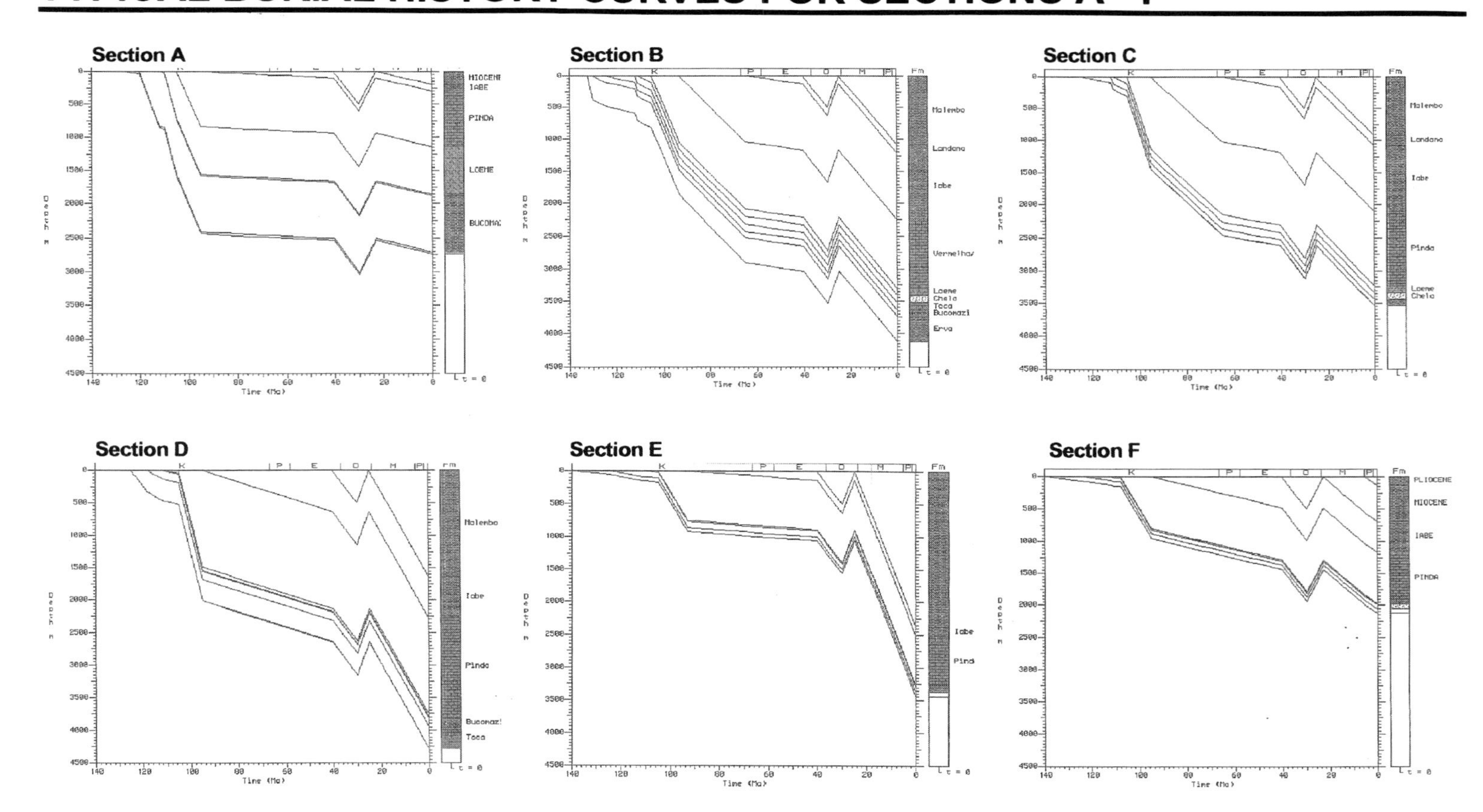

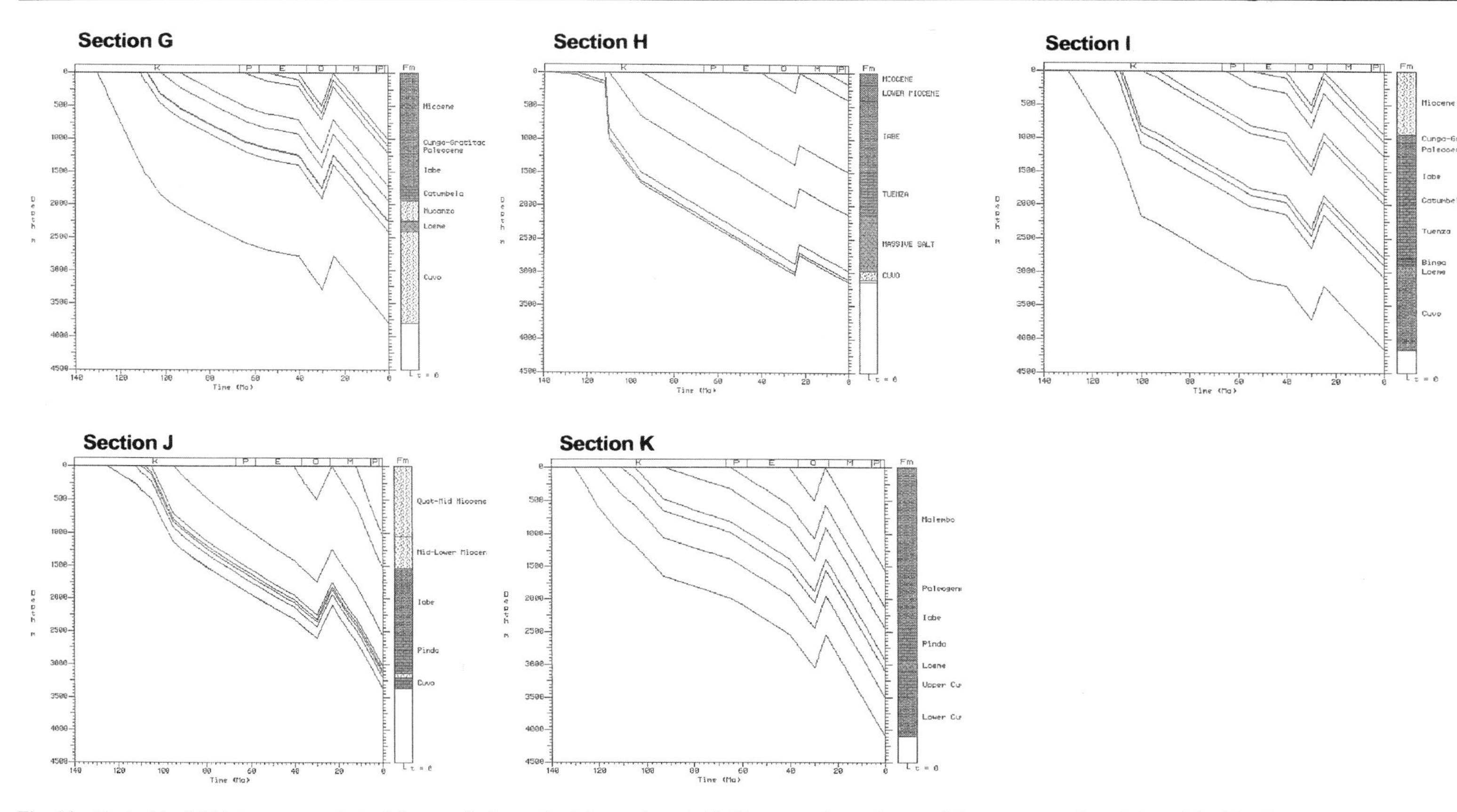


**Fig. 14.** Typical burial history curves derived from wells for each of the sections A–K. The age codes at the top of the curves are, from left to right: K = Cretaceous; P = Palaeocene; E = Eocene; O = Oligocene; M = Miocene; P = Pliocene.

These variations indicate that, regarding the post-salt structure, the continental margin is segmented into units with different tectonic patterns.

## Variation in the burial history of wells along the Angola–Zaire margin

To investigate the structural independence of tectonic units along the margin, variation in the subsidence of these units was examined by constructing burial history curves for 41 wells (Fig. 2), all of which have reached, or have come very close to, metamorphic basement. The data from these wells are proprietary and therefore unavailable for publishing. The stratigraphic tops used were the Malembo, Landana, Iabe, Pinda/Vermelha, Loeme, Chela, Bucomazi, Erva/Lucula and the metamorphic basement (Fig. 3). For calculation of the subsidence curves, BasinMod software was used and the sediments were assumed to be uncompacted. A constant 500 m of erosion in the Oligocene was also used. The wells are conveniently spread all along the Angola–Zaire Basins and their burial history curves indicate that significant variation in the subsidence history exists along the margin. Wells with similar burial history curves were then grouped together and it was found that these groups coincided areally with the tectonic sections. On the basis of subsidence history and structure, therefore, the Angola–Zaire margin was subdivided into 11 sections (A–K in Fig. 10). Representative well burial curves for each section are presented in Fig. 14.

In the Cabinda and Zaire areas (sections A–D), large uniform subsidence occurred during Cretaceous–Eocene, while the Miocene–Recent subsidence is relatively small in sections A–C and increases to the south (section D). Section E is markedly different from sections A–D, since it experienced a small subsidence during the Cretaceous and a very large one after the Miocene. Section F, on the contrary, shows a small Miocene–Recent subsidence. The Cretaceous–Eocene subsidence was quite large in sections G–K (Fig. 14) and the main difference observed in these sections is in the Miocene–Recent subsidence, which attains large values only off southern Kwanza Basin (section K). These variations provide an additional indication that the Angola–Zaire margin has not behaved tectonically as a single block, but rather is segmented into distinct units, each with its own structure and subsidence history. It is interesting to note that the Tertiary subsidence in sections C and D (Fig. 14) is smaller than in section E, although the former sections lie adjacent to the Zaire River estuary (Figs 1 and 10) and receive the major part of the sediment discharge. Similarly, the Tertiary subsidence in section I, which lies near the Kwanza River estuary (Figs 1 and 10), is much smaller than in section K, which receives only minor river discharge.

## Segmentation of the continental margin

Significant lateral differences in the subsidence and structure of segments along extensional margins have been observed in various other parts of the world. Some examples are the Campos and Santos basins, Brazil (Demercian *et al.* 1993); the Recôncavo–Tucano–Jotobá rift, Brazil (Milani & Davison 1988); the Roer Valley Graben, The Netherlands (Zijerveld *et al.* 1992); the Pannonian Basin (Tari *et al.* 1992); the Faeroe Basin (Rumph *et al.* 1993); the Danish sector of the North Sea (Cartwright 1987); the Western Anatolian province in Turkey (Sengör 1987); the eastern USA continental margin (Bally 1981); and the Bass, Gippsland and Otway basins of southeastern Australia (Etheridge *et al.* 1985). In all the above examples, apart from the Roer Valley Graben case, the authors have suggested that the boundaries of the segments are deep transfer zones which remained active throughout the extensional phase of the margin. Tectonic models, which make use of transfer fault tectonics, have been presented by Bally (1981), Etheridge (1986), Lister *et al.* (1986) and Cartwright (1992). In these models, deep faults, perpendicular to the general strike of the basin, dissect the margin into segments that exhibit distinct tectonic variations, such as changes in the throw of normal faults and alternation of basement highs and lows along the margin. Examples for the influence of transfer faults in modern rift basins can be seen in the East African Rift System (Morley *et al.* 1990), Dead Sea (Ben-Avraham 1992), central Greece (Gawthorpe & Hurst 1993) and the Gulf of Suez (Colletta *et al.* 1988).

Although there are strong indications from the present study that the Angola–Zaire margin is segmented, the presence of transfer faults as segment boundaries may be proposed only tentatively as a possible structural model. This is due to lack of along-strike seismic lines and detailed magnetic and gravity data from this study. Some indirect evidence for transfer zone existence is given by the fact that several segment boundaries (C–D, E–F, G–H, J–K) coincide with the offshore extension of basement transverse faults (Geological Map of Angola 1988) and oceanic fracture zones (Fig. 10).

## Discussion

The independent behaviour of the individual segments described above results in a complicated tectonic picture, which influences the hydrocarbon

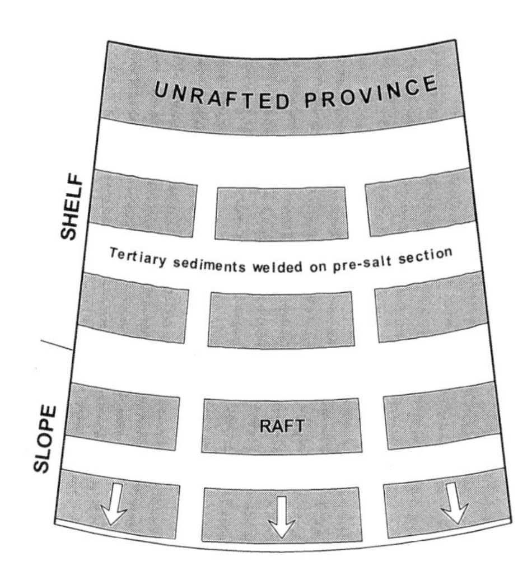


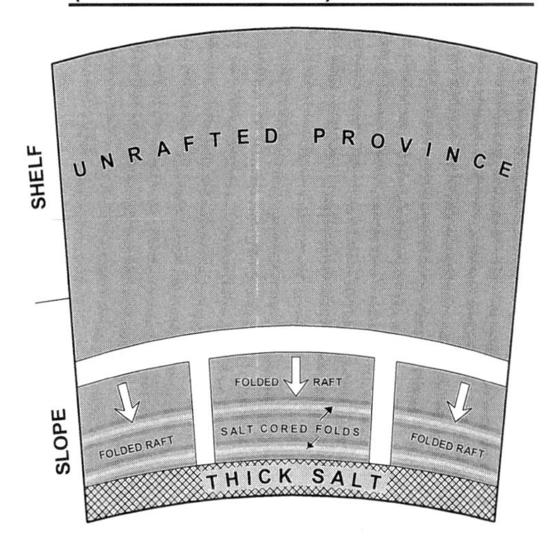


**Fig. 15.** Radial gliding of rafts on a convex-upward or concave-upward surface created by basement salients and re-entrants. This model, which can explain the changes in raft tectonics patterns, is a variation of the one discussed by Cobbold & Szatmari (1991).

prospectivity of the margin. The burial curves of each of the sections A–K indicate that periods of high and low subsidence rates occurred intermittently during the Neocomian to Recent time interval (Fig. 14). Rumph *et al.* (1993) suggested that this could be due to changes in the regional stress regime which, in the case of the West African margin, may have been caused by a change in the pole of rotation of the opening South Atlantic. The importance of the differential subsidence concept in oil exploration can be shown using as examples the subsidence histories of three oil-rich tectonic sections, A, B and E (Figs 10 and 14). The first two sections experienced a continuously large subsidence and behaved as depocentres throughout the Neocomian-Miocene period. This resulted in: (a) the deposition of a thick lacustrine anoxic layer of Neocomian–Barrremian shales which provide the source rock (Brice *et al.* 1982; Schlumberger 1991); and (b) the deposition of a thick Aptian–Albian sandstone unit (Vermelha Sandstone) which provides the main reservoir in Cabinda and Zaire (Brice *et al.* 1982; Schlumberger 1991). Section E, on the other hand, behaved as a relatively high-standing block during the Early Cretaceous–Early Tertiary period (Fig. 14) onto which the shallow-water, high-energy Aptian–Albian Pinda carbonates, which form the main reservoir off northern Angola, were deposited (Tillement 1987). The ensuing dramatic collapse of this section in the Miocene caused the maturation of the Upper Cretaceous Iabe source rocks which provided the hydrocarbon charge to a number of oil fields in the area (Edwards & Bignell 1988). It is evident, therefore, that knowledge of the differential subsidence along a margin provides valuable information about which hydrocarbon plays may be expected.

The difference in raft tectonics observed between the offshore southern Kwanza Basin (wide unrafted province, only one graben, rapid transition to compressional tectonics in the post-salt section) and the northern Kwanza and Lower Congo Basins (several rafts and grabens, mainly extensional post-salt tectonics) can be tentatively explained by the radial gravitational gliding model proposed by Cobbold & Szatmari (1991) for the southeast Brazilian basins. According to this model, convergent or divergent radial gliding of sediments on a décollement layer takes place in coastal re-entrants or salients, respectively. In that sense, a coastal re-entrant defines a concave-upward shape of basin, while a salient suggests that the basin is convex upward. In the first case (e.g. Santos Basin), the upslope extensional regime is followed by a pronounced contractional province at the lowermost part of the basin, with folds and anticlines cored by salt. In coastal salients (e.g. Campos Basin), the extensional regime holds throughout most of the extent of the basin.

The Kwanza Basin, with its large contractional downslope province (Figs 10 and 13), seems to be a case where convergent radial movement (re-entrant case) of the post-salt sedimentary load has taken place. It is important to note, however, that in this basin the re-entrant shape is formed by the basement outcrop rather than by the coastline, which, in fact, is a salient (Fig. 10). This suggests that, although the radial gliding model may be correct, it would be better to choose the basement outcrop shape to define salients and re-entrants (or, in other words, convex-upward or concave sub-basins) rather than the coastline (Fig. 15).

The lack of an unrafted province in the Kwanza Basin can be attributed to: (a) the large Late Cretaceous–Miocene uplift that affected this area (Sahagian 1988; Lunde *et al.* 1992); and (b) the great thickness of salt deposited in the basin (Brognon & Verrier 1966). The strong basinward tilting combined with the availability of a thick décollement layer resulted in the complete break-up of the unrafted province into large rafts and grabens, such as the Quenguela Graben (Verrier & Castello Branco 1972), and the Gaivota Graben (Duval *et al.* 1992).

## Conclusion

The seismic and well data that were available, together with the recent published laboratory studies of salt movements, have allowed a regional study of salt and raft tectonics in the Angola–Zaire basins to be carried out. The outcome of this work provides a new insight into thin-skinned tectonics and salt movements observable in these basins. The main results are as follows.

1. Raft tectonics has been the dominant process in shaping the post-salt structure of the Lower Congo and Kwanza Basins. This process has created two different stress regimes across the margin: (a) an extensional regime, located under the shelf and upper slope, where the Albian–Cenomanian carbonate platform and the overlying Upper Cretaceous–Early Tertiary sequence broke up in blocks and slid downslope on the salt décollement layer; and (b) a contractional regime, under the lower slope, where the previous extension is taken up by folding of the post-salt sequence.

2. Five different tectonic provinces can be identified on the basis of their raft and salt tectonics structure. The unrafted province comprises the region where the Albian–Early Tertiary sequence has been affected by listric faulting, without, however, complete separation of footwalls from hangingwalls. In the raft province, the thin-skinned extension has created isolated blocks of Albian–Early Tertiary age (rafts), separated by large grabens with Late Tertiary–Quaternary fill. In the salt diapir province, which essentially marks the beginning of the contractional stress regime, the post-salt sequence is folded and the salt has risen forming diapirs. The salt swell province is characterized by a very thick salt layer forming large swells under a strongly folded post-salt section. Thickness of salt continues to be great in the tabular salt province, but no significant folding is observed in the post-salt section, as most of the upslope extension has been taken up.

3. The areal distribution of the above provinces changes significantly along strike of the margin. Abrupt variations in the width of the unrafted and raft provinces, as well as in the style of raft tectonics, occur over quite small distances. Using these variations, the Angola–Zaire margin can be divided into distinct tectonic segments.

5. Differences in the subsidence of these segments, revealed by burial history curves, indicate that they behaved tectonically as independent blocks, at least during their post-salt (Aptian–Recent) history. The differential subsidence of the segments has significant consequences for the sedimentary history and hydrocarbon potential of each of them.

6. Intermittent periods of high and low subsidence rates occur within an individual section. Sections A and B, for instance, experienced a steady subsidence throughout their tectonic history, while section E stayed as a high block until the Miocene and then collapsed very quickly (Fig. 10 & 14). This variation may have been caused by changes in the regional stress regime due to changes in the pole of rotation.

8. The variation in the style of raft tectonics and in the extent of the structural provinces along the margin can be tentatively explained by using the radial gliding concept. According to this, the breadth of the extensional regime will depend on whether the post-salt sequence glides radially inwards or outwards on a concave or convex base, respectively.

The significance of lateral changes in the sedimentology and structure caused by along-strike segmentation of passive margins, and the question of the existence of transfer zones are subjects that merit further detailed studies, owing to their importance in hydrocarbon exploration in several basins.

I would like to express my gratitude to Amerada Hess Ltd for permission to publish this work and for the support that was offered to me. In particular, I would like to thank Quentin Rigby, Joe Lambiase and John Booth for their stimulating suggestions and encouragement, as well as Martin Turner for drawing the figures. I would also like to thank GECO-PRAKLA for permission to use several seismic lines from their GWA88 survey.

## References

BALLY A. W. 1981. Atlantic-type margins. *In: Geology of Passive Continental Margins*, American Association of Petroleum Geologists Education Course Note Series, **19**, 1–48.

BEN-AVRAHAM, Z. 1992. Development of asymmetric basins along continental transform faults. *Tectonophysics*, **215**, 209–220.

BOLLI, H. M., RYAN, W .B. F. ET AL. 1978. Angola continental margin – Sites 364 and 365. *In*: NATLAND J. H. (ed.), *Initial Reports of the Deep Sea Drilling Project*, US Govt. Printing Office, Washington, **40**, 357–390.

BRICE, S. E., COCHRAN, M. D., PARDO, G.& EDWARDS, A. D. 1982. *Studies in Continental Margin Geology*, American Association of Petroleum Geologists Memoir, **34**, 5–20.

BROGNON, G. P. & VERRIER, G. R.1966. Oil and geology in Cuanza Basin of Angola. *AAPG Bulletin*, **50**, 108–158.

BROWN, D. 1993. Raft tectonics mixes up geology – West Africa challenges explorers. *American Association of Petroleum Geologists, Explorer*, October 1993, 12–15.

CARTWRIGHT, J. A. 1987. Transverse structural zones in continental rifts – an example from the Danish sector of the North Sea. *In*: BROOKS, J. & GLENNIE, K. (eds) *Petroleum Geology of North West Europe*. Graham & Trotman, London, 441–452.

—— 1992. Fundamental crustal lineaments and transverse structural zones in continental rifts. *In*: MASON, R. (ed.) *Basement Tectonics*, **7**, 209–217.

COBBOLD, P. R. & SZATMARI, P. 1991. Radial gravitational gliding on passive margins. *Tectonophysics*, **188**, 249–289.

COLLETTA, B., LE QUELLEC, P., LETOUZEY, J. & MORETTI, I. 1988. Longitudinal evolution of the Suez rift structure (Egypt). *Tectonophysics*, **153**, 221–233.

DEMERCIAN, S., SZATMARI, P. & COBBOLD P. R. 1993. Style and pattern of salt diapirs due to thin-skinned gravitational gliding, Campos and Santos basins, offshore Brazil. *Tectonophysics*, **228**, 393–433.

DUVAL, B., CRAMEZ, C. & JACKSON, M. P. A. 1992. Raft tectonics in the Kwanza Basin, Angola. *Marine and Petroleum Geology*, **9**, 389–404.

EDWARDS, A. & BIGNELL R. 1988. Nine major play types recognized in salt basin. *Oil and Gas Journal*, **86**, 55–58.

ETHERIDGE, M. A. 1986. On the reactivation of extensional fault systems. *Philosophical Transactions of the Royal Society, London*, **A317**, 179–194.

——, BRANSON, J. C. & STUART-SMITH, P .G. 1985. Extensional basin-forming structures in Bass Strait and their importance for hydrocarbon exploration. *Australian Petroleum Exploration Association Journal*, **25**, 344–361.

GAULLIER, V., BRUN, J. P., GUÉRIN, G. & LECANU, H. 1993. Raft tectonics: the effects of residual topography below a salt décollement. *Tectonophysics*, **228**, 363–381.

GAWTHORPE, R. L. & HURST, J. M. 1993. Transfer zones in extensional basins: their structural style and influence on drainage development and stratigraphy. *Journal of the Geological Society, London*, **150**, 1137–1152.

*Geological Map of Angola* (Carta Geológica de Angola), Instituto Nacional de Geologia, Ministério da Indústria, 1988.

HONGXING, G. E., JACKSON, M. P. A. & VENDEVILLE, B. C. 1994. Experimental initiation of salt structures and emplacement of salt sheets by prograding sediment wedges. *Salt Tectonics Conference Abstracts*, Geological Society, London.

JENYON, M. K. 1986. *Salt Tectonics*. Elsevier, Barking.

LISTER, G. S., ETHERIDGE, M. A. & SYMONDS, P. A. 1986. Detachment faulting and the evolution of passive continental margins. *Geology*, **14**, 246–250.

LUNDE, G., AUBERT, K., LAURITZEN, O. & LORANGE, E. 1992. Tertiary Uplift of the Kwanza Basin in Angola. *In*: CURNELLER (ed.) *Géologie Africaine-Compte Rendu des Colloques de Géologie de Libreville*, 6–8 May 1992, Bull. Centres Rech. Expl.-Prod. ELF Aquitaine, Memoir, **13**.

LUNDIN, E. R. 1992. Thin-skinned extensional tectonics on a salt detachment, northern Kwanza Basin, Angola. *Marine and Petroleum Geology*, **9**, 405–411.

MILANI, E. J. & DAVISON, I. 1988. Basement control and transfer tectonics in the Recôncavo–Tucano–Jatobá rift, Northeast Brazil. *Tectonophysics*, **154**, 41–70.

MOHRIAK, W. U., MELLO, M. R., DEWEY, J. F. & MAXWELL, J. R. 1990. Petroleum geology of the Campos Basin, offshore Brazil. *In*: BROOKS, J. (ed.) *Classic Petroleum Provinces*. Geological Society, London, Special Publication, **50**, 119–141.

MORLEY, C. K., NELSON, R. A., PATTON, T. L. & MUNN, S. G. 1990. Transfer zones in the East African Rift System and their relevance to hydrocarbon exploration in rifts. *AAPG Bulletin*, **74**, 1234–1253.

RUMPH, B., REAVES, C. M., ORANGE, V. G. & ROBINSON, D. L. 1993. Structuring and transfer zones in the Faeroe Basin in a regional tectonic context. *In*: PARKER, J. R. (ed.) *Petroleum Geology of Northwest Europe: Proceedings of the 4th Conference*. Geological Society, London, 999–1009.

SAHAGIAN, D. 1988. Epeirogenic motions of Africa as inferred from Cretaceous shoreline deposits. *Tectonics*, **7**, 125–138.

SCHLUMBERGER. 1991. *Well Evaluation Conference*. Schlumberger, Angola.

SENGÖR, A. M. C. 1987. Cross-faults and differential stretching of hanging walls in regions of low-angle normal faulting: examples from western Turkey. *In*: COWARD, M. P., DEWEY, J. F. & HANCOCK, P. L. (eds) *Continental Extensional Tectonics*, Geological Society, London, Special Publication, **28**, 575–589.

SZATMARI, P., GUERRA, M. C. M. & PEQUENO, M. A. 1996. Genesis of large counter-regional normal fault by flow of Cretaceous salt in the South Atlantic Santos Basin – Brazil. *This volume*.

TARI, G., HORVÁTH, F. & RUMPLER, J. 1992. Styles of extension in the Pannonian Basin. *Tectonophysics*, **208**, 203–219.

TILLEMENT, B. 1987. Insight into Albian carbonate geology in Angola. *Bulletin of Canadian Petroleum Geology*, **35**, 65–74.

VENDEVILLE, B. C. & JACKSON, M. P. A. 1992*a*. The rise of diapirs during thin-skinned extension. *Marine and Petroleum Geology*, **9**, 331–353.

—— & —— 1992*b*. The fall of diapirs during thin-skinned extension. *Marine and Petroleum Geology*, **9**, 354–371.

VERRIER, G. & CASTELLO BRANCO, F. 1972. The Tertiary trough and oil accumulation of North Quenguela

(Cuanza Basin). *French Institute of Petroleum Review*, Jan/Feb 1972.

WALGENWITZ, F., PAGEL, M., MEYER, A., MALUSKI, H. & MONIE, P. 1990. Thermo-chronological approach to reservoir diagenesis in the offshore Angola Basin: a fluid inclusion, $^{40}Ar$–$^{39}Ar$ and K–Ar investigation. *AAPG Bulletin*, **74**, 547–563.

ZIJERVELD, L., STEPHENSON, R., CLOETINGH, S., DUIN, E. & VAN DEN BERG, M. W. 1992. Subsidence analysis and modelling of the Roer Valley Graben (SE Netherlands). *Tectonophysics*, **208**, 159–171.

# Influence of salt on fault geometry: examples from the UK salt basins

S. A. STEWART[1], M. J. HARVEY[2], S. C. OTTO[3] & P. J. WESTON[1]

[1]*Amerada Hess Ltd, 33 Grosvenor Place, London SW1X 7HY, UK*

[2]*Department of Geology, Imperial College, London SW7 2BP, UK*

[3]*Petroconsultants, 266 Upper Richmond Road, London SW15 6TQ, UK*

**Abstract:** The influence of salt layers on cover faulting during thick-skinned extensional faulting is examined using seismic examples from the salt basins offshore UK. The ratio of salt layer thickness to basement fault displacement is a key geometric parameter governing spatial offset between cover and basement fault segments. The influence of geological factors such as stratigraphic variation in salt thickness, basement fault zone geometry and basement fault growth through time on this parameter are individually examined. 3D models are constructed to illustrate the spectra of possible fault geometries which result from variations in these factors. Lateral stratigraphic variations within individual 'salt' layers are also considered, as are complications introduced by the addition of further salt layers into cover stratigraphy. Since diapiric intrusions occur after cover fault geometries are established and are localized by such faulting, an understanding of basement-cover fault relationships also illuminates salt diapir-basement fault relationships. An example from the Central North Sea diapir province is subject to such a genetic analysis, relating the diapir and its location to an underlying basement fault. The influence of salt on cover fault reactivation during basin inversion is discussed, focusing on the southern North Sea, where inhibition of reverse fault propagation from basement to cover is noted. It is suggested that analysis of local basement-cover relationships should form a first step in any attempt at a genetic classification of salt bodies.

Structural mapping in the sedimentary basins offshore UK is complicated in many areas by the presence of salt at one or more stratigraphic levels. Particular problems surround the interpretation of fault systems which interact with salt layers. For example, opinions have varied as to whether or not fault systems on the western margin of the southern North Sea link directly from strata above the salt to faults below the Zechstein salt (Christian 1969; Glennie & Boegner 1981; Walker & Cooper 1987; Arthur 1993). The issue is complicated because fault systems in the 'cover' above salt may be kinematically linked with 'basement' faults below salt, but spatial relationships may be difficult to identify due to offset between cover and basement via detachment in the salt layer (Nalpas & Brun 1993; Jackson & Vendeville 1994). For the purposes of this paper, 'cover' and 'basement' are mechanically distinct layers, which may be separated by a viscous (salt) layer (Fig. 1). Whilst it is true that in exclusively thin-skinned structural situations there need be no relationship between fault systems above and below salt (e.g. Crans *et al.* 1980; Duval *et al.*1992), significant basement faults have governed the location, geometry and evolution of the salt-bearing basins offshore UK. The purpose of this paper is to examine the geometries of the upwards continuations of such basement faults through salt layers into the cover. By considering fault systems in their entirety in the vertical sense, and the spatial variations in geometry in relation to geological circumstance within the UK basins, we hope to illustrate the impact of salt layers on the trajectories of thick-skinned extensional faults which post-date salt deposition.

## Review

Simple analog models demonstrate that salt layers have an important impact on the manner in which basement faults propagate upwards into cover (Fig. 1). It is easy to envisage a continuous series of models between those examples illustrated in Figs 1a and 1b with varying viscous layer thickness. Although a single fault strand does not always pass directly from basement to cover, the cover and basement fault elements are kinematically linked and could be considered to represent a single fault surface, composed of brittle and ductile portions (Fig. 1d). Furthermore, since salt layers in sedimentary basins are not infinite in lateral extent, one might expect line length balance between pre-rift beds above and below the salt (Nalpas & Brun 1993; Jackson & Vendeville 1994). A case can therefore be made for considering basement fault – detachment – cover faults as single fault surfaces on

*From* ALSOP, G. I., BLUNDELL, D. J. & DAVISON, I. (eds), 1996, *Salt Tectonics*, Geological Society Special Publication No. 100, pp. 175–202.

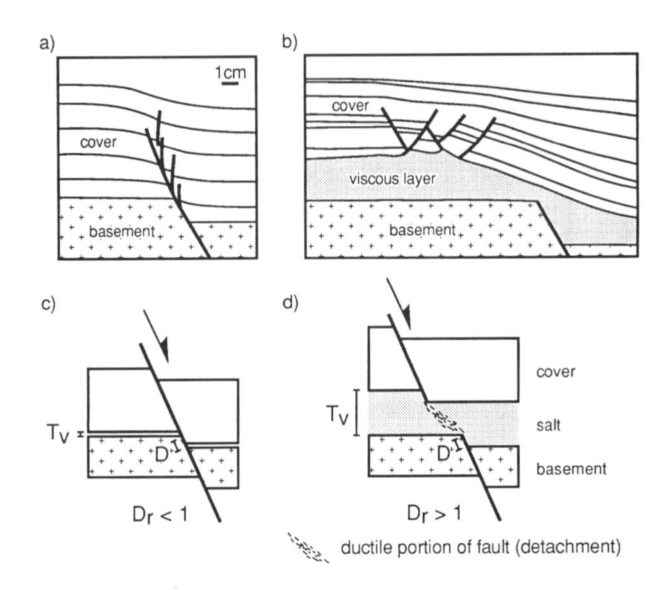


**Fig. 1.** Analog models illustrating the impact of a viscous layer on the upward propagation of a fault from 'basement' to 'cover', and the relationship of basement-cover fault linkage to displacement ratio. (a) 'Hard' linked, planar fault which has grown from basement to cover where no intervening detachment is present, after Withjack *et al.* (1990). (b) 'Soft' linked, ramp-flat extensional fault with planar, brittle elements visible in basement and cover, and a flat-lying, ductile detachment occupying the viscous layer in the zone offsetting the cover and basement ramp elements (after Larroque in Oudmayer & de Jaeger 1993). Displacement ratio ($D_r$) is defined in terms of $D$ (basement fault displacement) and $T_v$ (viscous layer thickness), $D_r = T_v/D$ (Koyi *et al.* 1993). (c) $T_v < D$, $D_r < 1$. Basement fault 'hard' linked to cover in a continuous brittle strand. (d) $T_v > D$, $D_r > 1$. Basement fault 'soft' linked to cover fault via ductile detachment in the salt, giving extensional ramp-flat geometry.

the basis of kinematic linkage, the variation in structural style reflecting the rock rheology.

Ramp-flat trajectories are considered to be characteristic of single fault surfaces passing through layered sequences of strongly contrasting rheology in both thrust systems (Boyer & Elliott 1982) and extensional systems (Peacock & Sanderson 1992; Peacock & Zhang 1994). Just as salt layers form important fault 'flats' in thrust systems (Davis & Engelder 1985), they may perform the same role in extensional systems, where the term 'detachment' is commonly used for the tectonically active portion of the salt layer.

Several geological parameters appear to influence the geometry of extensional ramp-flat systems. A review by Koyi *et al.* (1993) indicated that the magnitude, direction and rate of basement fault displacement are important, as are the thicknesses of the overburden and the salt layers. In particular, the magnitude of spatial offset between linked basement and cover fault systems has not yet been quantitatively described in terms of the above parameters. It is noted here that one of the key parameters may turn out to be salt layer top surface slope, which must control localized gravity sliding in the vicinity of a basement fault at the very first stages of fault growth. However, investigating the mechanics of this model is beyond the scope of this paper.

A single, fundamental, parameter can be defined, which encapsulates salt layer thickness and basement fault displacement. The parameter is the displacement ratio $D_r$ which is the thickness of the viscous layer $T_v$ divided by the basement fault displacement $D$ (Koyi *et al.* 1993). $D_r$ is a useful parameter because it embraces several geological controls. $T_v$ can vary spatially for stratigraphic or structural reasons, while $D$ can vary in space and time. One can envisage a special value of $D_r$, close to 1, at which the magnitude of basement fault displacement exceeds the thickness of the salt layer and a 'hard', direct fault linkage would be established between basement and cover (Fig. 1). This transition is illustrated in analog models presented by Oudmayer & de Jaeger (1993). In this paper, several examples from various basins on the UKCS are examined using seismic data. The aim in each example is to examine the influence of a single geological control, in terms of variation in observed fault system geometry and $D_r$.

Analogue modelling has shown that the locations of salt diapirs in extensional environments are

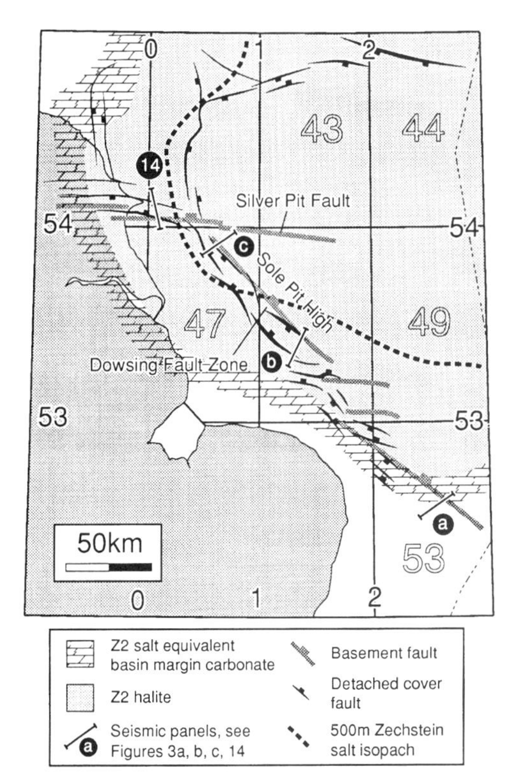


**Fig. 2.** West margin, southern North Sea Basin, showing location of sections presented in Fig. 3 and the spatial relationship between fault systems and Zechstein facies. Basement fault elements have not been mapped in detail during this study. The detached cover fault elements are only present within the Zechstein salt basin, but their location and orientation is controlled by basement structures.

predetermined by the locations of cover fault systems (Vendeville & Jackson 1992). The modelling also demonstrates that at least 20% extension of the cover ($\beta = 1.2$) via normal faulting must occur before diapirs are initiated. The geometry of a given linked basement – cover fault will already have been established at this stage and so the effects of diapirism in the examples discussed here will generally not be emphasized. By taking this approach we hope to illustrate the indirect control exerted by salt on diapir location during thick-skinned extension. The discussions presented here will make the assumption that basement fault movement has been dip-slip. Richard (1991) has shown that the obliquity (strike-slip / dip-slip) of basement fault movement will also affect basement-cover linkage geometry, a result which is relevant to North Sea geology as several tectonic syntheses have noted the importance of oblique slip on North Sea basement structures (e.g. Bartholomew *et al.* 1993).

## Stratigraphic variation in salt thickness ($T_v$) : west margin, southern North Sea

The west margin of the southern North Sea basin is structurally defined by the northeasterly dipping Dowsing Fault Zone (DFZ) which was episodically active as a major basement extensional fault from the middle Triassic to the lower Cretaceous (Cameron *et al.* 1992). Significant quantities of salt are present within the upper Permian Zechstein Supergroup, which accumulated in a flexural basin driven largely by thermal subsidence rather than active faulting. As a result, Zechstein salt isopachs and stratigraphic pinchout are not parallel in trend to the DFZ (Glennie & Boegner 1981; Fig. 2). Serial sections across the DFZ from southeast to northwest illustrate variation in style of the DFZ as it has grown upwards through a salt layer which thickens northwards. Seismic sections presented on Fig. 3 show the spatial variation in fault style. Towards the southeastern end of the DFZ, in the Hewett Shelf area, the Zechstein facies are basin margin carbonates and faults pass without deviation from basement to cover (Badley *et al.* 1989; Fig. 3a). Note the evidence for late (Cretaceous to Tertiary) inversion on these faults while Permian stratigraphic levels remain in net extension.

Proceeding northwestwards, the DFZ passes through the Zechstein salt pinchout in the northwest of Quadrant 53. Mismatch between the cover and basement fault systems in plan view has been noted in the Camelot area, blocks 53/1 and 53/2, where the Zechstein salt (the sum of Z2, Z3 and Z4 halites) is approximately 100 m thick (Holmes 1991). On the southeast margin of the East Midlands Shelf, with just over 100 m of Zechstein salt, there is clearly a degree of detachment occurring between basement and cover faults (Fig. 3b).

Several issues regarding the DFZ are raised on Fig. 3b. The net displacement on the faults at basement level does not balance the displacements seen across the cover faults. The Hewett Shelf section demonstrated that inversion has occurred on the basement DFZ faults. Uplift of the Sole Pit High (Fig. 3b) indicates that inversion also occurred where Zechstein salt is present. It is likely that a significant amount of extensional heave has been lost on most of the basement fault strands. However, reverse displacement has not passed upwards to cover fault segments which had been active during extension. The kinematics of this system during inversion will be examined later. Comparisons of heave across the cover fault system with likely pre-inversion heave across the basement faults (based on the thickness of sediments preserved in the Sole Pit High) suggest that the heave across the cover system may be greater than that which existed at basement level (Arthur 1993;

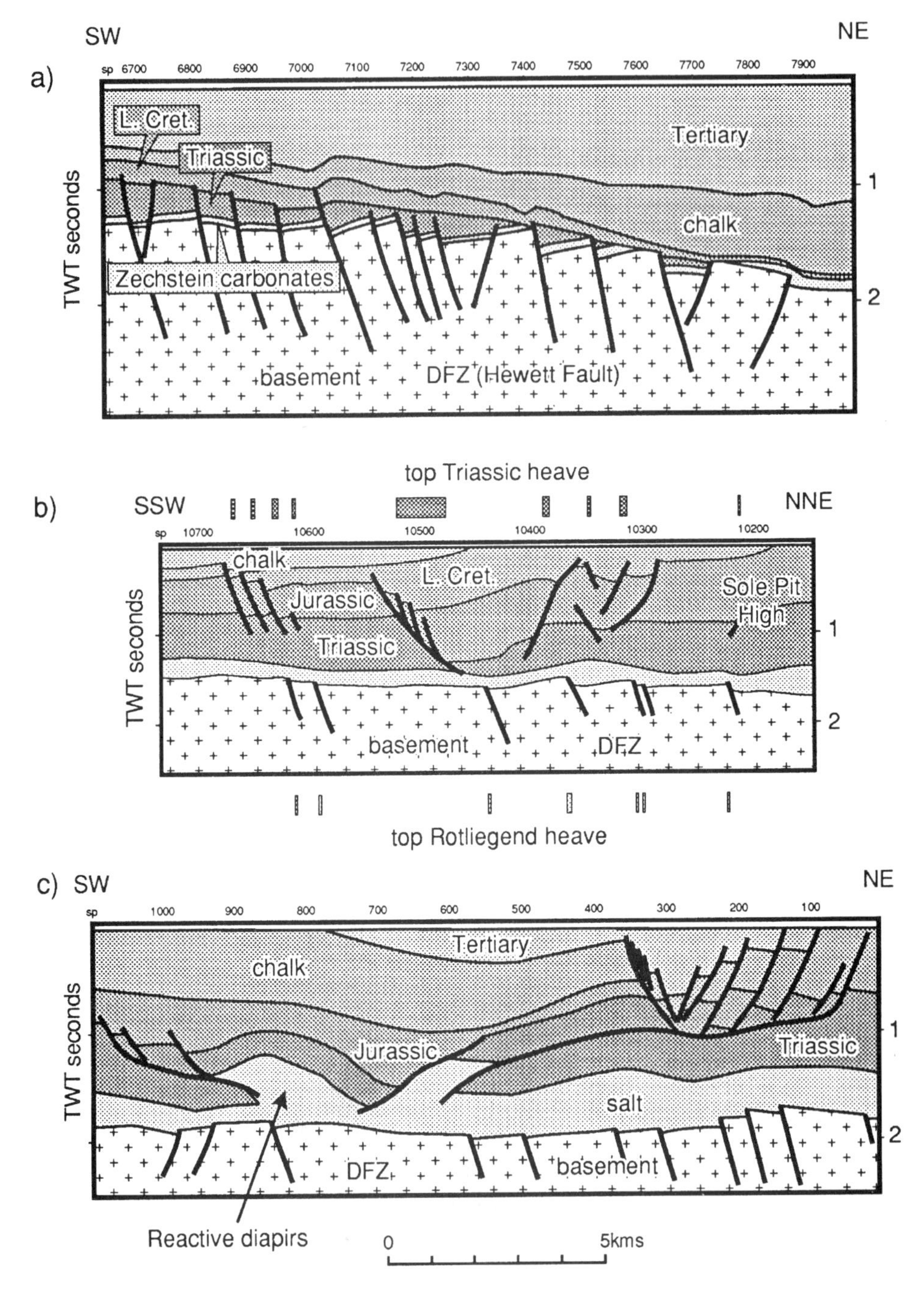


**Fig. 3.** Dowsing Fault Zone seismic examples and geoseismic interpretations (see Fig. 2 for locations) (a) Hewett Shelf; no Zechstein salt, hard linked faults pass directly from basement to cover. Inversion-related folds in hangingwalls to some faults present at base Chalk and top Chalk. (b) East Midlands Shelf; relatively thin Zechstein. Detachment between basement and cover faults. Approximately 2 km heave discrepancy between basement and cover fault systems. Note westerly facing monoclinal folding at top Jurassic and base Chalk, a ductile cover response to basement inversion. (c) East Midlands Shelf; relatively thick Zechstein. No clear relationship between basement and cover faults. Cover faults intruded by Zechstein diapirs. Significant bedding-parallel detachments on intra-Triassic salt layers.

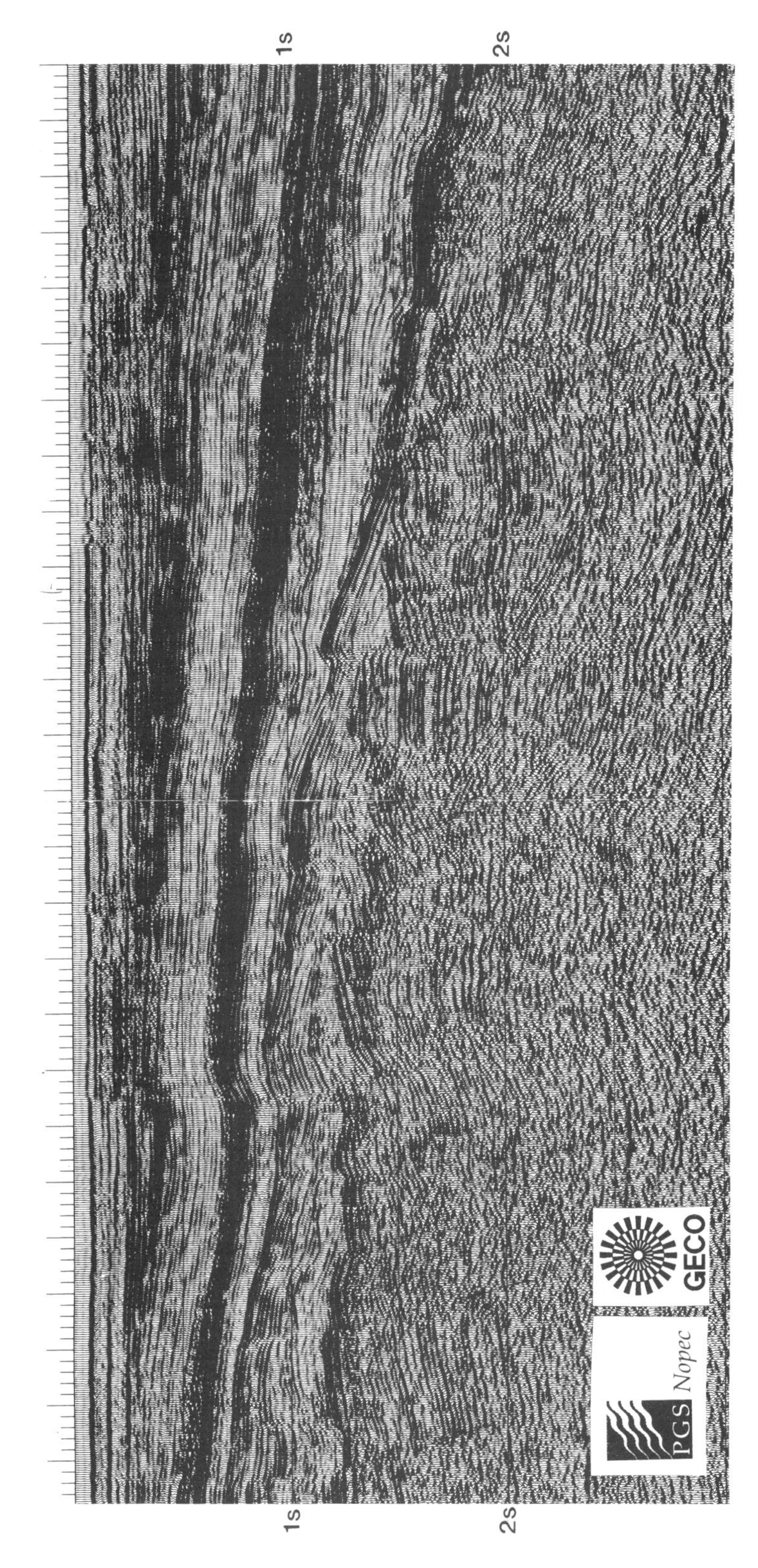


**Fig. 3a.** Seismic section.

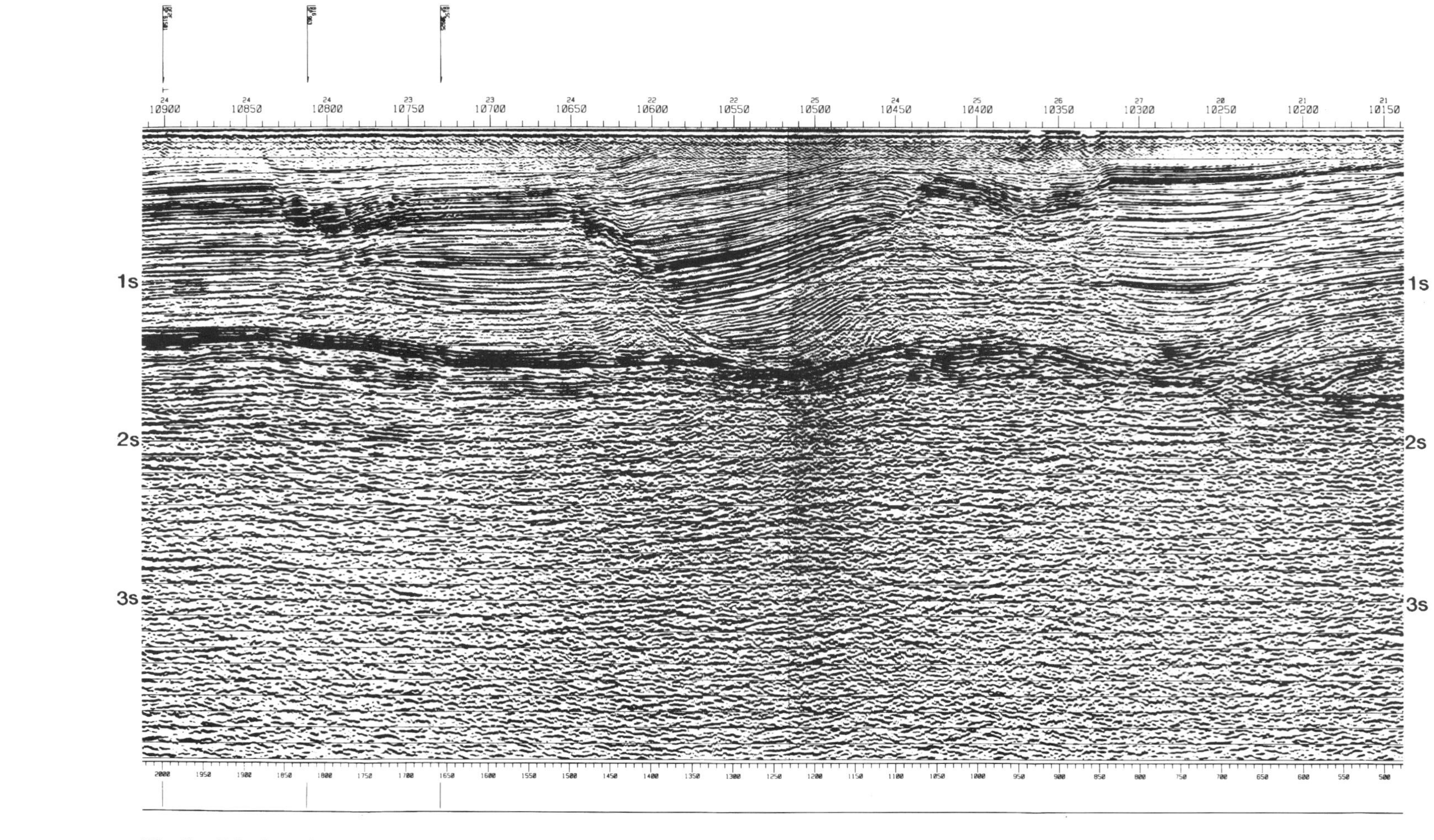


**Fig. 3b.** Seismic section.

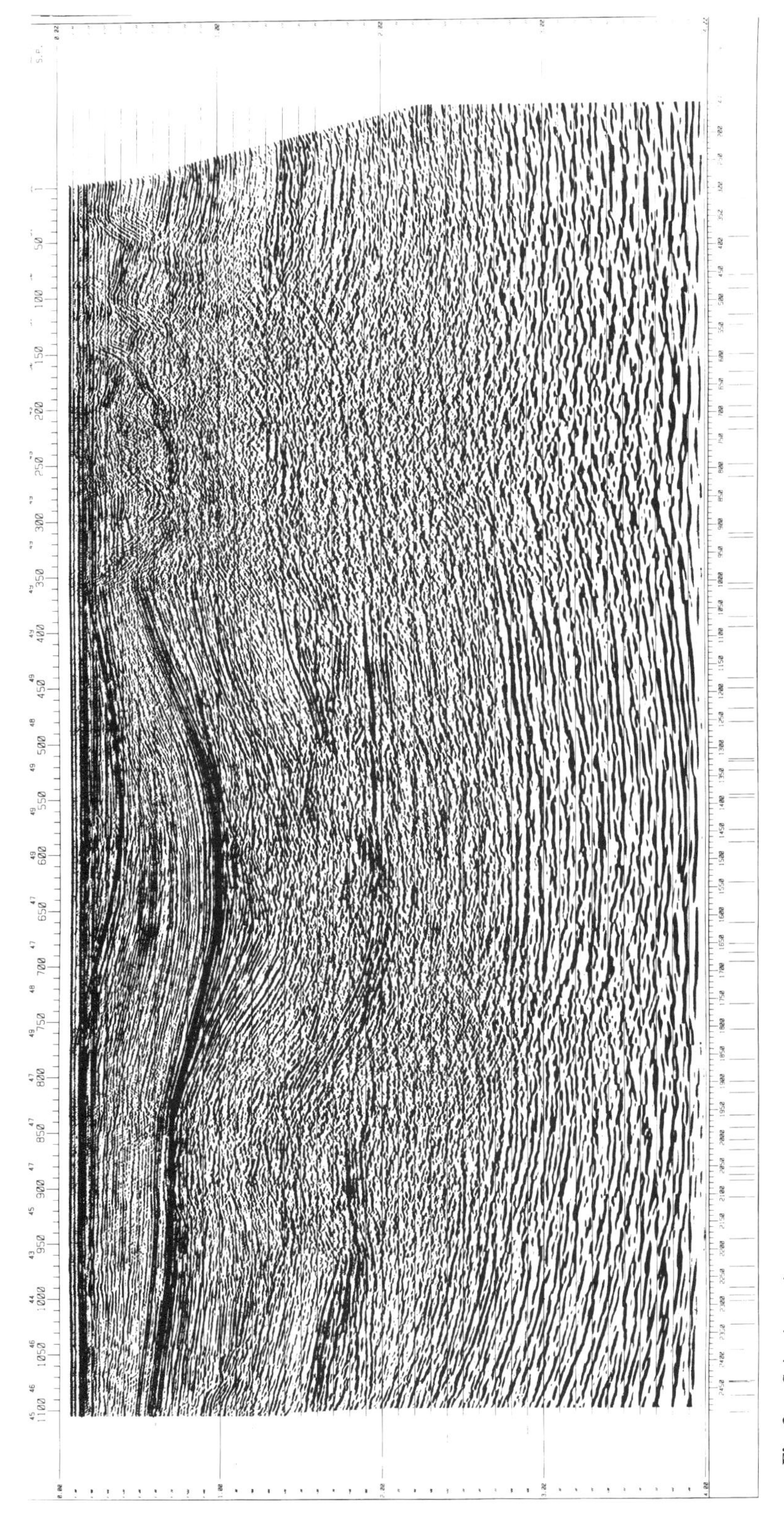

**Fig. 3c.** Seismic section.

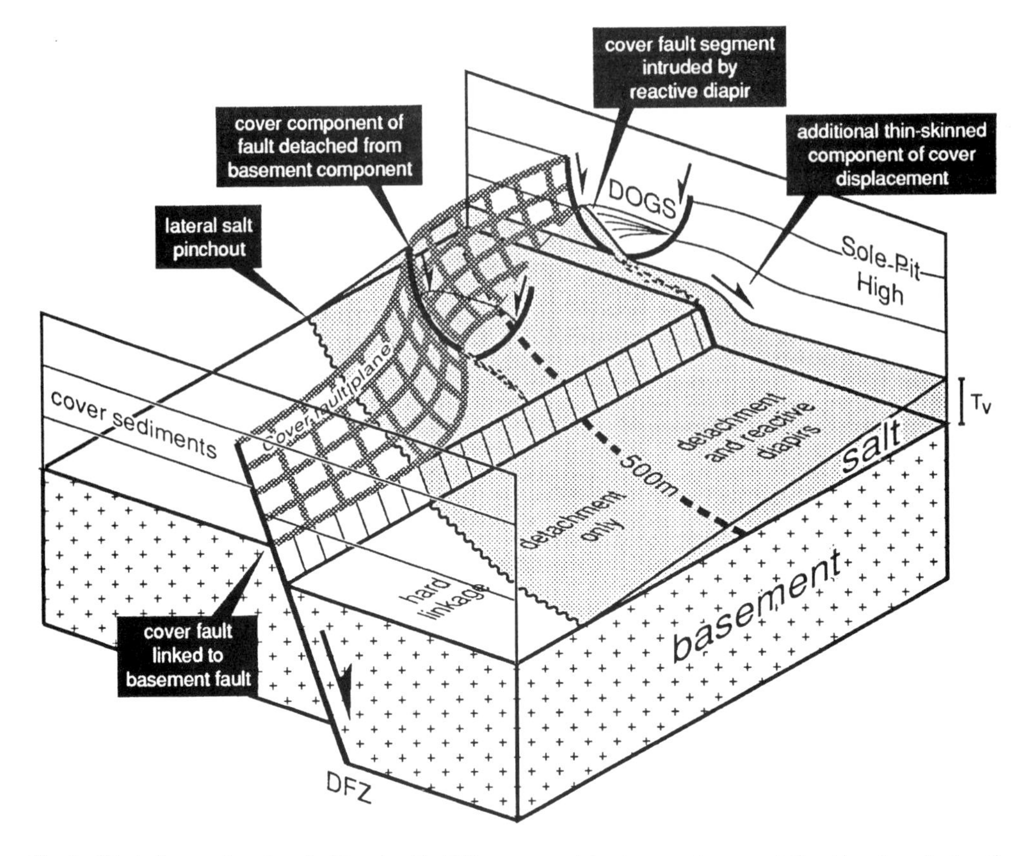


**Fig. 4.** Block diagram based on the Dowsing Fault Zone summarizing basement-cover relationships along the length of a major fault strand, illustrating the effect of spatial variation in salt layer thickness. Note the onset of diapirism within the 500 m Zechstein salt isopach.

Stewart & Coward 1995). This apparent anomaly can be explained by a component of thin-skinned gliding of the cover section eastwards towards the Sole Pit Trough, balanced presumably by thick-skinned basement extension or thin-skinned cover shortening within the basin. Thin-skinned extension of this nature has been described from the other side of the southern North Sea basin (Allen *et al.* 1994) and the central North Sea (Bishop *et al.* 1995). Given the likelihood that the cover fault system on the east margin of the East Midlands Shelf does not solely represent the upwards continuation of basement faults, reference to the cover fault system as 'Dowsing Fault Zone' could be misleading. Terming the basement faults DFZ and the cover faults Dowsing Graben System (DOGS) has been suggested (Stewart & Coward 1995).

The final example from the DFZ is taken from the northeast corner of the East Midlands Shelf, where the Zechstein Supergroup is approximately 800 m thick and is dominated by the Z2 Stassfurt Halite Formation (Fig. 3c). In this area the spatial relationship between cover and basement faults is far from clear although the DOGS is again present, below the base Cretaceous unconformity. Two important features on this section are: firstly, a major post-Cretaceous thin-skinned collapse graben detaching on upper Triassic evaporites, resulting from uplift of the Sole Pit High; secondly, the DOGS in this area is invaded by reactive diapirs (Vendeville & Jackson 1992), which become less common southeastwards where the salt is thinner.

The DFZ basement – cover relationships are summarized on Fig. 4, which attempts to portray the 3D geometry of a fault system with constant displacement passing through a salt layer which systematically varies in thickness. This illustrates the most basic effect of salt layers on growing extensional fault systems – the importance of the detachment is in proportion to the thickness of the detaching layer (Richard 1991). The examples in Fig. 3 also indicate that as salt increases in thickness along the DFZ, the onset of detachment precedes the onset of diapirism. A tentative conclusion from this observation would be that salt layers can constitute effective detachments though they are not thick enough to support diapirism. In this context it is noted that Price & Cosgrove (1990) demonstrated that the extrinsic mobility of salt in terms of $T_v$ is in proportion to the fourth power of $T_v$. They calculate that significant lateral salt movement such as that required to feed reactive and active diapirs can only occur if the salt layer is in excess of 500 m thick at temperatures likely to be encountered in sedimentary basins (Price & Cosgrove 1990). The 500 m salt isopach does indeed pass between the locations of the sections presented in Figs 3b and 3c (Figs 3 and 4).

## Tectonically-induced variation in salt thickness ($T_v$) : Forth Approaches Basin, central North Sea

In addition to initial, stratigraphic variations in salt layer thickness, there are several post-depositional processes, independent of basement tectonics, which may affect salt thickness. Examples include thin-skinned extension due to gravity gliding (Vendeville & Jackson 1992), salt redistribution due to buoyancy instability (Geil 1991; Zaleski & Julien 1992) and passive infill of the cores of contractional buckle folds (Koyi 1988). Tectonically-induced salt thickness variation in the North Sea can be found on the platforms either side of the Central Graben, for example the Forth Approaches Basin (Fig. 5).

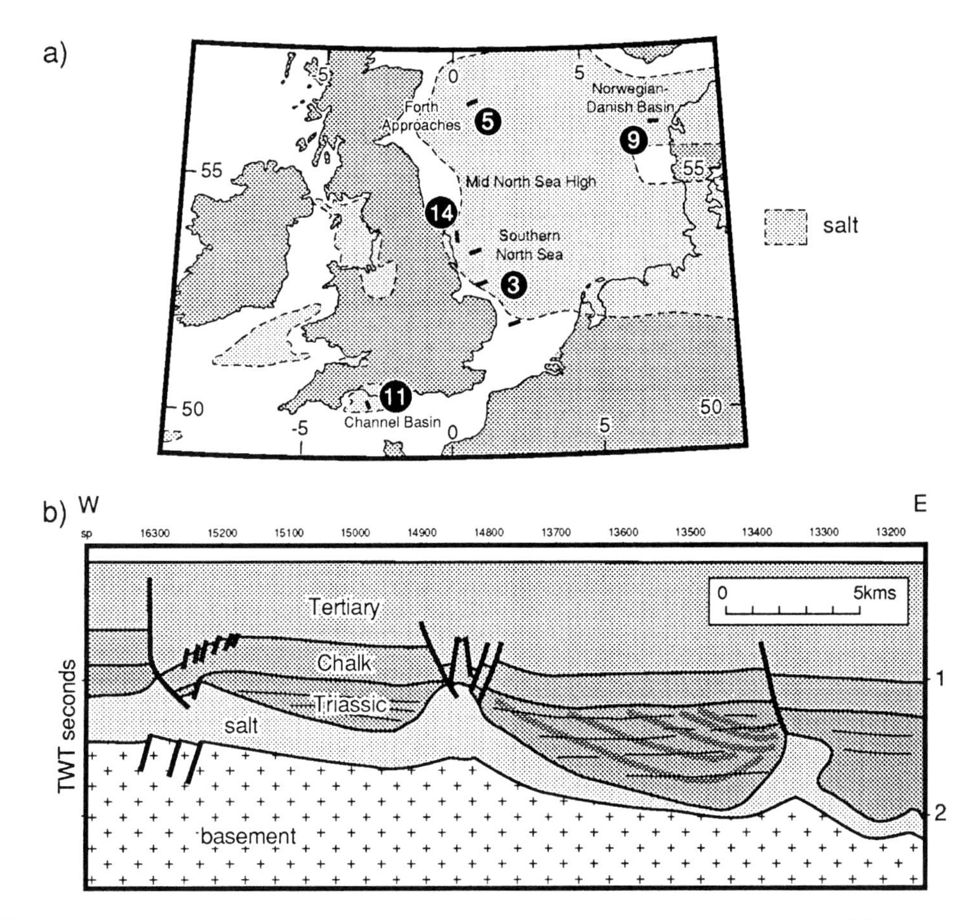


**Fig. 5.** Forth Approaches Basin seismic and geoseismic. (a) Location map for examples presented in this paper. Figure numbers circled. (b) Variations in salt and Triassic thickness below the base Chalk unconformity (relief at top basement due to velocity pull-up below salt highs). Tertiary thin-skinned extensional faults have been localised above the salt highs. Sub-horizontal reflectors are characteristic of the seismically transparent Triassic Smith Bank shales, regardless of folding recorded by the top Zechstein reflector. The sub-horizontal reflectors often suggest onlap relationships with the top Zechstein. Vestiges of reflectivity within some Triassic pods indicate that the Triassic beds may also be folded and that the sub-horizontal reflectors are multiple energy.

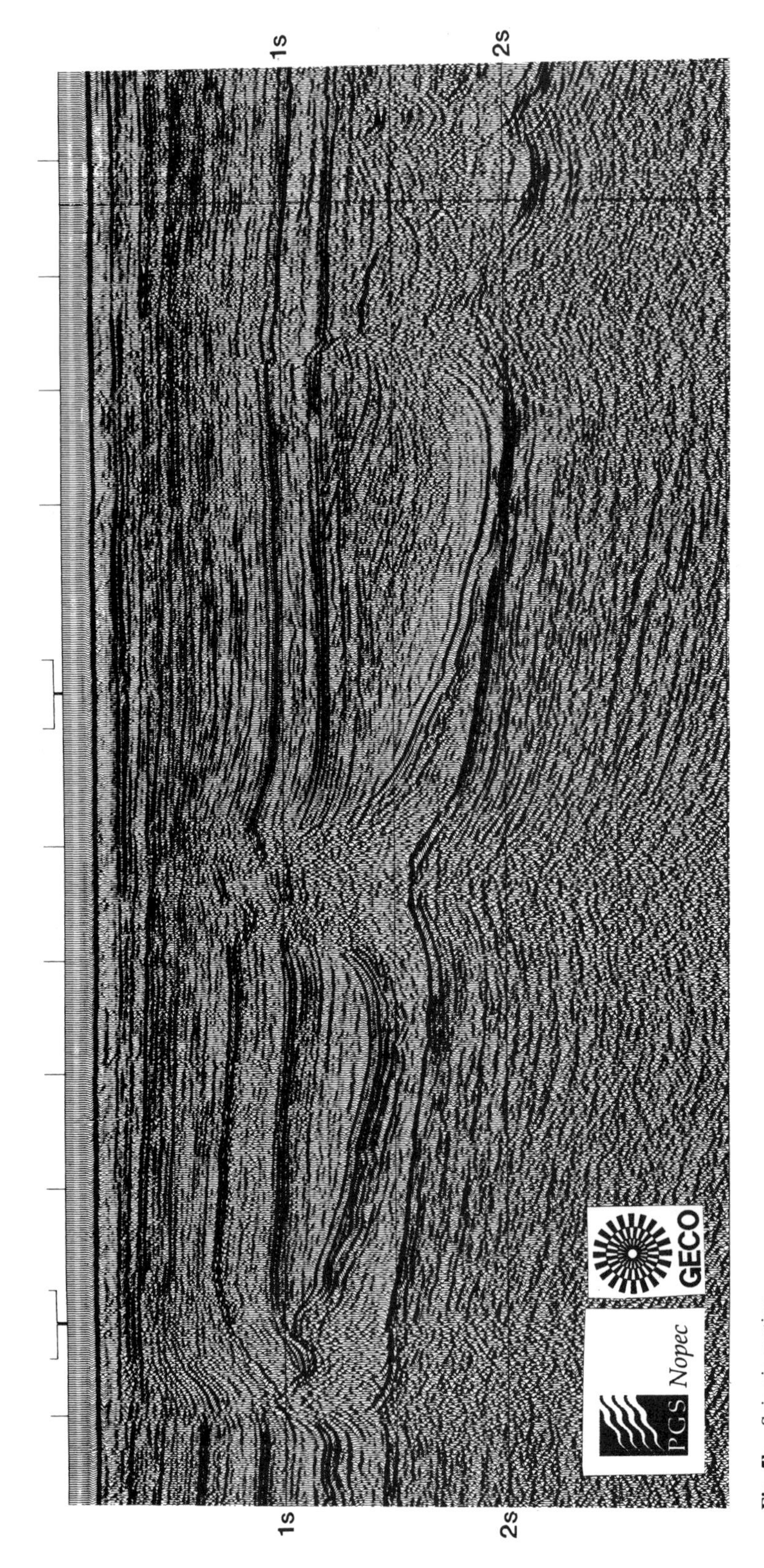


**Fig. 5b.** Seismic section.

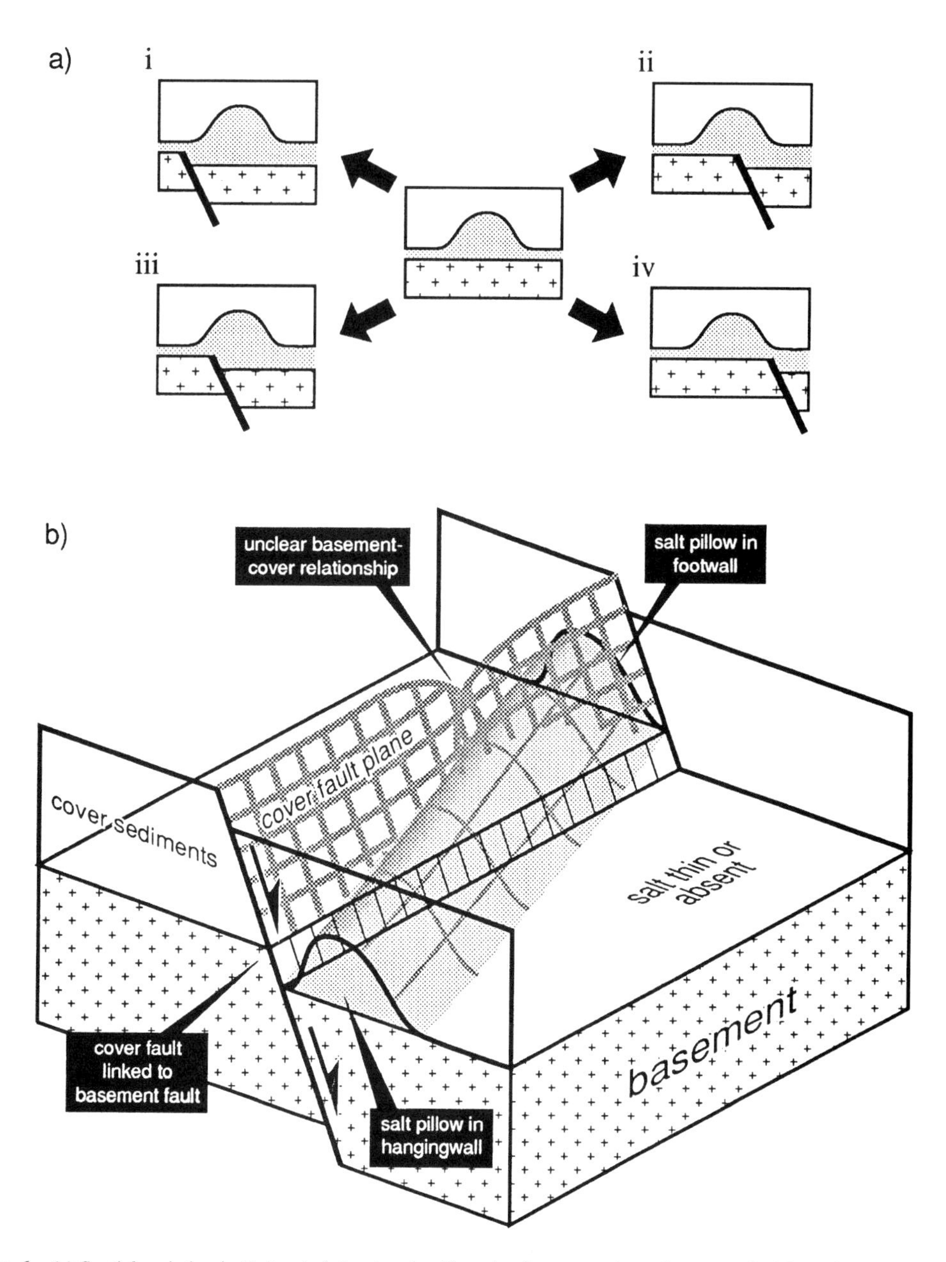


**Fig. 6.** (a) Spatial variation in $T_v$ (centre) due to salt pillow development prior to basement faulting, after Erratt (1993). The subsequent location of the basement fault will govern whether the pillow is preserved in the hangingwall (i) or footwall (iv) to the fault system. Prediction of basement – cover geometries in the intermediate cases (ii, iii) is not straightforward, as there may be cover structures from the time of pillow growth to inherit and reactivate. (b) 3D variation in $D_r$ along the length of a basement fault strand which has grown slightly oblique in trend to a pre-existing salt pillow. Cover fault is locally detached from basement strand.

Although the interpretation presented in Fig. 5 follows the thin-skinned raft tectonics interpretation of Penge *et al.* (1993), other workers have not yet discounted the possibility of salt redistribution driven by buoyancy instability as an important factor in this area (Hodgson *et al.* 1992; Smith *et al.* 1993). Either way, post-deposittional spatial variations in $T_v$ have been imparted to the Zechstein salt

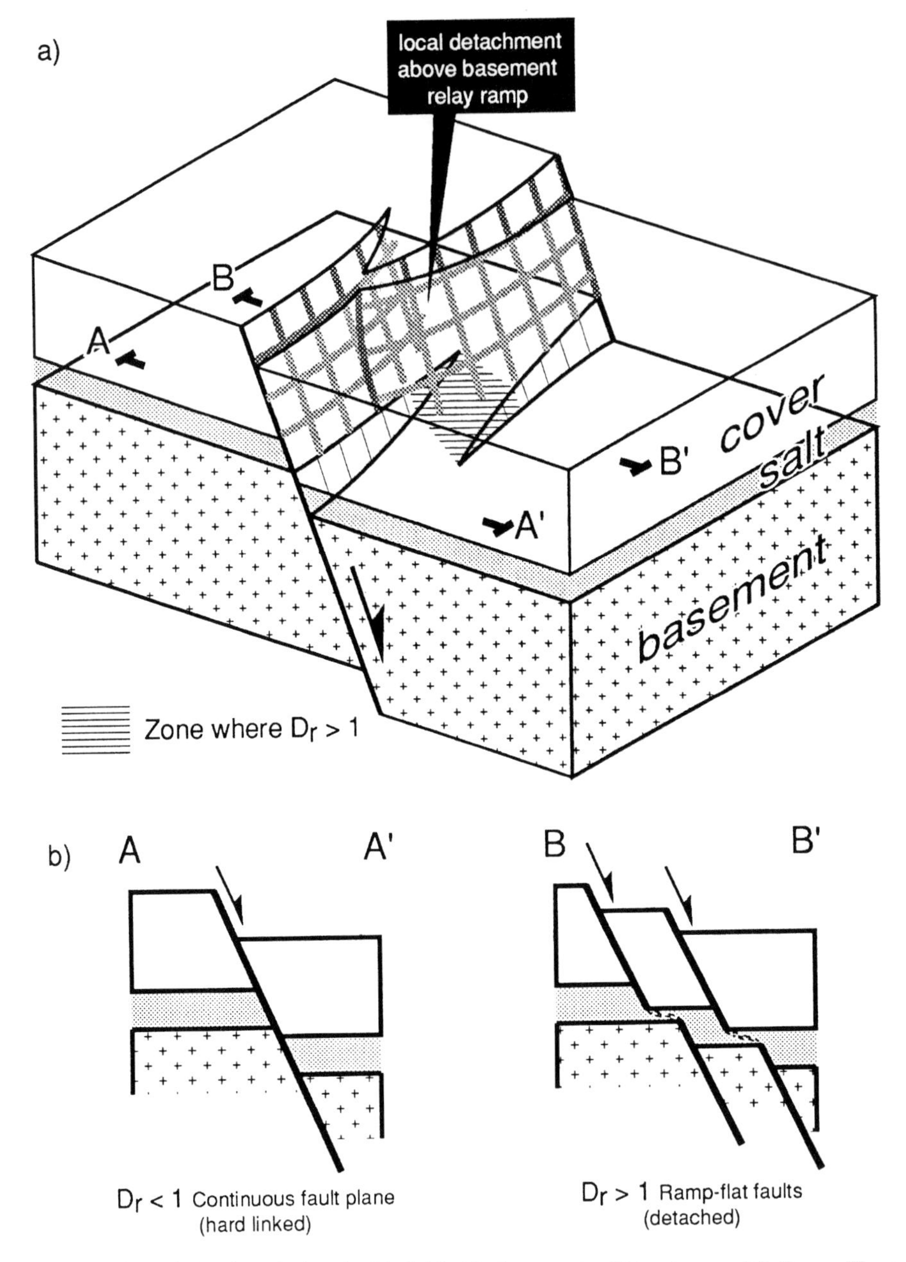


**Fig. 7.** (a) 3D variation in $D_r$ along the length of individual fault segments within a basement fault zone ($T_v$constant). $D_r < 1$ along most of the length of basement fault strands, but rises above 1 in the relay ramp zone.
(b) Detachment tectonics in the ramp zones of an otherwise hard-linked fault system are predicted - compare sections A–A' and B–B'.

and have preceeded the growth of major basement faults. Were a basement fault to grow now in the Forth Approaches Basin, the spatial location of this notional fault would clearly govern $D_r$ (Erratt 1993; Fig. 6a). The options presented on the 2D cartoons of Fig. 6a can also be considered in terms of variation along the length of a notional basement fault which grows along a trend slightly oblique to a pre-existing salt pillow (Fig. 6b). The salt pillow may be fossilized above the hangingwall or the footwall of

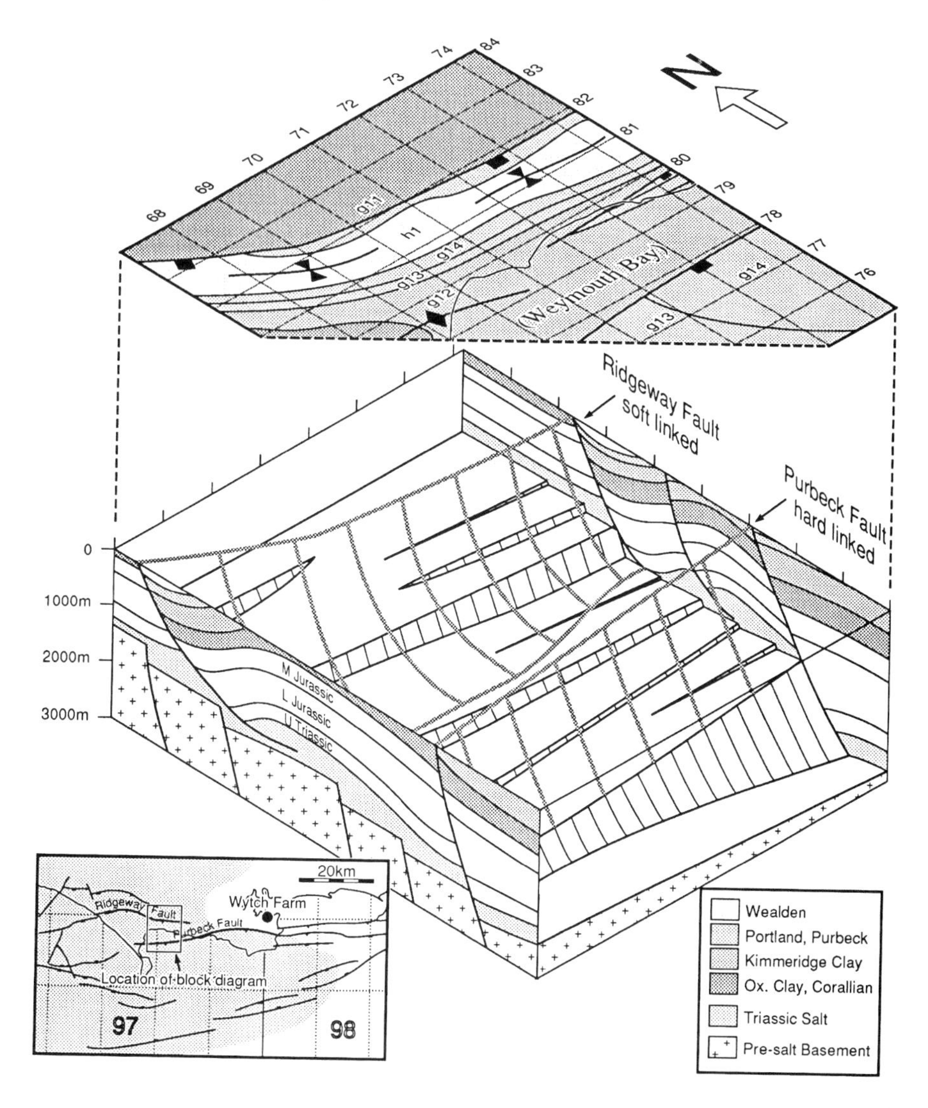


**Fig. 8.** 3D view of north margin, Channel Basin, showing part of a relay zone in the vicinity of Weymouth. Isometric view; grid on map is Ordnance Survey kilometre grid. Area lies fully within the Triassic salt basin (inset location map), $T_v$ is assumed to be constant over the relay zone. Block diagram is a restoration to Albian (pre-inversion) extensional geometry. Basement fault displacement is partitioned across several minor faults in the relay zone. D decreases westwards on the hard-linked Purbeck Fault ($D_r \approx 1$ on the Purbeck Fault on the western cross section). The Ridgeway Fault is detached in this part of the relay structure.

the basement fault - such geometries mimic those which have been discussed elsewhere in relation to salt movement during and after basement faulting (Roberts *et al.* 1990).

It would appear that variations in $T_v$ of the type observed in the Forth Approaches Basin occurred over a wide area north of the Mid-North Sea High (Fig. 5a) from the Triassic onwards (Hodgson *et al.* 1992; Smith *et al.* 1993), predating the major basement faulting episodes during the Jurassic which formed the Central Graben. Some of the wide variety of salt-related structures on the margins of, and within the Central Graben may be due in part to pre-rifting salt movement (Hodgson *et al.* 1992; Erratt 1993; Gowers *et al.* 1993; Høiland *et al.* 1993). The particular issue of the controls upon diapir location within the Central Graben will be discussed later.

## Spatial variation in basement fault displacement (*D*) : north margin, Channel Basin

In the same way that $D_r$ is sensitive to spatial variation in $T_v$, spatial variation in basement fault displacement $D$ will also impact $D_r$ and hence fault geometry. For instance, if a basement fault were to vary in displacement along strike while the thickness of the overlying salt layer remained constant, $D_r$ would vary in the same manner as already illustrated in the context of the Dowsing Fault Zone. This effect might become important where basement faults die out, losing their displacement to a spatially offset en echelon strand. Such relay zones maintain the overall displacement of the basement fault zone (Peacock & Sanderson 1994). If the total basement fault zone displacement is such that $D_r$ is less than 1, then $D_r$ will rise and exceed 1 towards the pins of individual fault segments, and local detachment tectonics in the vicinity of basement relay ramps would be predicted (Fig. 7).

Mapping in the Channel Basin has suggested that such localized detachments might be present around relay zones in the major basement fault system on the northern basin margin (Fig. 8). The Purbeck Fault, which defines the northern margin onshore in Dorset, trends from south of the Wytch Farm field in Bournemouth Bay (Block 98/11) 40 km westwards to Weymouth. As the fault dies out towards its western tip, displacement is transferred north to the Ridgeway Fault. In the part of the relay zone detailed in the pre-inversion restoration Fig. 8, the Purbeck Fault ($D_r < 1$) is still hard linked to its basement counterpart, while the Ridgeway Fault is detached. Further west, this situation is reversed. Tertiary basin inversion involved reverse movement on these faults and led to the growth of folds in the relay zone (e.g. Poxwell pericline). The pre-Albian subcrop pattern above the basement relay zone shows, however, that folds related to basin extension are spatially offset from the Tertiary structures (Ridd 1973). The geometries of these folds suggest that they are rollovers related to listric faults soleing on the Dorset Halite (which has been drilled by five adjacent onshore wells). Displacement at top basement level is partitioned between several small faults in the relay ramp, where $D_r > 1$. The 3D geometry of the Purbeck–Ridgeway extensional relay zone is constrained in the east by subcrop below the exposed Albian unconformity (Fig. 8, map and east side panel) and in the west by forward models of rollover geometry adjacent to the Ridgeway Fault (Fig. 8, west side panel). Cover faulting was probably localized above major basement faults in this area, as can be seen elsewhere in the Channel Basin (see below). A sharp swing in strike of the Ridgeway Fault in map view may indicate a second, northwesterly stepping relay ramp in the basement. The degree of Tertiary reversal on the Ridgeway Fault reflects variations in the degree of linkage between basement and cover during inversion, which may have been controlled by local salt welds in the hangingwall block.

## Temporal variation in basement fault displacement (*D*): Danish Basin

For constant $T_v$, variation in basement fault displacement $D$ due to fault growth may clearly lead to fault systems characterized by $D_r > 1$ early in their history changing to $D_r < 1$ at a later stage (Fig. 9a), indeed, this must be true of all fault strands with $D_r < 1$. It can be seen that cover components of major fault systems which pass through viscous layers may inherently be more numerous than their basement counterparts. A fault system which seems to display this evolutionary sequence is illustrated in Fig. 9. This example is taken from the Norwegian-Danish Basin, an area which has been the subject of several papers discussing relationships between salt and faults in the basement and cover (Hospers *et al.* 1988; Geil 1991; Madirazza 1992; Petersen *et al.* 1992). In Fig. 9 the Triassic package thickens from east to west across the basement fault. Divergent sedimentary packages can be seen in the hangingwalls of detached cover faults above the basement footwall. Balancing movement on the detached cover faults with early displacement on the basement fault as suggested in Fig. 9 accounts for the present-day displacements on the cover and basement fault segments and the thickness variations in Triassic strata across the basement fault zone. Similar geometries can be found on the west margin of the Jæren High (Høiland *et al.* 1993) and can be generated in analog models (Oudmayer & de Jaeger 1993).

## Shared detachment: impact on basement-cover fault linkage

The ramp-flat trajectories discussed so far in the context of single fault strands with $D_r > 1$ represent an idealized situation, applicable to single major faults. If on the other hand a basin is characterized by many basement fault strands, which individually have displacement ratios greater than 1, then the zones of detachment in the salt may spatially overlap (Fig. 10). A question then arises as to whether the fault trajectories separate again to proceed independently into the cover as depicted on Fig. 10. An example illustrating this problem can be found in the Channel basin (Fig. 11). In this area the Upper Triassic to Jurassic sediments lie in domino fault blocks defined by fault strands which seem to detach in the Upper Triassic Dorset Halite, while the tops of tilted fault blocks at top basement level can be seen below the evaporite layer (Fig. 11a, b). One can find in this area the opposite situation to that described in the context of temporal $D_r$ variation in that basement faults are more numerous than cover faults (Figs 8 & 11). The depth converted equivalent of this section (Fig. 11c) shows that cover fault segments which appear to follow ramp-flat trajectories in two-way time are planar in depth. The planar faults apparently continue down through the evaporite layer to link directly with basement faults. Of several issues raised by this depth section we first consider the significant discrepancies in displacement between the cover and basement elements of these apparently continuous strands. The same situation was noted on the west margin of the southern North Sea. In that area, the discrepancy was attributed to inversion of the basement faults and basinwards sliding of the thin-skinned cover. In the Channel Basin, one could also argue that the minor basement faults were older than the cover sediments. While none of these suggestions can be ruled out in the Channel Basin, we note that there is approximate balance of basement and cover fault displacements at the scale of individual cover fault blocks (Fig. 11c). The implication could be that the location of cover faults may be controlled by the most significant basement faults, but that the minor basement faults flatten upwards into the salt layer, contributing their displacement to an offset cover fault (Fig. 12). One could alternatively argue that the displacement discrepancy between cover and basement on the main fault strands could be due to salt migration from hangingwall to footwall, indeed, an element of salt redistribution is an inherent feature of the rigid fault block model shown in Fig. 12. If salt redistribution were the only cause of the basement-cover displacement anomaly, however, the displacement across the minor basement faults would not be accounted for.

If a basin were pervaded by many basement faults, each with a flat-lying portion in the same evaporite layer, then these flat detachments might overlap to the extent that they formed a basin-wide décollement. This situation would alias laterally extensive salt detachment characteristic of gravity

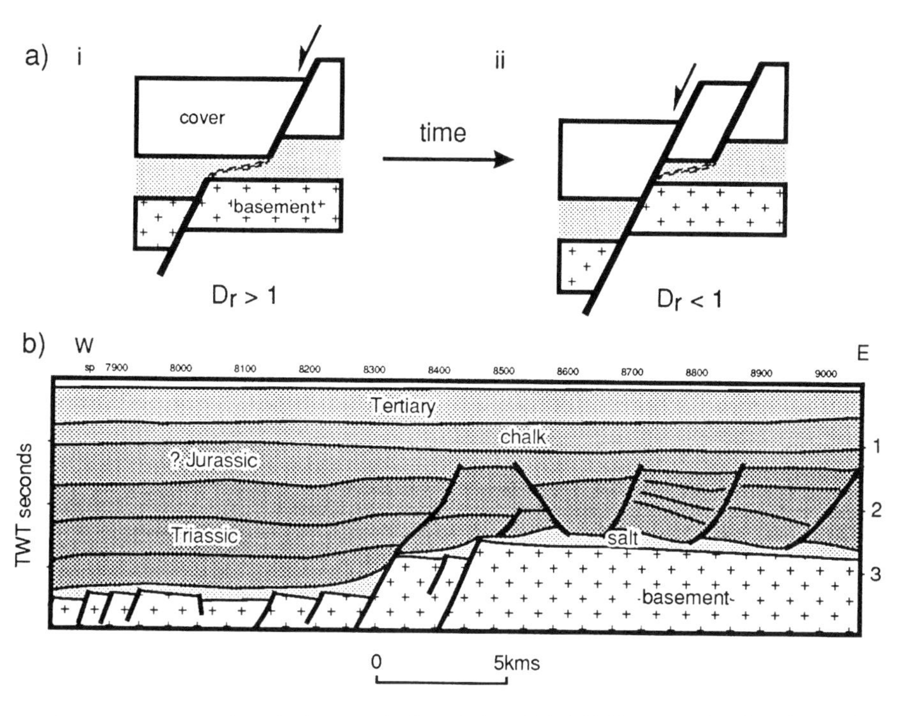


**Fig. 9.** (a) Increasing basement fault displacement D through time. $D_r$ is initially greater than 1, and ramp-flat linkage may occur (i). As the basement fault grows $D_r$ falls below 1 and hard linkage to a cover fault may occur (ii). The final fault geometry splays upwards from a single basement fault into two or more cover strands. The early spatially offset cover fault segment is 'fossilized' above the basement footwall. (b) Seismic and geoseismic from an example in the Norwegian–Danish Basin. Note thickening of the Triassic section across the basement fault and Triassic growth sequences in the detached faults above the basement footwall suggesting early ramp-flat geometry when $D_r > 1$.

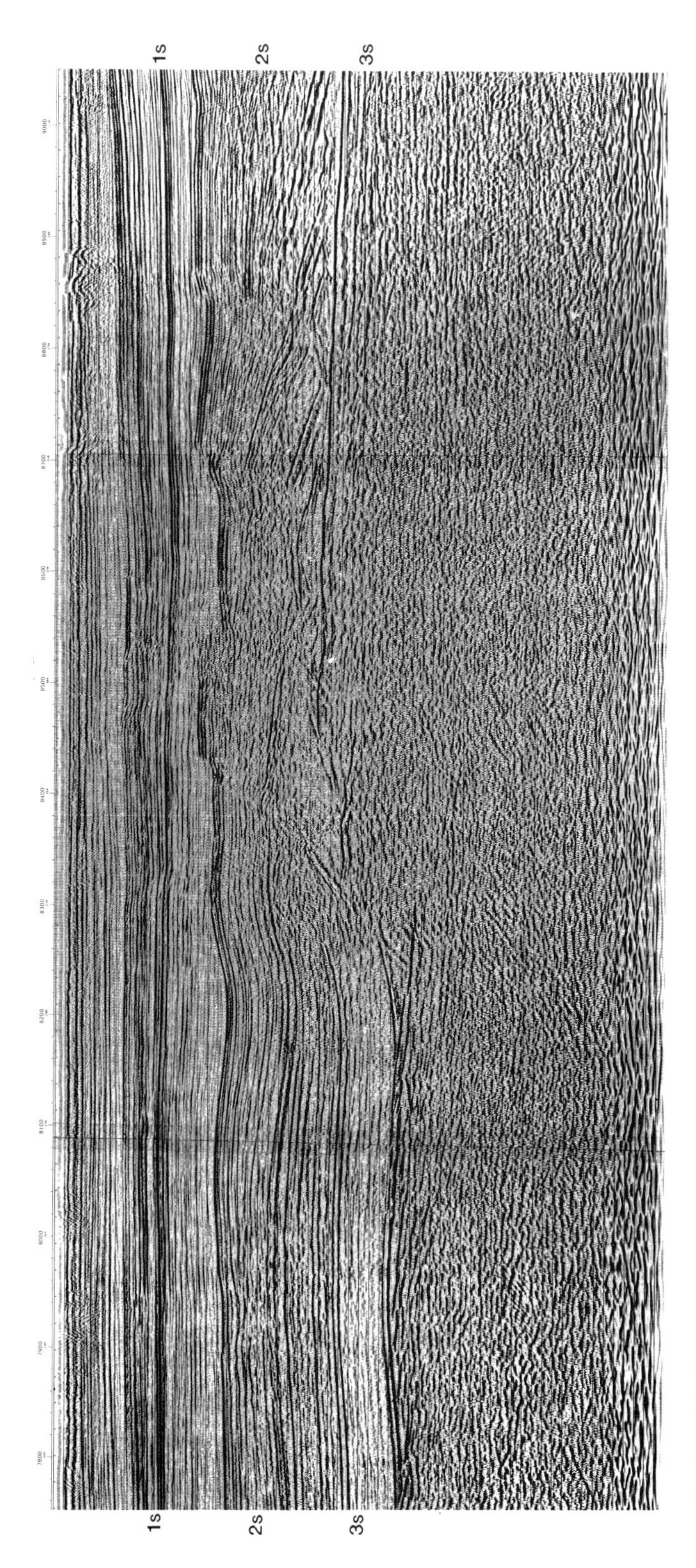


**Fig. 9b.** Seismic section.

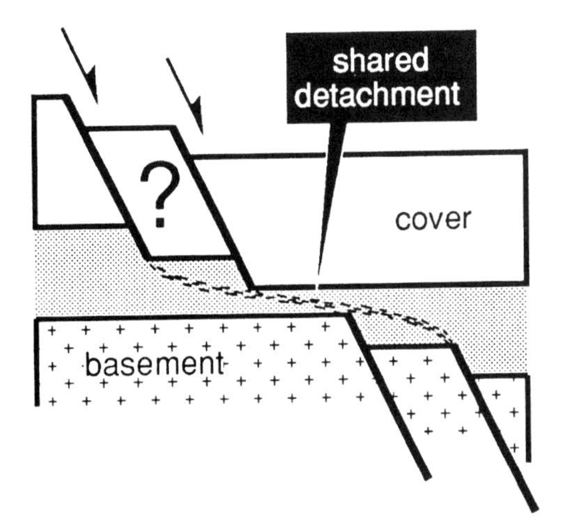


**Fig. 10.** Closely spaced basement faults which sole out into an overlying detachment. If the detachments overlap, do the individual fault trajectories diverge and proceed separately into the cover ?

gliding, and the fault style imparted to the cover sediments might be expected to be similar. The style of extensional thin-skinned deformation which characterizes gravity gliding is termed raft tectonics following type examples observed offshore Angola (Duval *et al.* 1992). During raft tectonics the cover is extended by normal faults whose location and spacing is largely determined by the material properties of the cover strata (Vendeville & Jackson 1992), although localization of cover faults will occur above pronounced, pre-existing edges (fault scarps) in basement topography (Gaullier *et al.* 1993). During thick-skinned extension it is traction at the base of the cover sequence which will drive cover extension, rather than tension in the cover due to basinward collapse. Of course, both processes could occur simultaneously. An important difference between the two processes is, however, that the former can generate cover raft tectonics in the absence of appreciable regional tilt. This also represents an instance of thick-skinned tectonics where, with the exception of anomalously large faults, no spatial relationship

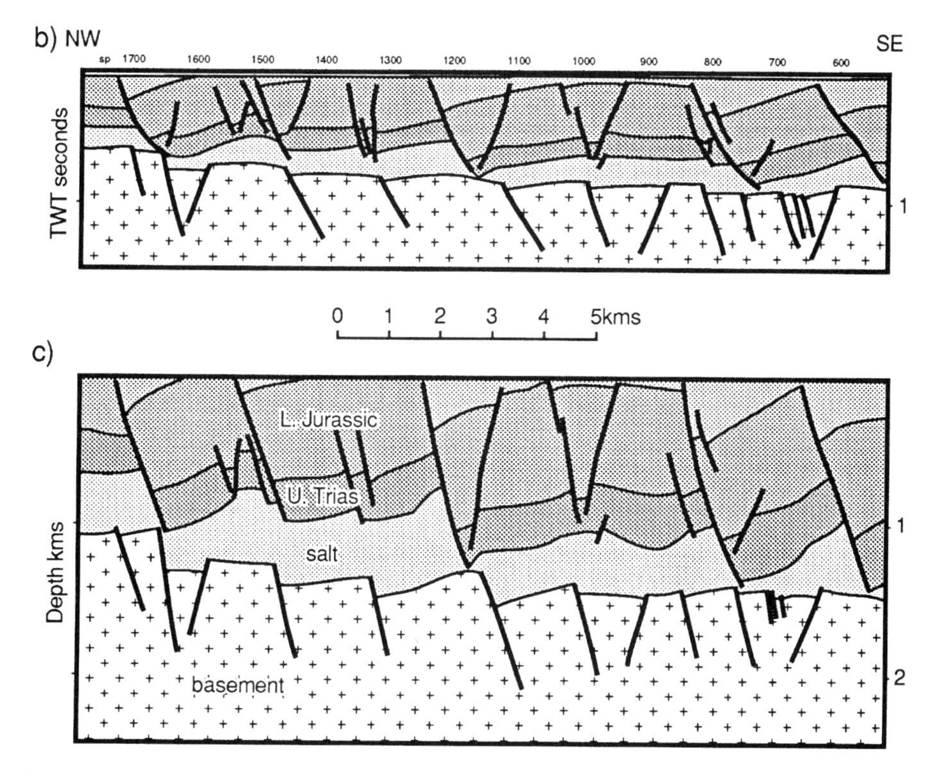


**Fig. 11.** North-south trending section from the Channel Basin. (a) time section. (b) geoseismic. The main faults are distorted into apparent ramp-flat trajectories by relatively high seismic velocity in the salt layer. (c) Depth converted section, vertical exaggeration = ×3. Main faults appear to be planar and continuous with basement strands. Note discrepancy in displacement between corresponding basement and cover strands.

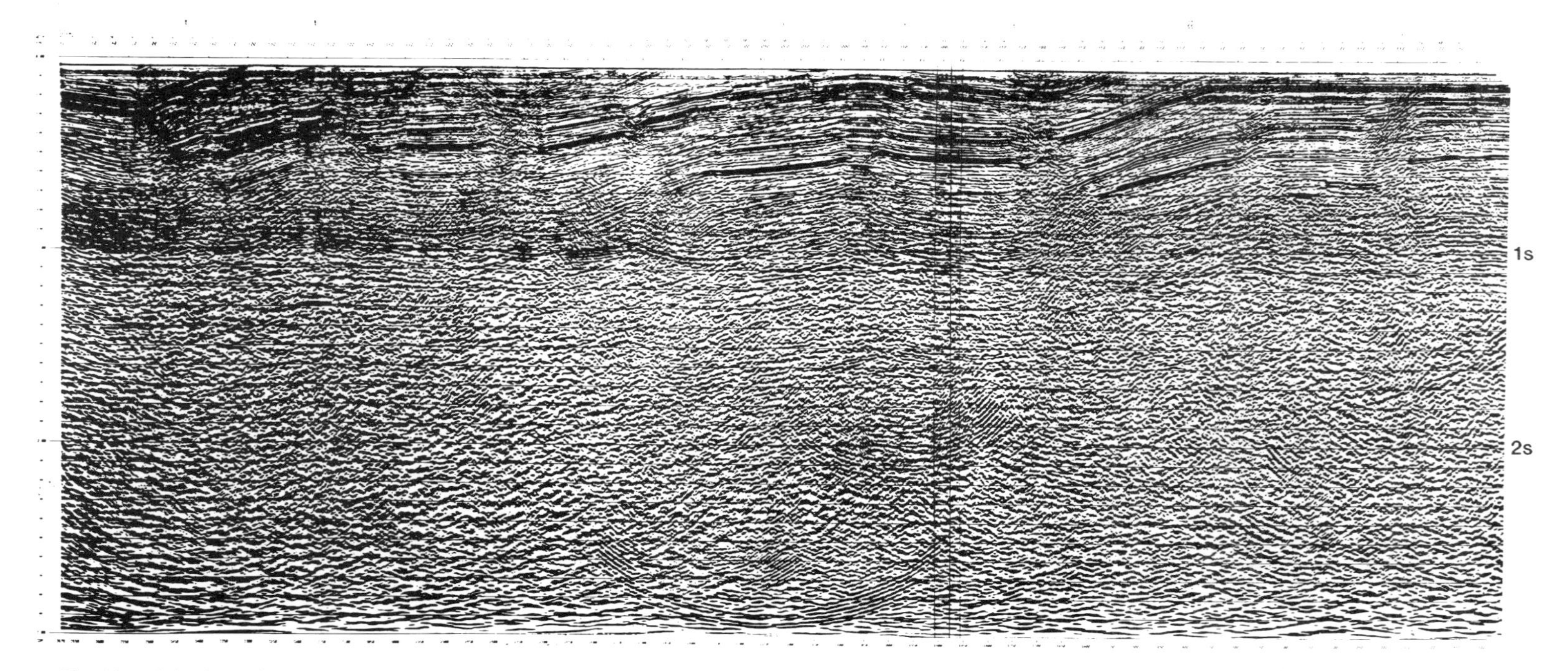


**Fig. 11a.** Seismic section.

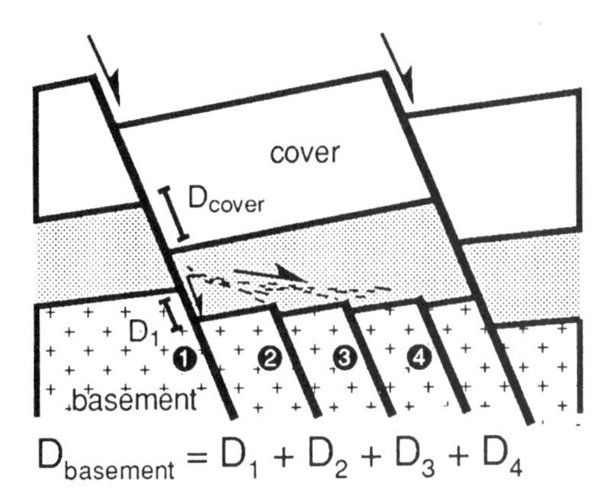


**Fig. 12.** Channel Basin kinematic model. Several closely spaced basement faults, which could be reactivations of older structures, share a detachment which transfers displacement onto a cover strand localized above the most prominent basement scarp. The rigid fault blocks depicted in this model imply some degree of salt redistribution away from each linked basement-cover strand.

need exist between cover and basement structure. The Whale–Horseshoe Basin system offshore Newfoundland is an example of raft tectonics in a thick-skinned system (Balkwill & Legall 1989; Fig. 4 in Jackson & Vendeville 1994).

## Defining effective salt layer thickness ($T_{ve}$)

With the exception of the first example discussed, salt layer thickness $T_v$ has been defined as the distance from the top of the 'basement' to the base of the 'cover'. This approach is perfect for analogue models which contain a single, pure, viscous layer but it is less satisfactory in real sedimentary basins where a salt 'Formation' may contain interbedded anhydrite, dolomite and so on (Taylor 1990). Such interbeds are much stronger than salt and if they are laterally extensive they will reinforce the otherwise weak salt interval. The question arises of how to define an effective $T_v$ value ($T_{ve}$) which could then be considered in terms of $D_r$ – compare Figs 1 and 13. Choices include using only the thickest salt

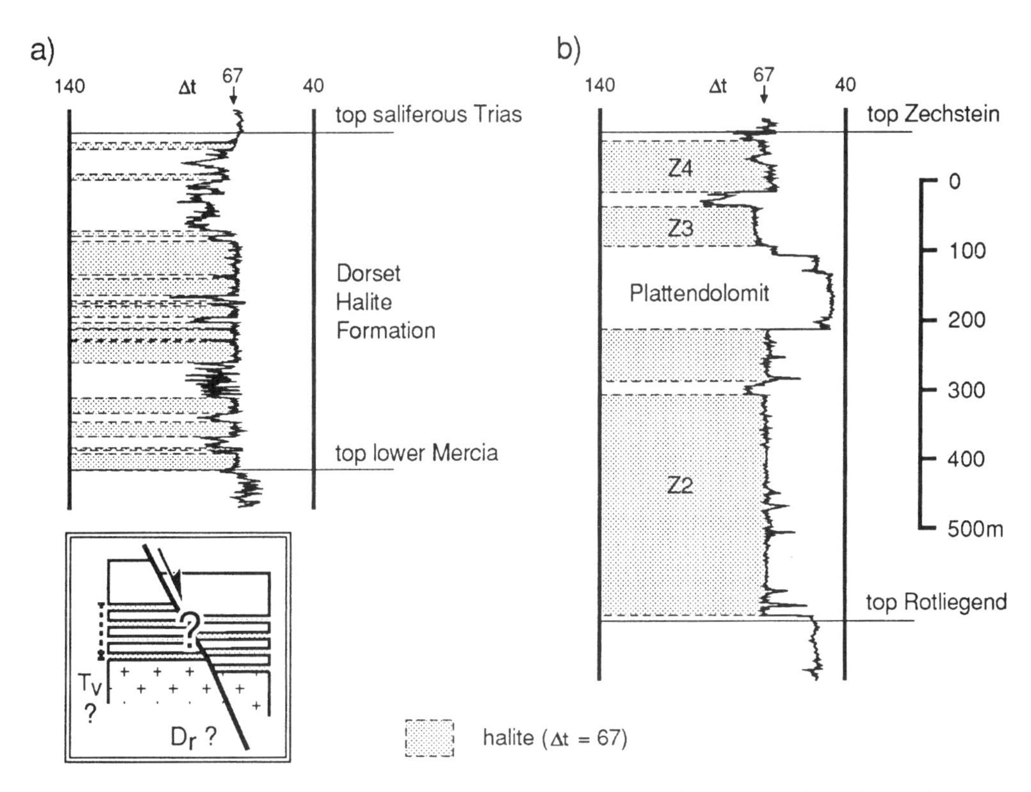


**Fig. 13.** Comparison of sonic logs of saliferous intervals from released wells in the Channel Basin (a) and the southern North Sea (b). $\Delta t$ : interval transit time (μs/foot). (a) Channel Basin well, Quadrant 97, showing Dorset Halite Formation (within the Triassic Mercia Mudstone Group). (b) Southern North Sea well, Quadrant 49 (after Taylor, 1990), showing the Permian Zechstein Supergroup. Inset box : if the salt 'layer' contains interbeds of strong, brittle material, what value should be used to represent $T_v$ when calculating $D_r$ and what is the fault trajectory?

layer; summing the salt layers to give the net salt thickness; averaging the salt layer thicknesses, etc. The Channel Basin example (Fig. 11a) has already shown that cover faults may link directly down onto basement faults at anomalously high displacement ratios ($D_r > 1$) if $T_{ve}$ is measured as the entire thickness of the Dorset Halite Formation. A comparison of typical sonic logs through the main saliferous sequences of the Channel Basin and the southern North Sea is shown on Fig. 13. Although the overall thicknesses of the saliferous sections are of the same order, the Channel Basin section contains far more interbeds and the most significant salt layer is an order of magnitude thinner than the thickest halite layer on the southern North Sea log (Z2 Stassfurt Halite Formation). The Channel Basin example indicates that the entire thickness of the saliferous series is an inappropriate choice for $T_{ve}$, as is the net thickness of salt. Only values close to (but not necessarily equal to) the thickness of the thickest individual salt layer give $D_r < 1$, and account for the observed fault geometry.

## Extensional fault trajectories in viscous – brittle multilayers

The examples illustrated in Fig. 13 could represent the same, notional sedimentary formation with a spatially varying stratigraphy. It is easy to imagine a spatially varying stratigraphic section passing from pure halite through an interbedded zone (such as Fig. 13a) to, with increasing thickness of some of the non-salt interbeds, a multilayer sequence containing salt layers separated by significant thicknesses of sediment. It is likely that real examples of fault planes passing through stratigraphies similar to any of these can be found, but the majority of modelling literature focuses on the 'single, pure layer' end of the spectrum (e.g. Koyi 1991; Vendeville & Jackson 1992; Zaleski & Julien 1992; Weijermars *et al.* 1993).

Having already considered examples here of faults passing through stratigraphies similar to those illustrated in Fig. 13, an example of fault trajectories through a cover sequence containing multiple salt layers is presented (Fig. 14). This example is located in the western part of the southern North Sea, where several significant halite layers are present in the middle and upper Triassic, in addition to the Permian Zechstein salt. The thickness of sediments between the lowermost Triassic evaporites and the Zechstein salt in this area is approximately 600 m (Cameron *et al.* 1992). The structure displayed is known as the 'Flamborough Disturbance', part of the Dowsing Fault Zone and its cover counterpart, both of which pass onshore into East Yorkshire (Kirby & Swallow 1987). The style of faulting present on both sides of this symmetrical graben system below the Cretaceous unconformity clearly consists of ramp-flat extensional faults which define domino fault blocks at several stratigraphic levels. Overall, the basement fault system (DFZ) detaches in the Zechstein, then ramps up in synthetic and antithetic systems through the Bacton Group (Lower Triassic) as it does further southeast (Fig. 3b), however the fault surface flattens again in each of the successive evaporite layers in the Triassic, stepping up in ramps which occur progressively further away from the basement fault scarp.

It is likely that the comments pertaining to basement-cover relationships made so far will also apply to each successive ramp-flat-ramp offset in multi-

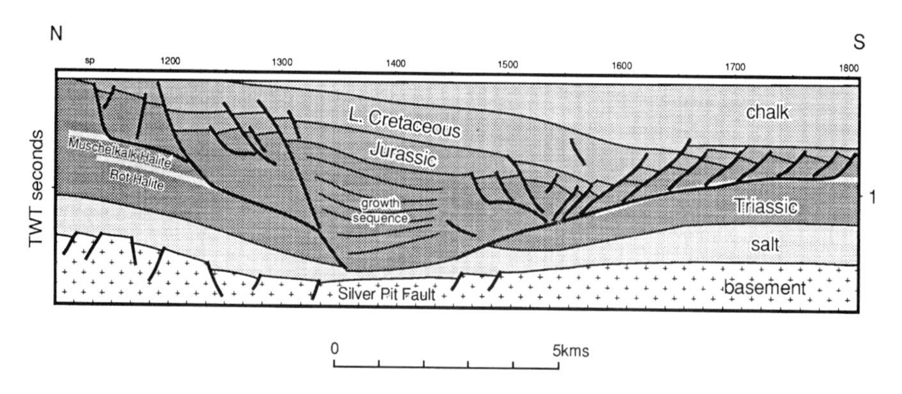


**Fig. 14.** Seismic profile offshore Flamborough Head, East Yorkshire. Multilevel ramp-flat fault profiles on both sides of the cover graben, detachment occurring on several intra-Triassic salts (Röt Halite, Muschelkalk Halite and Keuper Halite). Note the train of domino fault blocks above the Röt Halite on the south side of the graben, the kink-band style of folding, and southerly facing monoclinal folding at high stratigraphic levels (as seen on the east side of the East Midlands Shelf – Fig. 3b). The ramp-flat geometries in the cover continued to be active while $D_r$ fell below 1 (see text).

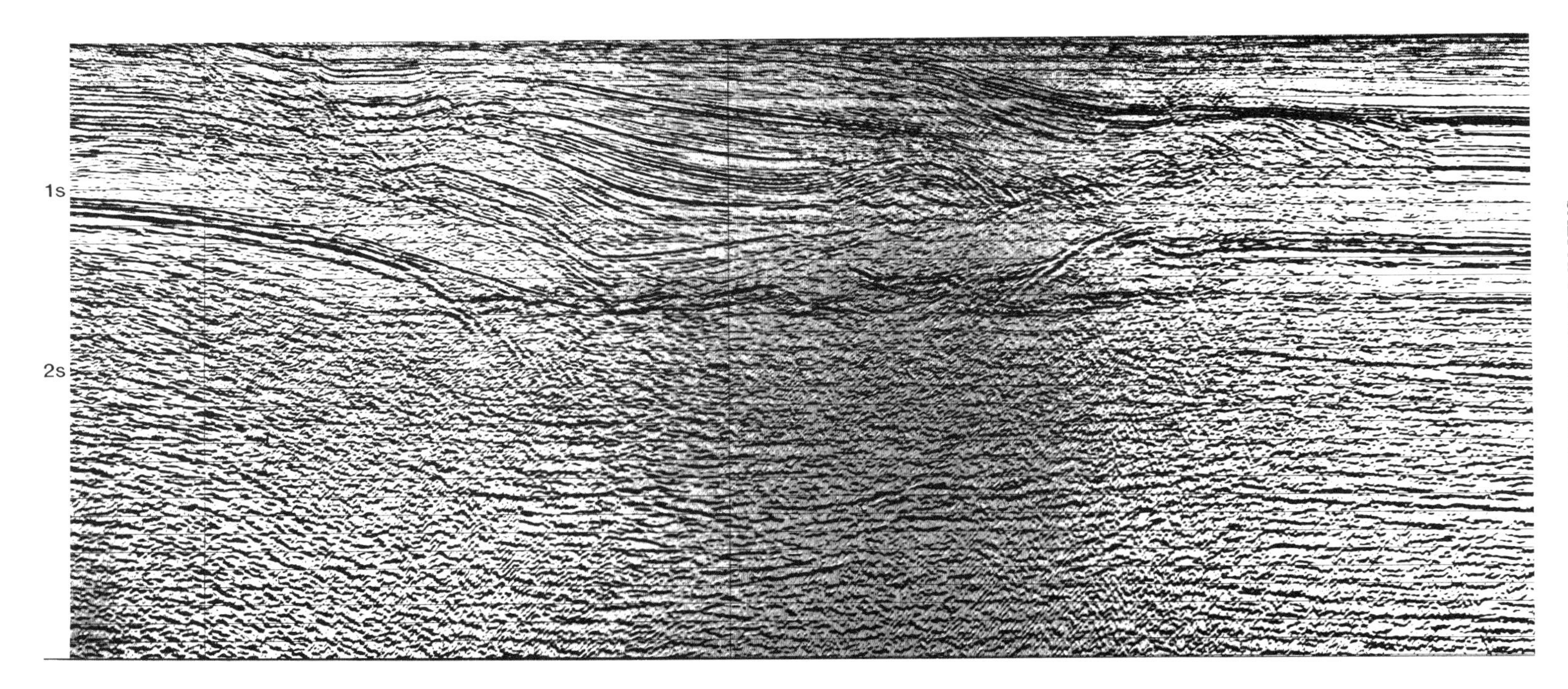


**Fig. 14.** Seismic section.

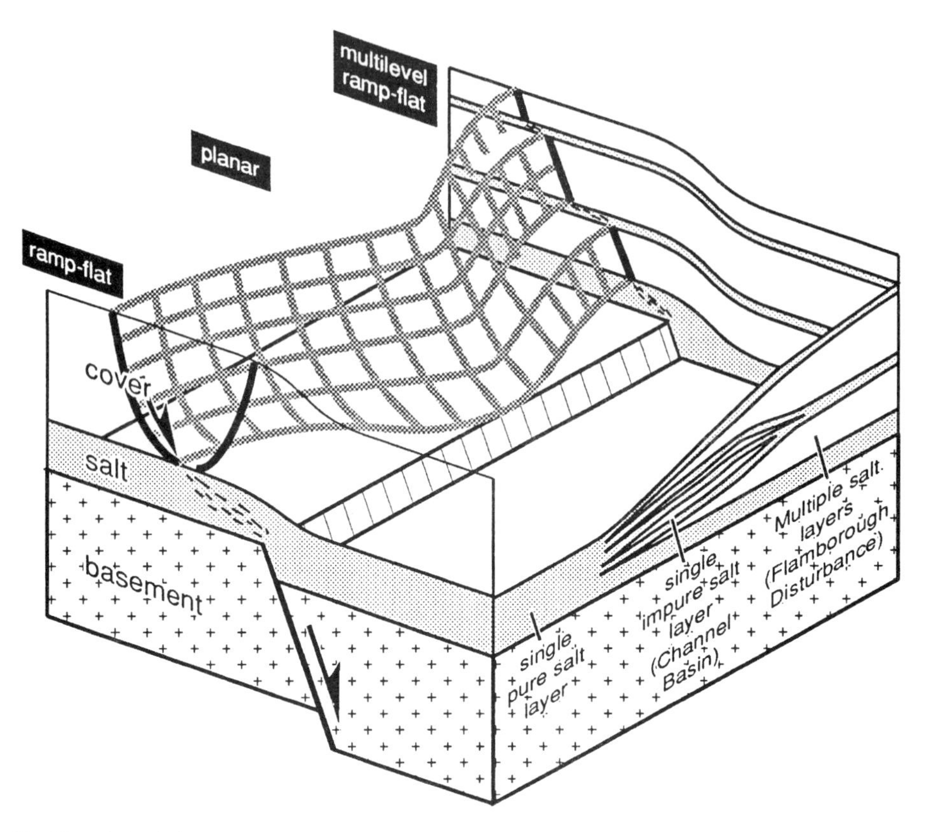


**Fig. 15.** 3D variation in basement-cover relationship due to intra-salt stratigraphic variation from a single, pure halite layer through interbedded salt and 'shales' to a multilayered salt / cover sequence. Based on analog models (Fig. 1), the Dowsing Fault system (Fig. 3), the Channel Basin (Fig. 11) and the Flamborough Disturbance (Fig. 14).

level stair-stepping fault trajectories. It is noted, however, that the ramp-flat geometries shown in Fig. 14 have persisted in spite of $D_r$ falling below 1 in the Triassic section as the system has grown (cf. Fig. 9). The survival of active ramp-flat surfaces when $D_r < 1$ may reflect a change in the behaviour of the multilayer as it becomes tilted (controlled in this case by Zechstein salt redistribution, Fig. 14), however the influence of multilayer dip is not explored further here. Although the graben system is broadly symmetric, the sub-Chalk Mesozoic sequence clearly thickens northwards, suggesting that the footwall to the system in basement and cover lies on the south side of the graben. This contrasts with the interpretation of Kirby & Swallow (1987) who infer the cover strata south of the graben to be allochthonous.

Ramp-flat geometries within the Triassic can be found in several other parts of the southern North Sea, for example in the crest of the Sole Pit High (Van Hoorn 1987; Walker & Cooper 1987) and on the flanks of major salt-cored fold structures (Owen & Taylor 1983; Jenyon 1988). These examples are entirely gravity driven and arise due to local uplift. The faults in these examples grow downwards towards the Zechstein salt and accommodate local gravity spreading. They are therefore genetically unrelated to the structures seen in association with the cover equivalent of the DFZ. The fault style observed in the Flamborough Disturbance can be considered alongside the patterns seen elsewhere on the DFZ and in the Channel Basin to construct the 3D geometry of a fault growing through a salt-cover stratigraphy which varies from a single to multiple viscous layers (Fig. 15). This figure emphasizes the importance of the purity of the salt layer and selection of appropriate $T_v$ values.

## *Cover delamination and salt sills in multilayer sequences*

Following sufficient cover extension, the Zechstein salt intrudes the cover fault planes as reactive

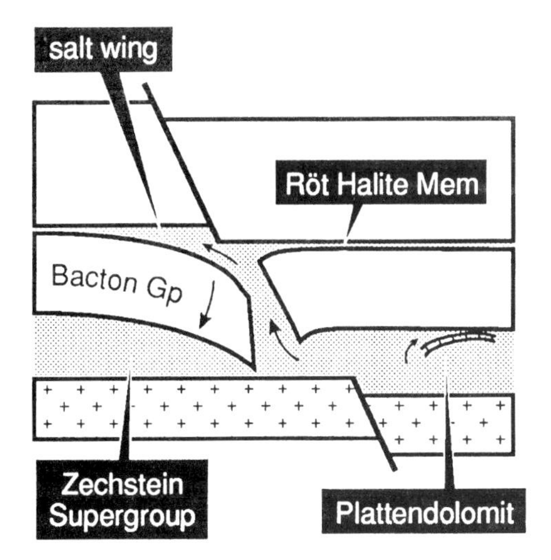


**Fig. 16.** Model of lower Triassic (Bacton Group) delamination facilitated by diapiric intrusion into a cover fault system. Lateral spreading of Zechstein salt at Röt Halite level is passive, a suction effect driven by unpeeling of Bacton Group into underlying Zechstein salt. Note that if Bacton Group delamination occurred in the hangingwall alone, with the Röt Halite fed by an out of section diapir, the geometry would ape sedimentary growth during Röt Halite times. The process of Bacton Group delamination is a larger-scale example of upper Zechstein carbonate (Z3 Plattendolomit) delamination, which is a common occurrence.

diapirs (Vendeville & Jackson 1992). Once these diapirs have intruded to the top of the Bacton Group, cohesion between the Bacton Group and the surrounding sediments is lost (the Bacton Group is overlain by the Röt Halite Member). With an abundant supply of Zechstein salt adjacent to the Röt Halite layer, the Bacton Group is free to delaminate from the rest of the cover sequence and collapse down, driven by gravity, into underlying Zechstein Supergroup, while mobile Stassfurt Halite feeds up the adjacent conduit to spread laterally into the space created above the Bacton Group (Fig. 16). This process represents localized redistribution driven by buoyancy, facilitated by reactive diapirism. It is a very similar process to that seen on a smaller scale within the Zechstein Supergroup, wherein the Z3 and Z4 carbonates often delaminate on the Z3 and Z4 halite members and sink down into the Z2 Stassfurt Halite, forming allochthonous rafts enveloped by Z2 salt (Jenyon 1989; Fig. 16). Indeed, such rafts can be carried up into the Triassic within reactive diapirs. It is noted that salt plays a passive role in this mechanism for Zechstein sill intrusion.

Zechstein sills may be lopolithic or laccolithic in form and are characterized by divergence of Triassic reflectors which can be misinterpreted as a sedimentary growth sequence. Examples of salt wings associated with lower Triassic delamination are common in the southern North Sea, Danish Central Graben (figure 11 in Sundsbø & Megson 1993) and the northwest German basin (Baldschuhn *et al.* 1991). Sag of the Bacton Group may be more extensive parallel to a cover fault trend than the adjacent reactive diapir conduit, so a given 2D section in the dip direction may contain an anomalously thick Röt Halite swell, apparently unconnected to the Zechstein, though the extra thickness at Röt Halite level is due to movement of Zechstein salt into the section.

## Influence of thick-skinned extension on subsequent diapirism

The discussion so far has focused on the geological criteria which control the initial geometry of extensional faults which grow from basement through a viscous layer to the cover. It has been argued that the geometry of such extensional faults is established before diapirism occurs. With continuing extension, salt intrudes the cover fault planes as reactive diapirs, which may be planar in 3D (Jackson & Vendeville 1994). This important period of diapir evolution occurs during active rifting (Late Jurassic in the North Sea) and is a period during which salt in the form of diapirs establishes discordant relationships with cover sediments and attains high, shallow, structural levels. Periods of basin extension via active basement faulting may be followed by periods of thermally driven subsidence (McKenzie 1978). This is the case in the North Sea, where flexural subsidence centred on the Central Graben during the Tertiary has led to the burial of the Jurassic rift system by up to 3 km of post-rift sediment. This second phase of basin subsidence is different from the first in that there is no inherent basement faulting. But since diapirs were already emplaced in association with fault planes at shallow levels, they are ideally placed to passively maintain their altitude relative to the geoid while the surrounding sediments 'downbuild', falling as the basin subsides (Jackson & Vendeville 1994). The conditions of active basement faulting in an area of thick salt, followed by thermal subsidence, are met in the Central Graben, where indeed many impressive salt diapirs can be found, often having 'risen' over 2km to reach juxtaposition with very young Tertiary strata. Some of these diapirs have been drilled and have been found to be capped by condensed sequences which are equivalent to the adjacent stratigraphic sections (Foster & Rattey 1993).

An understanding of the relationship between diapir location and basement structure can be

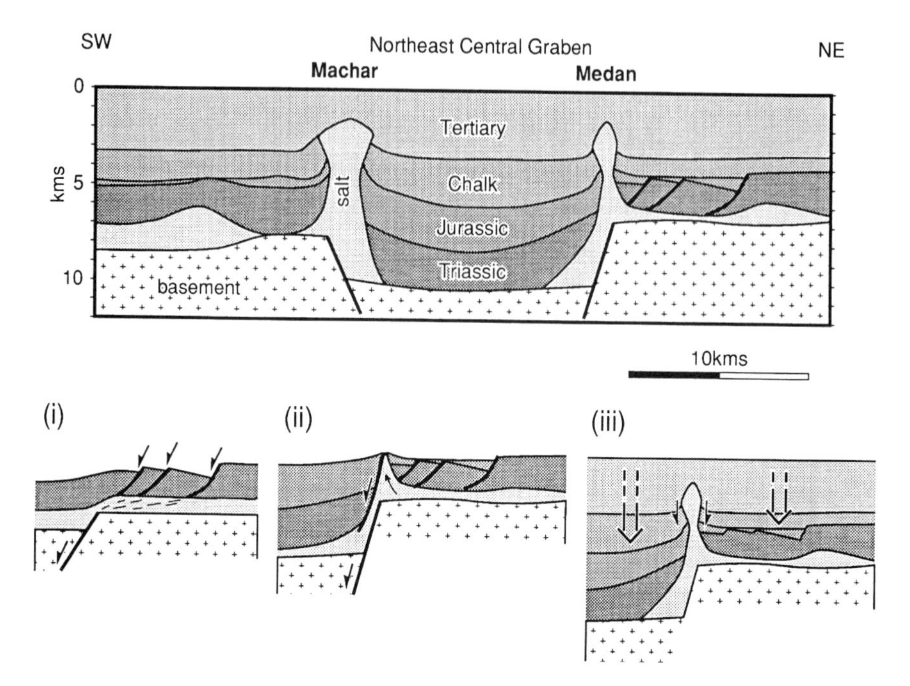


**Fig. 17.** Cross section across the Machar and Medan diapiric structures, Central North Sea, from Foster & Rattey (1993). The location and evolution of the Medan diapir is explained by a consideration of the basement fault history. (i) Initial basement fault growth ($D_r > 1$) links to cover via detachment in the salt, giving offset cover fault segments with Jurassic growth sequences (cf. Fig. 9). (ii) Continued basement fault growth until $D_r < 1$ leads to hard linkage of basement fault to cover, at which point the cover segment is intruded by a reactive diapir. (iii) Cretaceous and Tertiary basin subsidence carries the sedimentary sequence down around the diapir, which passively maintains its altitude relative to the geoid.

approached if one considers the early (pre-diapir intrusion) stages of fault evolution in terms of the geological criteria which have been discussed so far (Fig. 17). Interpretation of mapped spatial relationships is also possible, for example if a train of passive diapirs is spatially offset from a basement trend, they may have evolved from a salt wall which reactively intruded early cover faults offset from their basement roots. Diapirs overlying basement scarps may have intruded faults which attained displacement ratios less than 1, and so on. It should be borne in mind that basement-cover relationships may be modified after active basement faulting by thin-skinned gravity gliding, which would passively move any salt structure present in the cover relative to the basement. Local gravity gliding post-dating basement faulting has been noted on the west margin of the Central Graben (Erratt 1993; Rattey & Hayward 1993) and further west, on the West Central Shelf (Bishop *et al.* 1995).

## Influence of salt layers on inversion kinematics

Basin inversion is a common feature of the sedimentary basins offshore UK, particularly those in the south, such as the southern North Sea and Channel Basins (Roberts 1989). There may be many facets to the influence of salt layers on inversion styles, as there are during extension, but a collation of observations made on the west margin of the southern North Sea (Fig. 3) enables some key points to be made. Where no salt is present (Fig. 3a), reverse movements propagate directly into the cover. A key difference is noticed further northwest, where salt is present (Figs 3b, 3c) in that despite reverse movements at basement level (evidenced by monoclinal flexure of the Sole Pit High flanks) there is no reactivation of extensional fault segments in the cover. Given that the cover is certainly pinned to the basement further west at the salt pinchout (with no intervening thin-skinned compressional structures), the basement shortening

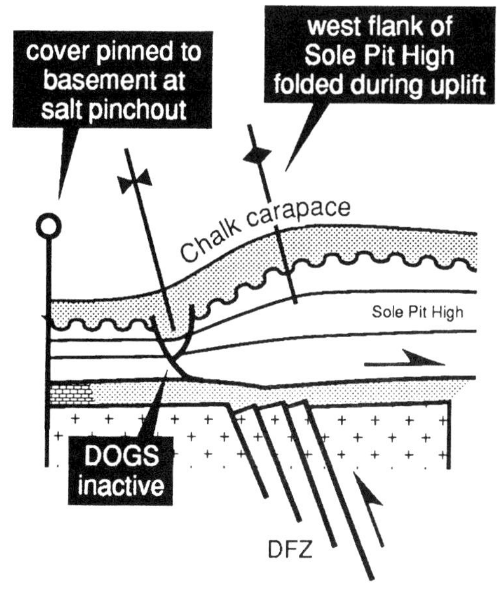


**Fig. 18.** Inversion kinematics of the Dowsing Fault Zone. Reverse movements on the basement faults are inferred from monoclinal folding of the cover sequence (and are observed along strike). The cover components of the Dowsing Fault system, which were active during extension (DOGS), are not reactivated during basement inversion. This implies basinwards movement of the cover (Sole Pit High) relative to the basement, given that the salt pinches out above the basement footwall. The basement hangingwall block is geometrically similar to a triangle zone.

can only be balanced by basinwards movement of the Sole Pit High relative to the basement hangingwall (Fig. 18). The line length mismatch between cover and basement is balanced by shortening in the cover on the east side of the Sole Pit High, where major salt-cored detached fold structures grew during the Palaeogene, contemporaneous with Sole Pit High inversion as dated by unconformable sediments further south (van Hoorn 1987; Badley *et al.* 1989; Stewart & Coward 1995). The spatial variation in cover fault reactivation is clearly influenced by the Zechstein salt (Fig. 19). This effect has also been noted in analog experiments (figure 7 in Richard 1991). The apparent difficulty in reactivating a stair-stepping fault trajectory during compression may underline an important role of detachments during shortening of multilayers.

## Discussion

Several geological criteria which control the geometry of faults which grow from basement to cover through salt layers have been examined using examples from offshore UK. These criteria include spatial variations in salt thickness and basement fault zone geometry. They are summarized by the Displacement ratio $D_r$ which has a key value of 1 marking the transition from soft-linked (detached) basement cover systems to hard-linked (continuous) strands. Examples of spatial coincidence and spatial offset between cover and basement fault elements can be found and accounted for. Although the geological criteria have been individually examined here, any one fault system is likely to be characterized in 3D by simultaneous variation of several criteria, giving complex variations in $D_r$. Most basin margins are characterized not by a single fault, but by fault systems comprised of several elements in basement and cover. Faults may splay upwards on passing through salt layers, or they may join upwards. In both cases the salt forms a detachment, the difference lying in the number of basement faults contributing displacement to the detachment. One of the most important, if least apparent, criteria is time. The role of salt in the evolution of a fault system changes from the earliest stages, when it contains ductile portions of the fault system, through later stages when it intrudes and deforms the upper portions of the system.

Experimental work has shown that diapir intrusion only occurs after significant amounts of extension on cover fault systems. Given the control of cover systems upon diapir initiation, an understanding of the spatial relationship between cover and basement fault segments should lead to an understanding of the spatial relationship between diapirs and basement faults. Diapir location may additionally reflect cover faulting related partially or wholly to thin-skinned tectonics. Although diapirs can be classified according to their ontogenic progression from reactive through active to passive (Vendeville & Jackson,1992), this does not identify their mode of origin. Many examples of diapirs related to thin-skinned tectonics have been described, as have diapirs related to thick-skinned tectonics (Jackson & Vendeville, 1994). The range of examples discussed here suggests that the problem of salt structure – basement structure relationship is tractable in many places within the UK offshore basins, and a genetic classification of salt structures as an adjunct to the ontogenic classification would be possible.

The authors acknowledge Geco-Prakla (UK) Ltd, PGS Nopec (UK) Ltd, Shelf Exploration Ltd, Western Geophysical, British Gas Exploration and Production, Kerr-McGee Oil (UK) Ltd., OMV (UK) Ltd, and Amerada Hess Ltd for permission to include seismic examples, and reviews by C. Smith, C. Elders and J. Hossack. The views expressed here are those of the authors and not necessarily those of Amerada Hess, British Gas, Kerr McGee or OMV.

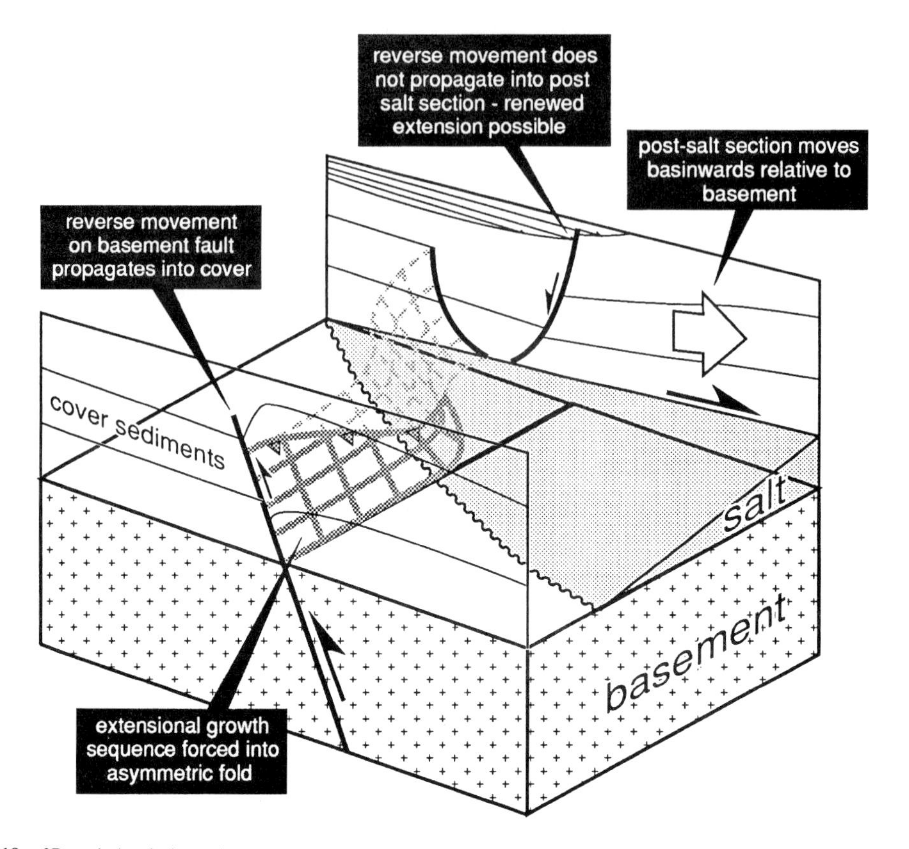


**Fig. 19.** 3D variation in inversion kinematics based on the Dowsing Fault System (Fig. 3). Without salt, the cover is hard-linked to basement and the cover extensional system is reactivated as a reverse fault, creating an asymmetric, hook-like hangingwall anticline structure. With a significant salt layer lying between basement and cover, reverse movement at basement level does not propagate into the extensional cover system. The basinwards movement and passive uplift of the cover can lead to minor extensional reactivation of the cover fault system. An element of rotation introduced into the cover between the hard-linked shortening zone and the detached zone may contribute to enigmatic basinal thin-skinned cover faulting.

## References


ALLEN, M. R., GRIFFITHS, P. A., CRAIG, J., FITCHES, W. R. & WHITTINGTON, R. J. 1994. Halokinetic initiation of Mesozoic tectonics in the southern North Sea: a regional model. *Geological Magazine*, **131**, 559–561.

ARTHUR, T. J. 1993. Mesozoic structural evolution of the UK Southern North Sea: insights from analysis of fault systems. *In*: PARKER, J. R. (ed.) *Petroleum Geology of Northwest Europe: Proceedings of the 4th Conference*. Geological Society, London, 1269–1279.

BADLEY, M. E., PRICE, J. D. & BACKSHALL, L. C. 1989. Inversion, reactivated faults and related structures: seismic examples from the southern North Sea. *In*: COOPER, M. A. & WILLIAMS, G. D. (eds) *Inversion Tectonics*. Geological Society, London, Special Publication, **44**, 201–219.

BALDSCHUHN, R., BEST, G. & KOCKEL, F. 1991. Inversion tectonics in the north-west German basin. *In*: SPENCER, A. M. (ed.) *Generation, Accumulation and Production of Europe's Hydrocarbons*, European Association of Geoscientists, Special Publication **1**, 149–159.

BALKWILL, H. R. & LEGALL, F. D. 1989. Whale Basin, offshore Newfoundland: extension and salt diapirism. *In*: TANKARD, A. J. & BALKWILL, H. R. (eds) *Extensional Tectonics and Stratigraphy of the North Atlantic Margins*. American Association of Petroleum Geologists, Memoir, **46**, 233–245.

BARTHOLOMEW, I. D., PETERS, J. M. & POWELL, C. M. 1993. Regional structural evolution of the North Sea: oblique slip and the reactivation of basement lineaments. *In*: PARKER, J. R. (ed.) *Petroleum Geology of Northwest Europe: Proceedings of the 4th Conference*. Geological Society, London, 1109–1122.

BISHOP, D. J., BUCHANAN, P. G. & BISHOP, C. J. 1995. Gravity-driven thin-skinned extension above Zechstein Group evaporites in the western central North

Sea: an application of computer-aided section restoration techniques. *Marine and Petroleum Geology*, **12**, 115–135.

BOYER, S. E. & ELLIOTT, D. 1982. Thrust systems. *AAPG Bulletin*, **66**, 1196–1230.

CAMERON, T. D. J., CROSBY, A., BALSON, P. S., JEFFREY, D. H., LOTT, G. K., BULAT, J. & HARRISON, D. J. 1992. *United Kingdom Offshore Regional Report: the Geology of the Southern North Sea*. London, HMSO for the British Geological Survey.

CHRISTIAN, H. E. 1969. Some observations on the initiation of salt structures of the southern British North Sea. *In*: HEPPLE, P. (ed.) *The Exploration for Petroleum in Europe and North Africa*. Institute of Petroleum, London, 231–248.

CRANS, W., MANDL, G. & HAREMBOURNE, J. 1980. On the theory of growth faulting : A geomechanical delta model based on gravity sliding. *Journal of Petroleum Geology*, **2**, 265–307.

DAVIS, D. W. & ENGELDER, T. 1985. The role of salt in fold-and-thrust belts. *Tectonophysics*, **119**, 67–88.

DUVAL, B., CRAMEZ, C. & JACKSON, M. P. A. 1992. Raft tectonics in the Kwanza Basin, Angola. *Marine and Petroleum Geology*, **9**, 389–405.

ERRATT, D. 1993. Relationships between basement faulting, salt withdrawal and Late Jurassic rifting, UK Central North Sea. *In*: PARKER, J. R. (ed.) *Petroleum Geology of Northwest Europe: Proceedings of the 4th Conference*. Geological Society, London, 1211–1219.

FOSTER, P. T. & RATTEY, P. R. 1993. The evolution of a fractured chalk reservoir: Machar oilfield, UK North Sea. *In*: PARKER, J. R. (ed.) *Petroleum Geology of Northwest Europe: Proceedings of the 4th Conference*. Geological Society, London, 1445–1452.

GAULLIER, V., BRUN, J. P., GUÉRIN, G. & LECANU, H. 1993. Raft tectonics: the effects of residual topography below a salt decollement. *Tectonophysics*, **228**, 363–381.

GEIL, K. 1991. The development of salt structures in Denmark and adjacent areas: the role of basin floor dip and differential pressure. *First Break*, **9**, 467–483.

GLENNIE, K. W. & BOEGNER, P. L. E. 1981. Sole Pit inversion tectonics. *In*: ILLING, L. V. & HUDSON, G. V. (eds) *Petroleum Geology of the Continental Shelf of North-West Europe*. Institute of Petroleum, London, 110–120.

GOWERS, M. B., HOLTAR, E. & SWENSSON, E. 1993. The structure of the Norwegian Central Trough (Central Graben area). *In:* PARKER, J. R. (ed) *Petroleum Geology of Northwest Europe: Proceedings of the 4th Conference*. Geological Society, London, 1245–1254.

HODGSON, N. A., FARNSWORTH, J. & FRASER, A. J. 1992. Salt-related tectonics, sedimentation and hydrocarbon plays in the Central Graben, North Sea, UKCS. *In*: HARDMAN, R. F. P. (ed.) *Exploration Britain*. Geological Society, London, Special Publication, **67**, 31–63.

HØILAND, O., KRISTENSEN, J. & MONSEN, T. 1993. Mesozoic evolution of the Jæren High area, Norwegian Central North Sea. *In*: PARKER, J. R. (ed.) *Petroleum Geology of Northwest Europe: Proceedings of the 4th Conference*. Geological Society, London, 1189–1195.

HOLMES, A. J. 1991. The Camelot fields, Blocks 53/1a, 53/2, UK North Sea. *In*: ABBOTTS, I. L. (ed.) *United Kingdom Oil and Gas Fields, 25 years Commemorative Volume*. Geological Society, London, Memoir, **14**, 401–408.

HOSPERS, J., RATHORE, J. S., JIANHUA, F., FINNSTRØM, E. G. & HOLTHE, J. 1988. Salt tectonics in the Norwegian-Danish Basin. *Tectonophysics*, **149**, 35–60.

JACKSON, M. P. A. & VENDEVILLE, B. C. 1994. Regional extension as a geologic trigger for diapirism. *Geological Society of America Bulletin*, **106**, 57–73.

JENYON, M. K. 1988. Overburden deformation related to the pre-piercement development of salt structures in the North Sea. *Journal of the Geological Society, London*, **145**, 445–454.

—— 1989. Plastic flow and contraflow in superposed Zechstein salt sequences. *Journal of Petroleum Geology*, **12**, 477–486.

KIRBY, G. A. & SWALLOW, P. W. 1987. Tectonism and sedimentation in the Flamborough Head region of north-east England. *Proceedings of the Yorkshire Geological Society*, **46**, 301–309.

KOYI, H. 1988. Experimental modeling of role of gravity and lateral shortening in Zagros Mountain Belt. *AAPG Bulletin*, **72**, 1381–1394.

—— 1991. Gravity overturns, extension, and basement fault activation. *Journal of Petroleum Geology*, **14**, 117–142.

——, JENYON, M. K. & PETERSEN, K. 1993. The effect of basement faulting on diapirism. *Journal of Petroleum Geology*, **16**, 285–312.

MADIRAZZA, I. 1992. Comment on 'The development of salt structures in Denmark and adjacent areas: the role of basin floor dip and differential pressure'. *First Break*, **104**, 134.

MCKENZIE, D. 1978. Some remarks on the development of sedimentary basins. *Earth and Planetary Science Letters*, **40**, 25–32.

NALPAS, T. & BRUN, J. P. 1993. Salt flow and diapirism related to extension at crustal scale. *Tectonophysics*, **228**, 349–362.

OUDMAYER, B. C. & DE JAEGER, J. 1993. Fault reactivation and oblique slip in the Southern North Sea. *In*: PARKER, J. R. (ed.) *Petroleum Geology of Northwest Europe: Proceedings of the 4th Conference*. Geological Society, London, 1281–1290.

OWEN, P. F. & TAYLOR, N. G. 1983. A salt pillow structure in the southern North Sea. *In*: BALLY, A. W. (ed.) *Seismic Expression of Structural Styles*. American Association of Petroleum Geologists, Studies in Geology Series, **15**, 2.3.2–7–10.

PEACOCK, D. C. P. & SANDERSON, D. J. 1992. Effects of layering and anisotropy on fault geometry. *Journal of the Geological Society, London*, **149**, 793–802.

—— & —— 1994. Geometry and development of relay ramps in normal fault systems. *AAPG Bulletin*, **78**, 147–165.

—— & ZHANG, X. 1994. Field examples and numerical modelling of oversteps and bends along normal faults in cross-section. *Tectonophysics*, **234**, 147–167.

PENGE, J., TAYLOR, B., HUCKERBY, J. A. & MUNNS, J. W.

1993. Extension and salt tectonics in the East Central Graben. *In*: PARKER, J. R. (ed.) *Petroleum Geology of Northwest Europe: Proceedings of the 4th Conference*. The Geological Society, London, 1197–1209.

PETERSEN, K., CLAUSEN, O. R. & KORSTGÅRD, J. A. 1992. Evolution of a salt-related listric growth fault near the D-1 well, block 5605, Danish North Sea: displacement history and salt kinematics. *Journal of Structural Geology*, **14**, 565–577.

PRICE, N. J. & COSGROVE, J. W. 1990. *Analysis of Geological Structures*. Cambridge, Cambridge University Press.

RATTEY, R. P. & HAYWARD, A. B. 1993. Sequence stratigraphy of a failed rift system: the Middle Jurassic to Early Cretaceous basin evolution of the Central and Northern North Sea. *In*: PARKER, J. R. (ed.) *Petroleum Geology of Northwest Europe: Proceedings of the 4th Conference*. Geological Society, London, 215–249.

RICHARD, P. 1991. Experiments on faulting in a two-layer cover sequence overlying a reactivated basement fault with oblique-slip. *Journal of Structural Geology*, **13**, 459–469.

RIDD, M. F., 1973. The Sutton, Poyntz, Poxwell and Chaldon Herring anticlines, southern England: a reinterpretation. *Proceedings of the Geologists' Association*, **84**, 1–8.

ROBERTS, A. M., PRICE, J. D. & SVAVA OLSEN, T. 1990. Late Jurassic half-graben control on the siting and structure of hydrocarbon accumulations: UK/ Norwegian Central Graben. *In*: HARDMAN, R. F. P. & BROOKS, J. (eds) *Tectonic Events Responsible for Britain's Oil and Gas Reserves*. Geological Society, London, Special Publication, **55**, 229–257.

ROBERTS, D. G. 1989. Basin inversion in and around the British Isles. *In*: COOPER, M. A. & WILLIAMS, G. D. (eds) *Inversion Tectonics*. Geological Society, London, Special Publication, **44**, 131–150.

SMITH, R. I., HODGSON, N. & FULTON, M. 1993. Salt control on Triassic reservoir distribution, UKCS Central North Sea. *In*: PARKER, J. R. (ed.) *Petroleum Geology of Northwest Europe: Proceedings of the 4th Conference*. Geological Society, London, 547–557.

STEWART, S. A. & COWARD, M. P. 1995. Synthesis of salt tectonics in the southern North Sea, UK. *Marine and Petroleum Geology*, **12**, 457–476.

SUNDSBØ, G. O. & MEGSON, J. B. 1993. Structural styles in the Danish Central Graben. *In*: PARKER, J. R. (ed.) *Petroleum Geology of Northwest Europe: Proceedings of the 4th Conference*. Geological Society, London,1255–1267.

TAYLOR, J. C. M. 1990. Upper Permian-Zechstein. *In*: GLENNIE, K. W. (ed.) *Introduction to the Petroleum Geology of the North Sea*. Blackwell, Oxford, 153–190.

VAN HOORN, B. 1987. Structural evolution, timing and tectonic style of the Sole Pit inversion. *Tectonophysics*, **137**, 239–284.

VENDEVILLE, B. C. & JACKSON, M. P. A. 1992. The rise of diapirs during thin-skinned extension. *Marine and Petroleum Geology*, **9**, 331–353.

WALKER, I. M. & COOPER, W. G. 1987. The structural and stratigraphic evolution of the northeast margin of the Sole Pit Basin. *In*: BROOKS, J. & GLENNIE, K. W. (eds) *Petroleum Geology of North West Europe*. Graham & Trotman, 263–275.

WEIJERMARS, R., JACKSON, M. P. A. & VENDEVILLE, B. 1993. Rheological and tectonic modelling of salt provinces. *Tectonophysics*, **217**, 143–174.

WITHJACK, M. A., OLSON, J. & PETERSON, E. 1990. Experimental models of extensional forced folds. *AAPG Bulletin*, **74**, 1038–1054.

ZALESKI, S. & JULIEN, P. 1992. Numerical simulation of Rayleigh-Taylor instability for single and multiple diapirs. *Tectonophysics*, **206**, 55–69.

# The development of the Gorleben salt dome (northwest Germany) based on quantitative analysis of peripheral sinks

MAX ZIRNGAST

*Federal Institute of Geosciences and Natural Resources (BGR), Stilleweg 2, D30631 Hannover, Germany*

**Abstract:** The structure of the most important stratigraphic horizons in the Gorleben salt dome area, northwest Germany, has been studied using reflection–seismic data and drilling results. Epeirogenic, tectonic and halokinetic movements in the area can be reconstructed on the basis of structure–contour maps, isopach maps, and palaeostructure–contour maps. For this purpose the increase of sediment volume in the rim synclines was calculated in three dimensions and compared with the results of the 'normal' subsidence. The additional subsidence in the rim synclines was used for calculation of salt flow by volume. By budgeting the salt movements, the flow rate of salt into the diapir and the salt loss by erosion and leaching were determined. The movement of the salt from the Late Buntsandstein (Röt) until today is depicted in 13 successive stages in three-dimensional diagrams using computer programs.

The objective of this study was to work out as detailed a description as possible of the history of development of the Gorleben salt dome and to make reliable estimates of the rate of flow of the salt.

Rates of uprise have been calculated for some of the salt domes in northwest Germany on the basis of various observations. The highest rates were determined for the relatively young Segeberg (Teichmüller 1948) and Lüneburg (Drescher *et al.* 1973) salt domes, which are still in the diapiric stage, and which have a maximum rate of uprise of 2 mm $a^{-1}$ (downbuilding in the sense of Jackson & Talbot (1991)). This would mean that radioactive waste disposed of in a repository there would rise 200 m in 100 000 years. These extremely high rates, however, are still disputed (Jaritz 1980).

Jaritz (1980) assessed the rates of uprise of several salt domes in northwest Germany on the basis of the thickness of the sedimentary fill in the rim synclines, and obtained a maximum rate of 0.5 mm $a^{-1}$. For salt domes in their post-diapiric stage, among which is the Gorleben dome, he obtained rates of a few hundredths of a millimetre per year or even less.

This would mean that radioactive waste disposed of in a salt dome still in the diapiric stage would rise 500 m in 1 million years, but in a salt dome in the post-diapiric stage, for example, it would rise only 10–20 m. Thus the uprise rates of salt domes during the diapiric phase are high enough to threaten the long-term safety of a permanent repository for radioactive waste, whereas the uprise rate during the post-diapiric phase is negligible.

Because of its significance for the final disposal of wastes, an attempt was made to quantify, as accurately as possible, salt movement or flow during and after the formation of the salt domes so that questions concerning the safety and thus suitability of the Gorleben salt dome could be satisfactorily answered.

In addition to the Gorleben 1984 seismic reflection survey, which was carried out specifically for the purpose of finding an answer to these questions, seismograms and drilling data obtained by companies exploring for hydrocarbons were made use of in this study. The author is indebted to the companies associated in the WEG (Wirtschaftsverband Erdöl- und Erdgasgewinnung e.V.) for allowing these results to be published.

## Regional structure and tectonic development of the area

Figure 1 shows the structures in this area, which belongs to the Pompeckj Block. The southwestern part of the area in Fig. 1 shows salt structures located on the NW and NNE trending edges of fault blocks; the northeastern part of the study area belongs to a region displaying isolated, halokinetic structures that are apparently independent of basement structures. Only the N–S trending Gülden–Braudel and Rosche–Thondorf structures in the western part of the study area follow a system of faults in the basement. The Gülden–Braudel structure is a part of a prominent fault system, the Gifhorn fault system, on which Early Triassic rifting may have occurred.

The area east of the Gülden–Braudel structure is characterized by intense halokinesis above a pre-Zechstein basement in which only minor faults occur. A large number of individual structures were formed here, almost all of which have reached the diapiric stage. In the area adjacent to the north and east, salt migration has taken place and most of the

*From* ALSOP, G. I., BLUNDELL, D. J. & DAVISON, I. (eds), 1996, *Salt Tectonics*, Geological Society Special Publication No. 100, pp. 203–226.

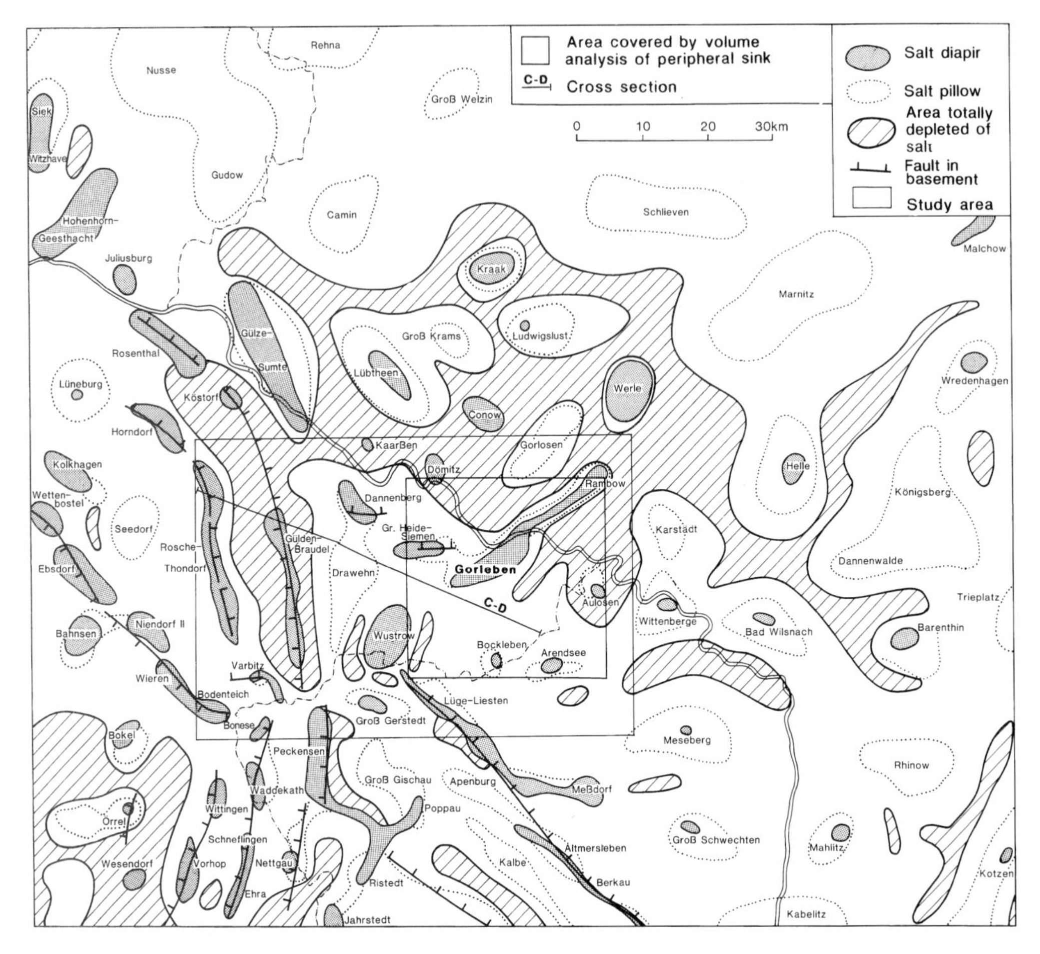


**Fig. 1.** Salt structures in the eastern part of Lower Saxony and western part of the former GDR, modified after Meinhold & Reinhardt (1967) and Kockel *et al.* (1989).

Zechstein has accumulated in the diapirs (Reinhardt 1967). In the north, along a zone where the base of the Zechstein dips to the north, there is an area where the Zechstein salt forms large salt pillows.

## Structural development of the Gorleben salt dome and surrounding area

The following sequence of structural development in the study area can be reconstructed on the basis of the thickness of the sedimentary strata and thus the age of the faults in the basement (Fig. 2). The most important fault in the basement beneath the Gülden–Braudel structure has been active, with a varying sense of displacement, since the Early Buntsandstein (darkest layer in Fig. 2). A rifting episode in the Trias caused a reactive stage of the salt dome and 'apparent downlap' (Jackson *et al.* 1994) on the western flank of the Gülden structure. During Keuper times, the Zechstein salt withdrew from the area around the structure almost completely. This process was completed at the end of the Keuper as is indicated by the regularity of the post-Keuper sedimentary cover above the Gülden salt structure, which does not show any variations of thickness due to salt movement. During the Keuper, early salt flow away from the area of the Gülden–Braudel structure initiated the formation of salt pillows in the area of Rosche–Thondorf and Drawehn. The Drawehn structure remained in the pillow stage, whereas in Rosche–Thondorf diapirism started during the Jurassic. This is due to the fact that only minor amounts of salt were available in this area, since some of the salt from beneath this area had already flowed away in the direction of the Dannenberg, Heide–Siemen and Wustrow

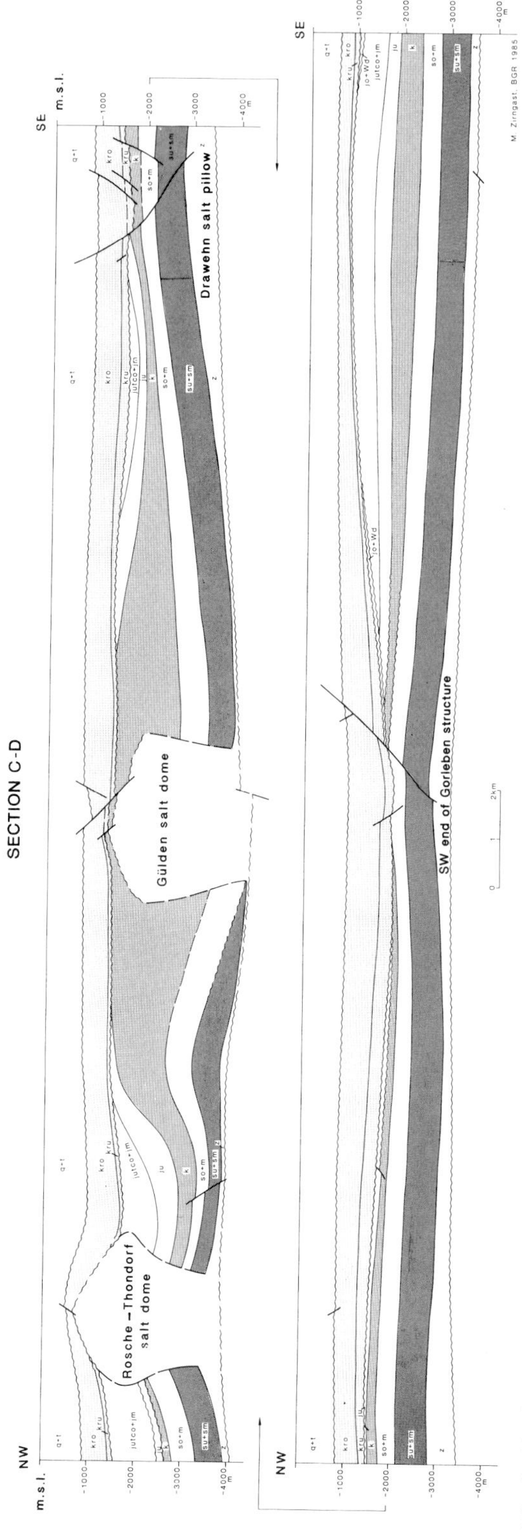


**Fig. 2.** Northwest–southeast section through the most important salt structures west of Gorleben. Line of section shown in Fig. 1.

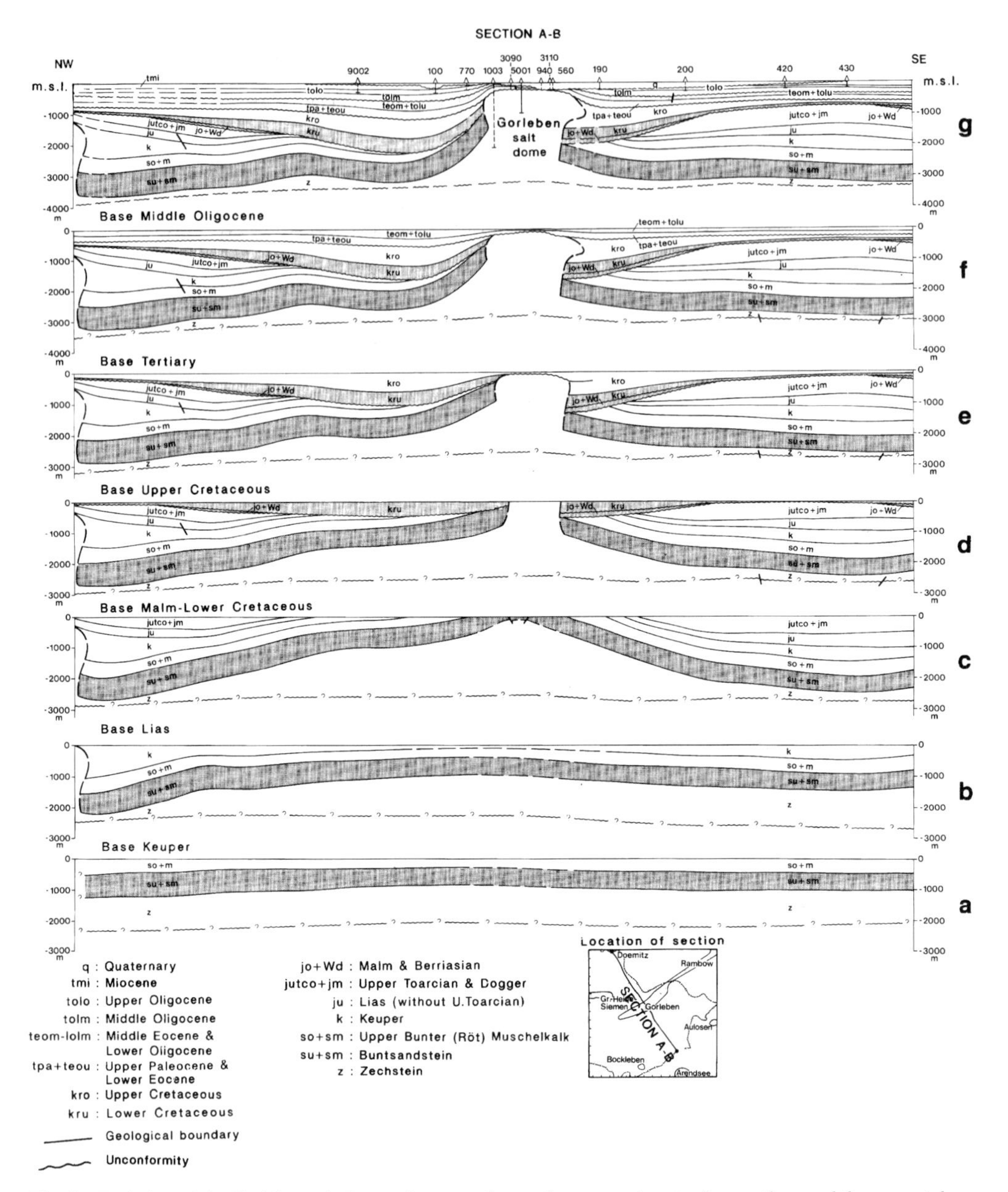


**Fig. 3.** Evolution of the Gorleben salt dome: six successive northwest–southeast palaeosections and the present-day section A–B.

structures. Early withdrawal of salt away from the neighbourhood of the Aulosen and Dömitz salt domes, which were formed during the Keuper, caused a large salt pillow to be formed in the area of the present-day Gorleben, Heide–Siemen, Wustrow and Dannenberg salt domes. Basement faults beneath the Dannenberg, Gr.-Heide–Siemen and Wustrow salt domes, whose age cannot be determined, and the basement high in the Gorleben area favoured the formation of individual salt structures in the area of the large salt pillow.

## Quantitative analysis of peripheral sinks

### *Fundamental principles*

The various stages of salt flow and accumulation can be reconstructed from the synsedimentary changes in the thickness of the strata overlying the Zechstein. Formation of salt pillows causes the sedimentary cover above the structure to thin out by non-sedimentation and erosion, and that above the areas from which salt withdrawal has taken place, i.e. the peripheral sinks, to increase in thickness. At the transition between the pillow and diapiric stages, the salt pierced through the sedimentary cover and large amounts of salt were leached at the surface. The salt at the margins of the partially depleted pillow also flowed into the diapir, causing a corresponding amount of surface subsidence. Thus, during the course of the diapiric stage, the secondary rim synclines were formed and their depocentres gradually moved inwards towards the salt dome.

The series of cross-sections of the Gorleben palaeostructure (Fig. 3) show the successive stages of salt withdrawal and accumulation, and development of the peripheral sink from the Zechstein to the present day (see also Jaritz 1980). The Buntsandstein and Muschelkalk display slight thinning on the flanks of the Gorleben salt dome, which was caused either by upwarping of the pre-Zechstein rocks during this time (as shown in Fig. 3, section 'a'; beginning of Keuper) or by early flow of salt. During the Keuper, salt flow and accumulation took place. Evidence is provided by the variable thickness of the Keuper in the area of the rim synclines and above the salt structure itself (Fig. 3, section 'b'; beginning of Lias). During the Lias and Dogger, salt flow increased, and an extensive, thick pillow with a distinct culmination was formed (Fig. 3, section 'c'; beginning of Malm–Lower Cretaceous). During this stage, the salt flowed in mainly from the southeast. Diapirism started during a regional uplift of the surrounding area (Pompeckj block) in Late Jurassic to Berriasian time, which led to truncation of about 600 m of the sedimentary cover (Jaritz 1980). Initially, salt was supplied from the southeast; in the course of the Cretaceous (Fig. 3, section 'd'; beginning of Upper Cretaceous) it started to flow in from the northwest. Remnants of Upper Cretaceous rocks in the roof of the salt dome indicate that the salt dome had a sedimentary cover at that time. Epeirogenic uplift and late minor flow of salt mainly from the northwest caused the top of the salt dome again to approach the surface and erosion to take place during Late Cretaceous and Palaeocene times. The late minor rise of salt continued during the Tertiary (Fig. 3, sections 'e' to 'g').

An accurate reconstruction of salt flow requires a detailed quantitative determination of the formation thicknesses in the peripheral sinks for certain selected periods of time. For this purpose, the following data must be determined:

1. The present positions and depths of all formation boundaries;
2. the thicknesses of the individual formations;
3. the area from which the salt in the salt dome was supplied, i.e. withdrew;
4. the amount of epeirogenic subsidence during given time intervals.

The method of analysing the formation thickness of the peripheral sinks is based on Rühberg (1976). The individual steps are briefly described below; in this way a reasonably accurate quantitative determination of the volume of sediments in the peripheral sinks can be obtained with a justifiable expenditure of effort.

## Steps necessary to determine the thickness of sediments in the peripheral sinks

### *Step 1. Compilation of structure contour maps*

All further work depends on producing an accurate map of the most important sedimentary units at depth. In 1984, PRAKLA-SEISMOS carried out a seismic reflection survey in the area of the Gorleben salt dome (Fig. 4) within the scope of the site investigation programme of the Federal Institute of Physics and Technology (PTB). In the seismograms recorded at that time, the reflectors were marked that correspond to the bases of the following units (stratigraphic symbols of the Lower Saxony Geological Survey (NLfB) are given in brackets):

Quaternary (q)
Miocene (tmi)
Upper Oligocene (tolo)
Middle Oligocene (tolm)
Middle Eocene (teom)
Palaeocene to Lower Eocene (tpa-teou)
Upper Cretaceous (kro)
Lower Cretaceous (kru)
Malm to Berriasian (jo-Wd).

Time/depth conversion of the selected reflectors was carried out using the Sattlegger program or the Migra program written by Buttkus, BGR. Using these methods, based on normal ray migration, the reflectors in the stacked sections corresponding to

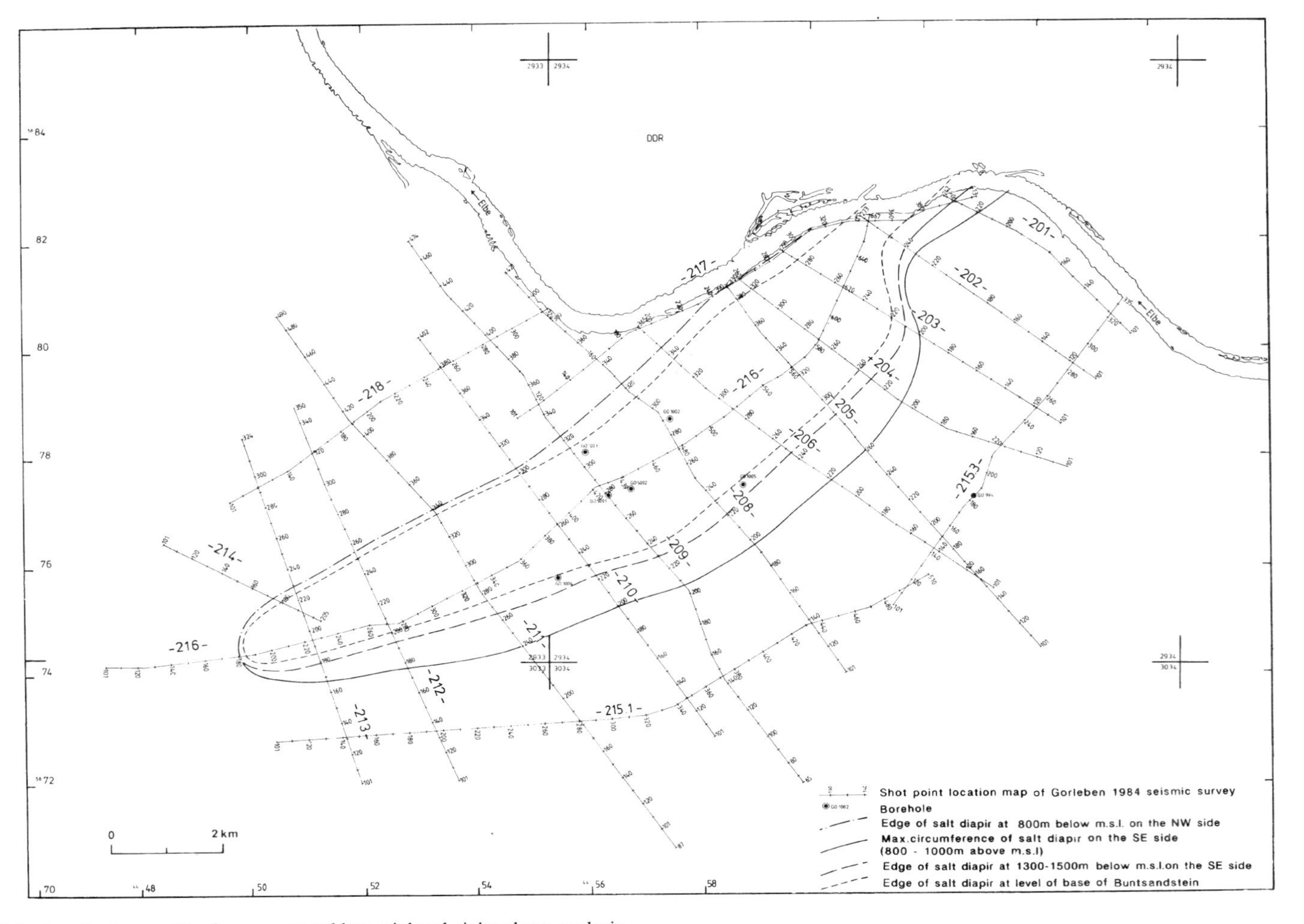


**Fig. 4.** Seismic reflection profiles in area covered by peripheral-sink volume analysis.

the lithological boundaries are digitized and, taking refraction at the velocity boundaries into consideration, converted into depths and represented as depth sections (Fig. 5).

The accurate determination of the depth of the individual horizons marked in the time sections is based on a seismic velocity concept developed by Jaritz *et al.* 1991). Jaritz *et al* determined optimum velocity gradients and values of 'initial velocities' (formation velocities at mean sea level) for ten stratigraphic sequences in northern Germany. These surface velocities were plotted in the form of isolines, taking into consideration geological, facies and structural features.

Thus, the seismic reflection profiles of the Gorleben 1984 seismic survey (Fig. 4) and the reflection seismic measurements made by companies exploring for hydrocarbons on the margins of the study area were evaluated and represented in the form of geological cross-sections. Using these cross-sections, as well as drilling data, it was possible to compile detailed structure contour maps (e.g. base of the Tertiary, Fig. 6).

The results of the survey are shown in block diagrams (Figs 7 and 8) and cross-sections (Figs 3 and 9). The data on parts of the surveyed areas in the former GDR were supplemented with information obtained from the literature (Beutler & Schüler 1978; Knape 1963; Putziger *et al.* 1966; Wienholz 1967); much less is known about these areas than about those in Lower Saxony.

The block diagram (Fig. 7) shows the subsurface structure of Zechstein to Cretaceous rocks in part of the study area. The surface shown here is the base of the Tertiary and the structures produced in it by rise of the Gorleben, Rambow, Heide–Siemen and Arendsee salt domes, since the Early Tertiary. The second block diagram (Fig. 8) shows the variation in thickness of the Tertiary sediments in the area of the Gorleben salt dome. The surface is the Quaternary base, which is irregular on account of channels deeply incised during the Elsterian glaciation. The cross-sections (Figs 3 and 9) show the varying thicknesses of the individual stratigraphic units in the study area. The sediments of Keuper to Tertiary age show considerable variation in thickness due to salt flow, whereas the Buntsandstein shows reduction of thickness only in the marginal zone of the salt dome due to truncation during the salt-pillow stage. Beginning with the Cretaceous, regional tilting of the subsurface caused reduction in thickness of the sedimentary beds from NW to SE. The tilting is clearly detectable in the SE of profile 3–3' shown in Fig. 9, where the Cretaceous wedges out and the Tertiary oversteps the Lower Jurassic.

### *Step 2. Construction of isopach maps*

Isopach maps are based on the structure contour maps compiled using data from seismic sections and well results; as an example, the base of the Tertiary is taken (Fig. 6). Then the depths of all individual stratigraphic units were taken at 2200 points on a 500 m grid and in marginal areas on a 1000 m grid (Fig. 10) and fed into a datafile. The difference between the values in the Middle Eocene base and Tertiary base datafiles corresponds approximately to the thickness of the Palaeocene plus Lower Eocene sediments. Figure 11 shows an isopach map compiled from these datafiles using the UNIRAS program. Towards the end of the study, the ISM program became available for calculation of sediment thicknesses and volumes. This program permits depth differences to be calculated from point data obtained by digitization of the structure contour maps. Comparison of the two methods showed good agreement.

### *Step 3. Determination of the amount of epeirogenic subsidence (i.e. 'normal' sediment thickness)*

To determine the proportion of sediment thickness in rim synclines that is due to halokinesis, the amount of epeirogenic subsidence (referred to here as 'normal' sediment thickness) must be deducted from the actual thickness calculated as above.

The 'normal' sediment thickness was determined in areas in the surroundings of the Gorleben salt dome in which salt withdrawal during the time intervals considered here is inferred to be zero, i.e. around the Gülden–Braudel (Fig. 2), Dömitz and Aulosen salt domes (Fig. 9). Formation of these particular salt domes was completed by the end of the Keuper, so that the thickness of the overlying post-Keuper strata corresponds approximately to the amount of epeirogenic subsidence.

In order to calculate the thickness of sediments in peripheral sinks that can be ascribed to salt withdrawal at depth, the amount of epeirogenic subsidence ('normal' sediment thickness) for each period was determined. For example, for the period Palaeocene–Early Eocene, the thickness in these areas is about 150 m. This amount is deducted from the total thickness of this stratigraphic sequence and yields the thickness of the sediments filling the peripheral sinks ascribable to salt withdrawal at depth (Fig. 12).

### *Step 4. Delineation of the source area of the salt in the Gorleben salt dome*

Since only those rim synclines from beneath which salt migrated towards the Gorleben salt dome could

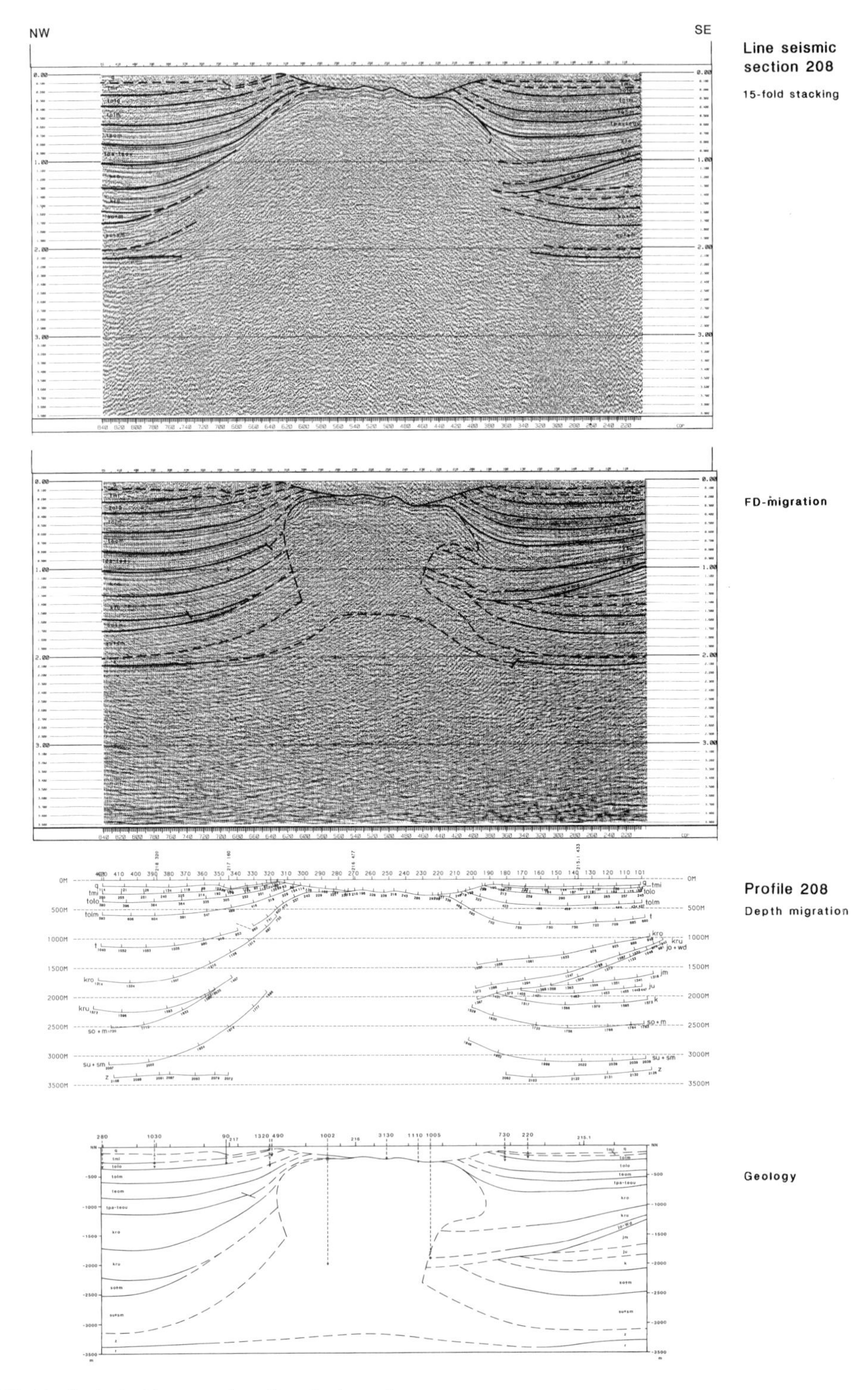


**Fig. 5.** Geological interpretation of a seismic section.

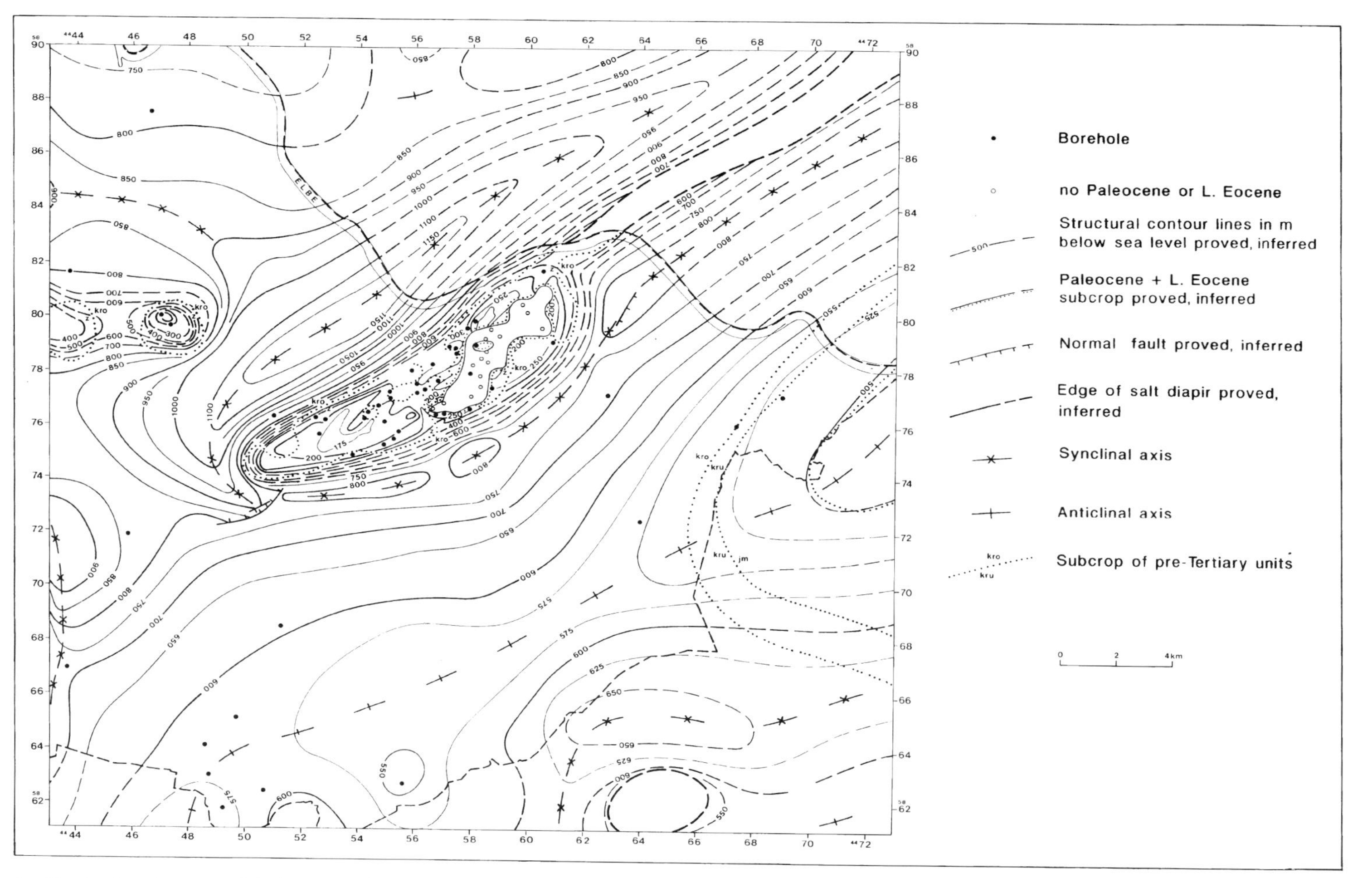


**Fig. 6.** Structural contour map of the Palaeocene/Lower Eocene unconformity.

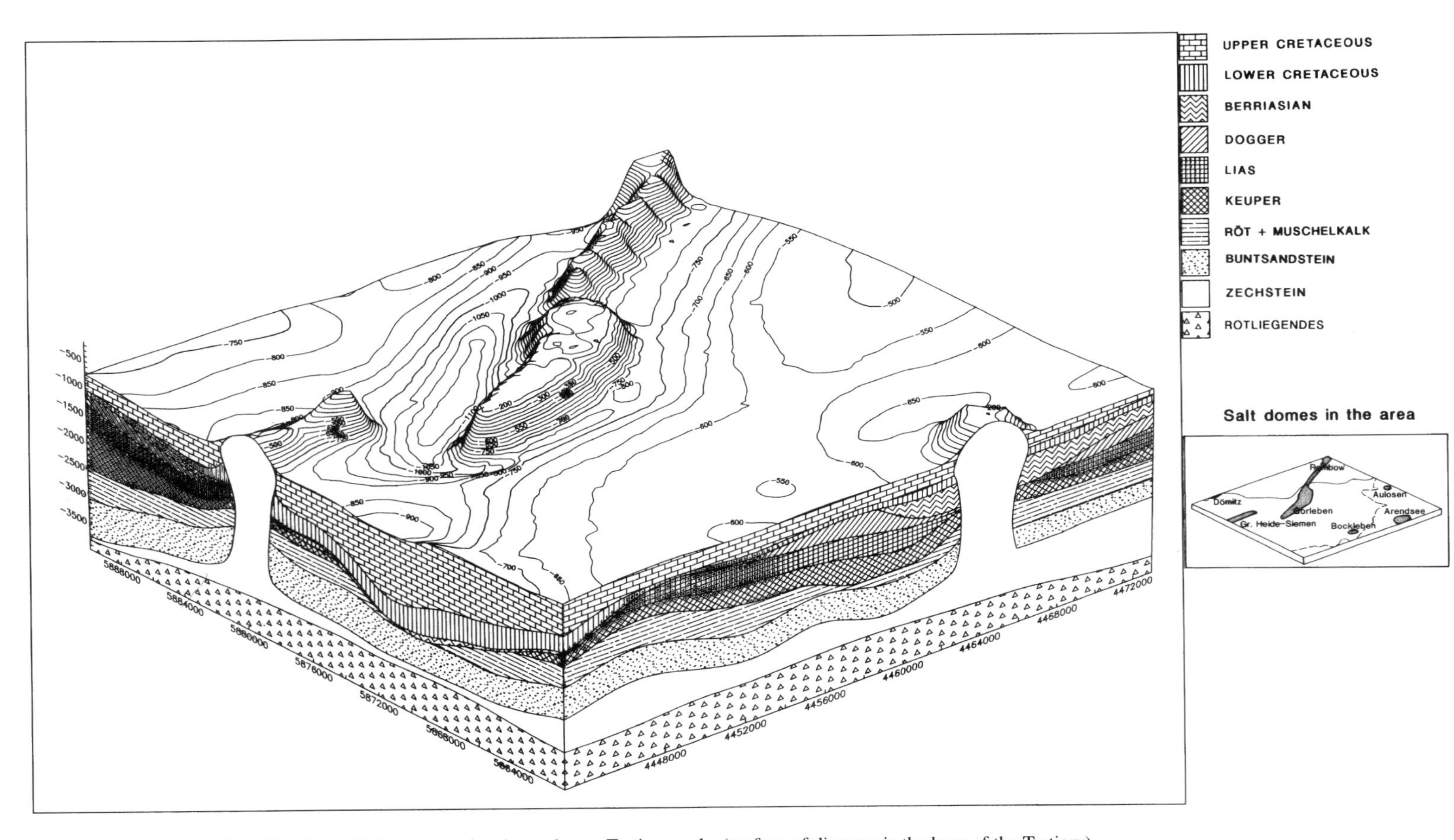


**Fig. 7.** Block diagram of the Gorleben salt dome area showing only pre-Tertiary rocks (surface of diagram is the base of the Tertiary).

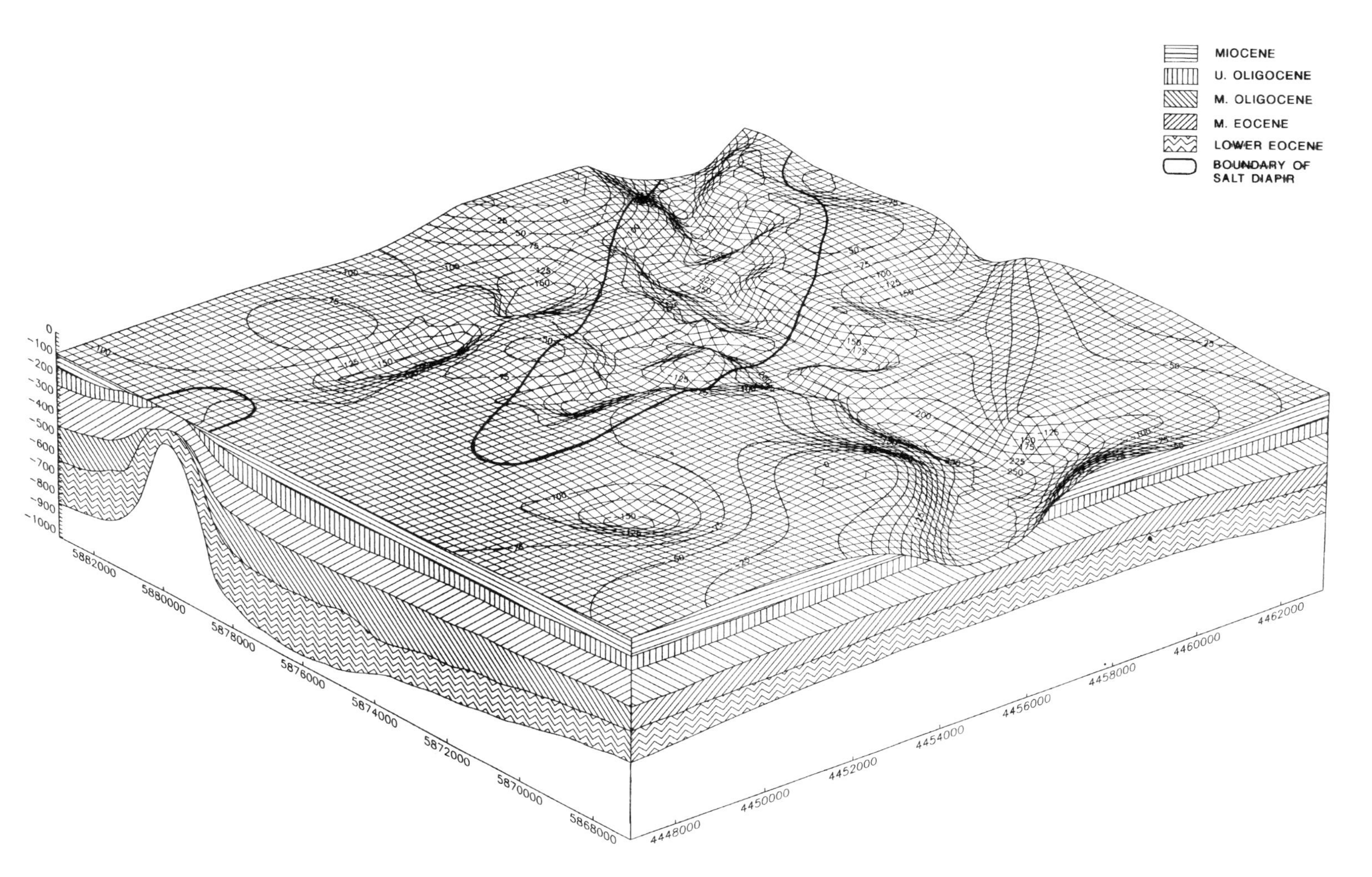


**Fig. 8.** Block diagram of the Gorleben salt dome area showing the Tertiary succession (surface of diagram is the base of the Quaternary).

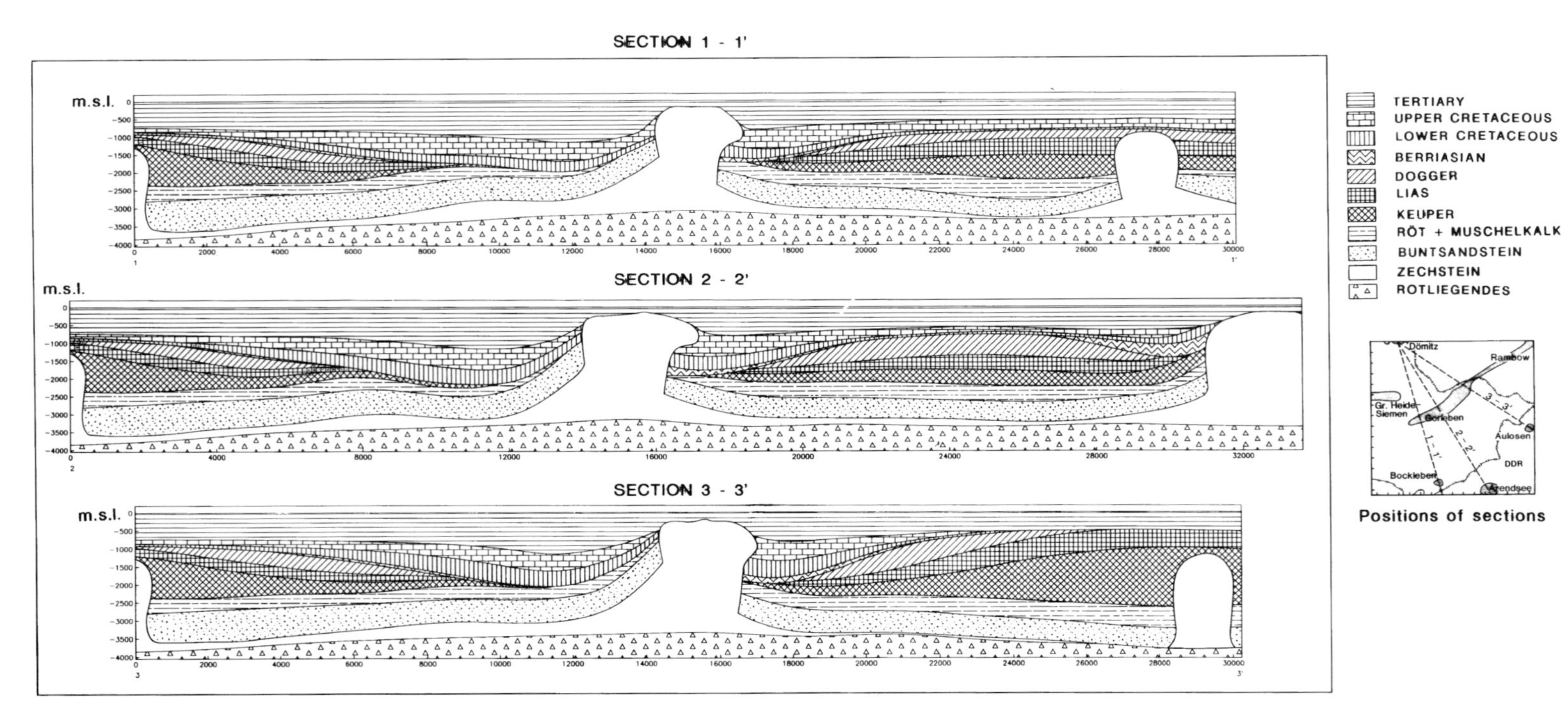


**Fig. 9.** Geological sections across the area covered by peripheral-sink volume analysis.

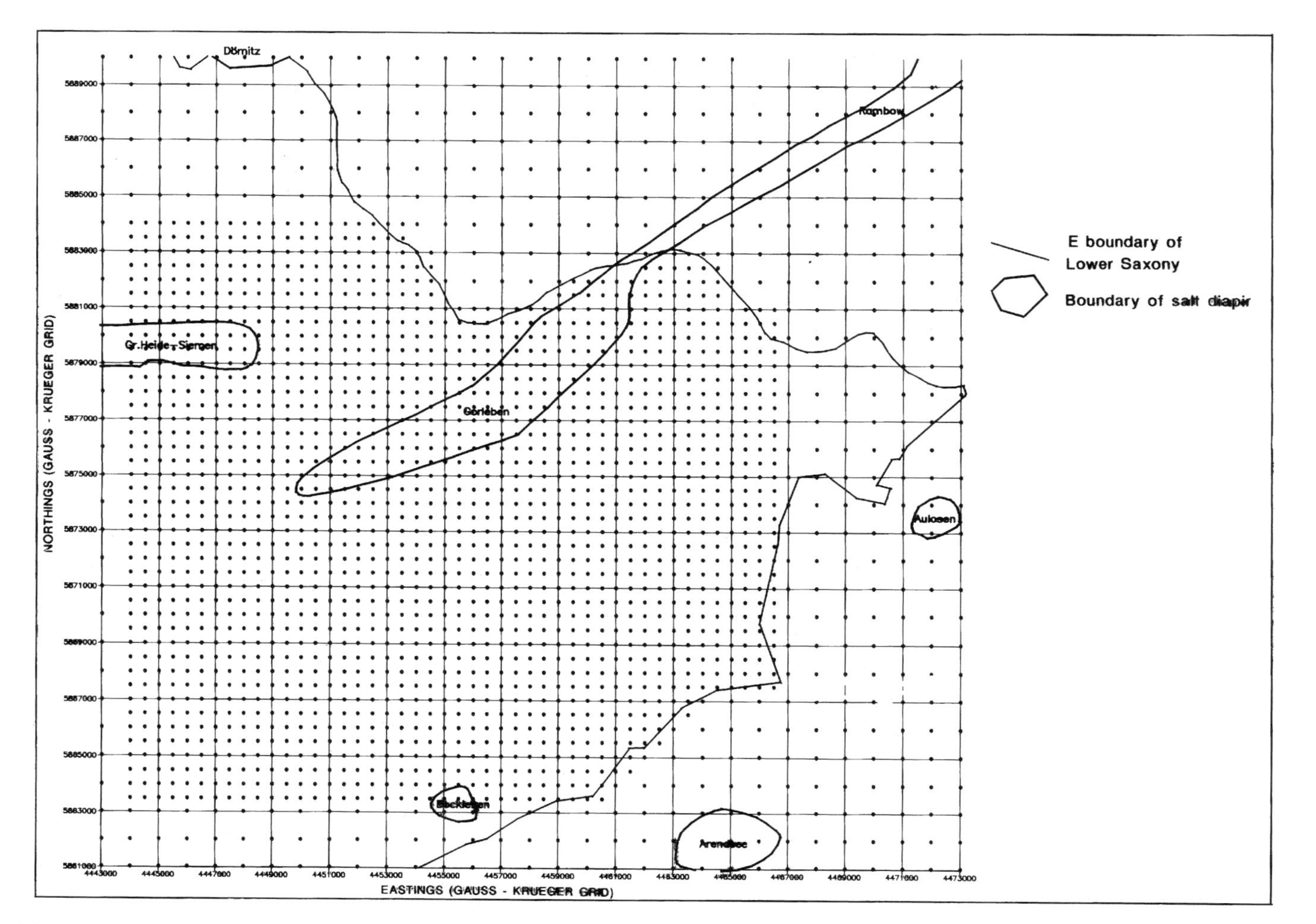


**Fig. 10.** Grid points used for construction of isopach maps.

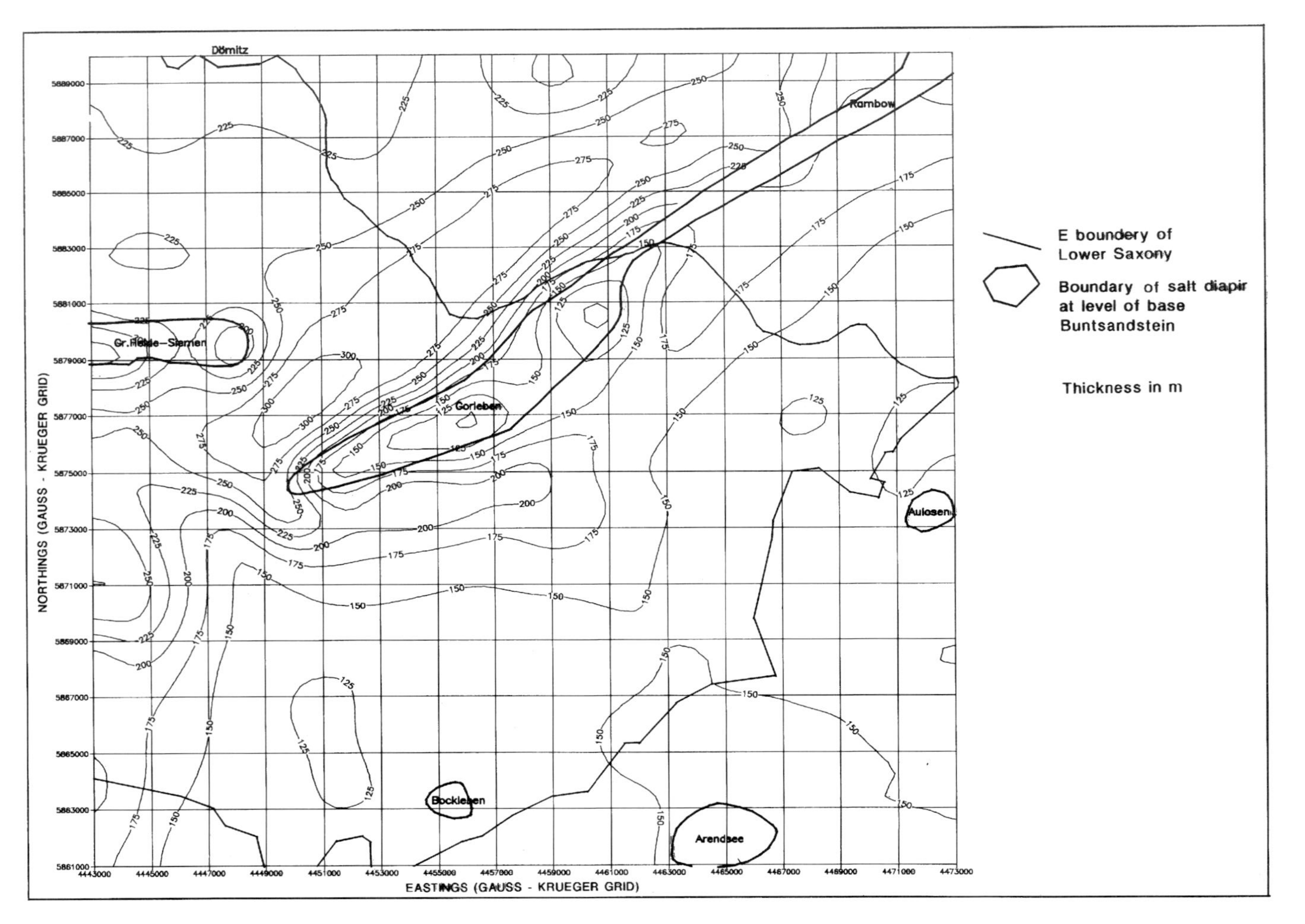


**Fig. 11.** Isopach map of strata between base Palaeocene and base Middle Eocene.

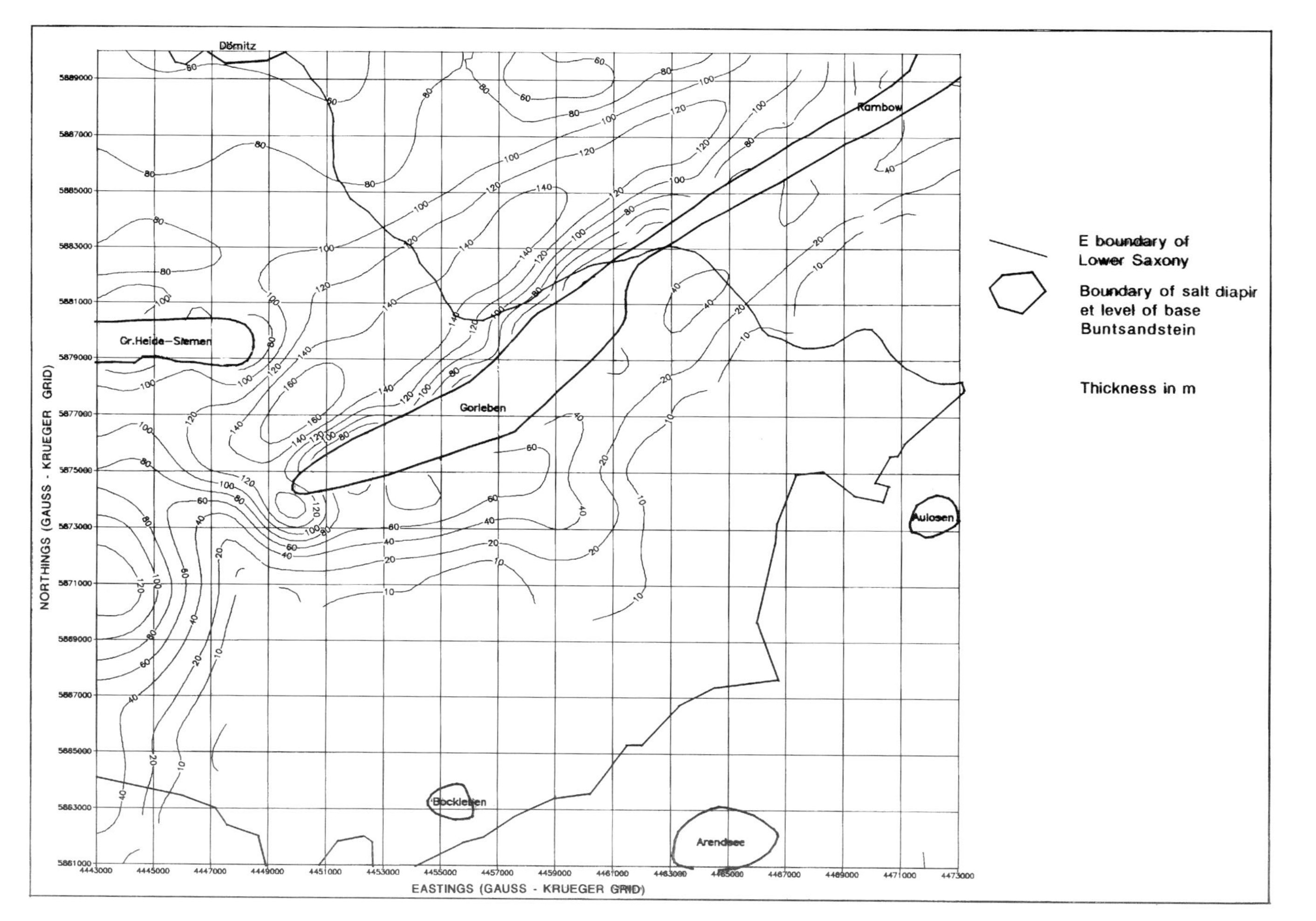


**Fig. 12.** Isopach map showing the thickness of sediments laid down in the peripheral sink during the Palaeocene and Early Eocene directly ascribable to salt movement at depth.

be used in the calculations, it was essential to delineate the source areas of the salt in the salt dome. Assuming that the salt always flows from a lower position to a higher position, the direction of flow can be derived from the position of the base of the Buntsandstein (i.e. top Zechstein) during a given period, in this example at the beginning of the Tertiary. Figure 13 shows the base of the Buntsandstein during the period from the Palaeocene to the Eocene, compiled on the basis of thickness datafiles.

The axes of the depressions in this surface around the Gorleben salt dome represent the boundaries of the source area of the salt. The boundary of the source area of the salt between the Gorleben and Rambow salt domes was assumed to run perpendicular to the contours at the northeastern end of the Gorleben salt dome, considering the elongated form of the entire structure.

The volume represented by the thickness of sediments (see Fig. 12) contained in the source area shown in Fig. 15 corresponds to the amount of salt that migrated into the salt dome during the period from the Palaeocene to the Early Eocene. Using this procedure, the volumes of salt that moved into the salt dome during eight different periods from the Berriasian to the present day were calculated.

## Rates of salt flow and salt dome uprise

The amounts of salt that migrated into the salt dome were derived by evaluation of the sedimentary fill of the peripheral sinks. They are listed in Table 1. Moreover, Table 1 provides all data necessary to determine the rate of flow and salt dome rise: absolute ages, geological time spans and epeirogenic subsidence.

For calculating the columns 3–5, the following horizontal sections of the salt dome were used: narrowest horizontal section (Fig. 13) = 26.9 $km^2$ (at base-Buntsandstein level); and largest horizontal section (Fig. 6) = 44.2 $km^2$ (at Late Cretaceous level).

In the left-hand column of Table 1, eight periods from the beginning of diapirism (Berriasian) to the present day are listed. The bottom line provides average figures for the entire period. The amounts of epeirogenic subsidence per period are given in column 1. In the case of the Lower and Upper Cretaceous, it can be seen from the maximum, minimum and mean values that epeirogenic subsidence varies considerably due to synsedimentary tilting of the basement. Column 2 gives the volume of the sedimentary fill accommodated by subsidence of the peripheral sink due to halokinesis; this corresponds to the amount of salt that flowed into the salt dome. Altogether, about 280 $km^3$ of salt has flowed into the salt dome since the beginning of diapirism (bottom line). If this salt volume is converted into a column with a section area equal to the largest horizontal section of the salt dome (see above), and if the total amount of epeirogenic subsidence is subtracted from this column, then one obtains that portion (given in column 3 of Table 1) of the salt column which theoretically rose above the Earth's surface and was removed by subsurface solution (subrosion) and/or by surface erosion. The maximum amount of salt removed during the time elapsed since diapirism began corresponds to a column about 7 km high. During salt movement, the highest flow rates occurred in the part of the salt dome with the smallest horizontal cross-section. These maximum flow rates were calculated twice, using two different geological time scales (IUGS 1989; Harland *et al.* 1990); they are shown in column 4 of Table 1. Salt flow reached its highest rate (a maximum of 0.14 mm $a^{-1}$) during the Late Cretaceous; the lowest rate was obtained for the most recent period from the Miocene to the Quaternary. Taking the largest horizontal section in the zone of the salt overhang as a basis for calculating the rate of salt flow, the theoretical rate of uprise of the salt dome surface is obtained. These values were again calculated using two geological time scales and are listed in column 5 of Table 1. The salt dome reached its maximum rate of uprise of 0.088 mm $a^{-1}$ during the Late Cretaceous phase of diapirism; the minimum rate (0.018 mm $a^{-1}$) was reached during the Miocene and Quaternary.

The rates calculated for the individual geological periods are shown in Fig. 14. Depending on which of the two geological time scales (IUGS 1989; Harland *et al.* 1990) calculation of the rate was based, different rates of uprise and flow rates were obtained. As some of them differ considerably, both values are shown. Conspicuous in this figure are the high rates that were calculated for the Middle Oligocene on the basis of the time scale of Harland *et al.* (1990), and for the Late Oligocene based on the IUGS (1989) time scale. One of the reasons for the apparently elevated rate of uprise during the Middle Oligocene could have been the high rate of epeirogenic subsidence during this time (0.037 mm $a^{-1}$ according to the time scale of IUGS (1989) and 0.047 mm $a^{-1}$ according to the time scale of Harland *et al.* (1990)). However, this was not corroborated by the fact that during the Middle Oligocene the salt dome was covered by Rupelian clay. As a result of this cover, only minor subsurface solution (subrosion) took place, so that the actual uplift of the salt dome must have been approximately equal to the calculated uprise of the salt. This is contradicted by the fact that the calculated salt uprise of 195 m during the Middle Oligocene exceeds the rate of epeirogenic subsidence during this period of time (150 m). The presence of Middle Oligocene cover,

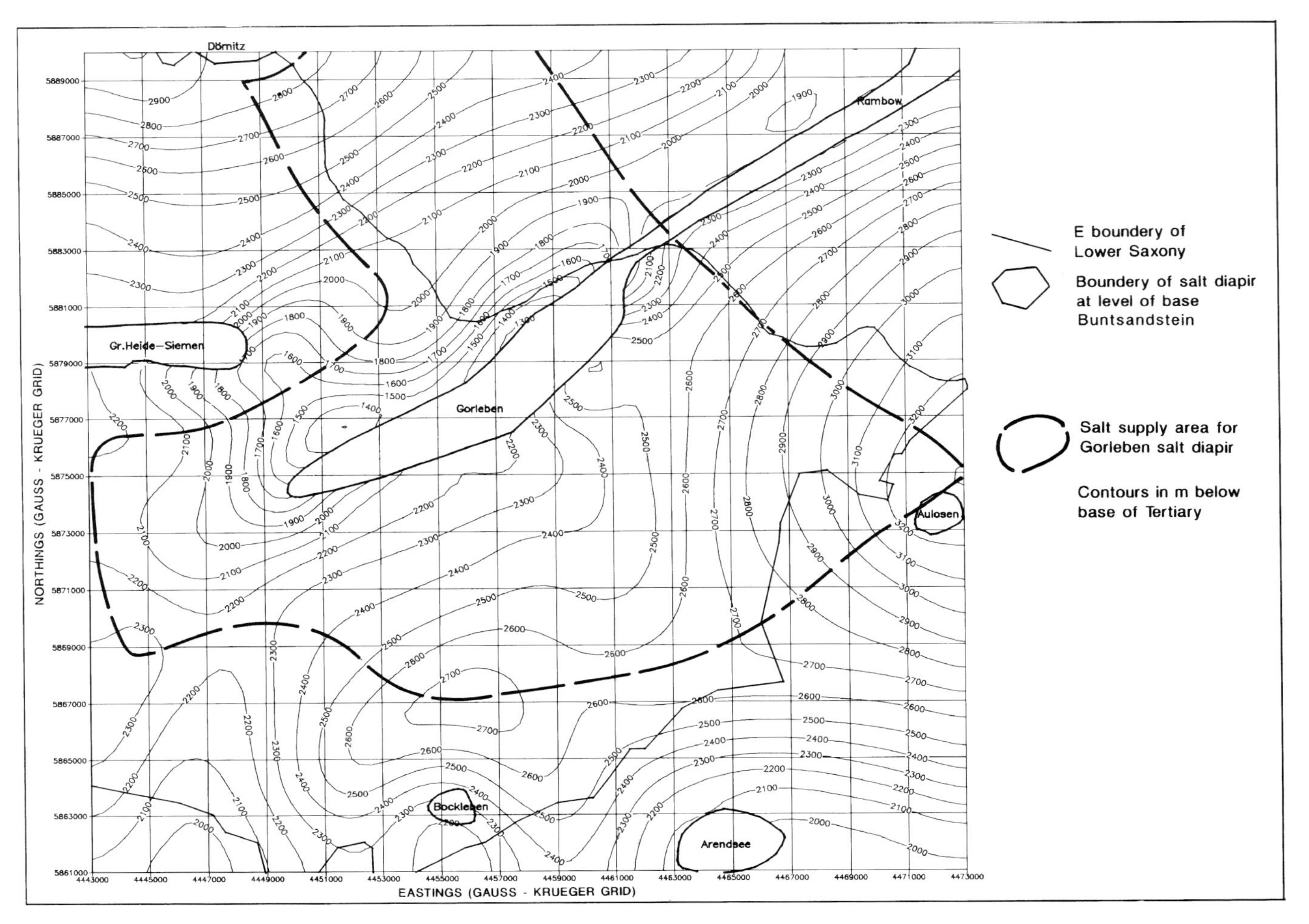


**Fig. 13.** Depth of base Buntsandstein below the Tertiary at the beginning of Tertiary times. The boundary of the area that supplied salt for the Gorleben salt diapir is also shown.

**Table 1.** *Results of structural analysis of the rim syncline*

| | 1 Epeirogenic subsidence [m] | | 2 Volume of salt moved into the diapir $[km^3]$ | | 3 Height of column of dissolved salt | | | | 4 Velocity of salt flow in the narrowest part of the diapir | | | | | | | | 5 Rate of uprise of surface of diapir | | | | | | | |
|---|---|---|---|---|---|---|---|---|---|---|---|---|---|---|---|---|---|---|---|---|---|---|---|---|
| | | | | | using smallest diapir diameter [km] | | using largest diapir diameter [km] | | using time scale of HARLAND et al.(1990) | | | | using time scale IUGS (1989) | | | | using time scale of HARLAND et al.(1990) | | | | using time scale of IUGS (1989) | | | |
| | | | | | | | | | Time [Ma] | | Velocity [mm/a] | | Time [Ma] | | Velocity [mm/a] | | Time [Ma] | | Velocity [mm/a] | | Time [Ma] | | Velocity [mm/a] | |
| Miocene – Quaternary | 100 | | 18.80 | | 0.599 | | 0.323 | | 23.3 | | 0.03 | | 23 | | 0.030 | | 23.3 | | 0.018 | | 23 | | 0.018 | |
| | | | | | | | | | 23.3 | | | | 23 | | | | 23.3 | | | | 23 | | | |
| Late Oligocene | 55 | | 11.01 | | 0.354 | | 0.193 | | 6 | | 0.068 | | 4 | | 0.102 | | 6 | | 0.041 | | 4 | | 0.062 | |
| | | | | | | | | | 29.3 | | | | 27 | | | | 29.3 | | | | 27 | | | |
| Middle Oligocene | 150 | 305 | 8.65 | 42.53 | 0.173 | 1.277 | 0.046 | 0.654 | 3.7 | 27.2 | 0.087 | 0.058 | 4 | 22 | 0.081 | 0.072 | 3.7 | 27.2 | 0.053 | 0.035 | 4 | 22 | 0.049 | 0.043 |
| | | | | | | | | | 33 | | | | 31 | | | | 33 | | | | 31 | | | |
| Middle Eocene - Early Oligocene | 100 | | 22.87 | | 0.750 | | 0.415 | | 17.5 | | 0.049 | | 14 | | 0.061 | | 17.5 | | 0.029 | | 14 | | 0.037 | |
| | | | | | | | | | 50.5 | | | | 45 | | | | 50.5 | | | | 45 | | | |
| Late Paleocene - Early Eocene | 150 | | 21.39 | | 0.645 | | 0.332 | | 7.5 | | 0.106 | | 10 | | 0.079 | | 7.5 | | 0.064 | | 10 | | 0.048 | |
| | | | | | | | | | 58 | | | | 55 | | | | 58 | | | | 55 | | | |
| Late Cretaceous (Cenomanian – Early Paleocene) | 100 | Min. 0<br>Max. 150 | 114.57 | 152.08<br>97.49 | 4.159 | 5.653<br>3.475 | 2.480 | 3.425<br>2.05 | 39 | | 0.109 | 0.145<br>0.093 | 40 | | 0.106 | 0.140<br>0.091 | 39 | | 0.066 | 0.088<br>0.056 | 40 | | 0.065 | 0.086<br>0.051 |
| | | | | | | | | | 97 | | | | 95 | | | | 97 | | | | 95 | | | |
| Early Cretaceous (Hauterivian – Albian) | 50 | Min. 0<br>Max. 75 | 71.37 | 90.5<br>62.89 | 2.603 | 3.364<br>2.288 | 1.557 | 2.038<br>1.341 | 38 | | 0.07 | 0.089<br>0.062 | 25 | | 0.106 | 0.135<br>0.095 | 38 | | 0.042 | 0.054<br>0.037 | 25 | | 0.064 | 0.082<br>0.057 |
| | | | | | | | | | 135 | | | | 120 | | | | 135 | | | | 120 | | | |
| Berriasian - Valanginian - | 0 | | 10.79 | | 0.401 | | 0.243 | | 10.6 | | 0.038 | | 15 | | 0.026 | | 10.6 | | 0.023 | | 15 | | 0.016 | |
| | | | | | | | | | 145.6 | | | | 135 | | | | 145.6 | | | | 135 | | | |
| Cretaceous – Quaternary | 705 | Min. 555<br>Max. 780 | 279.45 | 336.09<br>253.89 | 9.684 | 11.93<br>8.76 | 5.59 | 7.01<br>4.92 | 145.6 | | 0.071 | 0.086<br>0.065 | 135 | | 0.077 | 0.092<br>0.070 | 145.6 | | 0.044 | 0.052<br>0.039 | 135 | | 0.047 | 0.056<br>0.042 |

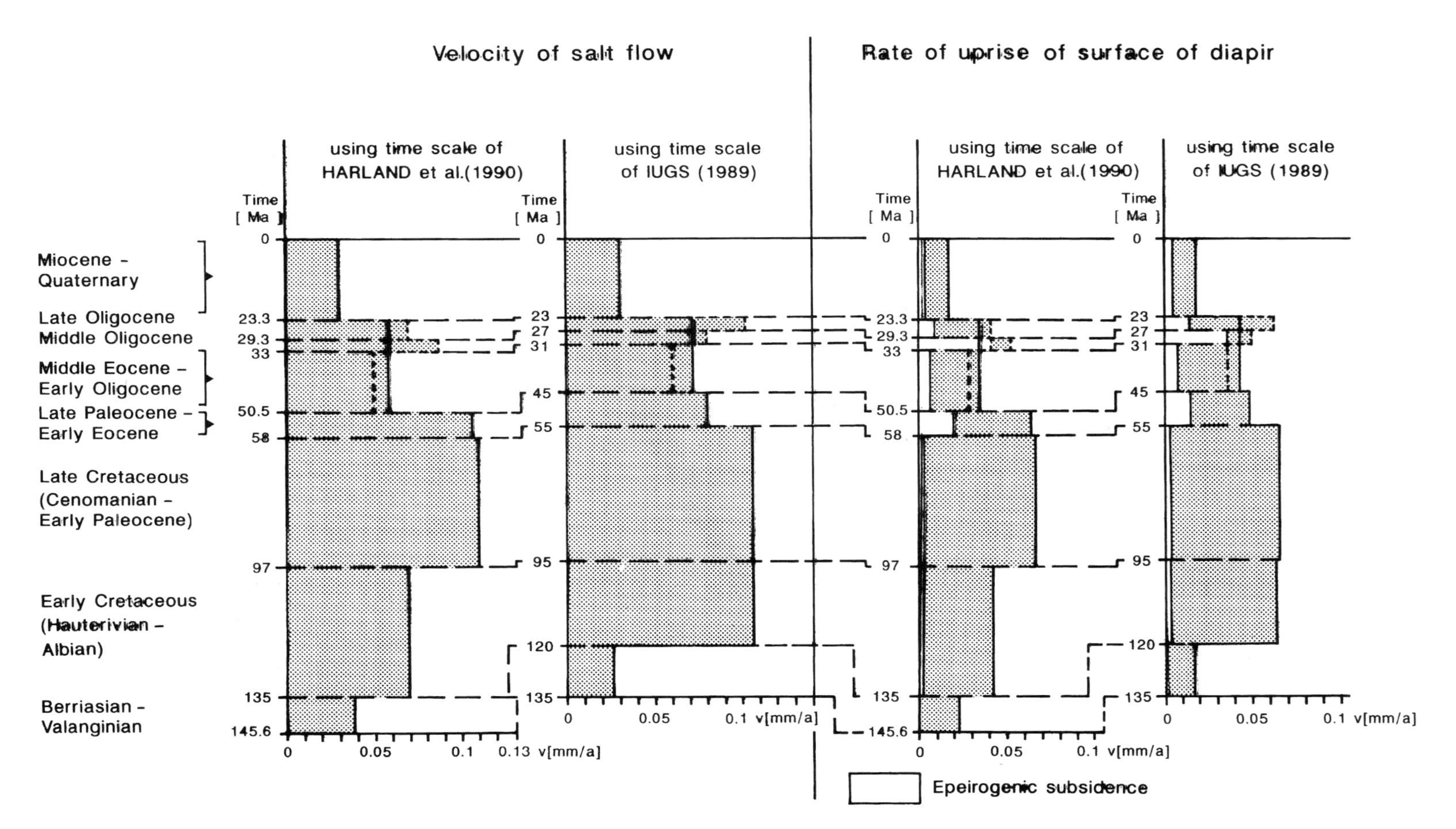


**Fig. 14.** Velocity ($v$) of salt flow in the narrowest part of the diapir and the resulting mean rate ($v$) of uprise of the salt at the top of the salt diapir.

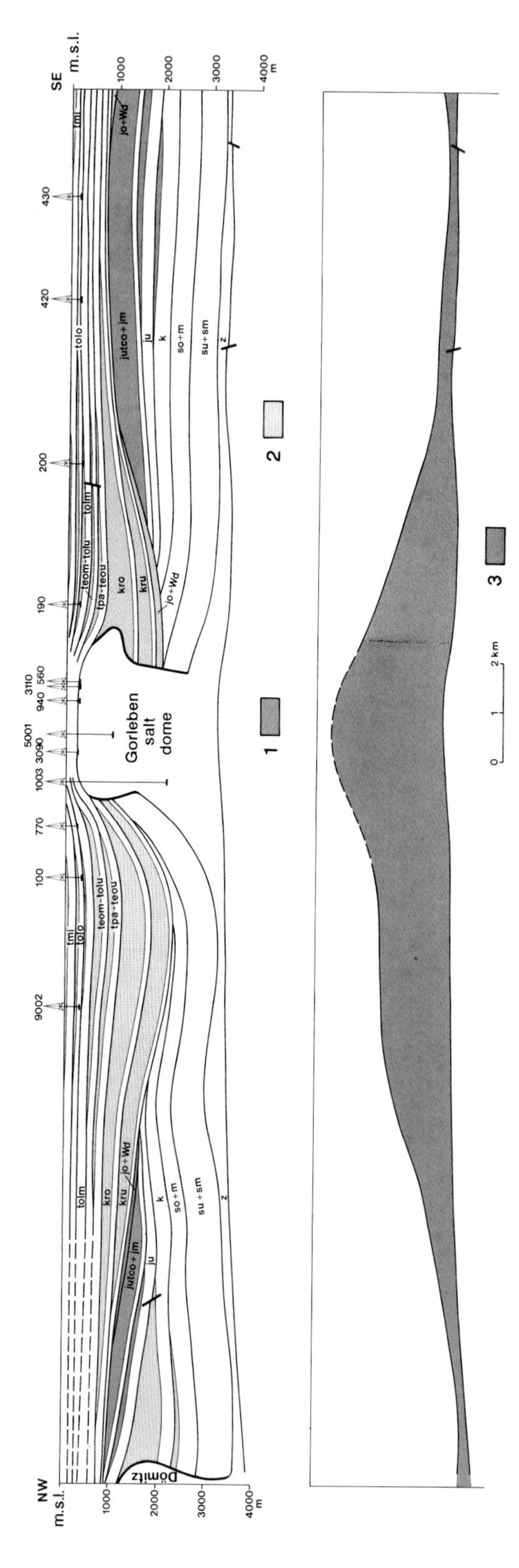


**Fig. 15.** Reconstruction of the former salt pillow at the site of the Gorleben salt dome. 1, additional thickness of sediment in the primary peripheral sinks due to halokinesis. 2, additional thickness of sediment in the secondary and tertiary peripheral sinks due to halokinesis. 3, thickness of salt pillow derived from the additional thickness of sediment in the secondary and tertiary peripheral sinks due to halokinesis.

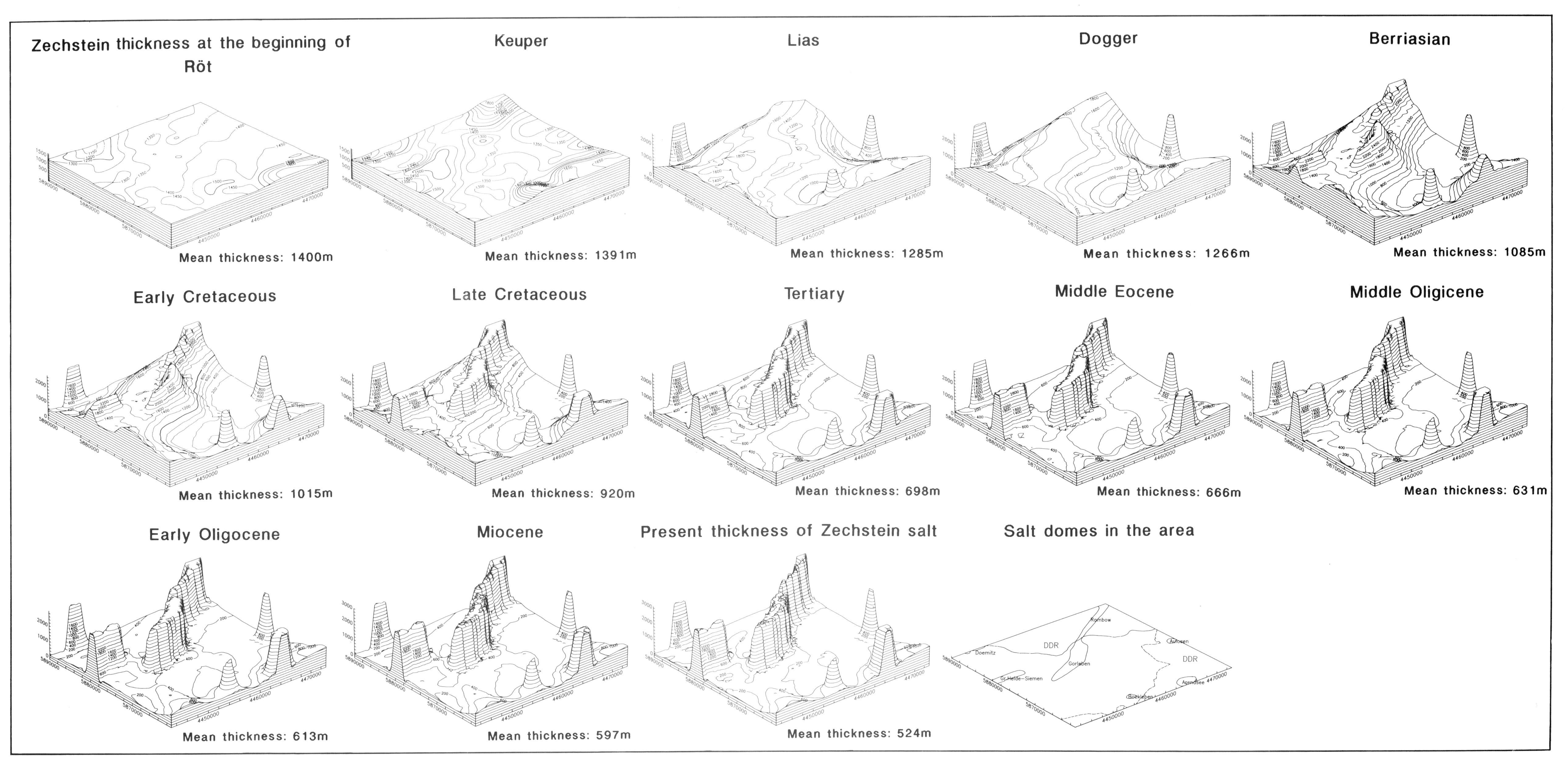


**Fig. 16.** Halokinetic evolution of the Zechstein evaporite.

however, shows that the uprise of the salt dome is less than the epeirogenic subsidence, i.e. the calculated uprise rates for this time span are too high. It is understandable that erroneously high rates were obtained for the comparatively short periods of the Middle Oligocene and/or Late Oligocene when some of the absolute ages given by Harland *et al.* (1990) differ from those of IUGS (1989) by as much as 50%.

If the Eocene and Oligocene are considered as one large unit (see Table 1 and Fig. 14), these extreme variations in rate do not occur. In this case, the curve for the rate of salt uprise shows an increase from the Berriasian to a maximum during the Late Cretaceous and/or Early Eocene. Afterwards, the rate decreases and reaches a minimum in the most recent of the periods studied here, i.e. Miocene to Quaternary.

## Primary thickness of the Zechstein

Figure 15 shows a reconstruction of the original salt pillow at the site of the Gorleben salt dome using the method described by Rühberg (1976). The thickness of sediment in the secondary peripheral sinks due to salt flow is added to the thickness of Zechstein salt remaining outside the salt dome. In this way one obtains the shape of the salt pillow just prior to the moment of piercement. The actual position of the diapir is interpolated. Assuming that before and during the salt-pillow stage no salt is leached away, the volume of the salt pillow can be taken as equivalent to the volume of salt originally present in the source area. If this calculation is done in three dimensions for the Gorleben salt dome, then the original thickness of the Zechstein source layer amounts to 1150 m. However, if this primary Zechstein thickness is used to reconstruct the history of Zechstein salt flow throughout the whole study area, as shown in Fig. 16, and not only in the source area of the Gorleben salt dome, then it appears that the Zechstein salt could have been even thicker than this, at least around the Dömitz and Aulosen salt domes, which were formed during the Keuper. A reconstruction of the course of evolution of the Zechstein up to diapir formation in the Berriasian was obtained by adding amounts equivalent to the 'normal' thicknesses of the Buntsandstein and Muschelkalk formations to the primary thickness of the Zechstein of about 1150 m. The present thicknesses of these two units were then subtracted from the total thickness calculated above. The result represents the thickness of the Zechstein at the beginning of the Keuper. This method of calculation can be used for all units for which the 'normal' thickness due to epeirogenic subsidence, and the present thickness are known. If the Zechstein thickness is reconstructed in this way up to the end of the Keuper (beginning of the Lias), i.e. the original thickness of the Zechstein is added to the 'normal' thickness of the Buntsandstein, Muschelkalk and Keuper and the present thicknesses of these three units subtracted from it, then the calculated base of Buntsandstein in the area of the large Keuper peripheral sinks of the Dömitz and Aulosen diapirs would sink beneath the level of the base Zechstein. Assuming that no local movements of the base Zechstein took place and that the geological data on which the maps are based is reliable, this means that the inferred thickness of the Zechstein (1150 m) is not large enough. Thus it is necessary to assume that the original thickness of the Zechstein was c. 1400 m.

## The salt budget

In Table 2, the computed volumes of salt remaining in the withdrawal area and the volumes of salt lost either by erosion or leaching are compared with the volumes of salt originally present. Since the beginning of the diapiric phase in the Berriasian, about 248 km$^3$ of Zechstein salt has risen to the Earth's surface. The actual volume of salt leached away is about 9 km$^3$ more than this since the present-day salt table lies about 200 m below sea level. Depending on whether one uses the figure of 1400 m for the original thickness of the Zechstein salt, as must be assumed for the whole study area, or only 1150 m, as was calculated for the source area of the Gorleben salt dome, one obtains a figure of 47% or 57%, respectively, for the amount removed by leaching. This means that the remaining salt is only 53% or 43%, respectively, of the original volume.

Assuming that an original thickness of 1400 m of Zechstein salt was present in the source layer beneath the Gorleben salt dome – although this figure is difficult to determine accurately – then *c.* 18% (about 93 km$^3$) of the salt originally present must have been removed by subsurface solution (subrosion), in spite of the fact that there appears to be no equivalent thickness of sediment in the peripheral sinks today. Jaritz (1980) explains that in the case of the Zwischenahn salt dome, whose penetration phase was between Malm and Hauterivian (as in the case of the Gorleben salt dome), epeirogenic uplift took place at this time, and the amount of this must be taken account of by adding a correction figure. In the Gorleben area, uplift during the Late Jurassic and Berriasian was about 600 m. During this time span, salt loss may have occurred from the top of the salt dome by erosion or subrosion, and the equivalent amount of sedimentation in the peripheral sink may have been reduced by uplift, or sediments may have been removed by erosion. Thus it follows that the original Zechstein thickness of 1150 m calculated on the basis of

**Table 2.** *Zechstein salt budget in the source area of the Gorleben salt dome*

| | Salt moved into diapir ($km^3$) | Height of dissolved salt above diapir (m) | Volume of dissolved salt ($km^3$) |
|---|---|---|---|
| Miocene – Quaternary | 18.8 | 323 | 14.3 |
| Upper Oligocene | 11.1 | 193 | 8.6 |
| Lower Oligocene | 8.6 | 46 | 2.0 |
| Middle and Upper Eocene | 22.9 | 415 | 18.4 |
| Palaeocene – Lower Eocene | 21.4 | 332 | 14.7 |
| Upper Cretaceous | 114.6 | 2480 | 110.1 |
| Lower Cretaceous | 71.4 | 1557 | 69.1 |
| Berriasian | 10.8 | 243 | 10.8 |
| | 279.6 | 5589 | 248.0 |

Source area: 376.4 $km^2$
Salt in the source area at present: 185.6 $km^3$
Dissolved salt + remnant salt = original volume of Zechstein salt
248.0 $km^3$ + 185.6 $km^3$ = 433.6 $km^3$
which means an original thickness of 1150 m
Dissolved salt: 57%
Salt moved into the salt dome: 63%

sediment thicknesses in the peripheral sinks is only a minimum value.

## Error analysis

Our knowledge about the deep geological structure of an area depends on the amount of investigation done, i.e. on the number of boreholes and seismic reflection profiles. Errors may occur during conversion of times to depths in seismic reflection profiles and during the stratigraphic interpretation of drilling data. Jaritz (1980) considers that the most frequent sources of error in the peripheral sink analysis include determination of the 'normal' stratigraphic thicknesses in the source area of the salt, the complicated reconstruction of salt dome cross-sections at various geological times, and the inaccuracy of absolute dating of geological periods.

For peripheral sink analysis, it is assumed that all the salt that migrates during the individual periods considered can be derived from, i.e. is equivalent in volume to, the volume of sediment in excess of the 'normal' in the peripheral sinks. Short phases of uplift or stationary phases within the single periods considered do not affect the final result. There is, however, one exception to this: the Late Jurassic phase of uplift and erosion, when about 600 m of sediment were eroded off. Any rock salt (e.g. the top of the salt pillow) included in the eroded material would not be accounted for in the salt budget, since it is assumed that no salt losses occurred before formation of the salt diapir. Any salt eroded in this way would amount to a maximum of 18 $km^3$. It can be assumed that no larger losses of salt took place due to extrusion of salt onto the surface during this phase of uplift. If this had taken place then it would have led to the development of peripheral sink thickness in excess of the 'normal' Jurassic thickness.

To estimate the effect of erroneous calculation of the thickness of the sedimentary fill in the peripheral sinks and of erroneous ages on the rate of salt uprise, comparative calculations using various 'normal' thicknesses and ages were carried out (see Table 1). Synsedimentary tilting caused the 'normal' thickness of Upper Cretaceous sediments to vary between 0 m in the southeastern part of the study area and 150 m in the northwest. Both these extreme values of the 'normal' thickness are given in column 1 ('epeirogenic subsidence') in Table 1.

The rates of uprise of the salt dome derived from these two values are 0.088 mm $a^{-1}$ and 0.056 mm $a^{-1}$ (column 5), respectively. This means that these extremely different thickness values cause the rate of uprise to change by only 36%. By comparison, the rate of uprise is much more affected by variations in the age, since it is inversely proportional to the age. The duration of the Late Oligocene, for example, is given by IUGS (1989) as 4 million years, whereas Harland *et al.* (1990) give 6 million years (Table 1). If these figures, which differ by a factor of 1.5, are used to calculate the rate of uprise, then we obtain values which also differ by a factor of 1.5.

Another source of error lies in the fact that compaction of sediments has been neglected in the calculations, since the compaction behaviour of the beds in the study area is not known. Studies by Matthesius (1974), of sedimentary rocks in the Gifhorn trough, show that the Cretaceous sediments are reduced in thickness by compaction by an amount that corresponds approximately to the additional thickness ascribed to subsidence due to compaction of the underlying beds. Thus, we can assume that the present-day thickness of the Cretaceous rock corresponds to the sum of epeirogenic subsidence and the thickness of salt that flowed into the salt domes during the Cretaceous. From the Tertiary onwards, the additional thickness of sediments deposited as a result of compaction of the underlying strata may even be larger than the reduction in thickness due to later compaction of these strata. This means that the calculated volumes of fill in the peripheral sinks are not only a function of salt withdrawal and epeirogenic downwarp, but also depend partly on compaction of the underlying strata. Calculations carried out by Matthesius (1974) on an 80 m thick Quaternary bed show that compaction has so far caused a reduction in thickness of 10 m, whereas the additional thickness ascribed to compaction of the underlying strata amounted to 21 m. Hence it follows that, if compaction is taken into account, the calculated salt flow rates would, in the case of the more recent, mainly argillaceous sequences, be slightly lower (maximum 10%), whereas slightly higher rates would be obtained for the older, more sandy-silty parts of the sequences (maximum 20%).

Consequently, using this method to determine the rate of salt uprise, the most marked error is due to different absolute age scales for the sedimentary sequences studied here; in the case of short stratigraphic intervals they may lead to a maximum error of 50%. In an earlier paper on this study area (Zirngast 1991), in which the older time scales of Harland *et al.* (1982) and Odin (1982) were used to calculate the rates of rise of the salt dome, the maximum error was as high as 100%. On the other hand, the calculated rates of salt flow are only slightly affected by uncertainties in the determination of the 'normal' thicknesses and the volume of sediments filling the peripheral sinks. The error can be estimated as between -10% and +20%.

## Reconstruction of the halokinetic movements of the Zechstein salt layer

With the help of the computer program ISM, plots were made of the successive stages of Zechstein salt withdrawal and formation of the salt domes between the beginning of the Röt and the present day (Fig. 16). The shapes of the plotted salt bodies do not correspond exactly to the true shapes of the salt domes, because the program used for the construction cannot portray overhangs in three dimensions. The shape at the beginning of the Berriasian, i.e. at the end of the salt-pillow stage and beginning of the diapiric stage, was obtained by adding all secondary peripheral sink sediment thicknesses of halokinetic origin to the thickness of the Zechstein salt still present today. The later block diagrams up to the beginning of the Miocene were obtained by subtracting the amounts of subsidence of the peripheral sinks corresponding to the successive time intervals.

Up to the beginning of the Keuper, there was little change in the shape of the top of the Zechstein in the area of the present Gorleben salt dome. The large difference in thickness between the NE and SE is possibly due to our poor knowledge of these areas. During the Keuper, the relatively small Dömitz, Aulosen and Bockleben diapirs (see Fig. 1) were formed in a comparatively short period of time.

In the area between these three salt domes, a large salt pillow formed and subsequently gave rise to the Gorleben, Heide–Siemen and Wustrow salt domes. In the south, another pillow formed, i.e. the Arendsee structure. At the beginning of the Berriasian, the Gorleben, Heide–Siemen and Wustrow structures started to develop as separate salt domes. Here and in the Arendsee area, this led into the diapiric stage. During the Early and Late Cretaceous, the flanks of the pillow retained their outline but became thinner. At the beginning of the Tertiary, the diapirs had almost the same shape as today. Salt withdrawal still took place southeast of Gorleben during the Tertiary; the decrease in size of the root of the salt dome provides evidence of this.

At the beginning of the Late Oligocene, the part of the pillow that remained northwest of the salt dome was distinctly reduced in size. The last block diagram (Fig. 16) shows the present-day Zechstein salt distribution. The thickness of the salt remaining in the source layer around the salt dome varies between 100 m to the southeast of the diapir and 400 m in the northwest. In the entire study area, the present-day salt volume within and around the salt dome corresponds to an average thickness of 524 m, i.e. about 37% of the primary Zechstein thickness of 1400 m.

## Conclusions

The calculations demonstrated an original thickness of 1150 m for the Zechstein salt in the source area of the Gorleben salt dome. About 63% by volume of this salt moved upwards into the Gorleben salt stock after the pillow stage. Altogether, about 248 km$^3$ of salt, i.e. about 57% of the Zechstein salt originally present in the supply area, rose to the Earth's surface and was eroded or dissolved by groundwater; 6% remains in the top of the present-day structure due to downbuilding after the pillow stage. The salt flow velocity in the area of the narrowest horizontal section of the salt stock varied between a maximum of 0.14 mm $a^{-1}$ during the Late Cretaceous and a minimum of 0.03 mm $a^{-1}$ during Miocene–Quaternary. The resulting rate of elevation of the surface of the salt column varied from 0.08 mm $a^{-1}$ in the Late Cretaceous to 0.02 mm $a^{-1}$ in the Miocene–Quaternary.

## References


BEUTLER, G. & SCHÜLER, F. 1978. Über altkimmerische Bewegungen im Norden der DDR und ihre regionale Bedeutung. *Zeitschrift der geologischen Wissenschaften*, **6**, 403–420.

DRESCHER, J., HILDEBRAND, G. & SCHMIDEK, R. 1973. *Bodensenkungen in der Lüneburger Altstadt: Vorschläge zur baulichen Sanierung*. Deutsche Gesellschaft für Erd- und Grundbau, Symposium Erdfälle und Bodensenkungen, **T4-G**.

HARLAND, W. B., ARMSTRONG, R. L., COX, A. V., CRAIG, L. E., SMITH, A. G. & SMITH, D. G. 1990. *A Geological Time Scale* 1989. Cambridge University Press.

——, COX, A. V., LLEWELLYN, P. G., PICKTON, C. A., SMITH, A. G. & WALTERS, R. 1982. *A Geological Time Scale*. Cambridge University Press.

IUGS. 1989. Supplement to *Episodes* **12**, 2.

JACKSON, M. P. A. & TALBOT, C. J. 1991. *A Glossary of Salt Tectonics*. Geological Circular **91-4**, The University of Texas at Austin.

——, VENDEVILLE, B. C. & SCHULTZ-ELA, D. D. 1994. Salt related structures in the Gulf of Mexico: A field guide for geophysicists. *The Leading Edge*, **13**(8), 837–842

JARITZ, W. 1980. Einige Aspekte der Entwicklungsgeschichte der Nordwestdeutschen Salzstöcke. *Zeitschrift der Deutschen Geologischen Gesellschaft*, **131**, 387–408.

——, BEST, G., HILDEBRAND, G. & JÜRGENS, U. 1991. Regionale Analyse der seismischen Geschwindigkeiten in Nordwestdeutschland. *Geologisches Jahrbuch*, **E 45**, 43–57.

KNAPE, H. 1963. Tektonischer Bau und Strukturgenese im nordwestlichen Vorland des Flechtinger Höhenzuges. *Geologie*, **12**, 505–760.

KOCKEL, F., BALDSCHUHN, R., BEST, G. *et al.* 1989. Geotektonischer Atlas von NW-Deutschland – Übersicht über die Strukturen des Oberbaus 1:300000. *Nachrichten der Deutschen Geologischen Gesellschaft*, **41**, 103–104.

MATTHESIUS, G. 1974. *Vertikale Dichte-, Porenanteil- und Druckdifferenzprofile an Sedimentgesteinen des Nordwestrandes des Gifhorner Troges*. Dissertation Technische Universität Braunschweig (unpublished).

MEINHOLD, R. & REINHARDT, H.-G. 1967. Halokinese im Norddeutschen Tiefland. *Berichte der deutschen Gesellschaft für geologische Wissenschaften*, **A12**, 329–353.

ODIN, G. S. 1982. The phanerozoic time scale revisited. *Episodes*, 3–9.

PUTZIGER, K., REINHARDT, H.-G. & WEGERT, F. 1966. Geophysikalische Beiträge zur Erkundung des regionalen geologischen Baues in Norddeutschland. *Geophysik und Geologie*, **8**, 3–20.

REINHARDT, H.-G. 1967. Hinweise der Prospektionsseismik auf rezente Salzbewegungen im Norden der DDR. *Geologie*, **16**, 95–100.

RÜHBERG, N. 1976. Probleme der Zechsteinsalzbewegung. *Zeitschrift für angewandte Geologie*, **22**, 413–420.

TEICHMÜLLER, R. 1948. Das Oberflächenbild des Salzdoms von Segeberg in Holstein. *Zeitschrift der Deutschen Geologischen Gesellschaft*, **98**, 7–92.

WIENHOLZ, R. 1967. Über den geologischen Bau des Untergrundes im Norddeutschen Flachland. *Jahrbuch für Geologie*, **1**, 1–87.

ZIRNGAST, M. 1991. Die Entwicklungsgeschichte des Salzstocks Gorleben. *Geologisches Jahrbuch*, **A132**.

# Physical modelling of fold and fracture geometries associated with salt diapirism

G. IAN ALSOP

*Department of Geology, Royal Holloway, University of London, Egham, Surrey, TW20 0EX, UK*

**Abstract**: The generation of salt diapirs is associated with a protracted deformation history in the surrounding overburden, which may be difficult to image seismically due to the highly inclined and fractured nature of sediments adjacent to the salt. To further the understanding of overburden deformation related to diapir development, a series of physical models has been produced incorporating a viscous SGM polymer source layer which is overlain by a granular, glass bead overburden. Model diapirs are initially triggered (to the pillow stage) via a basal indentor, but subsequently develop by a passive downbuilding process. The maximum widths of withdrawal basins and diapir-related upturn are approximately equivalent to one another. Experiments in which the salt analogue is allowed to extend laterally result in listric normal faulting of the overburden, to define graben structures centred on the diapir. Injection of the source layer into the footwall of normal faults results in reactive diapirism on the flanks of the major structure, whilst the adjacent central sections illustrate geometries which could be attributed to both active and passive diapirism. Multiple generations of normal faults produce graben with an inward-stepping sequence of fault propagation towards the central block. Complete evacuation of the source material at the keel of the normal fault simulates localized salt welds and results in extensional displacement being transferred into the hangingwall block. Sand–clay mix layers act as relatively competent horizons and display radial fracture patterns which may also be associated with a central ring fault and circular collapse graben above the diapir. Normal faults defining graben structures converge on the radial fracture pattern overlying the diapir. Circular thrust faults surrounding the diapir are developed when more competent horizons temporarily arrest vertical growth resulting in diapiric 'ballooning'.

This study was designed to analyse and interpret the geometry and kinematics of folding and faulting generated in the sedimentary overburden adjacent to diapiric structures. The steeply dipping attitude of rotated bedding, coupled with lateral facies variation, typically results in poor seismic resolution around diapirs, although improved techniques do now enable clearer imaging in some instances (e.g. Ratcliff *et al.* 1992).

Physical analogue modelling as an aid to interpretation and understanding of geological problems has been undertaken in a wide variety of studies, including granite tectonics (e.g. Ramberg 1970, 1981; Castro 1987), development of Calderas and volcanism (e.g. Marti *et al.* 1994) and especially salt tectonics (e.g. Parker & McDowell 1955; Lemon 1985; Cobbold *et al.* 1989; Vendeville & Jackson 1992*a,b*). The series of experiments described below examines the geometries of primary and secondary rim synclines developed adjacent to diapirs formed by passive downbuilding. The models were also designed to investigate the role and possible control exerted by extensional faulting on (reactive) diapiric evolution, together with analysis of the geometry and pattern of diapir-related fracturing in competent overburden. It should be noted that the experiments on passive diapirism were not developed to study the internal deformation of salt, nor to investigate possible triggering mechanisms of diapirism.

## Experimental materials

In previous experiments, layers of sand and clay were used to simulate the sedimentary overburden, and a viscous transparent polymer (SGM-36 polydimethylsiloxane) with a density of 965 kg m$^3$ and a viscosity of $5 \times 10^4$ Pa s was used to simulate the mobile salt layer (see Weijermars 1986*a,b*; Vendeville & Jackson 1992*a,b*). However, such experiments have failed to reproduce a notable upturn in the granular overburden adjacent to diapirs, a frequently observed phenomenon in both outcrop (Davison *et al* 1996*a,b*), and seismic studies (Ratcliff *et al.* 1992). These observations suggest that, despite its widespread use, sand may be too strong an analogue material to accurately model overburden deformation associated with salt (SGM) diapirs. Use of a more viscous diapiric source layer which could more effectively penetrate the sand overburden would require an unfeasibly long experimental time frame. This conclusion was originally reached by Parker & McDowell (1955,

*From* ALSOP, G. I., BLUNDELL, D. J. & DAVISON, I. (eds), 1996, *Salt Tectonics*, Geological Society Special Publication No. 100, pp. 227–241.

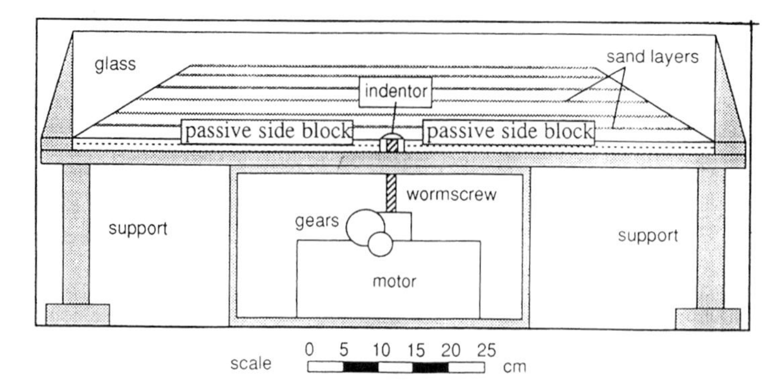


**Fig. 1.** Line drawing of open-ended rig used in this series of experiments. The wooden indentor is driven upwards into the polymer by a motor-driven wormscrew. Granular overburden is deposited during diapiric rise to maintain a horizontal top surface.

p. 2401), who considered that the shear strength of sand was too great to be scaled up to 'real' examples.

As a consequence of these modelling constraints, this series of experiments used a relatively weaker overburden of glass beads instead of sand. The beads may deform by both pervasive granular flow and more discrete faulting, and allow diapiric penetration through overburdens (up to 11 mm thick) within an acceptable scaled time span. The glass beads are spherical, with a diameter of 230–320 μm, a coefficient of friction of 15.3°, and a pseudocohesion of 200 Pa (E. Thompson pers. comm. 1993). They have a bulk density of 1910 kg $m^{-3}$ and an average porosity of 29%.

Although the smooth upturn of overburden

**Table 1.** *Summary of modelling parameters from Experiments A3–A28. Thickness of the premovement overburden, polymer and amount of indentation are shown to the nearest millimetre. The diapir height/width ratio relates to the parameters outlined in fig 2.*

| Experiment number | Thickness premovement overburden (mm) | Thickness of polymer (mm) | Amount of indentation (mm) | Diapir height/ width ratio | Fracture pattern |
|---|---|---|---|---|---|
| A3 | 10 | 14 | 23 | 3.50 | Hub, radial |
| A4 | 9 | 12 | 40 | 2.22 | No hub, radial |
| A5 | 8 | 10 | 23 | 1.14 | Linear, radial |
| A6 | 8 | 11 | 48 | 1.7 | Hub, radial |
| A7 | 7 | 11 | 45 | 1.65 | Linear, radial |
| A8 | 6 | 14 | 23 | 2.4 | Linear, asymm. |
| A9 | 3 | 10 | 23 | 1.7 | No hub, radial |
| A10 | 16 | 10 | 40 | – | Hub, radial |
| A11 | 3 | 10 | 7 | 3.3 | No faulting |
| A12 | 9 | 10 | 23 | 3.7 | Hub, radial |
| A13 | 14 | 13 | 45 | – | No hub, radial |
| A14 | 4 | 10 | 23 | 3.1 | Hub, radial |
| A15 | 3 | 15 | 6 | 3.29 | Linear, radial |
| A16 | 3 | 10 | 6 | 3.43 | Linear, radial |
| A17 | 3 | 13 | 6 | 3.50 | Hub, radial |
| A18 | 5 | 8 | 9 | 3.00 | Hub, radial |
| A19 | 5 | 11 | 9 | 3.01 | Linear, radial |
| A20 | 4 | 10 | 9 | 3.12 | Hub, radial |
| A21 | 6 | 8 | 23 | 2.4 | No faulting |
| A22 | 5 | 10 | 18 | 2.0 | Linear, radial |
| A23 | 5 | 10 | 18 | – | Linear, radial |
| A24 | 7 | 3 | 13 | 1.67 | No faulting |
| A25 | 7 | 9 | 18 | 2.08 | Linear, radial |
| A26 | 15 | 8 | 27 | 1.00 | No faulting |
| A27 | 12 | 9 | 31 | 1.58 | – |
| A28 | 10 | 10 | 29 | 1.18 | – |

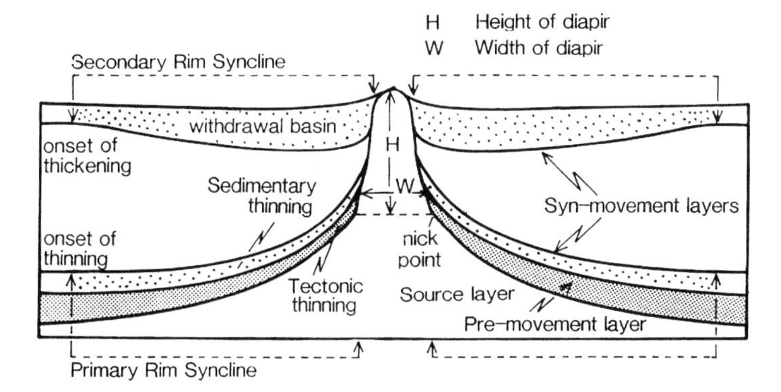


**Fig. 2.** Diagram illustrating the geometries and definitions associated with salt diapirism. A primary rim syncline is defined as the onset of sedimentary thinning, whilst a secondary rim syncline is considered as the onset of sedimentary thickening associated with a withdrawal basin.

adjacent to diapirs in seismic sections may be partially accommodated by pervasive minor fracturing, the bulk structural geometry is one of gentle to moderate folding (Alsop *et al.* 1995). A clear geometric similarity between natural diapirs and this series of models suggests that the scaling associated with the experiments represents a reasonable approximation. Glass beads are therefore considered a suitable analogue with which to model both the overall fault patterns and gross fold and flexure geometries observed adjacent to salt diapirs.

## Scaling

The models are scaled at an approximate ratio of 1:100 000, where 10 mm dimension in the model represents 1km in natural examples. The strengths of the materials are scaled so that the polymer: rock salt strength ratio is approximately $1{:}10^{14,}$ assuming that rock salt has a viscosity of $10^{18}$ Pa s. (Van Keken *et al.* 1993), and that the polymer has a viscosity of $5 \times 10^4$ Pa s (Weijermars 1986*a,b*). The time scaling factor is determined by this ratio and the ratio of the stresses. The stress ratio of model to real diapirs is approximately $5 \times 10^{-6}$ (derived from a scale ratio of $10^{-5}$ and density ratio of polymer to salt of approximately 0.5). Therefore the time scale ratio of model to real diapirs is $10^{-9}$ so that 1 h of modelling time is very approximately equivalent to 0.1 Ma.

## Experimental procedure

An approximate thickness of 10 mm of SGM polymer was allowed to settle (and air bubbles dissipate) on the floor of the rig, which had previously been sealed by Plasticene around the edges and by a thin, greased plastic sheet at the base to prevent escape and leakage of polymer. A central reservoir of polymer measuring 40 mm $\times$ 40 mm with a depth of 50 mm was set into the base of the rig and this was underlain by a sealed, flat-topped wooden indentor. The 40 mm diameter indentor was driven upwards into the well of viscous polymer by a motor-driven wormscrew regulated at a constant 9 mm $h^{-1}$, which represented a convenient and reasonable rate for the development of diapiric models. Significantly, however, the motorized indentation (which displaced a corresponding volume of polymer upwards out of the well) was entirely contained within the well and did not directly intrude into the overlying granular overburden. It should be noted that the rigid indentor (which has no 'natural' equivalent in salt diapirs) was typically only used for small amounts of indentation at the beginning of experiments to trigger diapirism, and once this was achieved motorized indentation ceased (Fig. 1, Table 1).

As noted previously, this study is concerned with overburden deformation adjacent to diapirs and is not intended as an investigation into the causes of salt diapirism. However, the downbuilding technique employed in the models, where sediment is deposited around the developing diapir to encourage further passive growth, is considered to be a viable and widespread mechanism of diapiric development (e.g. Jackson & Talbot 1991,1993). The resulting diapiric geometries and patterns of overburden deformation observed within the models are therefore a consequence of viscous polymer flow and interaction with the granular overburden, i.e. a passive downbuilding process of diapiric growth rather than active intrusion as modelled by Davison *et al.* (1993) (Fig. 2).

The present study involves two basic sets of

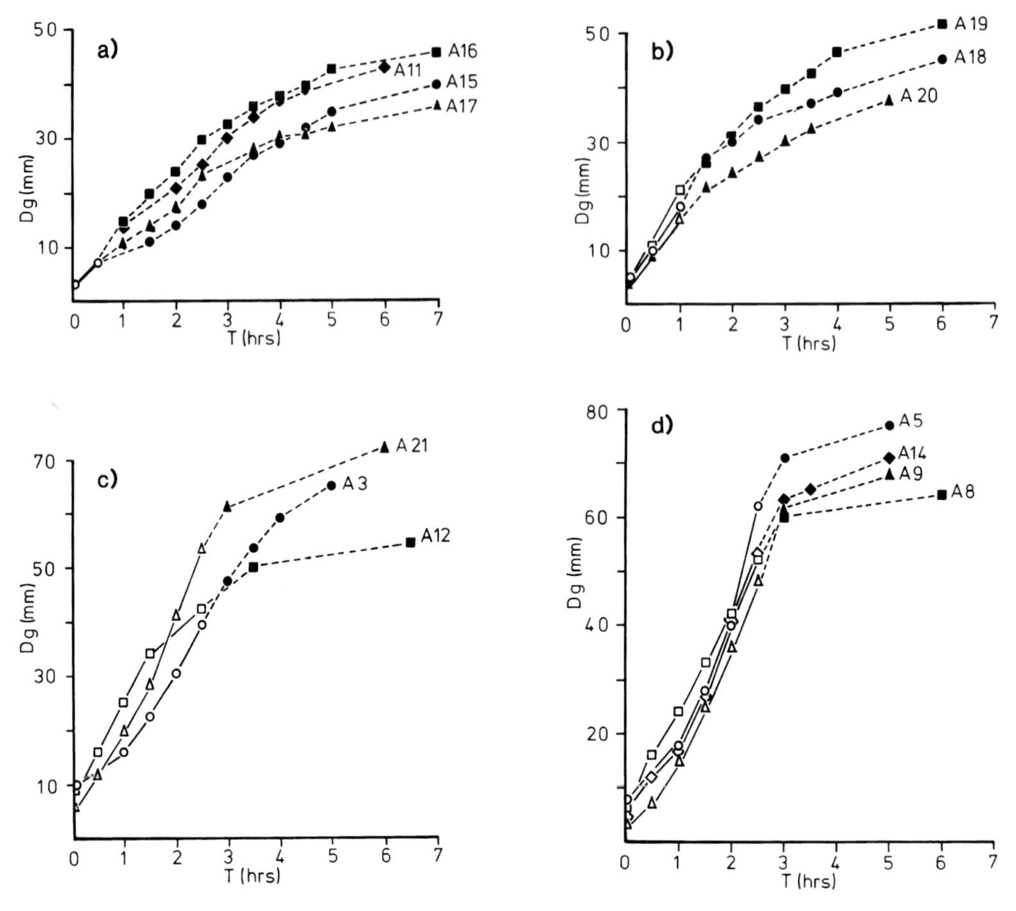


**Fig. 3.** Graphs of diapiric growth (Dg) plotted against experimental time (T). Open symbols connected by solid lines represent diapiric growth during motor-driven indentation (9 $mmh^{-1}$) of the circular boss, whilst solid symbols represent post-indentation growth. The intersection with the vertical axis marks the thickness of the premovement horizons. Experimental numbers are shown on the right. (**a**) and (**b**) illustrate the reduction in diapiric growth rate after an initial phase of indentation with a laterally constrained source layer. (**c**) and (**d**) illustrate the more rapid nature of diapiric growth during prolonged indentation associated with source layer extension. Refer to Table 1 for further details of experimental parameters.

experiments. In one set the source layer is laterally constrained by the Plasticene end walls (Experiments A3,10–13,15–21), whilst in the other set the polymer is allowed to flow laterally (Experiments A4–9,14,22–28). Both sets of experiments were run with variable overburden thickness and amounts of basal indentation (Table 1). In the series of experiments designed to investigate the role of source layer extension on the development of diapiric structures, the lateral flow within the polymer was driven entirely by overburden loading. Deposition of granular overburden results in loading, which leads to lateral extrusion of polymer over either one or both of the Plasticene end walls. The amount and direction of source layer flow may be controlled by blanketing one of the end walls with sand to prevent polymer escape. This technique is considered to reflect the development of diapirs during thin-skinned regional extension (see Jackson & Vendeville 1994).

Prior to initiation of indentation, a thin layer of barium powder was sprinkled onto the top surface of the polymer to facilitate its separation from the granular overburden during final preservation of the model. A constant, thin unit of glass beads with blue sand marker layers (which highlight the overburden structure during subsequent deformation) was then deposited over the barium and polymer. These

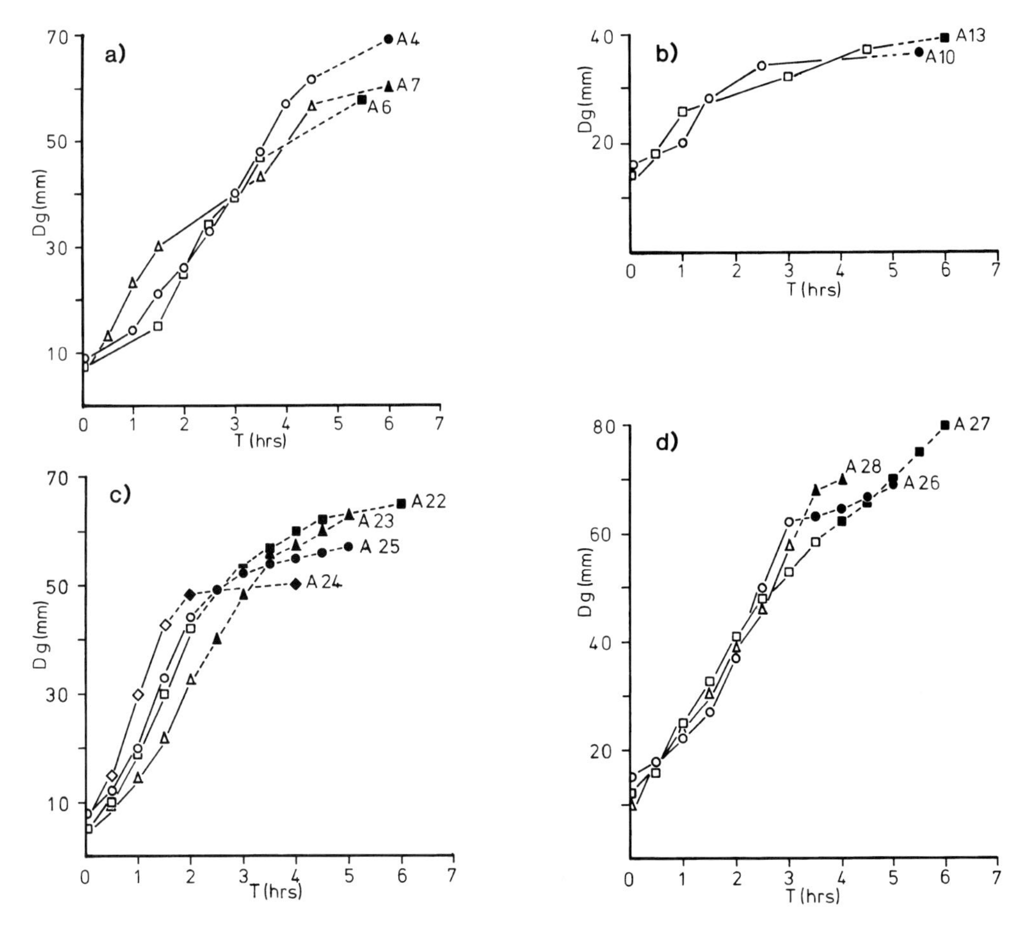


**Fig. 4.** Graphs of diapiric growth (Dg) plotted against experimental time (T). Open symbols connected by solid lines represent diapiric growth during motor-driven indentation (9 mm $h^{-1}$) of the circular (**a,b,c**) or elliptical (**d**) boss, whilst solid symbols represent post-indentation growth. The intersection with the vertical axis marks the thickness of the premovement horizons. (**a**) and (**b**) illustrate the slower nature of diapiric growth rate with a thicker premovement overburden during source layer extension (**a**) and with a laterally constrained source (**b**). A thick (15 mm) premovement layer will stifle pillow to diapir evolution. (**c**) and (**d**) illustrate similar growth profiles produced during source layer extension with a circular indentor (**c**) and an elliptical indentor (**d**). Refer to Table 1 for further details of experimental parameters.

premovement glass beads were layered only over the critical central 20 cm of the model (10 cm on either side of the indentor) and the rest of the flanks were infilled with sand, as these regions do not contain structures which are directly related to diapiric development.

During initial indentation and subsequent diapiric growth, glass bead layers with red sand markers were deposited onto the top surface at 30 minute intervals. These synmovement horizons infill any topography and maintain a horizontal upper surface which encourages further passive diapiric growth by downbuilding. It should be noted that whilst the extremely thin (<1 mm) sand marker layers may introduce some limited heterogeneity into the overburden, their sole purpose is to define structural geometries. Importantly, the sand markers have no apparent influence on the rate and nature of diapiric development.

The marker layers of sand have an average grain size of 250 μm (similar to the glass beads), an internal angle of friction of 31°and a density of 1580 kg $m^{-3}$ (see Weston *et al.* 1993). More competent layers within the synmovement overburden are represented by thin (2 mm) units of 80–20% sand–clay mix in which fracture patterns are developed directly above the diapir head during active piercement. Sand–clay mixes typically have angles

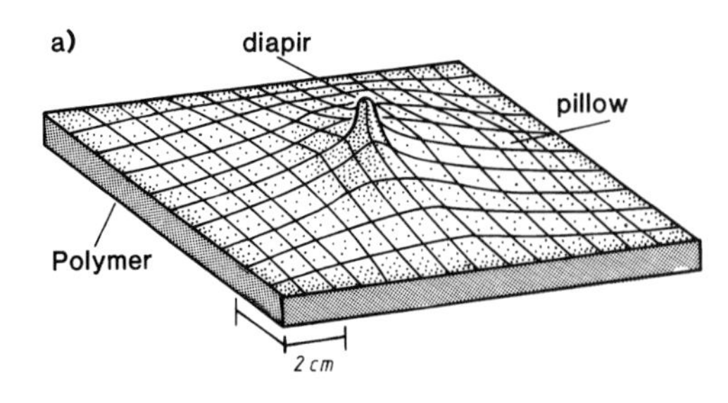


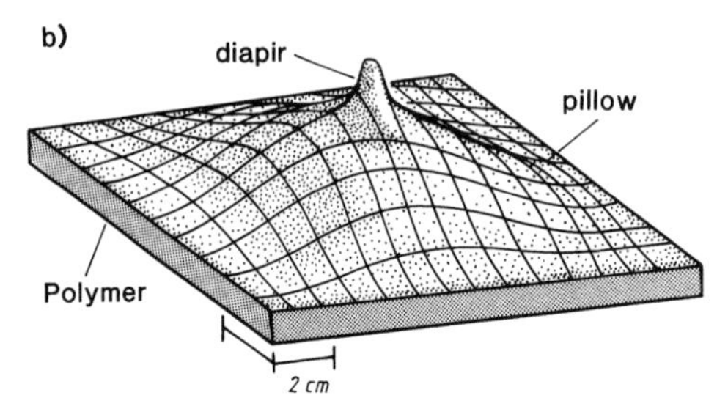


**Fig. 5.** Three-dimensional net diagrams of the top polymer surface in Experiments A13(**a**) and A10(**b**). The diagrams illustrate the geometry of broad pillow structures prior to the development of pronounced diapirs.

of friction of 33°, a cohesion of 180 Pa, and develop discrete fault zones approximately 0.5 mm wide (Davison *et al.* 1993). Sand–clay horizons represent units such as chalk, which may behave in a relatively competent manner within the deforming overburden (see Weston *et al.* 1993).

On completion of an experiment, the model was preserved via impregnation with hot gelatine solution. Once cooled and only slightly hardened, the model was carefully removed from the experimental rig and vertically sliced into a series of approximately 10 mm thick serial sections to allow detailed three–dimensional analysis of the preserved diapiric structure.

## The role of indentation on diapiric growth rates

Analysis of diapiric growth rates highlights the importance of motorized indentation, the thickness of premovement layers, and source layer extension during the development of diapirs.

Experiments with limited (6–7 mm) basal indentation (and a constrained source layer) demonstrate that a reduction in diapiric growth rate occurs as the diapir evolves (Fig. 3a). This is considered to relate to the increasing influence of viscous drag forces between source material and the flanks of the growing diapir. Experiments with slightly greater amounts (9 mm) of basal indentation (and a constrained source layer) display relatively rapid diapiric growth rates during indentation (Fig. 3b). However, when indentation ceases, a similar reduction in growth rate occurs producing an overall sigmoidal growth curve.

Experiments with moderate indentation (23 mm) during source layer extension display initial rapid growth rates (Fig. 3c). When basal indentation ceases (after 23 mm) and source layer extension continues, the rate of diapiric growth markedly decreases (Fig. 3d). Diapirs which developed most rapidly during the early growth history display the most abrupt reduction in growth when indentation ceases (Fig. 3d). Experiments in which a thicker (approximately 10 mm) premovement overburden is associated with moderate (40 mm) indentation and source layer extension do not display as rapid a growth profile as those experiments with thinner premovement overburdens (Fig.4 a,c). This conclusion is highlighted by extremely thick (15 mm) premovement overburdens which stifle the development of diapirs despite significant (40–45 mm) indentation (Fig. 4b).

The use of an elliptical basal indentor (axial ratio 1:1.66) during moderate (27–31 mm) indentation coupled with source layer extension (Fig. 4d) results in similar growth profiles to diapirs developed under similar conditions with the standard circular indentor (Fig. 4c).

## Experimental results.

### *Primary rim syncline geometries*

Primary rim synclines are typically concentric asymmetric synforms associated with increasing structural dips on the flanks of growing salt pillows (Trusheim 1960; Jenyon 1986) (Fig. 2). Salt pillows are typically associated with gentle dips forming broad culminations in the prediapiric stage (Fig. 5). The geometry of these structures is particularly relevant to hydrocarbon exploration as they define the limits of structural closure associated with individual salt diapirs. Within this series of experiments, primary rim synclines are characterized by a tectonic attenuation of concordant premovement layers coupled with a sedimentary thinning of the synmovement units towards the diapiric flanks (Fig 6).

The geometry and scale of the primary rim synclines was found to be influenced by two major factors: the thickness of the premovement horizons deposited prior to diapiric growth, and the volume of source material entering the system (controlled by the amount of motorized indentation) (Fig. 7). Increasing either or both of these parameters

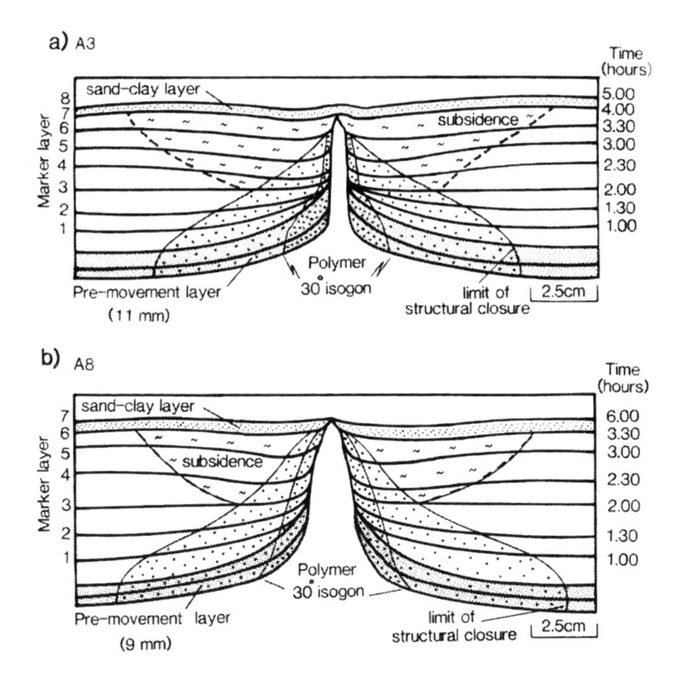


**Fig. 6.** Line drawings of sections through preserved models in Experiments A3(**a**) and A8(**b**). The diagrams highlight the limit of structural closure where beds start to upturn towards the diapir (spaced stipple), and also the extent of moderate dips immediately adjacent to the diapir (close stipple). The limit of downturn and thickening of beds associated with subsidence and development of withdrawal basins is shown by the dashed line.

typically results in wider rim syncline geometries associated with a broader structural closure, and accentuated thinning of synmovement horizons towards the diapir. Whilst the geometry and spacing of pillow structures in nature may be controlled by the thickness of the salt horizon (Hughes & Davison 1993), the thickness of the polymer layer in models does not significantly influence the geometry of the pillow structures due to the underlying well of polymer (Table 1). The width of the primary rim syncline is greatest in the premovement and lowermost synmovement horizons and progressively diminishes upwards towards the top of the diapir. The overall triangular geometry of the structural closure associated with the primary rim syncline is associated with concave-up flanks which typically steepen adjacent to the upper portions of the diapir (Fig. 6).

## *Secondary rim syncline geometries*

Secondary rim synclines are produced by diapirs withdrawing source material to feed further diapiric growth, thus resulting in peripheral subsidence and corresponding sediment thickening to fill localized basins (Figs 8, 9). The extent of the secondary rim syncline should therefore correspond to the limits of source layer withdrawal and consequently 'sphere of influence' of individual diapirs. The study of such withdrawal basins is particularly relevant as they provide a detailed sedimentological account of the history, timing and kinematic evolution of the adjacent diapir. Secondary rim synclines are characterized by the thickening of synmovement horizons into peripheral basins, coupled with the development of gentle diapir-directed dips which progressively diminish upwards through the withdrawal basin (Figs 6 & 8).

Within this series of experiments, secondary rim synclines are most clearly developed in the upper synmovement horizons deposited when diapiric growth was most vigorous. Secondary rim synclines are widest in the uppermost synmovement layers and narrow downwards towards the lower portion of the diapir (Fig. 6). Typically, the maximum widths of both the primary and secondary rim synclines are approximately equivalent to one another. The correspondence in the extent of the withdrawal basin associated with the evacuation of source material and the width of the pillow structure suggests that the diapir is entirely fed from the associated pillow. In general, the maximum radii of primary and secondary rim synclines was found to be approximately three to four times larger than the diameter of the associated diapir.

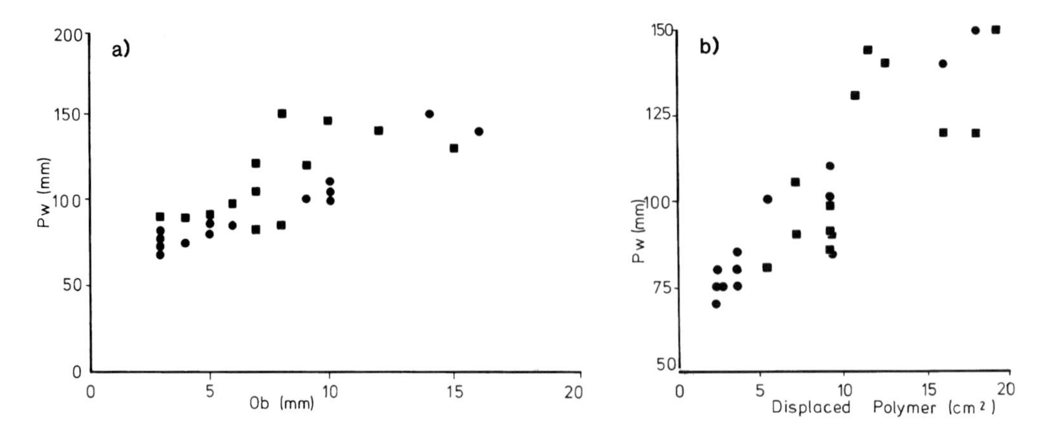


**Fig. 7.** (**a**) Graph of premovement overburden (Ob) thickness plotted against pillow width (Pw). (**b**) Graph of displaced polymer (cross-sectional area related to motorized indentation) plotted against pillow width (Pw). The graphs illustrate the positive correlation between these two variables. Solid circles relate to experiments with a constrained source layer whilst squares represent experiments with source layer extension.

## *Diapir-related fracturing*

Competent horizons may accommodate diapiric growth by brittle failure and fracturing, which is distinct in both style and geometry from fracture patterns initiated during regional extension (Withjack & Scheiner 1982; Jenyon 1986; Brewer & Groshong 1993). Importantly, models provide an opportunity to analyse and interpret, in isolation from regional tectonics, the fracture patterns and geometries produced by active diapiric growth.

Competent units, such as chalk, are simulated in models by 80–20% sand–clay mix horizons which are layered directly over the top of the developing diapir. Diapiric penetration through the competent horizon is achieved entirely by forceful intrusion of the overburden associated with active diapirism. Competent horizons above a developing pillow are arched into a flat-topped dome which displays a polygonal system of irregular, tensile fractures (Fig. 10e). Tensile fractures which initiate perpendicular to sedimentary layering and in plan view typically

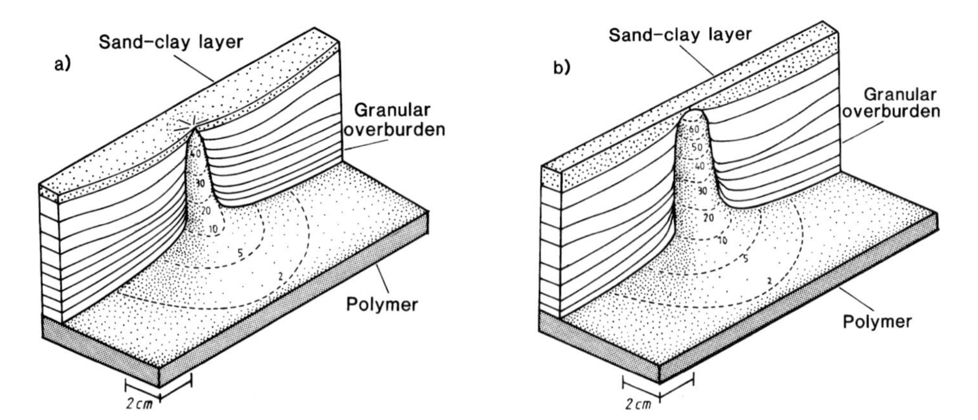


**Fig. 8.** Three-dimensional diagrams of the top polymer surface and overburden geometries in Experiments A3(**a**) and A8(**b**). The diagrams illustrate the broad pillow structures and nature of associated diapirs. The top polymer surface is contoured in millimetres (dashed lines). The diagrams highlight the geometry of secondary rim synclines developed in the overlying sedimentary overburden and also fracture patterns in sand–clay horizons (stippled).

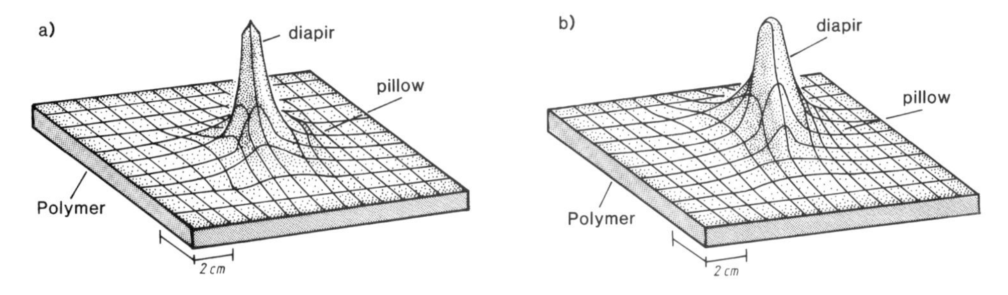


**Fig. 9.** Three-dimensional net diagrams of the top polymer surface in Experiments A3(**a**) and A8(**b**). The diagrams illustrate the evolution of broad pillow geometries and initiation of diapirism within the polymer source layer.

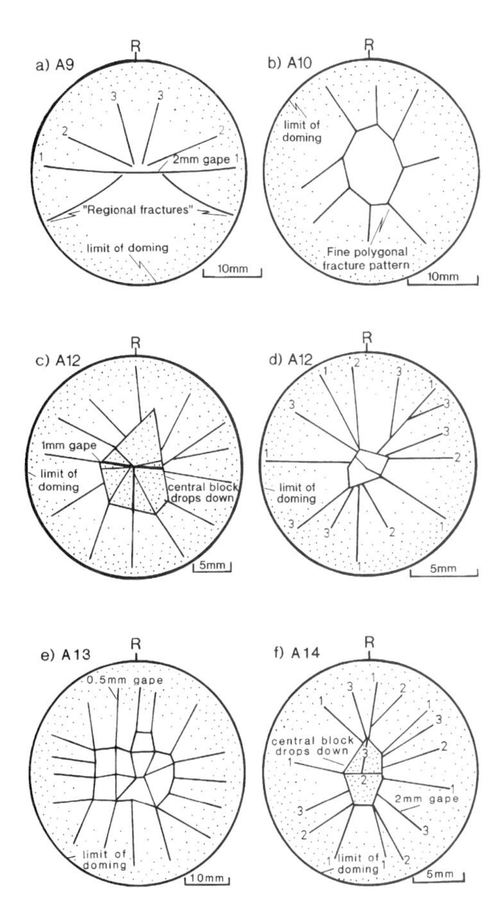


**Fig. 10.** Plan view line drawings illustrating the pattern and order of development of fractures in the sand–clay mix horizon directly above the diapir. Fracture patterns are typically radial and may incorporate a central ring fault with downthrown hub. The polygonal fracture pattern developed in (**e**) is a consequence of broad doming above a pillow structure (A13).

form a triple junction are developed directly above the diapiric head (Fig.10). The three major arms of the fracture system form at 120° to one another and progressively open and propagate outwards as diapir-driven doming continues. Subsequent sets of fractures sequentially develop between the major fracture arms to define an overall radial pattern. Propagation of fractures from a central area directly above the diapiric head notably contrasts with recent experiments by Marti *et al.* (1994), in which fractures developed above an expanding balloon were observed to nucleate at the periphery of the dome and gradually propagate both inwards and outwards. This difference in fracture propagation is thought to reflect the variation in geometry between the expanding balloon and the diapir crest.

In some experiments, the fracture arms propagate from a central hub which is surrounded by a ring fault. The central hub remains relatively undeformed and is simply dropped down relative to the updomed flanks (Fig. 10). The hub is subsequently carried passively upwards on the roof of the growing diapir, as radial fractures continue to gape. In section, the downthrown central portions are similar to graben geometries, although it should be noted that they are restricted to the diapiric crest and are not laterally extensive. The competent horizon may behave as a rigid layer and be rotated upwards as two flaps above the diapir, as shown in finite element modelling by Schultz-Ela *et al.* (1993). Injection of diapiric source material into the fracture system may result in complete detachment of overburden blocks and diapiric growth via a stoping mechanism.

## *Diapiric ballooning*

The growth of a diapir may be temporarily arrested by a relatively competent horizon within the overburden, resulting in *in situ* ballooning of a bulbous

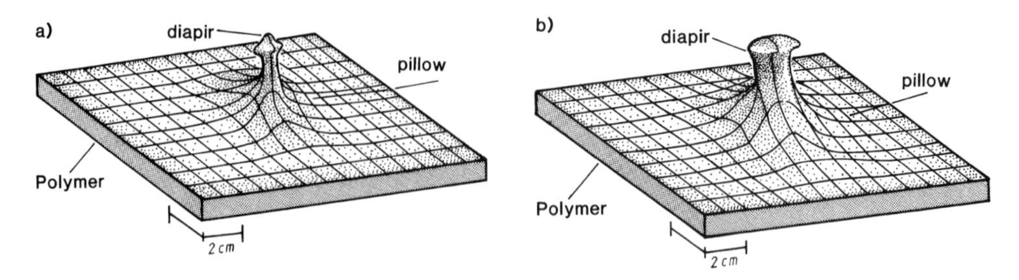


**Fig. 11.** Three-dimensional net diagrams of the top polymer surface in Experiments A15(**a**) and A22(**b**). The diagrams illustrate the geometry of *in situ* ballooning of the diapiric head above a relatively narrow stalk.

diapiric head (Fig. 11). The generation of ballooning diapirs is considered a consequence of the upward bouyancy force exerted by the diapir being approximately equal to the strength of the overlying competent horizon. The analysis of the geometry and kinematics of ballooning diapirs is particularly relevant to the hydrocarbon potential, as overhangs associated with swollen diapiric heads may be difficult to image seismically, but may provide important structural traps.

Within this series of experiments, ballooning diapirs have been successfully produced on several occasions, with expansion occurring within glass bead units directly beneath relatively competent sand-clay horizons (Fig. 12). Ballooning of the diapiric head or 'bulb' (Jackson & Talbot 1986) is associated with lateral expansion and production of overhangs, particularly along the sand–clay/glass bead interface, and also between sand–clay horizons which differ in competency as a result of varying proportions of clay (Fig. 12). Radial expansion of the diapiric head results in doming of the overlying sand–clay unit which deforms by brittle fracturing and faulting.

Diapiric expansion notably results in contractional reverse and low-angle thrust faulting around the periphery of the balloon. The occurrence of thrust faulting may therefore provide an indication of the presence of the underlying actively ballooning diapir. Directly above the diapiric head, however, the competent unit deforms by tensile fractures developed perpendicular to the layering and defining a radial pattern in plan view. Both tensile fracturing and contractional thrust faulting allow the source layer to inject along the dislocation and thereby facilitate ultimate penetration by the diapir of the competent unit (Fig. 12). Radial ballooning of the diapir is typically uniform, with no marked asymmetry in the direction of lateral expansion. However, in experiments where the source layer is allowed to extend in a single direction (see below), ballooning diapirs may become notably asymmetric with enhanced expansion in the direction of source layer flow (Figs 11 & 12).

### *Role of sediment progradation*

A series of three experiments has been run in order to evaluate and assess the control exerted on diapir evolution by prograding sediment wedges and the resulting differential loading. In each experiment, a premovement overburden of glass beads was deposited as an even, horizontal layer over the polymer, with subsequent layers of glass beads deposited around the growing diapir so that the top surface (immediately above the diapir) was horizontal. On the left-hand flank of the diapir, a greater thickness of overburden (glass beads and sand) was deposited to simulate differential loading associated with a prograding sediment wedge.

Experimental results indicate that differential loading of the laterally constrained source layer does not produce a markedly asymmetric diapir (Fig. 12). The diapiric walls remain perpendicular to the distal, horizontal flank. However, if the differential load is located at greater distances from the diapir, rotation of the source layer/overburden interface by angles of up to 5° from the horizontal may occur. The form and density of the prograding sediment wedge is critical in determining the amount and extent of rotation on diapiric flanks. Differential loading does not apparently produce a marked asymmetry in the development of rim synclines, although the true geometry of both the primary and secondary rim synclines is masked by the prograding sedimentary wedge. Differential loading does not produce a markedly different fracture pattern in competent sand–clay horizons overlying the diapir. Radial fracture systems are consistently developed throughout the prograding sedimentary wedge experiments.

## Role of asymmetric source layer flow

Both thin-skinned (Vendeville & Jackson 1992*a,b*) and crustal-scale, thick-skinned extension have recently been invoked as a major control on the initiation and evolution of salt structures (Nalpas & Brun 1993; Jackson & Vendeville 1994). Regional

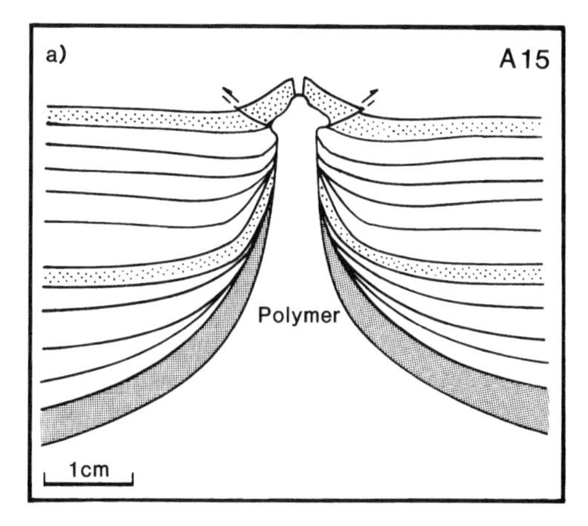


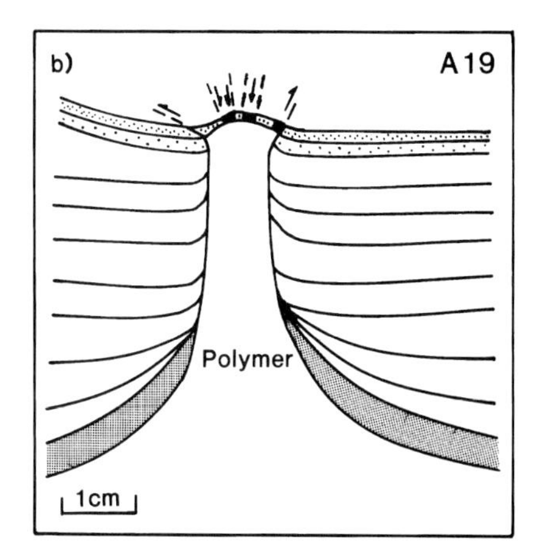


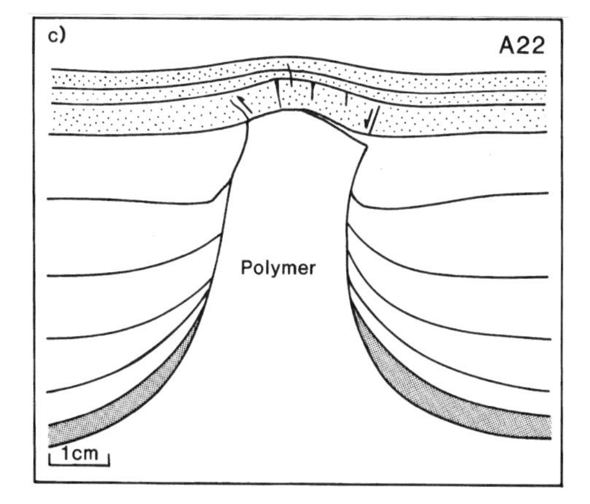


**Fig. 12.** Line drawings of central sections through preserved models. Premovement layers are shaded whilst the more competent sand–clay horizons are stippled. The sections illustrate the nature of *in situ* ballooning of the diapiric head or 'bulb' beneath competent horizons.
(**a**) Experiment A15 Overburden fracturing is associated with both peripheral thrust faults and overlying extensional fractures forming a graben.
(**b**) Experiment A19. This section illustrates the nature of *in situ* ballooning of the diapiric head or 'bulb' between two sand–clay horizons of varying competency. (**c**) Experiment A22. Section illustrating asymmetric ballooning with overlying tension fractures.

extension is considered to represent an effective mechanism for triggering salt diapirism, as the overburden is not only attenuated and weakened by normal faulting, but also topography generated by faulting results in differential loading of the underlying salt. Numerous physical modelling experiments have been conducted in which granular (sand) overburden has been extended at a fixed rate via mechanically driven end plates (e.g. Nalpas & Brun 1993) or by gravitational flow down a gentle gradient (e.g. Vendeville & Jackson 1992*a,b*; Jackson & Vendeville 1994).

The present series of experiments differs from those previously cited in the manner in which extension is generated, and by the use of glass beads rather than sand in the overburden. Within the models, extension and lateral flow within the polymer source layer was driven entirely by overburden loading (gravity spreading). Deposition of granular overburden results in loading, which leads to lateral extrusion of polymer over either one or both of the Plasticene end walls. The amount and direction of source layer flow may be controlled by blanketing one of the end walls with sand to prevent polymer escape. This technique is considered to simulate lateral withdrawal and flow of salt during thin-skinned extensional tectonics.

## *Patterns of overburden faulting*

The overall pattern of faulting developed within the granular overburden is governed by the nature and extent of lateral flow within the underlying polymer. The source layer behaves as a layer parallel décollement in which displacement is concentrated, whilst deformation in the passive overburden is in response to such displacements within the underlying detachment. Horizontal translations within the source layer are accommodated in the granular overburden by normal faulting, with concave-up listric geometries converging towards the source layer. The vertical displacement of the extensional fault is accommodated within the mobile layer,

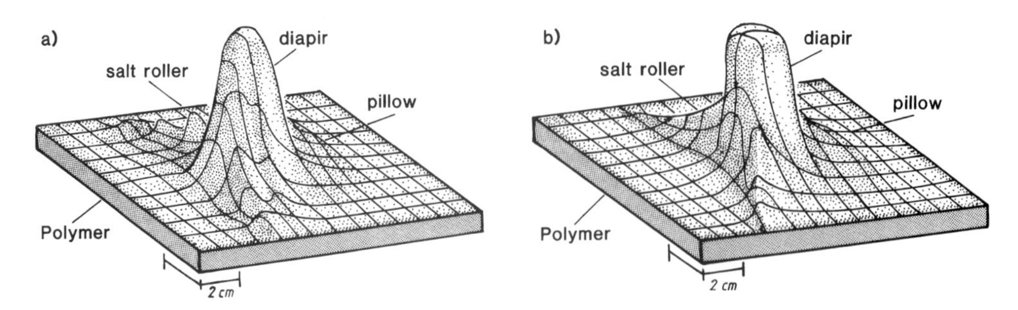


**Fig. 13.** Three-dimensional net diagrams of the top polymer surface in Experiments A27 (**a**) and A28 (**b**). The diagrams illustrate the geometry of diapirs developed during source layer extension. 'Salt rollers' developed in the footwalls of normal faults converge towards the diapiric crest.

resulting in significant relief along the source layer–overburden interface and development of simulated salt rollers (Fig. 13). Originally horizontal layers within the overburden may be significantly rotated by extensional faulting at angles of up to 30°.

Unidirectional lateral flow resulting from polymer extrusion at one end wall, typically leads to the formation of an asymmetric graben system, whilst polymer escape at both end walls produces a symmetric graben centred on the diapir. Normal faulting within the overburden is typically developed perpendicular to the trend of polymer flow, although in plan view fault planes are markedly curved towards the diapiric crest (Figs 13 & 14). Lateral extrusion of polymer results in a lowering of the source layer–overburden interface adjacent to the diapir. Total evacuation of the source layer occurs initially along the keels of downthrown fault blocks and simulates salt welds (Jackson & Talbot 1993). Continued lateral flow of polymer results in displacement being transferred from the now-locked normal fault into the hangingwall block. This produces an overstepping structural sequence of normal faults in the direction of source layer extension, i.e. structurally higher faults form sequentially later.

Central sections through diapirs frequently reveal little or no evidence for normal faulting and graben development within the overburden, despite significant extensional faulting in adjacent sections through diapiric flanks (Fig. 14). Extensional deformation within the overburden has been accommodated within the diapir itself. Such reactive diapirs may be associated with asymmetric primary rim synclines, with greater amounts and widths of upturn on one diapiric margin being representative of a 'regional' extensional fault. Further evidence of diapiric 'masking' of extensional fault systems is the downthrow and offset of the source layer/overburden interface across the central section of the diapir.

### *Extension and diapiric geometries*

Source layer extension associated with polymer escape induces fracturing of the overburden which encourages reactive diapirism (noted above). Diapirs associated with extension are characterized by more rapid growth and are typically larger than both passive and active diapirs produced when the source layer is constrained. The geometry of the diapir is also found to vary significantly, becoming relatively wider during source layer extension. Broader diapirs are considered to be a consequence of the overburden being extended by the flowing source layer, with the diapiric material simply responding passively to overburden geometry and fracturing (Talbot 1993; Jackson & Vendeville 1994).

Extensional faults result in keels of overburden within the hangingwall of the normal fault being displaced downwards into the source layer, which occupies the adjacent footwall. Salt walls in the footwalls of regional normal faults may thus extend at right angles to regional flow. Multiple sets of extensional faults associated with graben formation lead to the development of reactive diapirism on the flanks of the major diapiric structure (Fig. 14). The convergence of regional faults towards the diapir is associated with the convergence of 'salt walls' onto the diapir crest. Such structures produce complex three-dimensional geometries which vary considerably across the diapir (Fig. 15).

In summary, source layer extension is shown to be a major factor in controlling the geometry and evolution of diapiric structures.

## Conclusions

Analysis of diapiric growth curves shows that growth may accelerate during constant (9 mm $h^{-1}$) indentation (Figs 3c, 3d, 4c & 4d). This suggests that the increasing thickness of synmovement overburden, deposited during passive downbuilding,

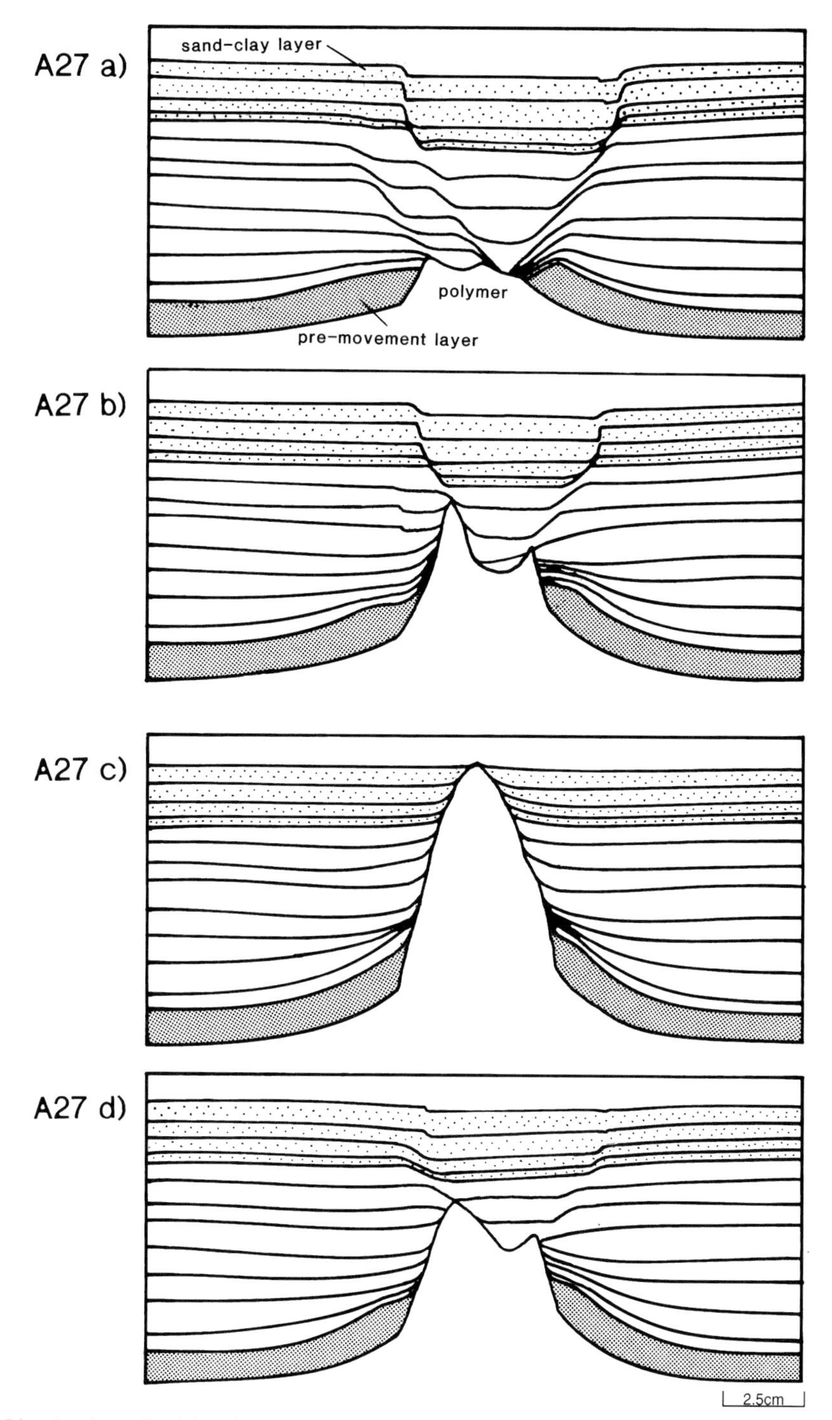


**Fig. 14.** Line drawings of serial sections through preserved model from Experiment A27. The spacing between each section is 10 mm. The diagrams highlight the extreme variation in geometry across the diapiric structure from the margins (**a**) to the centre (**c**).

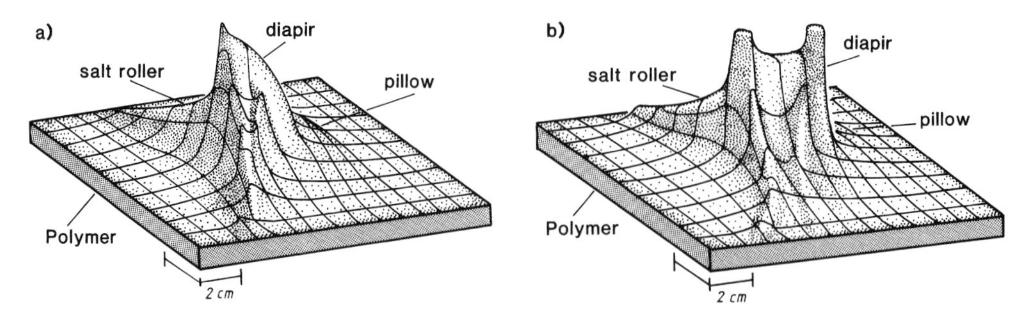


**Fig. 15.** Three-dimensional net diagrams of the top polymer surface in Experiments A25 (**a**) and A26 (**b**). The diagrams illustrate the geometry of diapirs developed during source layer extension. 'Salt rollers' developed in the footwalls of normal faults converge towards the diapiric crest.

encourages rapid growth. The typical reduction in diapiric growth rates at the cessation of indentation may relate to the increasing influence of drag forces within polymer in both the vertical diapiric 'pipe' and the increasingly depleted horizontal source layer. The development of 'salt welds', where faulted overburden comes into contact with the base of the source layer, will ultimately lead to source layer exhaustion.

Overall pillow dimensions (width and height) are proportional to one another and are dependent on the overburden thickness. Relatively wider pillows are, however, produced during source layer extension. Source layer flow into the diapir also results in localized secondary rim synclines, which display gentle profiles with diapir-directed dips progressively diminishing up through the structure. The width of the secondary rim syncline is greatest in the uppermost synmovement horizons and progressively diminishes downwards. The maximum extents of the primary and secondary rim synclines are approximately equivalent to one another, and this relationship relates to the developing diapir being directly sourced from the underlying pillow. The correspondence in the extent of the withdrawal basin associated with the evacuation of source material and the width of the pillow structure suggests that the diapir may be entirely fed from the associated pillow. Importantly, secondary rim synclines do not form when new source material is continually introduced into the diapiric system (via indentation).

The development of diapirs results in characteristic fracture patterns in overlying competent overburden. Competent horizons deposited over developing pillow structures display tension fractures which typically form polygonal patterns in plan view. Progressive diapiric doming produces a coherent and ordered sequence of faulting, with the development of a triple fracture system, coupled with minor fractures, sequentially forming between master faults.

*In situ* ballooning is associated with asymmetric swelling of the diapiric head coupled with a marked upturned zone of the adjacent overburden. Ballooning is typically developed between horizons of marked competency contrast (directly beneath the more competent layer). Forceful intrusion associated with ballooning of the diapiric head results in circular, peripheral thrust faults with associated hangingwall extension above the crest of the diapir.

Differential loading associated with a prograding sediment wedge does not produce asymmetric diapirs, as the diapiric walls remain perpendicular to the distal flank. Overburden fracture patterns are identical to those generated with an overburden of uniform thickness.

Source layer extension is clearly an important factor in controlling the generation and evolution of diapiric structures. Diapirs developed within such regimes are found to vary considerably from the flanks of the central section, highlighting the importance of three-dimensional analysis of diapiric structures.

This research was funded by Amoco (UK), Conoco (UK.), Enterprise Oil (UK), Hardy Oil & Gas, Ranger Oil (UK), Santos Europe Ltd and Saxon Oil (UK) within the Royal Holloway Salt Tectonics Project. I would especially like to thank Ian Davison for his encouragement and advice in the modelling programme together with T. Nalpas and M. Insley for careful and constructive reviews.

## References

Alsop, G. I., Jenkins, G. & Davison, I. 1995. The nature and geometry of drag zones adjacent to salt diapirs. *In*: Travis, C. (ed.) *Salt, Sediment and Hydrocarbons*. GCSSEPM Foundation Sixteenth Annual Research Conference.

Brewer, R. C. & Groshong, R. H. 1993. Restoration of

cross sections above intrusive salt domes. *AAPG Bulletin*, **77**, 1769–1780.

CASTRO, A. 1987. On granitoid emplacement and related structures. A review. *Geologische Rundschau*, **76**, 101–124.

COBBOLD, P., ROSSELLO, E. & VENDEVILLE, B. C. 1989. Some experiments on interacting sedimentation and deformation above salt horizons. *Bulletin of the Geological Society, France*, **8**, 453–460.

DAVISON, I., ALSOP, G. I. & BLUNDELL, D. J. 1996*a*. Salt tectonics: some aspects of deformation mechanics. *This volume*.

——, BOSENCE, D., ALSOP, G. I. & AL-AAWAH, M. 1996*b*. Deformation and sedimentation around active Miocene salt diapirs,on the Tihama Plain, northwest Yemen. *This volume*.

——, INSLEY, M., HARPER, M., WESTON, P., BLUNDELL, D., MCCLAY, K. R. & QUALLINGTON, A. 1993. Physical modelling of overburden deformation around salt diapirs. *Tectonophysics*, **228**, 255–274.

HUGHES, M. & DAVISON, I. 1993. Geometry and growth kinematics of salt pillows in the southern North Sea. *Tectonophysics*, **228**, 239–254.

JACKSON, M. P. A. & TALBOT, C. J. 1986. External shapes, strain rates and dynamics of salt structures. *Bulletin of the Geological Society of America*, **97**, 305–323.

—— & —— 1991. *A Glossary of Salt Tectonics*. Geological Circular **91-4**, Bureau of Economic Geology, University of Texas at Austin.

—— & —— 1993. Advances in salt tectonics. *In*: HANCOCK, P. L. (ed.) *Continental Deformation*. Pergamon, Oxford, 159–179.

—— & VENDEVILLE, B. C. 1994. Regional extension as a geologic trigger for diapirism. *Bulletin of the Geological Society of America*, **106**, 57–73.

JENYON, M. K. 1986. *Salt Tectonics*. Elsevier, Barking.

LEMON, N. M. 1985. Physical modelling of sedimentation adjacent to diapirs and comparison with late Precambrian Oratunga Breccia Body in Central Flinders Ranges, South Australia. *AAPG Bulletin*, **69**, 1327–1338.

MARTI, J., ABLAY, G. J., REDSHAW, L. T. & SPARKS, R. S. J. 1994. Experimental studies of collapse calderas. *Journal of the Geological Society, London*. **151**, 919–929.

NALPAS, T. & BRUN, J. P. 1993. Salt flow and diapirism related to extension at crustal scale. *Tectonophysics* **228**, 349–362.

PARKER, T. J. & MCDOWELL, A. N. 1955. Model studies of salt dome tectonics. *AAPG Bulletin*, **39**, 2384–2470.

RAMBERG, H. 1970. Model studies in relation to intrusion of plutonic bodies. *In*: NEWALL, G. & RAST, N. (eds) *Mechanisms of Igneous Intrusion*. Seel House Press, Liverpool, 261–286.

—— 1981. *Gravity, Deformation and the Earths Crust in Theory, Experiments and Geological Application*, 2nd edn. Academic, London.

RATCLIFF, D. W., GRAY, S. M. & WHITMORE, N. D. 1992. Seismic imaging of salt structures in the Gulf of Mexico. *The Leading Edge* **11**, 15–31.

SCHULTZ-ELA, D. D., JACKSON, M. P. A. & VENDEVILLE, B. C. 1993. Mechanics of active salt diapirism. *Tectonophysics*, **228**, 275–312.

TALBOT, C. J. 1996. Moulding of salt diapir profiles by stiff overburdens. *In*: JACKSON, M. P. A., ROBERTS, D. G. & SNELSON, S. (eds) *Salt Tectonics: A Global Perspective for Exploration*. American Association of Petroleum Geologists, Memoir.

TRUSHEIM, F. 1960. Mechanism of salt migration in northern Germany. *AAPG Bulletin*, **44**, 1519–1540.

VAN KEKEN, P. E., SPIERS, C. J., VAN DEN BERG, & MUYZERT E. J. 1993. The effective viscosity of rock salt: implementation of steady state creep laws in numerical models of salt diapirism. *Tectonophysics*, **225**, 457–476.

VENDEVILLE, B. C. & JACKSON, M. P. A. 1992*a*. The fall of diapirs during thin-skinned extension. *Marine and Petroleum Geology*, **9**, 354–371.

—— & —— 1992*b*. The rise of diapirs during thin-skinned extension. *Marine and Petroleum Geology* **9**, 331–353.

WEIJERMARS, R. 1986*a*. Finite strain of laminar flows can be visualised in SGM-36 Polymer. *Naturwissenschaften* **73**, 333–334.

—— 1986*b*. Flow behaviour and physical chemistry of bouncing putties and related polymers in view of tectonic laboratory applications. *Tectonophysics*, **124**, 325–358.

WESTON, P. J., DAVISON, I. & INSLEY, M. W. 1993. Physical modelling of North Sea salt diapirism. *In*: PARKER, J. R. (ed.) *Petroleum Geology of Northwest Europe: Proceedings of the 4th Conference*. The Geological Society, London, 559–567.

WITHJACK, M. O. & SCHEINER, C. 1982. Fault patterns associated with domes – An experimental and analytical study. *AAPG Bulletin*, **66**, 302–316.

# Salt flow by aggrading and prograding overburdens

H. KOYI

*Bureau of Economic Geology, Applied Geodynamics Laboratory, The University of Texas at Austin, Box X, Austin, 787 13 Texas, USA. Present address: Hans Ramberg Tectonic Laboratory, Institute of Earth Sciences, Norbyvägen 18B, S-752 36 Uppsala, Sweden*

**Abstract**: The mechanisms responsible for segmentation of salt sheets and their emplacement into higher stratigraphic levels are not separable and act simultaneously. Three sets of centrifuge models with strongly planar anisotropic (anisotropy, δ, the ratio between the effective viscosities in pure and simple shear, ranging between 4.8 and 16.4) microlaminate overburdens are used to study the effect of aggradation and progradation on segmentation and emplacement of allochthonous salt sheets.

In the first set of models, a tabular buoyant source layer was overlain by tabular anisotropic overburden simulating aggradation. During centrifuging, the underlying ductile source layer was segmented into individual wall-like diapirs by the subsiding blocks formed due to extension and faulting of the overburden. The extensional zone in these models started at the spreading edge (free face) of the model and migrated backwards.

In the second set of models, a tabular buoyant source layer was overlain successively by wedges of anisotropic overburdens simulating progradation. During centrifuging, the buoyant layer was displaced from the back of the model, where loading was higher, towards the free face in the front of the model, where overburden units were thinner. Overburden units extended at the back and the middle of the model while contractional structures dominated at the front, where asymmetric diapirs formed overhangs that spread 'basinward' to form 'salt' sheets.

In the third set of models, a wedge-shaped buoyant source layer was overlain successively by wedges of anisotropic overburden simulating progradation. The overburden wedge created a lateral pressure gradient ranging from 144 Pa at the back of the model to 80 Pa at the front when deformed in the centrifuge. In these models, as in the second set of models, the underlying buoyant mass was displaced 'basinward' by the subsiding thicker overburden units at the back of the model. Contractional structures dominated the deformation at the leading edge of the wedge.

Comparison of model results suggests that progradation (as in the second and third sets of models) loads underlying 'salt' differentially, displaces it downdip and segments it. As it segments at the back, the 'salt' flows laterally to areas of lower loading by intruding through the thinner overburden units and forming secondary 'salt' sheets at the front. On the other hand, aggradation of uniform overburden segments a buoyant sheet into two-dimensional salt walls or stocks, as in the first set of models.

Extrusive ductile salt spreads broad overhangs up to 80 km long (e.g. Worrall & Snelson 1989; Jackson & Talbot 1991) and up to 6 km thick (Simmons 1991) and forms salt sills (Nelson & Fairchild 1989), salt sheets (Curtis *et al.* 1988) or salt canopies when several of these overhangs coalesce (Jackson & Talbot 1986, 1991; Talbot & Jackson 1987).

Examples of salt sheets, which have been studied by many researchers (Worrall & Snelson 1989; West 1989; Wu *et al.* 1990a; Liro 1989, 1992; Seni 1992, 1994; Seni & Jackson 1989; Talbot 1992; Hudec *et al.* 1993; McGuinness & Hossack 1993; Koyi 1993), are present in the Gulf of Mexico. Wu *et al.* (1990*a*) used seismic data to describe the evolution of allochthonous salt in the northeastern Gulf of Mexico and concluded that autochthonous salt, allochthonous salt, and detached allochthonous salt are typical stages of the evolution. Wu *et al.* (1990*b*) distinguished five phases of progradation, starting with the Smackover carbonate unit (150–144 Ma BP), which loaded the Louann salt basinward. However, Wu *et al.* (1990*b*) emphasized that the rapid aggradation and progradation of enormous amounts of clastic sediments in the deep-water area since 30 Ma BP has led to the formation of the allochthonous salt structures in the northeastern Gulf of Mexico.

Jackson & Talbot (1991) defined segmentation of a salt sheet as 'deformation of a salt sheet acting like a second-cycle diapiric source layer, by means of differential loading or burial below the level of neutral buoyancy'. It is not clear from this definition whether a lateral flow of salt is also a part of the process. Therefore, in this study, the term 'segmentation' is used to describe partitioning of a salt layer into smaller entities (salt stocks and rollers) by aggradation, as in the first set of models. On the

*From* ALSOP, G. I., BLUNDELL, D. J. & DAVISON, I (eds), 1996, *Salt Tectonics*, Geological Society Special Publication No. 100, pp. 243-258.

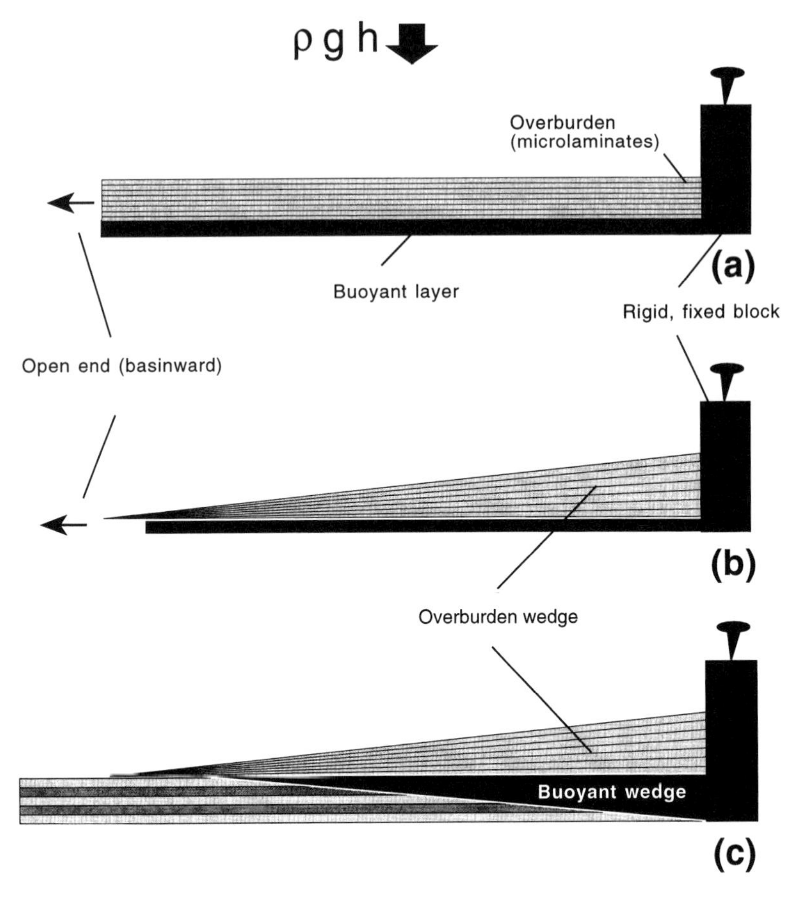


**Fig. 1.** Schematic diagram of model set-up before deformation. (**a**) First set of models with constant thick anisotropic overburden and a tabular buoyant layer. (**b**) Second set of models with an anisotropic overburden wedge overlying a tabular buoyant layer. (**c**) Third set of models with an anisotropic overburden wedge overlying a wedge-shaped buoyant layer. Thick arrow shows the direction of the centrifugal force which simulates the gravity force in nature. Small arrow shows the spreading direction during deformation in the centrifuge.

other hand, the term 'emplacement' is used to describe the process of salt flow laterally and across stratigraphy at the front of a salt sheet and its partitioning at the back due to progradation, as in the second and third sets of models (C. J.Talbot & M. P. A. Jackson, pers. comm.).

Recently, Talbot (1992) used viscous materials to build physical analogues to study the evolution of the salt structures in the Gulf of Mexico and concluded that progradation is driving the salt structures much further horizontally than they vertically rise. Koyi (1992, 1993) suggested that progradation and the presence of a slope at the base of a buoyant layer are essential for the formation and emplacement of a salt sheet. Ge *et al.* (1994) modelled the initiation and growth of salt structures induced solely by sediment progradation. They studied the effects of a flat salt base versus a stepped salt base on the deformation of the ductile layer. Cohen & Hardy (1996) emphasize the significance of differential loading in their two-dimensional numerical models of delta progradation over a mobile substrate. In their models, they assessed the effect of variations in mobile layer viscosity and thickness, progradation rate and sea-level on rate of movement of the mobile layer. In this paper, three physical analogues with anisotropic overburdens are used to study the effects of aggradation and progradation on segmentation of salt sheets, formation of secondary sheets and the deformation of the overburden units (Fig. 1).

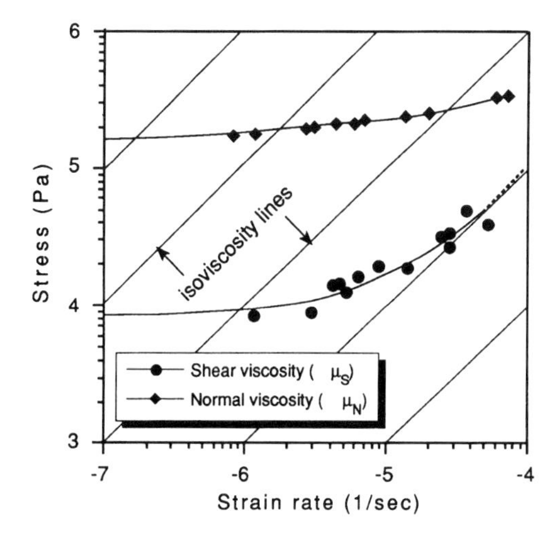


**Fig. 2.** Flow curves of microlaminates used as overburden in the models. The curves show that microlaminates possess different normal and shear viscosities, indicating anisotropic rheology.

## Modelling

### *Model materials*

In the models, rock salt was simulated using opaque silbione silicon putty, which is a Newtonian viscous fluid that behaves on short time scales like natural rock salt on geological time scales. The effective viscosity of the silbione silicone putty is $2 \times 10^4$ Pa s at the experimental strain rate of $10^{-2}$ $s^{-1}$, and its density is 1090 kg $m^{-3}$ (Weijermars 1986).

In contrast to previous studies (e.g. Talbot 1992; Ge *et al.* 1994), overburden units in these models were anisotropic microlaminates that simulate anisotropy likely to be present in nature and mimic faulting in natural overburdens. Microlaminates consist of alternating layers of two or more components with different effective viscosities. Microlaminates used in this study consisted of two main components, Dow Corning silicone and Plasticene, which were interlayered equally (microlayers were 0.2 cm thick). Dow Corning silicone (density 1300 kg $m^{-3}$) is a non-Newtonian silicone polymer whose effective viscosity varies from power-law at low strain rates to linear at higher strain rates (Dixon & Summers 1985). Plasticene (density 1700 kg $m^{-3}$) is a hygienic, non-toxic modelling material with a power-law stress exponent varying between 6 and 9 (McClay 1976; Weijermars 1986). Microlaminates can simulate anisotropy likely to occur in stratified sedimentary rock sequences when deformed in upper crustal conditions. Preparation of microlaminates has been described in detail by Dixon and Summers (1985).

Planar anisotropy ($\delta$) is the ratio between the effective viscosities in pure and simple shear, $\mu_N$ and $\mu_S$ respectively (Weijermars 1992)

$$\delta = \mu_N / \mu_S. \qquad (1)$$

Anisotropy of a viscous material is determined by the ratio between its viscosity in shortening/extension parallel to layering and its viscosity in shear parallel to layering. Anisotropy of a layered unit is dependent on strain rate because its normal and shear viscosities are strain-rate dependent. At high strain rates, if the effective shear and normal viscosities approach unity, the material becomes isotropic. The microlaminates used in these models have a strong anisotropy ranging between 4.8 and 16.4 at strain rates of $10^{-4}$ to $10^{-5}$ $s^{-1}$ (Fig. 2). The components of the microlaminate can be changed to simulate lower anisotropies if necessary.

The microlaminate overburden wedges of the second and third sets of models were prepared either by rolling tabular microlaminates until the thickness of the individual laminates at one end was thinner than their equivalents at the other end, or by cutting a wedge from the microlaminate stack. When placed on the buoyant layer, the overburden wedges formed an upper slope (Fig. 1b,c).

### *Scaling and limitations*

In the models described here, the mechanical properties of the anisotropic microlaminate overburden, rather than those of the low viscosity Newtonian source layer, influenced the deformation style. Therefore, the scaling procedures are approximated for the overburden units rather than the source layer.

It is essential that analogue models are scaled to their prototypes and geometric, kinematic and dynamic similarities are fulfilled. In the current models, geometric similarity is represented by a length ratio of $5 \times 10^{-6}$ when 1 mm in the models simulates 200 m in nature. Comparison of West's (1989) data, for a sedimentary wedge thinning from 3.6 km to 2 km in the Gulf of Mexico basin, with the overburden wedges used in the models gives a length ratio of $2 \times 10^{-6}$, which is very close to the value used for the model ratio (Table 1). Kinematic similarity, on the other hand, is approached by ensuring that time sequences simulating aggradation or progradation in the models were analogous to the time sequences of the same processes in their prototypes.

Dynamic similarity requires that ratios of forces acting in nature are equal to ratios of similar forces acting in the models. However, material properties determine how a model behaves to an applied stress. Therefore, to fulfil dynamic similarity, the ratio between the anisotropy factors of model materials

**Table 1.** *Scaling parameters for the overburden units of the models*

| Quantity | Model | Prototype | Model ratio |
|---|---|---|---|
| Thickness | | | |
| (overburden) | 5 mm | 1 km | $5 \times 10^{-6}$ |
| (source) | 4 mm | 0.8 km | $5 \times 10^{-6}$ |
| Density ($g\,cm^{-3}$) | | | |
| (overburden) | 1.4–1.45 | 2.4–2.5 | $(5.6\text{–}6) \times 10^{-1}$ |
| (source) | 1.090 | 2.2 | $5 \times 10^{-1}$ |
| Acceleration | 2500 *g* | 1 *g* | $2.5 \times 10^{3}$ |
| Time (s) | $9 \times 10^{2}$ | $9.4 \times 10^{8}$* | $9.5 \times 10^{-7}$ |
| Anisotropy at $(10^{-4}\text{–}10^{-5})s^{-1}$ | 4.8–16.4 | 4.8–16.4 † | 1 |

* Time for formation of allochthonous salt sheets due to enormous amounts of clastic sediments in the northeastern Gulf of Mexico, from Wu *et al.* (1990*b*)
† Suggested for prototype simulated by the models

to that of their natural prototypes should be equal or close to one. However, anisotropy of clastic rocks is very difficult to measure in nature. There are few laboratory measurements of anisotropic rocks (e.g. phyllite, by Paterson & Weiss (1966)). Measurements of anisotropy produced due to interlayering of different lithologies are rare. Therefore, it is difficult to compare the dynamics of model deformation to nature. However, in this study it is assumed that the overburden units in the natural prototypes have a similar anisotropy factor to the model overburdens. The microlaminates used in these models possess anisotropies ranging from 4.8 to 16.4 at strain rates of $10^{-4}$ to $10^{-5}$ $s^{-1}$, respectively. The anisotropic microlaminates used in the models could simulate interbedded limestone and shale, or sandstone and shale in nature (Dixon & Summers 1985).

Compared to nature, the model aggradation rate was very fast because each microlaminate overburden wedge (with thicknesses changing from 2 mm at the front to 5 mm at the back) were placed on top of the buoyant layer as one lump. However, rapid sedimentation in different parts of the Gulf of Mexico has been reported (West 1989; Wu *et al.* 1990*b*). West (1989) documented a gradual basinward thinning of the Upper Mexican Ridges seismic stratigraphic unit from 3.6 km in the depocentres to 2 km in the deep Gulf of Mexico basin during 30 Ma. Also, Wu *et al.* (1990*b*) stated that during the Late Oligocene–Neogene phase of rapid sedimentation, more than 5 km of clastic sediments accumulated due to shifting of the depocentres towards the northeastern Gulf of Mexico.

After running the model in the centrifuge for 5 minutes, cutting a section, and photographing, a new overburden was placed on top of the model. All models were run at 2500 *g* for 15 min. Unlike sand models, where sand grains could be poured on a larger model at variable rates, it is difficult to put very thin (0.5–1 mm thick) layers of microlaminate overburden periodically on a small centrifuge model to simulate a slower rate of aggradation. However, running the model continuously for 7 min in the centrifuge after placing each overburden wedge corresponds to the same time interval necessary to deposit wedges of similar thickness in nature. The time model ratio ($9.5 \times 10^{-7}$) is quantified by assuming 30 Ma for deformation of the Louann salt in the Gulf of Mexico (Wu *et al.* 1990b) versus 15 min for model deformation (Table 1). However, if the time of initiation of salt movement (144 Ma BP (Wu *et al.* 1990*b*)) is included in the calculations, the time model ratio is $1.8 \times 10^{-7}$, which is less than one order of magnitude different from the value used in this study (Table 1).

The models described here do not simulate any

**Fig. 3.** First set of models. (**a**) Top view and (**b**) profile of the model after placing the first microlaminate overburden and centrifuging for 7 min at 2500*g*. The diapirs of the buoyant layer (arrowed) rise through the faulted zones in the layered overburden at the front part of the model. Note that the back part of the model (right) is relatively undeformed compared to the front. This indicates that the deformation is concentrated at the spreading edge of the model (left). (**c**) Top view and (**d**) profile of the same model after addition of a second overburden microlaminate and centrifuging for an additional 7 min. The new overburden is extended more in the middle and the front part of the model than in the back of the model. The buoyant source layer is segmented into individual diapirs by the subsiding overburden blocks The diapirs rise through the faulted overburden along the fault zones. Note that the diapirs are buried by the new overburden as their supply is exhausted.

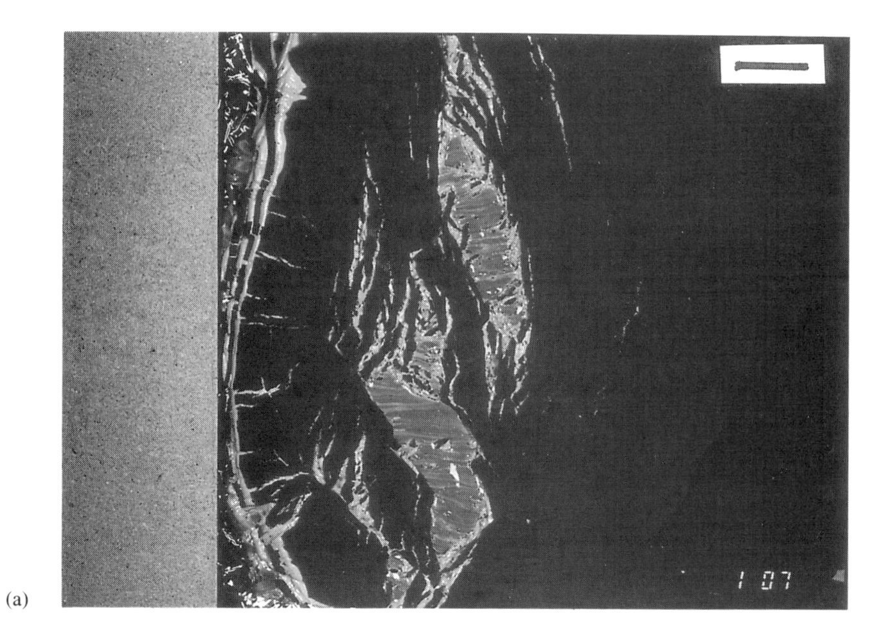


(a)

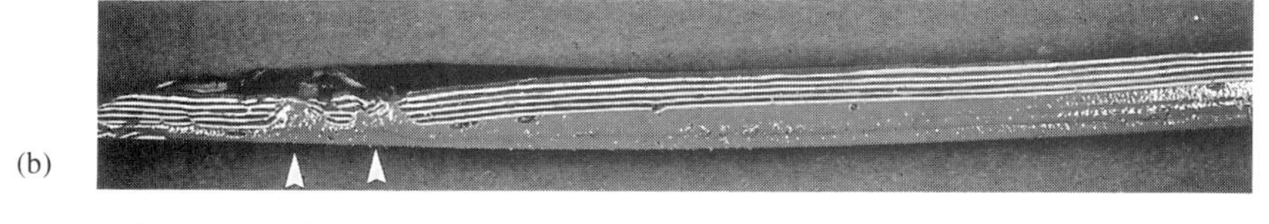

(b)

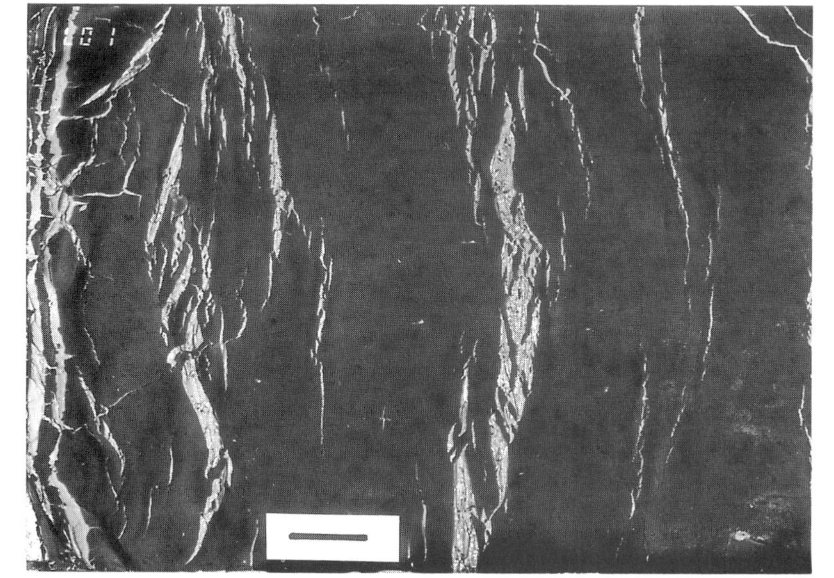

(c)

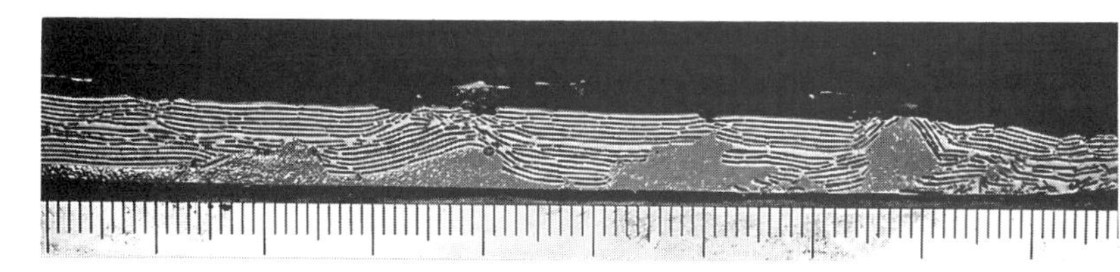

(d)

**Fig. 4(a).**

particular area in nature. However, they simulate prototypes where a salt layer is buried beneath an interlayered wedge that possesses a strong anisotropy (Table 1). These models are not dynamically scaled to the Gulf of Mexico, but it was used as a general guide.

## *Model preparation and kinematics*

Three sets of models were prepared to study segmentation and emplacement of salt sheets. These models were deformed at 2500*g* in a centrifuge at the Applied Geodynamics Laboratory at Austin, Texas. In all three sets of models, similar microlaminates were used for overburden units, and silicone putty simulated salt. There were two main differences between these three sets of models: (1) the geometry of the overburden units (tabular in the first set of models and wedge-like in the second and third sets of models); and (2) the basal slope of the source layer (horizontal in the first and second sets of models, but inclined in the third set of models).

The first set of models comprised a uniformly thick (4 mm) buoyant silicone layer (simulating rock salt in nature) that was resting on a horizontal lower boundary and was overlain by a uniformly thick (5 mm) layer of an anisotropic overburden of microlaminates (Fig. 1a). The models were allowed to extend incrementally at an open end (the front corresponding to the basinward direction) during centrifuging at 2500*g* (Fig. 3a,b). After photographing and adding another thick (5 mm) tabular layer of overburden, the model was deformed in the centrifuge (Fig. 3c,d). These models were centrifuged for a total of 15 minutes and extended up to 40%. Profiles were cut for photographing after each run (Fig. 3b,d). Extension was achieved in the centrifuge by spreading the buoyant and less viscous material beneath the subsiding overburden units towards the open end.

The second set of models consisted of a tabular source layer (4 mm thick) underlying a wedge-shaped microlaminate overburden (Fig. 1b). At first, the buoyant layer in these models consisted of two layers of silbione (each 2 cm thick), one of which was passively coloured in order to exhibit internal

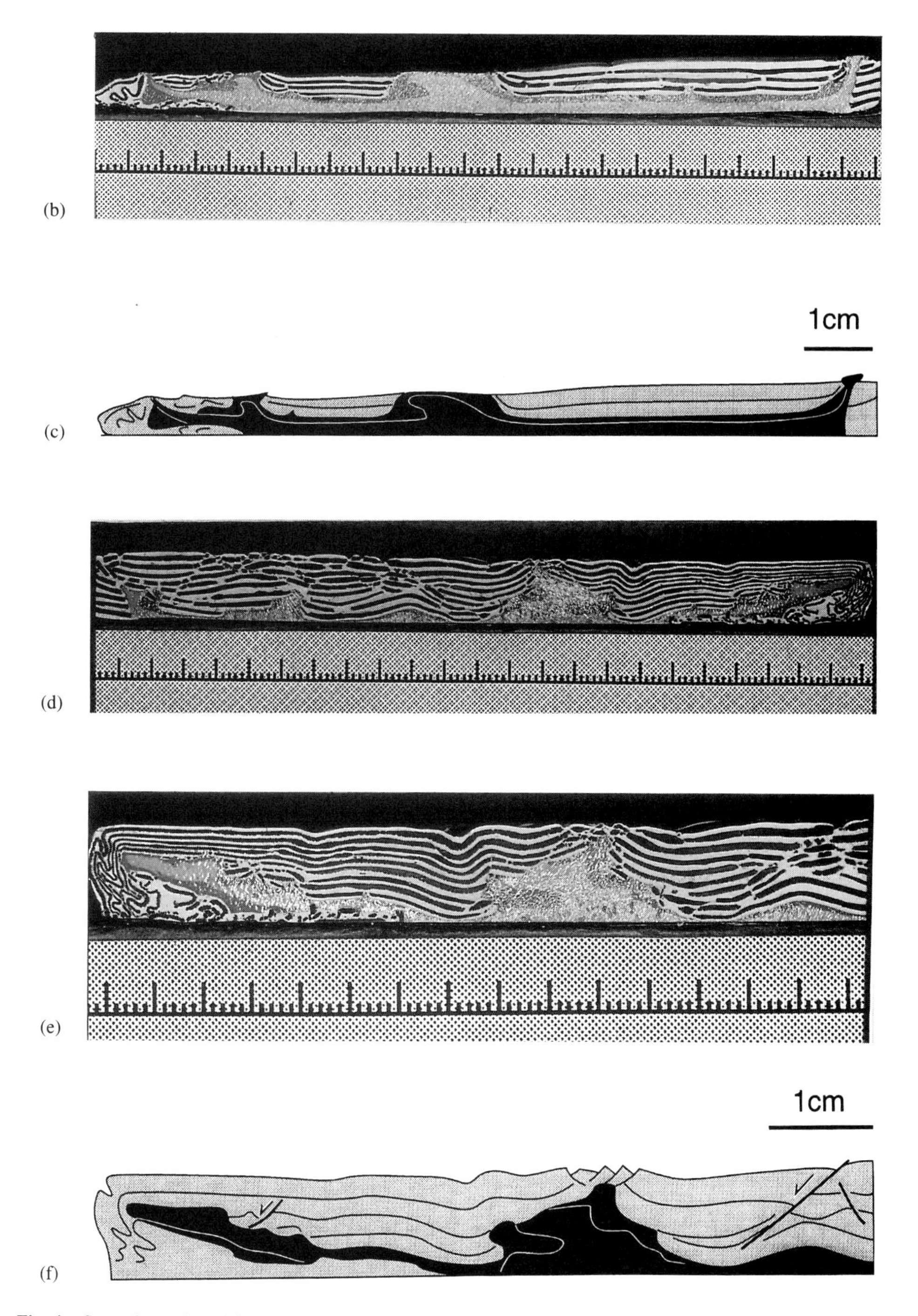


**Fig. 4.** Second set of models. (**a**) Top view and (**b**) profile after burial by one overburden wedge. (**c**) Line drawing of (**b**). (**d**) Profile after burial by two overburden wedges. (**e**). Close-up of the front area of (**d**). (**f**) Line drawing of (**e**). The white line in (**c**) and (**f**) marks the passive contact between the two buoyant layers. The overburden wedge is extended in the middle and back of the model while it is shortened at the front part of the model. The overturned overburden layers act as a ramp for the asymmetric diapirs to climb up the stratigraphy and form a buoyant sheet at the front area of the model. Observe the asymmetric basinward flow of the buoyant layer illustrated by the white line. The 'salt' sheet in (**b**) is pushed forward and upward across the stratigraphy due to further loading in (**d**) and (**e**).

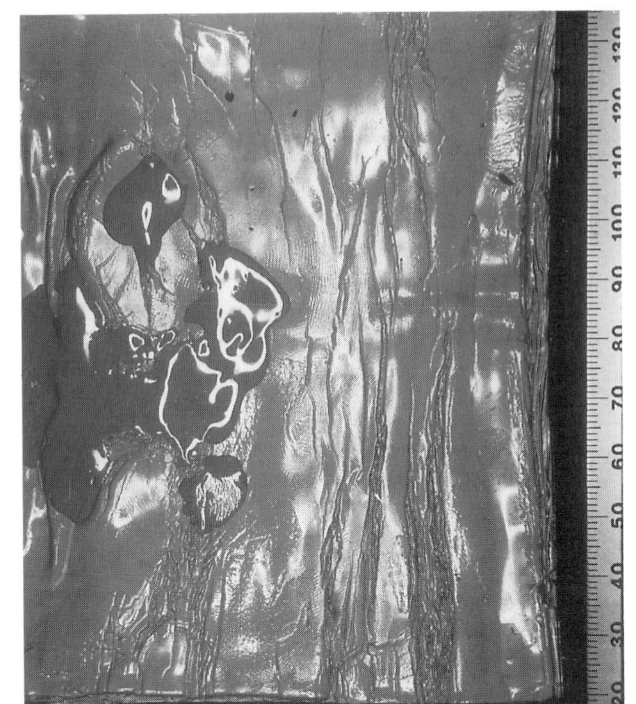

(a)

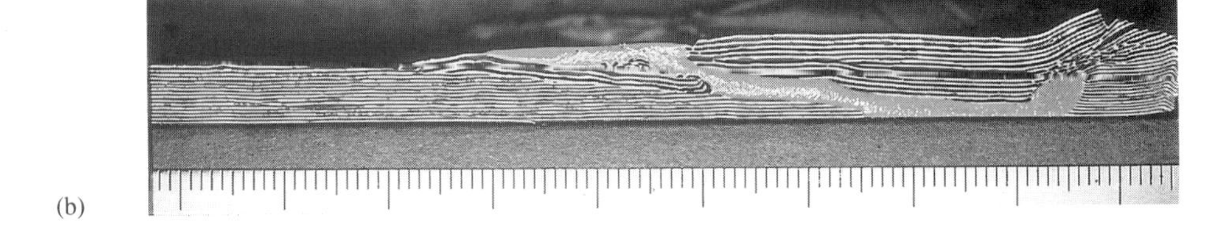

(b)

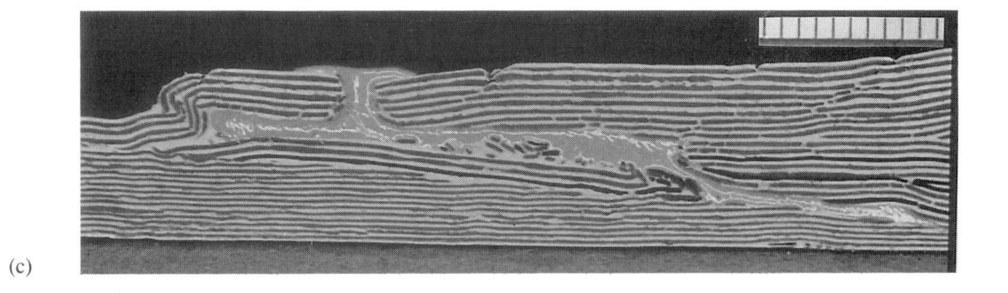

(c)

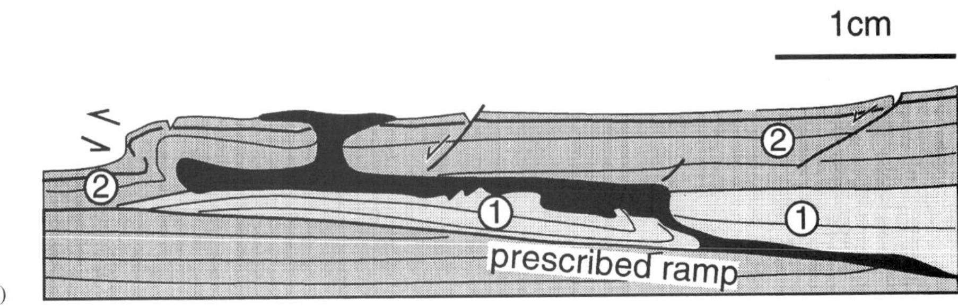


(d)

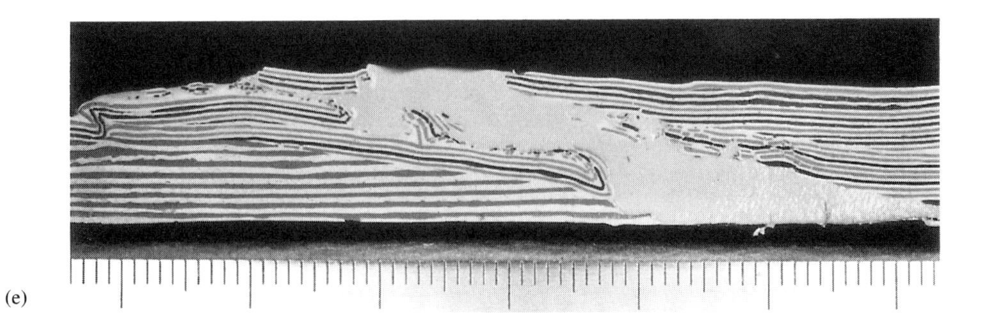

(e)

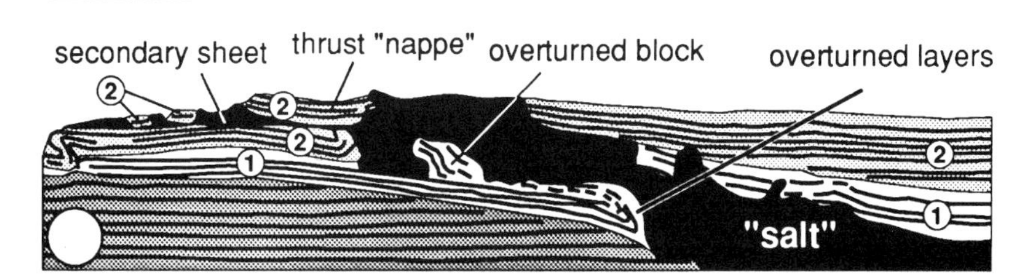
1 cm
secondary sheet
thrust "nappe"
overturned block
overturned layers
2
2
2
1
2
1
"salt"

(f)

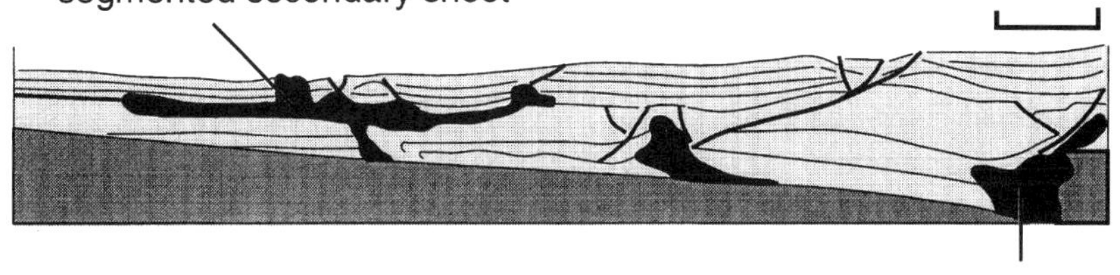
segmented secondary sheet
salt stock

(g)

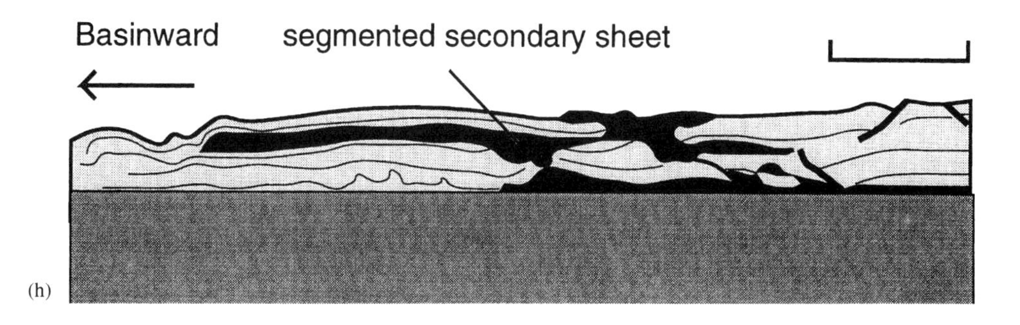
Basinward
segmented secondary sheet

(h)

deformation during flow of the buoyant layer (Fig. 4b–f). Two overburden wedges were placed on the buoyant layer at different times. During centrifuging, the overburden units produced a lateral pressure gradient that displaced the source layer to flow towards the open end of the model where overburden was thinner. Two profiles were cut for photographing before placing a new overburden wedge, which simulated additional progradation, on top of the model. The models were run for a total of 15 minutes at 2500*g* in the centrifuge.

Finally, the third set of models consisted of a buoyant 'basinward' thinning wedge of silicone with a landward dipping lower boundary (Fig. 1c). This buoyant wedge was overlain episodically by two wedges of anisotropic microlaminate units simulating prograding sediments (Fig. 1c). The overburden wedges produced a lateral pressure gradient ranging from 144 Pa at the back to 80 Pa at the front. Sequential propagation of overburden wedges resulted in basinward flow of the buoyant layer up through the stratigraphic section, and the formation of secondary sheets basinward. These models were also run for 15 minutes at 2500*g* in the centrifuge.

## Model results

Structures formed during centrifuging of these three sets of models were different. In the first set of models with tabular overburden, during centrifuging the overburden was faulted into two-dimensional blocks that subsided into the underlying source layer (Fig. 3b). Consequently, the viscous layer was segmented by the subsiding overburden blocks to diapiric walls trending perpendicular to the extension direction (Fig. 3a). These diapirs rose through the faulted overburden between the subsiding blocks (Fig. 3b) as long as the material supply was continuous from below. Further aggradation of the model buried most of these diapirs, which lacked supply from below (Fig. 3d). In map view, overburden faults formed orthogonally to the extension direction (Fig. 3a,c). In general, faulting in the overburden started at the spreading open 'basin' end of the model and progressively migrated to the back of the model (Fig. 3a,c).

In the second set of models (tabular source layer overlain by an overburden wedge), the overburden wedge displaced the source layer at the back towards the open end of the model where it formed asymmetric (basinward) diapirs and buoyant sheets that overrode the thinner overburden (Fig. 4b). Here, overburden units extended and faulted at the back and in the middle of the model, when it was folded and thrust at the front (Fig. 4). The overburden was faulted by basinward dipping normal faults at the back while it was stretched in the middle of the model (Fig. 4b). Addition of a new overburden wedge resulted in segmentation of the source layer into individual asymmetric diapirs at the back and the middle part of the model, while buoyant sheets formed at the front of the model (Fig. 4b–f). These sheets overrode their former overburdens where the latter were overturned (Fig. 4d–f).

In the third set of models, with both wedge-shaped overburden and source layer (Fig. 1c), the overburden wedge subsided at the back of the

**Fig. 5.** Third set of models. (**a**) Top view of the model after centrifuging for 7 minutes at 2500*g*. The extensional faults at the back (right) are due to subsidence of the overburden and flow of the underlying buoyant material forwards (basinwards) to rise as diapirs and form secondary sheets. Note that the faults at the back of the model extend regionally along strike while the faults close to the diapirs have limited extent. (**b**) A profile cut after the first stage of deformation showing the subsidence of the first overburden wedge at the back of the model where it has segmented the buoyant wedge and displaced parts of it forward to form a buoyant sheet that spread 'basinward'. (**c**) Close-up of the model frontal part after addition of a second overburden wedge and centrifuging. (**d**) Line drawing of (**c**). The buoyant sheet in (**b**) is displaced basinward and segmented by the second overburden wedge. Note that the overburden slab basinward to the diapir is about to thrust as it overrides the buoyant sheet. (**e**) A close-up of the frontal area of the same model after further centrifuging (without addition of any further addition of overburden wedges). (**f**) A line drawing of (**e**). This profile was cut through an area in the model away from the profile in (**c**). With further centrifuging, the overburden slab thrusts as it overrides the buoyant sheet. Below the 'salt', overburden units (layer 1) are overturned and are overridden by the forward spreading buoyant sheet. A detached block is overturned beneath the buoyant sheet. Similar areas in nature could form potential hydrocarbon traps. Part of layer 2 of the overburden is detached and forms a thrust nappe overriding the spreading buoyant sheet at the front. The extension at the back is accompanied by folding of the layered overburden at the front. (**g**) A profile cut at a later stage of deformation of a third type model showing segmentation of an initially buoyant wedge (black) by episodic addition of several overburden wedges. Overburden units at the back (right) subside and fault to push the buoyant sheet forwards and up in the stratigraphy, leaving some material behind to rise through fault zones in the overburden. Note that the faults in the overburden sole into the buoyant materials. Also, the secondary buoyant sheet acts as a new site for deformation of new overburdens. The salt stock lacking supply at the back of the model is buried by the new overburden wedge. The thrust beneath the secondary sheet represents an earlier leading edge of the sheet before it was emplaced by a new overburden wedge. (**h**) Profile of another model showing the frontal area, away from the prescribed slope, where three horizons of salt sheets are formed.

model and displaced the buoyant material forwards (basinwards) up the prescribed ramp of microlaminates (Fig. 5b,c). Subsidence of the overburden units formed regional (basinward dipping) normal faults which soled into the buoyant material and segmented it (Fig. 5b, c, d, g). Faults in the overburden formed at the back of the model at the start. With further centrifuging, the deformational front migrated basinward and younger faults formed on the basinward side of the older faults (Fig. 5 a, c, d, g). Downslope (basinward) flow of buoyant material from the back of the model to the front was accompanied by folding of the overburden in front of the leading edge of the buoyant sheet (Fig. 5c,d). These asymmetric folds evolved into thrusts with further deformation (Fig. 5e,f). New overburden wedges displaced the buoyant mass further basinward where they intruded their thinner overburden and formed secondary allochthonous sheets (Fig. 5c–h). These secondary sheets acted as new deformational sites where additional overburden faulted (Fig. 5c–h). This deformational cycle repeated itself during episodic addition of new overburden units after each centrifuging period. During each period, some of the buoyant material rose to a younger stratigraphic level and formed a step-like distribution of the buoyant material (Fig. 5h). Consequently, a basinward shallowing of the buoyant mass built up.

At the back, overburden units displayed extensional faults that soled into the buoyant mass (Fig. 5b,g). These extensional faults formed where thicker overburden sunk into and displaced the underlying buoyant material. At the front, overburden units displayed compressional structures represented by large thrust nappes overriding the forward-flowing buoyant mass at shallow levels (Fig. 5e,f). The overburden was stretched and faulted above these secondary sheets, forming detached slabs that thrust as they were overridden by the spreading sheet (Fig. 5d). Stretching of the overburden above the sheets created space for the buoyant mass to intrude as diapirs (Fig. 5c,d). During late stages of model deformation, extensional faults advanced into formerly compressed areas.

## Discussion

### *Salt structures*

Deformation styles in the models were governed by the geometry of the overburden, i.e. whether the microlaminate overburden was tabular or wedge shaped. In the first model, the horizontal buoyant sheet was segmented by the package of tabular overburden into individual wall-like diapirs that rose along fault zones in the overburden (Fig. 3). Here, no secondary buoyant sheets developed due to the segmentation of the initial buoyant layer. New overburden layers later failed above the pre-existing structures as long as flow of buoyant material continued. The lack of buoyant sheets in these models may be due to the fast aggradation rate relative to spreading rate of the diapir overhangs. However, even if the aggradation rate had been low, the diapirs would have developed symmetric instead of asymmetric overhangs that could spread basinward, driven by thicker sediments at the back. A fast spreading rate relative to aggradation rate would have resulted in formation of stacks of symmetric sheets spread at different stratigraphic levels located vertically above each other.

In the second set of models, the diapirs that formed in the middle of the model were asymmetric basinwards. The diapirs that formed at the front of the model were highly asymmetric with large overhangs, which spread basinward and up the stratigraphy (Fig. 4b–f). Lacking a prescribed slope at the base of the source layer in these models, the buoyant sheets were using their former overburdens to flow to higher stratigraphic levels (Fig. 4b–f).

Seni and Jackson (1992) and Seni (1994) distinguished four types of salt structures in the Green Canyon area that formed due to segmentation of a salt sheet: massifs, stocks, remnant-salt structures, and sheets. In the models where the 'salt' layer was prograded (second and third sets of models), two groups of diapirs were recognized according to their location in the models. Here, all the structures (stocks, massifs and remnant-salt structures) that form at the trailing edge of model salt sheets are called the first group of diapirs to separate them from the structures forming at the leading edge (second group of diapirs). As described for the structures in the northeastern Green Canyon area by Seni (1994), the first group of diapirs is separated from the trailing margin of the 'salt' sheet when differentially loaded by the overburden wedge (Fig. 4b,c, Fig. 5b,g), while secondary sheets form at the front of the model as the salt climbs up the stratigraphy and spreads 'basinward' (Fig. 4d,e, Fig. 5c,e, g, h, Fig. 6c).

The first group of diapirs, which rises along faults in the overburden units, is likely to consist of two-dimensional walls at deeper levels, but stocks at shallower levels. Because of limited material supply, these diapirs may be buried during further sedimentation and never reach the surface. These diapirs are generally bounded by a fault on their basinward side.

The second group of diapirs consists of those which rise from the leading edge of the sheets (Fig. 5b–h) and are located at shallow levels. They rise at the leading edge of the models in the compressional zone, and commonly spread secondary sheets (Fig. 4, 5 and 6). These secondary sheets are not *in situ* and could be detached from their feeder: they climb

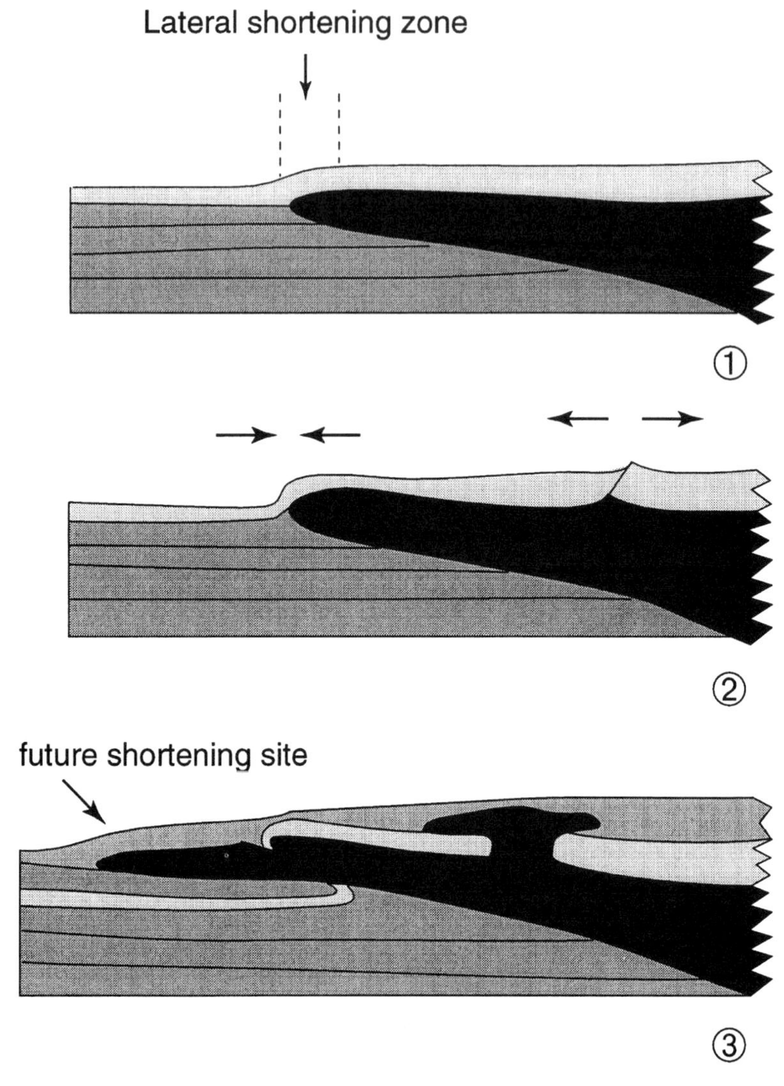


**Fig. 6.** Schematic illustrations based on model results and reflection seismic data, showing the sequence of deformation at the leading edge of a salt sheet. Overburden units extend and fault at the middle and back of the sheet while their continuation (stage 2) fold and (stage 3) thrust at the leading edge.

upwards in the stratigraphy, but down the slope laterally. Model 'salt' sheets have formed in a similar way to the model proposed by West (1989), who suggested that once a major sheet has formed, progradational loading will push out salt of the updip end of the sheet where it spreads downdip due to gravity acting on the sloping sea-level. In this way, the sheet continues to extend downdip, gradually cutting up-section as it does so. Worrall & Snelsen (1989) suggested that a significant volume of salt flows laterally over long distances due to gravity spreading.

Additional overburden wedges placed on top of the second and third sets of models loaded the buoyant material remaining beneath the back to flow further 'basinward' towards the front. These new overburden wedges covered and suppressed diapirs whose supply was exhausted and which could not overcome the yield strength of the new overburden units at the back (Fig. 5g). At the back of the model, deformation of the new increments of the overburden ceased as a weld formed when the buoyant material was displaced basinward. Similar structures in nature can be found where younger, undeformed sediments overly older, deformed sediments in areas which hosted a layer of salt at some stage.

### *Faulting*

Model results show that the two mechanisms (growth faulting and differential loading) proposed recently (Jackson & Talbot 1991) may act simultaneously to emplace a buoyant sheet. This is because growth faulting causes differential loading, and differential loading may cause faulting. Wedge-shaped overburden units fault due to subsidence of the thicker units at the back which drives the buoyant sheet to flow to areas of lower loading at the front. A wedge offers both a lateral thickness change (i.e. differential loading) and a slope, which could drive an underlying buoyant layer. The effects of slope and different models of differential loading are described in detail by Nelson (1991).

Wu *et al.* (1990a,b), Hudec *et al.* (1993) and McGuinness & Hossack (1993) distinguished two deformation zones above a salt sheet: an extension zone at the back of the sheet, and a zone of compression at the front. Seismic images in the Gulf of Mexico display these two zones associated with shallow salt sheets (Fig. 7a,b). The second and third sets of models displayed these two zones and the two groups of diapirs associated with them (Fig. 5b, c, d, g, h).

In the first set of models, overburden faulting started at the spreading edge at the front (basinward) part of the model with the least lateral support, and younger faults formed towards the 'landward' end of the model (Fig. 3). The deformation of this model is similar to the spreading of a nappe resting on a viscous substrate, where flow velocity is highest at the open end of the system (Ramberg 1986).

In the third set of models, the source layer was segmented by basinward flow. Extensional faulting started at the back of the model due to subsidence of the thicker parts of the overburden, which differentially loaded the buoyant mass and displaced it basinward. The majority of the faults dipped basinward (regional) and soled into either the mother source layer or the buoyant sheets formed due to its emplacement. The deformation front generally migrated basinward so that younger overburden faults formed basinward of the older faults. This happened when, after displacing the buoyant material, subsiding overburden units were welded with pre-'salt' strata. Seismically defined salt welds have been documented from the Gulf of Mexico by various workers (Jackson & Talbot 1991; Seni & Jackson 1992; McGuiness & Hossack 1993; Seni 1994).

Model 'regional' faults mark differential subsidence of the overburden relative to the underlying salt, which is displaced to areas of lower loading in the front part of the model. In these models, landward dipping (counter-regional) faults formed as antithetic faults to the regional ones (Fig. 5g). Basinward dipping faults are older than their antithetic landward dipping faults, and accommodate rotation of faulted overburden blocks that segment the underlying buoyant wedge. As suggested by Nelson (1991) for natural examples, rotation of model overburden units along regional faults and displacement of the underlying 'salt, alter the surface slope inherited from the initial wedge geometry of the overburden.

Model 'regional' faults may evolve and persist as growth faults. On map view, these 'regional' faults extend regionally along strike, but their equivalents close to the diapirs have limited extent around the diapirs (Fig. 5a). These faults outline the withdrawal basins at the back of the salt structures. Because 'salt' withdrawal at the middle and front of the model is more irregular and partitioned compared to a more equally distributed "basinward" flow at the back of the model, the faults here do not extend regionally.

The area between the compressional and extensional zones is where overburden units passively ride over the basinward flowing buoyant sheet and undergo no significant deformation (Fig. 5f, Fig. 6c). The movement along the down-to-the-basin master growth faults in the Gulf of Mexico is compensated mainly by salt withdrawal and partly by basinward shortening (Fig. 6; Wu *et al.* 1990b, figs 16 to 18).

In the models, the lateral shortening zone is localized above the leading edge of the 'salt' sheet (Fig. 5c, Fig. 6b). This zone is the border between overburden units that rest on the mobile 'salt' sheet and the overburden resting on non-mobile units in front of the leading edge of the sheet (Fig. 6a,b). The overburden units above the leading edge of the sheet glide basinward relative to their continuation above the non-mobile units in front of the sheet (Fig. 6b). This leads to folding of the overburden units immediately above the leading edge of the sheet (Fig. 6b). The asymmetric fold, with basinward vergence, amplifies until the overturned limb thrusts (Fig. 6c). Simultaneously, when this process occurs at the leading edge of the sheet, overburden units resting on the back of the sheet extend and fault (Fig. 6b,c). As a result, block-like nappes of overburden override the secondary sheets at the front, whilst extension dominates at the back (Fig. 5e, Fig. 6c). These structures result in repetition of the anisotropic overburden in the frontal area. Parts of the same overburden "stratigraphy" are separated by a thin layer of the buoyant material. Eventually, these nappes might sink into the secondary sheets and form a step in front of it.

Repetition of the stratigraphy by thrusting has been reported from the Gulf of Mexico by many workers (Hudec *et al.* 1993; McGuinness &

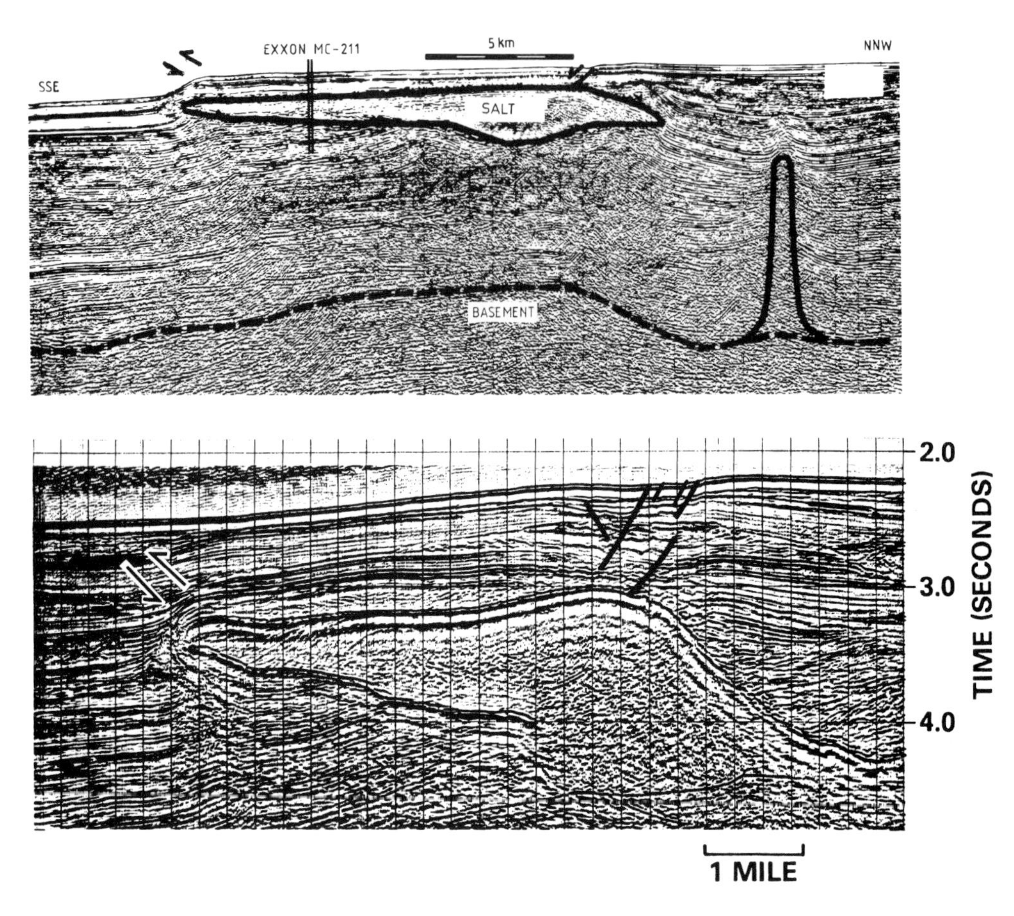


**Fig. 7.** (**a**) An orthogonal reflection seismic line, recently released by Exxon, showing a salt sheet in the Gulf of Mexico. The shallow, allochthonous salt sheet has not started to segment possibly due to relatively thin overburden with a constant thickness (after Simmons (1991)). (**b**) A seismic section across a salt sheet in offshore Louisiana. Note the extensional faults at the back of the salt sheet and the contractional structure at its leading edge (modified from Nelson, 1991).

Hossack, 1993). Moreover, Seni (1994) documented the presence of thrusts associated with remnant salt at deeper levels (5–6 s) in seismic lines of the northeast Green Canyon area. Seni (1994) interpreted these thrusts to have occurred during early development of the salt structures, which may have formed barriers to downslope movement of sediments along the glide plane of the regional faults. In the models, however, thrusts located at deeper levels (Fig. 5g) represent an earlier leading edge of the 'salt' sheet before it was emplaced basinward by further loading.

## *Base slope and sedimentation*

Thickness of overburden and the overall geometry of salt sheets influence the mode of emplacement and segmentation. Rapid and uniform accumulation of sediments results in fast reactivation of the salt sheet and formation of fewer salt stocks. Non-uniform sedimentation (turbidites and mass-flow deposits) and slow progradation of sediments may result in formation of many salt stocks. Emplacement and segmentation of salt sheets are processes closely connected in time and space. The boundary between these two processes is very diffuse because they occur simultaneously in the same allochthonous sheet.

The three models described here show that segmentation of salt sheets could be due to both progradation and/or aggradation. However, progradation of thick overburden units drives buoyant material laterally away from areas where loading is greatest towards the front of the slope, where it climbs up the stratigraphy and spreads under the

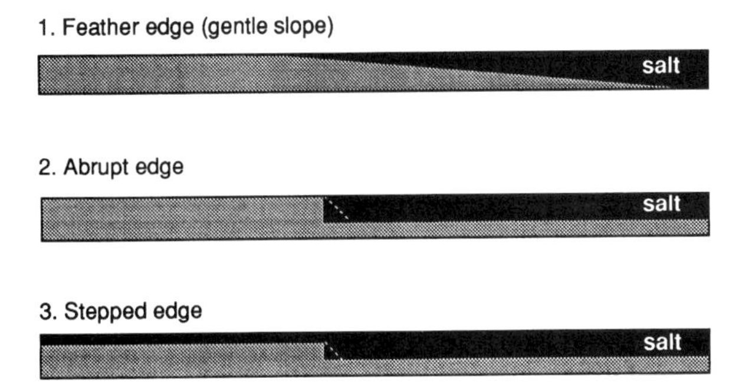


**Fig. 8.** Schematic illustrations showing three possible boundary effects that might force salt to climb up stratigraphy when loaded differentially from one end (right). Compiled from works by Jackson *et al* (1988), Koyi (1992), Koyi *et al.* (1993) and Ge *et al.* (1994).

force of gravity. As the leading edge of the sheet is emplaced, the trailing edge is segmented and welded.

Progradation of a wedge is more likely to load an underlying buoyant layer differentially than uniform sedimentation formed due to aggradation. Differential loading due to deposition of a sedimentary wedge triggers diapirs with asymmetric overhangs in an underlying salt sheet which, if it climbs up the stratigraphy, may later develop secondary sheets. The second and third sets of models suggest that differential loading is essential for the formation of secondary sheets.

The presence of a slope at the base of a buoyant sheet (as in the third set of models) or abrupt truncation of the lateral extent of the sheet promotes its emplacement (Jackson *et al.* 1988; Koyi 1992; 1993; Ge *et al.* 1994) more than in the absence of the prescribed slope (as in the second set of models). The slope promotes flow of buoyant material across the stratigraphic layering. This slope could form when salt rises and spreads faster than sedimentation that periodically truncates its lateral flow (Jackson & Talbot 1991). Thrust segments of the overburden that glide basinward over the spreading 'salt' sheet may form steps in front of a spreading sheet (Fig. 5e, Fig. 6c). Ge *et al.* (1994) emphasized the effect of a stepped salt base on the deformation style of the salt layer loaded by prograding sediments. In the flat-base salt model of Ge *et al.* (1994) the 'salt' layer extended along the model. As the overburden wedges loaded the 'salt' layer at the back, a rolling monocline accompanied and pushed the salt basinward (Ge *et al.* 1994, fig. 1a). However, if the 'salt' layer in Ge *et al.*'s (1994) model was of limited extent, the rolling monocline could have 'migrated' only to the edge of the 'salt' layer, where it would have eventually folded or thrust, for it could no longer glide on the mobile salt. In general, in the absence of an edge or a built-in step (Fig. 8), overburden units located at the edge of a 'salt' layer form a natural step along which the 'salt' climbs across the stratigraphy (Fig. 4b,d). In all the modelling attempts of salt sheets (Jackson *et al.* 1988; Koyi 1992, 1993; Talbot 1992; Ge *et al.* 1994), the presence of an edge (Fig. 8) has been a decisive factor in forcing the 'salt' to climb up through the stratigraphy. That edge could occur in nature due to thickness or facies change with the salt formation or a fault at the base of the salt layer (Koyi *et al.* 1993).

With their low permeability and large areas, allochthonous sheets of rock salt can seal huge traps. As they spread at or close to the surface or sea bottom, hydrocarbon traps associated with salt sheets are likely to be shallow. Recent discoveries of hydrocarbons beneath salt sheets of the deep Gulf of Mexico raise hopes for the future (Fig. 7a).

## Conclusions

1. Model results suggest that progradation is essential for upward movement of the sheet and formation of secondary sheets higher in the section. Presence of a slope or a step at the lower boundary of a salt sheet promotes flow of the sheet up the stratigraphy.
2. Regional faults in the overburden are more likely to form when wedge shaped overburden units subside and differentially load the buoyant material at the back.
3. Model counter-regional faults formed as antithetic faults to the regional faults.

Thanks are due to Martin Jackson for reading and commenting on an earlier version of this work and for introducing me to microlaminates, Christopher Talbot for commenting on the manuscript, and John Holder for help with measurements of material properties. Sincere thanks to Drs Mike Hudec and John Dixon for their critical review of this manuscript and many insightful suggestions. This work was carried out at the Applied Geodynamics Laboratory, Bureau of Economic Geology at Austin, and was funded by the Swedish Natural Research Council (NFR).

## References

COHEN, H. A. & HARDY, S. Numerical modelling of stratal architecture, resulting from differential loading on a mobile substrate. *This volume.*

CURTIS, M. P., FOX, J. F., RACE, R. & SENI, S. J. 1988. Deep-water Gulf of Mexico: Integrating recent ideas on salt tectonics and petroleum geology. *AAPG Bulletin*, **72**, 175.

DIXON, J. M. & SUMMERS, J. M. 1985. Recent development in centrifuge modelling of tectonic process: equipment, model construction techniques and rheology of model materials. *Journal of Structural Geology*, **7**, 83–102.

GE, H., JACKSON, M. P. A. & VENDEVILLE, B. C. 1994. Experimental initiation of salt structures and emplacement of salt sheets by prograding sediment wedges. *Salt Tectonics Meeting*, Geological Society, London, Abstracts, 51–52.

HUDEC, M. R., FLETCHER, R. C. & WATSON, I. A. 1993. The composite salt glacier: extension of the salt glacier model to post-burial conditions. *American Association of Petroleum Geologists Hedberg Research Conference on Salt Tectonics*, Bath, UK.

JACKSON, M. P. A. & TALBOT, C. J. 1986. External shapes, strain rates, and dynamics of salt structures. *Geological Society of America*, **97**, 305–323.

—— & —— 1991. *A Glossary of Salt Tectonics*. Bureau of Economic Geology, University of Texas at Austin, Geological Circular **91-4**.

——, —— & CORNELIUS, R. R. 1988. *Centrifuge Modelling of the Effects of Aggradation and Progradation on depositional Salt Structures*. Bureau of Economic Geology, The University of Texas at Austin, Report of Investigations, **173**. 93.

KOYI, H. 1992. Modelling of segmentation and emplacement of salt sheets. *Geological Society of America, Abstracts with Programs*, A363.

—— 1993. Modeling of segmentation and emplacement of salt sheets in anisotropic overburden. *Gulf Coast Section, SEMP Foundation, 13th Annual Research Conference, Proceedings*, 135-142.

——, JENYON, M. K. & PETERSEN, K. 1993. The effect of basement faulting on diapirism. *Journal of Petroleum Geology*, **16**, 285–312.

LIRO, L. M. 1989. Seismic investigation of salt deformational styles, lower slope of the northern Gulf of Mexico, extended and illustrated abstracts. *Gulf Coast Section, SEMP Foundation, Houston, 10th Annual Research Conference Proceedings*, 94–100.

—— 1992. Distribution of shallow salt structures, lower slope of the northern Gulf of Mexico, USA. *Marine and Petroleum Geology*, **9**, 433–451.

MCCLAY, K. R. 1976. The rheology of plasticine. *Tectonophysics*, **33**, T7–T15.

MCGUINNESS, D. B. & HOSSACK, J. R. 1993. The development of allochthonous salt sheets as controlled by the rates of extension, sedimentation, and salt supply. *Gulf Coast Section, SEMP Foundation, 14th Annual Research Conference, Rates of Geologic Processes*, 127–139.

NELSON, T. H. 1991. Salt tectonics and listric normal faulting. *In*: Salvador, A. (ed.), *The Geology of North America, The Gulf of Mexico Basin*, Geological Society of America, 73-89.

—— & FAIRCHILD, L. H. 1989. Emplacement and evolution of salt sills in the northern Gulf of Mexico. *AAPG Bulletin, Abstracts*, **73**, 393.

PATERSON, M. S. & WEISS, L. E., 1966. Experimental deformation and folding of phyllite. *Bulletin of the Geological Society of America*, **77**, 343–374.

RAMBERG, H. 1986. The stream function and Gauss' principle of least constraint: two useful concepts for structural geology. *Tectonophysics*, **131**, 205–246.

SENI, S. J. 1992. Evolution of salt structures during burial of salt sheets on the slope, northern Gulf of Mexico. *Marine and Petroleum Geology*, **9**, 452–468.

—— 1994. *Salt Tectonics on the Continental Slope, Northeast Green Canyon Area, Northern Gulf of Mexico: Evolution of Stocks and Massifs from Reactivation of Salt Sheets*. Bureau of Economic Geology, The University of Texas at Austin, Report of Investigations, **212**.

—— & JACKSON, M. P. A. 1989. Counter-regional growth faults and salt sheet emplacement, Northern Gulf of Mexico, extended and illustrated abstracts. *Gulf Coast Section, SEMP Foundation, Houston, 10th Annual Research Conference Proceedings*, 116–121.

—— & ——. 1992. Segmentation of salt allochthons. *Geology*, **20**, 169–172.

SIMMONS, G. 1991. Study suggests salt traps in deep Gulf. *American Association of Petroleum Geologists Explorer*, **12**, 4.

TALBOT, C. J. 1992. Centrifuged models of Gulf of Mexico profiles. *Marine and Petroleum Geology*, 9, 412–432.

—— & JACKSON, M. P. A. 1987. Internal kinematics of salt diapirs. *AAPG Bulletin*, **71**, 1068–1093.

WEIJERMARS, R. 1986. Flow behavior and physical chemistry of bouncing putties and related polymers in view of tectonic laboratory applications. *Tectonophysics*, **124**, 325–358.

—— 1992. Progressive deformation in anisotropic rocks. *Journal of Structural Geology*, **14**, 723–742.

WEST, D. B. 1989. Model for salt deformation on deep margin of central Gulf of Mexico basin. *AAPG Bulletin*, **73**, 1472–1482.

WORRALL, D. M. & SNELSON, S. 1989. Evolution of the northern Gulf of Mexico, with emphasis on Cenozoic growth faulting and the role of salt. *In*: BALLY, A. W. & PALMER, A. R. (eds), *The Geology of North America – An Overview*, Geological Society of America, 97–138.

WU, S., BALLY, A. W. & CRAMEZ, C. 1990*a*. Allochthonous salt, structure and stratigraphy of the north-eastern Gulf of Mexico. Part II: Structure. *Marine and Petroleum Geology*, **7**, 334–370.

——, VAIL, P. R. & CRAMEZ, C. 1990*b*. Allochthonous salt, structure and stratigraphy of the north-eastern Gulf of Mexico. Part I: Stratigraphy. *Marine and Petroleum Geology*, **7**, 318–333.

# Genesis of large counter-regional normal fault by flow of Cretaceous salt in the South Atlantic Santos Basin, Brazil

P. SZATMARI, M. C. M. GUERRA & M. A. PEQUENO

*PETROBRÁS Research Centre, Quadra 7, Cid. Univ., Ilha do Fundão, 21949-900, Rio de Janeiro, RJ, Brazil*

**Abstract**: The outstanding feature of the Santos Basin is a gigantic landward-dipping listric fault, about 300 km long, which trends NE parallel to the coast and is soled in the Aptian salt. This fault forms the seaward limit of a half-graben, up to 50 km wide and restricted to the post-salt sediments. In the present work we show, by physical modelling, that such a fault can result from the flow of salt in response to progradation of sediments over a spreading salt layer.

The Santos Basin is located on the continental shelf of southeastern Brazil (Fig. 1), between 23° and 29°S, occupying an area of approximately 350 000 km$^2$. This portion of the continental margin is unique because: (1) it forms the southern end of the early Cretaceous South Atlantic salt basin (Leyden *et al.* 1976); and (2) the salt basin in this segment is anomalously wide because, perhaps owing to an

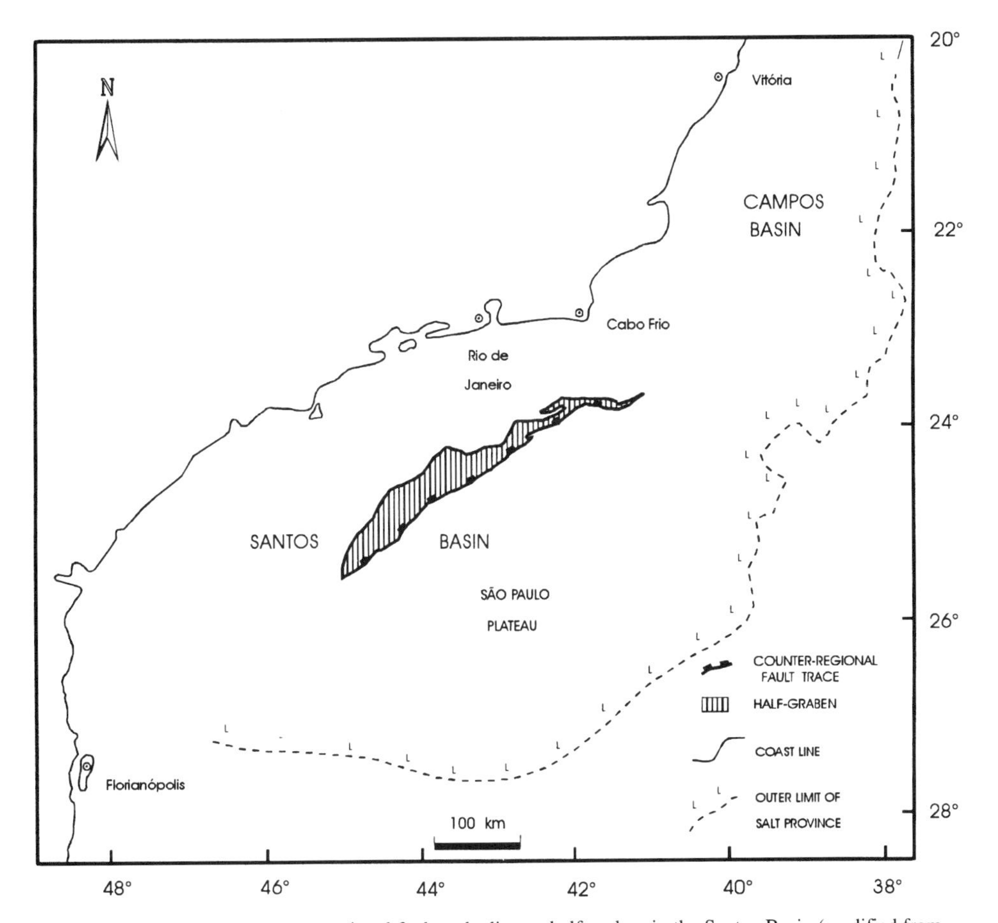


**Fig. 1.** Location of the giant counter-regional fault and adjacent half-graben in the Santos Basin (modified from Szatmari & Demercian 1993).

*From* ALSOP, G. I., BLUNDELL, D. J. & DAVISON, I. (eds), 1996, *Salt Tectonics*, Geological Society Special Publication No. 100, pp. 259–264.

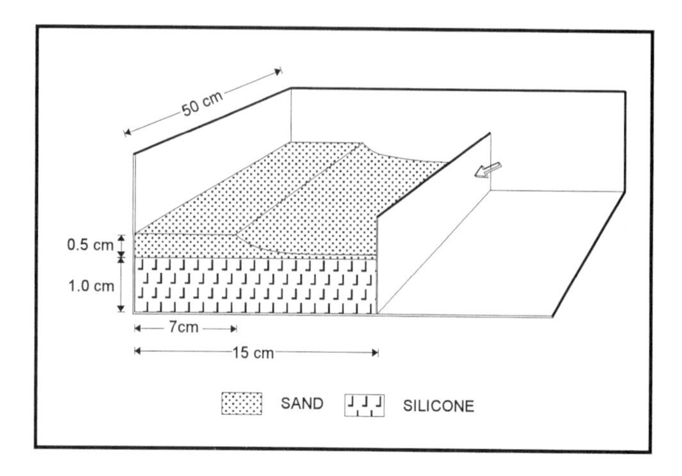


**Fig. 2.** Sketch of the model. Wall shown by arrow was removed after deposition of first sand layer.

early ridge jump during its evolution (Cande & Rabinowitz 1978; Emery & Uchupi 1984), there is no corresponding evaporite basin on the opposing margin of Africa.

The original thickness of the Lower Cretaceous (Aptian) evaporites in the Santos Basin is estimated at about 1000 m, possibly reaching 2000 m in the depocentres (Pereira & Macedo 1990). From Albian times, the salt has migrated basinwards (Cobbold & Szatmari 1991). Two distinct zones formed as a result of this flow (Demercian *et al.* 1993): a near-shore extensional zone, characterized in the overburden by synthetic and antithetic (counter-regional) normal faults, and a far-shore compressional zone, characterized by chevron folds, growth folds and reverse faults. This pattern is complicated by along-strike shortening caused by a general trend towards convergent flow in the basin (Cobbold & Szatmari 1991).

One of the most imposing features revealed by salt tectonic studies in the Santos Basin (Demercian *et al.* 1993; Szatmari *et al.* 1993*a*,*b*; Mohriak *et al.* 1996) is an ENE trending counter-regional normal fault, subparallel to the trend of the shoreline (Fig. 1). This fault, which soles in the salt, is unique to the Santos Basin, both in its length (about 300 km) and in the width of the half-graben (50 km) of which it forms the southeastern border fault.

The half-graben is restricted to the Upper Cretaceous through to Recent sedimentary sequence that lies above the salt. Growth in the thickness of sediment layers towards the border fault in the southeast shows that sedimentation was syntectonic; disturbance of the sea-floor, where intersected by the fault, shows that fault activity and sediment accumulation in the half-graben still continue today.

## Physical modelling

### *Objectives*

The occurrence of straight, extensive counter-regional faults of such length (several hundred kilometres) over a salt bed is unusual. Several hypotheses were proposed to account for its occurrence in the Santos Basin. Some of these hypotheses, including pre-existing relief on a basinward tilted salt base and reactivation of basement faults after salt deposition, were tested by physical modelling (Rizzo *et al.* 1990), but did not accurately reproduce the giant counter-regional fault.

Oceanward gliding of the overburden over salt is controlled primarily by the slope of the sediment–water interface. If the salt bed dips in the same direction as the slope, the dominant extensional structures are basinward dipping listric normal faults. In much of the Santos Basin, as documented by extensive seismic surveys, the sediment surface slopes basinward but the base of the salt is horizontal. The shape of seismic reflectors near the giant counter-regional fault suggests basinward expulsion and possibly dragging of the salt by the load of Upper Cretaceous and Tertiary sediments, up to 10 km thick, prograding from the continent.

Therefore we decided to test a model based on the oceanward escape of salt from beneath a prograding and aggrading wedge of sediments. The base of the salt layer was kept horizontal; the salt was free to spread in one direction.

### *Methodology*

Two materials were used in this modelling. The sedimentary cover was represented by dry quartz

**Fig. 3**. Plan view of top surface of model showing incipient fault (1) in the slope toe region, and small discontinuous synthetic faults (2) in the proximal region, formed at the beginning of the experiment.

sand layers (grain size < 0.4 mm, bulk density $\rho$ = 1600 kg $m^{-3}$) dyed in contrasting colours. The dry sand follows the Mohr–Coulomb failure criterion, with negligible cohesion and with an internal friction angle of 30°, which makes it a suitable analogue for brittle rocks. The evaporites were simulated by a silicone putty of almost Newtonian behaviour (GS1R Rhône-Poulenc, density $\rho = 1120$ kg $m^{-3}$, viscosity $\mu = 4 \times 10^4$ Pa s) which reacts to stress by creep, like most ductile rocks. The use of this silicone putty and sand for tectonic modelling is described in greater detail by Vendeville & Cobbold (1987) and Szatmari & Aires (1987).

A 1 cm thick silicone layer was placed over an area 50 cm long and, initially, 15 cm wide. The length and the width were chosen so as to reduce border effects and permit spreading. Over the silicone, a sand layer was deposited so as to cover the proximal half of the silicone layer with a thickness of 0.5 cm, and the distal half with a thickness of 0.2 cm. The slope between the two halves corresponds to the top of a prograding sediment wedge with an initial dip of 10–15° (Fig. 2).

After the deposition of this first sand layer, the wall from the distal, 'seaward' edge of the silicone was removed, permitting free spreading. The base of the silicone was kept horizontal so as to isolate the overburden effect by removing any flow that might be due to dip of the silicone layer. New sand layers were deposited, covering the previous sand layers and filling depressions created by extension. Like the first sand layer, the successive layers were kept thick (about 0.5 cm) over the proximal and thin (about 0.2 cm) over the distal half of the model, with a slope in between. The slope was moved slightly 'seaward' in each successive layer, simulating progradation. Total aggradation was 3.5 cm on the proximal and 1.2 cm over the distal portion of the silicone layer; the total slope progradation was 9.0 cm.

## *Experimental results*

As the silicone started spreading, discontinuous synthetic faults, with curved traces, developed in the proximal region where the sand was thicker. At the same time, a long, continuous straight fault formed at the toe of the slope, and parallel to it, but with no significant throw (Fig. 3). In the successive sediment layers, as the slope prograded basinwards, this long straight fault also migrated basinwards (7.5 cm in total), at all times forming at the toe of the new slope (Fig. 4). As the experiment progressed, displacement on the counter-regional fault increased until a maximum horizontal separation of 6.5 cm had developed (Fig. 5).

At the end of the experiment, the model was sectioned into a series of vertical slices perpendicular to the strike of the slope. The sections showed a proximal region with listric normal faults, bordered by a central region at the slope toe, where the large counter-regional fault developed. All these structures were underlain by silicone antiforms that varied from high to low amplitude pillows throughout the model, witnessing lateral flow. Under the prograding wedge, the thickness of the silicone was reduced to about a quarter of its original value. Faulting was accompanied by growth in the sedimentary cover and by block rotation (Fig. 6).

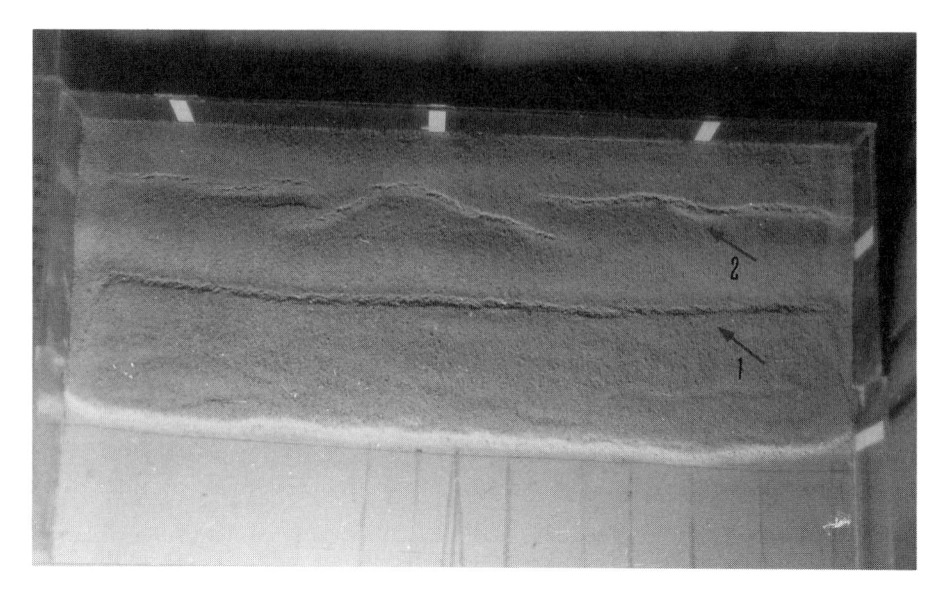


**Fig. 4.** Plan view of more evolved fault. See Fig. 3 for meaning of arrows 1 and 2.

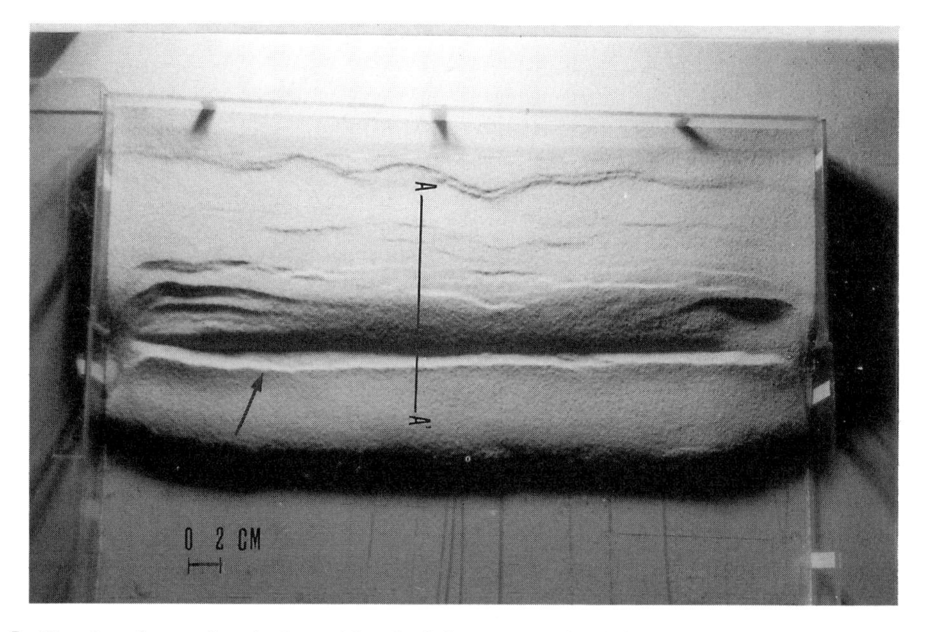


**Fig. 5.** Plan view of top surface. As the model evolved, the counter-regional character of the fault at the slope toe (arrow) became distinct. Section A–A′ is shown on Fig. 6a.

## Discussion and conclusions

In the model, the main counter-regional fault resulted from silicone flow from under the prograding wedge towards the regions where the load was smaller. The flow induced rotation of the overlying layers which became vertical in some places. Small synthetic and antithetic faults developed late in the uppermost layers of the model to accommodate this rotation.

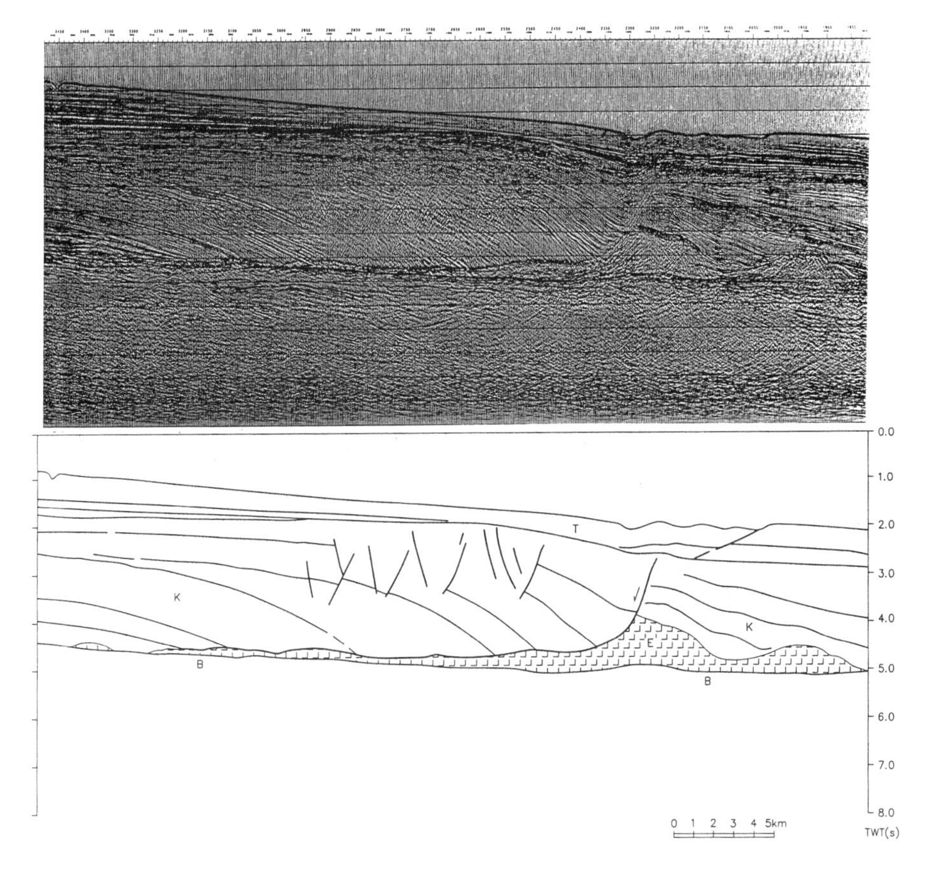


**Fig. 6.** Excellent correspondence is found between section A–A′ of physical model (**a**) (see Fig. 5 for location of section) and reflection seismic line (**b**) across the giant antithetic fault in the Santos Basin. Scale of seismic line (**b**) is given on line drawing (**c**) (after Demercian *et al.* 1993). T, Tertiary; K, Upper Cretaceous; E, Aptian evaporites; B, salt base.

The final form of the model reproduced the giant counter-regional fault observed on seismic profiles in the Santos Basin, even when taking into account the probable vertical exaggeration (1.5–2 times) of the seismic profiles. The most striking features reproduced by the model include the wide gap caused by extension in the basal layers, the steep dip and increased thickness of the layers next to the fault, the salt pillow underlying this fault, and the presence of minor synthetic and antithetic faults in the uppermost layers (Fig. 6).

The excellent match between the model and the geophysical data supports our initial hypothesis that this huge counter-regional fault formed as a result of horizontal seaward flow of a ductile salt layer from beneath a prograding wedge towards a region where the load was smaller. Once formed, the fault would act as a barrier to the seaward transport of the prograding sediments, with the sediments captured in the half-graben on the landward side of the fault maintaining its activity.

We would like to thank PETROBRÁS for permission to present and publish this paper, and referees I. Davison and M. O. Withjack for their useful comments on the manuscript.

## References

CANDE, S. C. & RABINOWITZ, P. D. 1978. Mesozoic seafloor spreading bordering conjugate continental margins of Angola and Brazil. *10th Annual Meeting, Offshore Technical Conference, Houston, Texas*, 1869–1876.

COBBOLD, P. R. & SZATMARI, P. 1991. Radial gravitational gliding on passive margins. *Tectonophysics*, **188**, 249–289.

DEMERCIAN, L. S., SZATMARI, P. & COBBOLD, P. R. 1993. Style and pattern of salt diapirs due to thin-skinned gravitational gliding, Campos and Santos basins, offshore Brazil. *Tectonophysics*, **228**, 393–433.

EMERY, K. O. & UCHUPI, E. E. 1984. *The Geology of the Atlantic Ocean*. Springer, New York.

LEYDEN, R., ASMUS, H., ZEMBRUSCKI, S. & BRYAN, G. 1976. South Atlantic diapiric structures. *AAPG Bulletin*, **60**, 196–212.

MOHRIAK, W. U. & THE CABO FRIO PROJECT WORK GROUP. 1996. Salt tectonics and structural styles in the deep water province of the Cabo Frio frontier region, Rio de Janeiro, Brazil. *In*: JACKSON, M. P. A., ROBERTS, D. G. & SNELSON, S. (eds) *Salt Tectonics: A Global Perspective for Exploration*. American Association of Petroleum Geologists, Memoir.

PEREIRA, M. J. & MACEDO, J. M. 1990. A Bacia de Santos: perspectivas de uma nova província petrolífera na plataforma sudeste brasileira. *Boletim Geociências Petrobras*, **4**, 3–11.

RIZZO, J. G., MOHRIAK, W. U., AIRES, J .R. & BARROS, A. Z. N. 1990. Modelagem física de falhamentos antitéticos em águas profundas da região de Cabo Frio na Bacia de Campos, RJ. Congresso Brasileirade Geologia, 37, Natal, 1990. *Anais. Natal, Sociedade Brasiliera Geologia*, **5**, 2238–2249.

SZATMARI, P. & AIRES, J. R. 1987. Experimentos com modelagem física de processos tectônicos no Centro de Pesquisas da Petrobras. *Boletim Geociências Petrobras*, **1**, 13–24.

—— & DEMERCIAN, L. S. 1993. *Halocinese na Bacia de Santos*. PETROBRÁS/CENPES, Relatório Interno **673.14051** (unpublished internal report).

——, ——, PEQUENO, M. A., GUERRA, M. C. M. & CONCEIÇÃO, J. C. J. 1993*a*. Origin, development and salt tectonics of the South Atlantic salt basin, Brazil. *AAPG Hedberg Conference on Salt Tectonics*, Bath, UK (Abstracts).

——, GUERRA, M. C. M. & PEQUENO, M. A. 1993*b*. Modelagem física da Falha antitética de Santos. PETROBRÁS/CENPES. Comunicação Técnica SETEC 22/93.

VENDEVILLE, B. & COBBOLD, P. R. 1987. Glissements gravitaires synsédimentaires et failles normales listriques: modèles expérimentaux. *Compte Rendu Academie Science Paris*, Sér. II: **305**, 1313–1319.

# Numerical modelling of stratal architectures resulting from differential loading of a mobile substrate

HARVEY A. COHEN[1] & STUART HARDY[1,2]

*Fault Dynamics Project, Department of Geology, Royal Holloway University of London, Egham Hill, Egham, Surrey, TW20 OEX, UK*

[1]*Present address: Institute of Earth Sciences (Jaume Almera), Consejo Superior de Investigaciones Cientificas (CSIC), Marti i Franques, s/n 08028 Barcelona, Spain*

**Abstract**: A numerical model is presented in which a viscous layer flows in response to differential loading by deltaic sedimentation. No density effects are considered; the mobile layer and overlying sediments deform solely in response to differential thicknesses of sedimentary overburden. Modelling results include the following key features: (1) sedimentary depocentres are asymmetric and spoon-shaped, with flat-lying shelf strata grading laterally into foreset slope strata at a prograding shelf-slope break; (2) the locus of deformation is controlled by the position of the shelf-slope break; and (3) a displaced bulge of mobile substrate migrates seaward faster than the shelf-slope break progrades. Using viscosities characteristic of natural salt ($\eta = 10^{16}$ to $10^{19}$ Pa s), the sensitivity of the model to several parameters is examined. Increasing the mobile layer thickness increases the rate of substrate withdrawal in a manner similar to decreasing its viscosity. A shorter sediment transport distance results in steeper depositional slopes and faster substrate withdrawal. Comparison with a Miocene example from offshore Texas demonstrates that structures qualitatively attributed to differential loading can be successfully modelled using reasonable values for sediment supply, relative sea-level rise, and substrate thickness.

Recent work on salt tectonics has been largely concerned with the behaviour of salt following its burial and mobilization in response to regional tectonics (e.g. Jackson & Vendeville 1994; Vendeville & Jackson 1992a,b) Whilst this work has added appreciably to our understanding of salt/overburden interactions, it has tended to minimize the significance of salt-related deformation during shallow burial. Natural salt does not require deep burial to acquire mobility (Weijermars *et al.* 1993) and may flow during its early burial history. Similarly, muds may become overpressured at relatively shallow depths, facilitating ductile deformation (Rieke & Chilingarian 1974; Jones 1994). Such syndepositional deformation will strongly influence the distribution of sedimentary facies and establish structures that may become modified during subsequent tectonism.

Differential loading of a mobile substrate has long been recognized as a possible mechanism for the large-scale shale tectonics observed in deltas (Bruce 1973; Dailly 1976; Evamy *et al.* 1978). In regions of salt tectonics, the presence of spectacular diapiric structures has tended to draw attention away from more subtle features; differential sedimentary loading is considered insufficient to initiate diapirs except under special conditions (Jackson & Vendeville 1994). Nonetheless, the ability of differential sedimentary loading to displace ductile substrata has been widely recognized (Jackson & Talbot 1986; Nelson 1991; Worrall & Snelson 1989). To date, there has been little rigorous investigation of how the mechanisms of deltaic progradation and salt flow might interact to influence the architecture of deltas. In this paper, we present the results of a numerical model which takes into account rates of delta progradation, thickness and viscosity of the mobile substrate, overburden deformation and sea-level variations. We assess the types of architectures that may be expected to form in response to differential loading of a ductile substrate under a variety of conditions, and compare our results to natural examples and conceptual models. Although applicable to both salt and shale tectonics, we emphasize the former, since the rheological behaviour of salt is better understood.

## Differential loading

Although it was intended solely as a model for movement of overpressured shales, Dailly's (1976) conceptual model of a progradational spoon (Fig. 1A) incorporates the key elements which characterize differential loading by delta progradation. In this model, a proximal wedge of 'sand' progrades over a distal wedge of ductile muds. The higher-density sands subside into this material and a deformed zone of ductile material is developed beneath and in

*From* ALSOP, G. I., BLUNDELL, D. J. & DAVISON, I. (eds), 1996, *Salt Tectonics*,
Geological Society Special Publication No. 100, pp. 265–273.

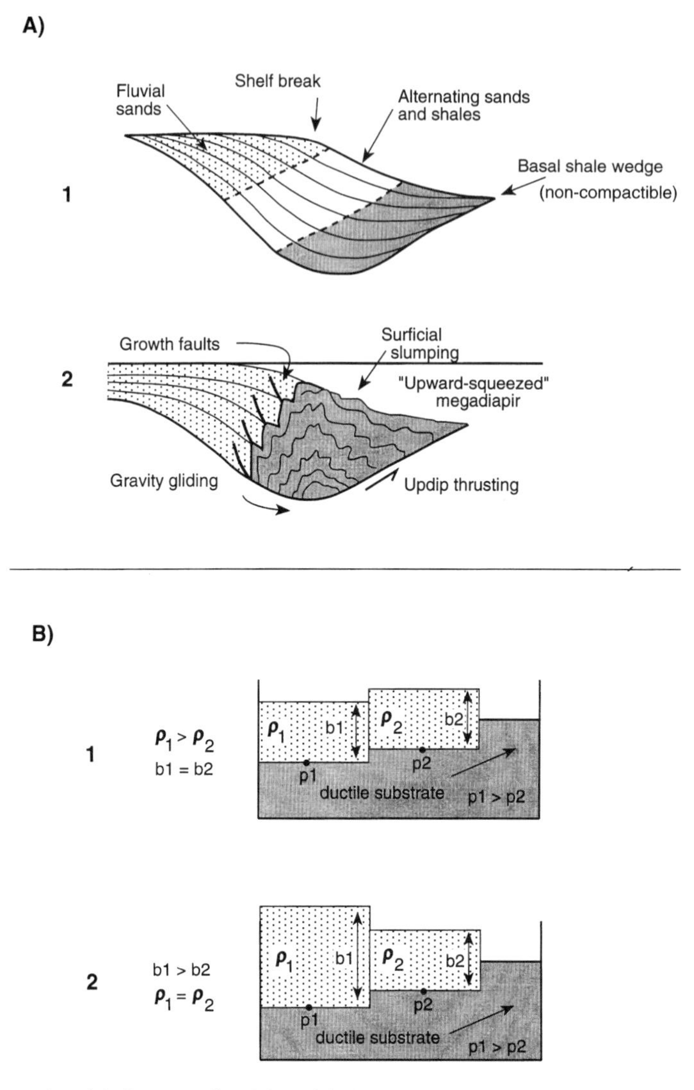


**Fig. 1.** (**A**) Conceptual model of a prograding delta, with higher density sands overtaking lower-density overpressured shales (after Dailly 1976). (**B**) Schematic diagram illustrating the two components of differential loading: (1) differential density load and (2) differential thickness of overburden. $\rho$ = density; $b$ = thickness of block. Other symbols are defined in the text.

front of the shelf-slope break. The resultant bulge is characterized by diapirism and gravity sliding, and is displaced laterally over time.

The lateral pressure differences that may cause flow of a ductile substratum can be attributed to two types of differential loading: (1) that due to a varying density of overburden, and (2) that due to a variable thickness of overburden (Fig. 1B). The classical approach to understanding salt tectonics has been primarily concerned with the former situation, in which an overburden of higher density triggers diapirism from underlying evaporites. The importance of the latter mechanism was noted by Jackson & Talbot (1986), and later demonstrated with physical models by Vendeville & Jackson (1992*b*). In these models, overburden of variable thickness caused salt analogues to flow, even when the substratum was of a greater density than the overlying sediment. This is because a Newtonian fluid with zero yield strength will flow in response to any differential pressure. A variable topographic load, e.g. a delta profile, will exert a laterally

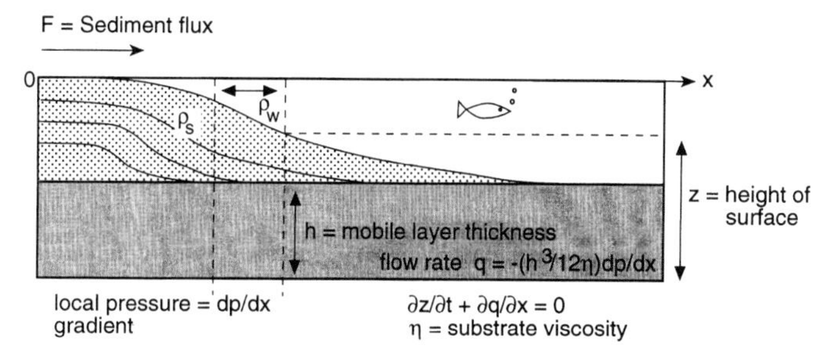


**Fig. 2.** Schematic cross-section illustrating the modelling approach. Details are given in the text.

varying load and cause a substrate of low shear strength (e.g. natural salt) to flow down even a small pressure gradient. Because the total load on the substrate is equivalent to the weight of sediment and water column, any sediment denser than the overlying water column can create a differential load when its thickness varies laterally.

Mandl (1988) showed that given a spatially varying load, if the thickness of an underlying fluid layer remains much smaller than the typical lateral dimensions of the problem and the radius of curvature of the fluid/overburden interface, then the local flow rate of the ductile material ($q$) is given by

$$q = -(h^3/12\eta)\mathrm{d}p/\mathrm{d}x \qquad (1)$$

where $h$ is the thickness of the mobile layer (in metres), $\eta$ is the viscosity of a linearly viscous mobile layer (Pa s), $p$ is the local pressure in the mobile layer (Pa) and $x$ is a horizontal coordinate. The local pressure is equal to the weight of the material column, consisting of mobile material, overburden and water column. Because it is our goal to examine differential loading rather than density inversion effects, in this analysis we consider the overburden and mobile material to have the same density. This is a reasonable assumption, as at shallow depths the density difference between normal and overpressured muds will be minimal, and typical sediments do not attain a density inversion with salt until at least 0.5 km of burial (Jackson & Talbot 1986).

Equation (1) may be combined with the equation of conservation of volume for an incompressible fluid

$$\partial z/\partial t + \partial q/\partial x = 0 \qquad (2)$$

to give

$$\partial z/\partial t = \partial((h^3/12\eta)\mathrm{d}p/\mathrm{d}x)/\partial x \qquad (3)$$

where $z$ is the height of a surface (in metres) at a fixed horizontal location and $t$ is time. Equation (3) gives the rate of change of height (i.e. velocity of the top surface) of a mobile substrate in response to differential loading. In the modelling described below, we use densities of 2200 kg m$^{-3}$ for overburden and mobile layer and 1000 kg m$^{-3}$ for water. The gravitational constant, $g$, equals 9.81 m s$^{-2}$.

## Modelling deltaic sedimentation and salt movement

In order to model deltaic sedimentation and substrate movement, we use the general tectono-sedimentary forward modelling equation of Waltham (1992):

$$\partial z/\partial t = \underbrace{[\, s - \partial F/\partial x \,]}_{\text{Sedimentary processes}} + \underbrace{[v - u\partial z/\partial x]}_{\text{Tectonic processes}} \qquad (4)$$

where $s$ is a source term (m a$^{-1}$), F is the sediment flux (m$^2$ a$^{-1}$) , $x$ is a horizontal coordinate, $v$ is the vertical velocity (m a$^{-1}$; +, uplift; –, subsidence), and $u$ is the horizontal velocity (m a$^{-1}$; +, to right; –, to left).

Waltham's (1992) equation is incorporated into a model of deltaic sedimentation previously described by Hardy & Waltham (1992) (Fig. 2). In this model, there is a flux of sediment to the sea. At sea-level the sediment settles out over a characteristic distance ($X$), giving rise to a delta profile; $X$ is the distance over which sediment load decreases by a factor of $1/e$ in the absence of erosion. Surfaces lifted above sea-level may be eroded, leading to a drop in the surface height and to an increase in the sediment load being transported. These processes are modelled in a very simple way. The sediment settling mechanism will give rise to a positive rate of change in height, $z$, which will be proportional to the sediment load; erosion will give rise to a negative rate of change in height equal to the erosion rate. Thus, in mathematical terms

$$\partial z/\partial t = aF - E \qquad (5)$$

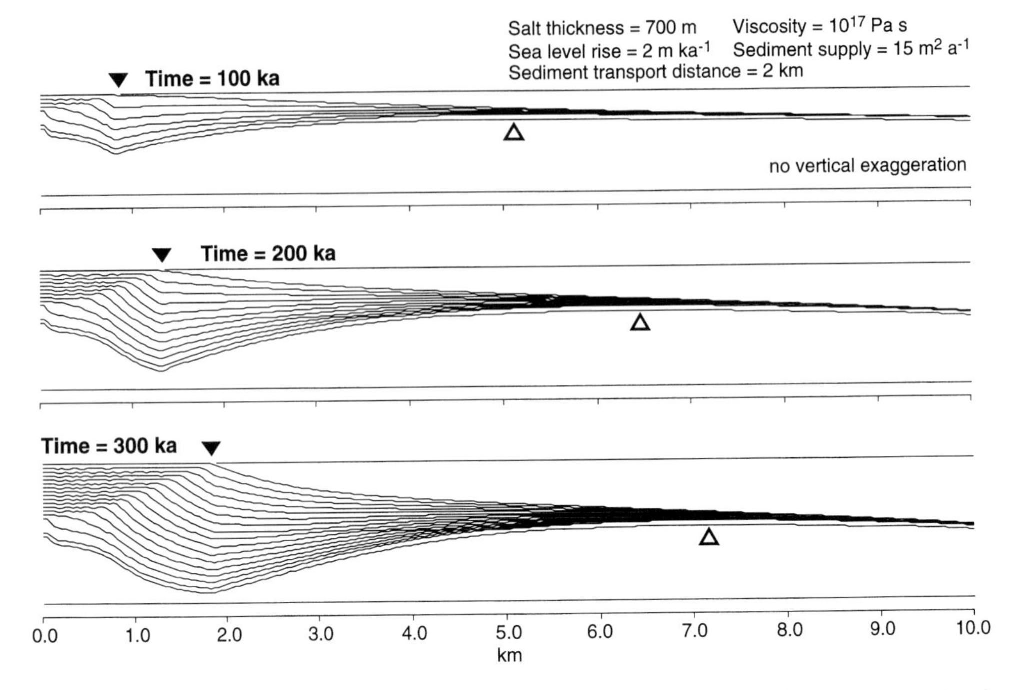


**Fig. 3.** Example of stratal evolution over 300 ka for model run with initial salt thickness of 700 m and viscosity of $10^{17}$ Pa s. Chronostratigraphic markers displayed every 20 ka. Filled triangles mark the position of the shelf-slope break; open triangles mark the apex of peripheral substrate bulge. The maximum sedimentation rate is calculated to be about 0.5 cm $a^{-1}$ at the shelf-slope break, a reasonable value for rapidly subsiding deltas.

where $a$ is a constant of proportionality and $E$ is the erosion rate. Both of these quantities are functions of position, $x$. The constant $a$ is the reciprocal of the characteristic transport distance, $X$. When the only process being considered is movement of sediment around the model, then from equation (4), if the source term $s = 0$,

$$\partial z/\partial t = -\partial F/\partial x \quad (6)$$

Combining equations (5) and (6) then yields

$$\partial F/\partial x = E - aF \quad (7)$$

Equation (7) is solved numerically to give $F$ which is inserted into equation (4) to model the evolution of the surface due to deltaic sedimentation. This sedimentation imposes a differential load upon a mobile substrate which deforms as described by equation (3). Equation (3) is substituted for the vertical velocity term in equation (4). In response to substrate flow, the overburden deforms by vertical simple shear. No horizontal motions are considered in the overburden and there are no source terms (background sedimentation). Thus we are merely considering the change in height of a surface due to sedimentation and/or substrate flow. Equation (4) is solved using explicit finite difference techniques. Compaction of the sediments with burial is not considered.

## Model results

A typical model run output is presented in Fig. 3. The physical parameters chosen for this run (e.g. viscosity of salt = $10^{17}$ Pa s and initial salt thickness 700 m) were chosen to be within the range of reported values (Jackson & Vendeville 1994; van Keken *et al.* 1993). The sediment influx (15 $m^2$ $a^{-1}$) and rate of relative sea-level rise (2 m $ka^{-1}$) are high, but were chosen to emphasize key features of the model. Similar, although more subdued, stratal architectures are produced by smaller magnitude sedimentation rates and sea-level changes (see section on natural examples).

Key features of the model output are: (1) an asymmetric, prograding sedimentary depocentre with an overall spoon shape; (2) flat-lying shelf strata which grade into foreset slope strata at a shelf-slope break; and (3) a seaward-migrating bulge of mobile substrate in front of the slope. Owing to the dependence of flow on slope angle, the locus of deformation is controlled by position of the shelf-slope break. The progradation rate of the

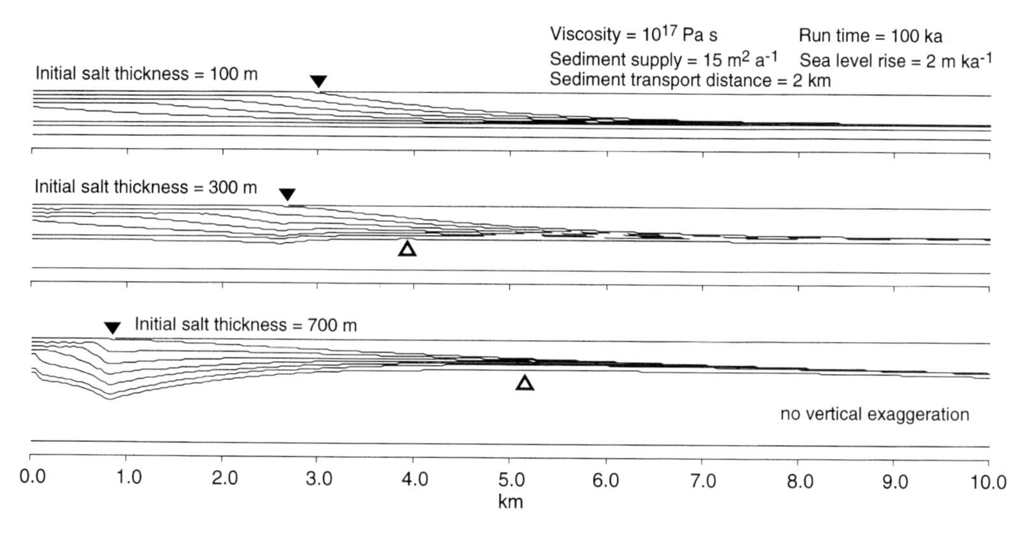


**Fig. 4.** Three model runs demonstrating the effect of variable initial salt thickness on stratal architecture. Chronostratigraphic markers displayed every 20 ka. Filled triangles mark the position of the shelf-slope break; open triangles mark the apex of the peripheral substrate bulge.

shelf-slope break decreases with time, because of the trapping of a constant sediment flux on a continually widening shelf. In contrast, the substrate flow rate depends purely upon the overlying slope and substrate thickness, and will not decrease as rapidly. Thus progradation of the shelf break lags behind migration of the mobile substrate bulge. Interaction between these two rates results in a seaward migration of the locus of subsidence and a pronounced asymmetry in the depocentre (Fig. 3).

Sediment geometries developed at each time step are determined by interplay between the sediment distribution and substrate flow algorithms. Thus the overall slope and shape of the deltaic sequence adjust to flow in the underlying substrate. For example, the upper depositional slope steepens over time (Fig. 3). Although this slope is initially determined by the sediment distribution algorithm, substrate withdrawal, which is dependent upon slope angle and substrate thickness, causes steepening of the overlying sediments. Therefore, even though the sediment distribution parameters do not change, the depositional geometry may vary with time.

## *Sensitivity analyses*

*Mobile layer thickness.* Fig. 4 shows the sensitivity of the model to variations in the initial thickness of mobile substrate. Thicker mobile layers flow more quickly than thinner layers over a given time period. Mathematically, this reflects the dependence of $\partial z/\partial t$ on $h^3$ in equation (3). In terms of final geometry, thicker initial salt layers result in greater trapping of sediment in the slope area and a more pronouncedly spoon-shaped depocentre. Note that for a mobile layer with an initial thickness on the order of 100 m, there is little or no influence upon sedimentary architecture because the flow rate is too slow relative to the sedimentation rate. The progradation rate of each delta is, however, a function of both initial mobile layer thickness and elapsed time: as salt is withdrawn from under the slope, its flow rate decreases so that the subsidence rate also decreases. Consequently, the influence of the mobile substrate on architecture of the shelf-slope break area decreases with time.

*Mobile layer viscosity.* Decreasing the mobile layer viscosity has a similar effect to increasing its thickness. This is illustrated in Fig. 5, which shows typical model geometries for the range of viscosities quoted for natural salt: $10^{16}$ to $10^{18}$ Pa s (van Keken *et al.* 1993). For highly viscous examples ($10^{18}$ Pa s), the final geometry is determined almost exclusively by the sediment supply, whereas with decreasing viscosity, deltaic progradation is progressively hindered by subsidence into the substrate. Sedimentary packages accumulating in this type of setting would tend to feature vertically stacked facies.

*Sediment transport distance.* In the model, the delta slope is controlled by a characteristic sediment transport distance. The influence of this parameter on model results is shown in Fig. 6. Shorter sediment transport distances, typical of coarser sediments, result in steeper slopes, a greater differential

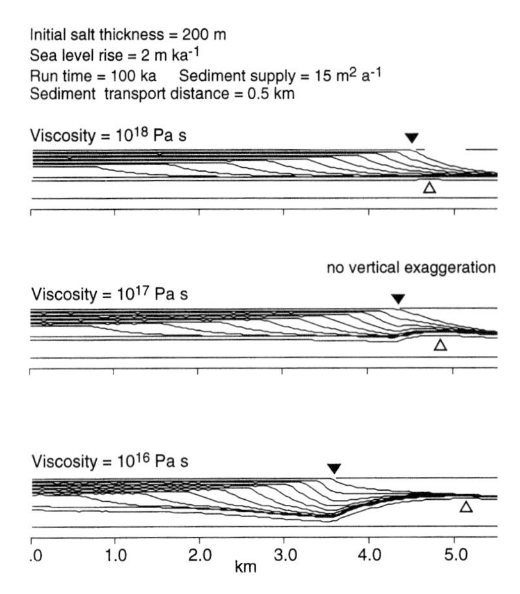


**Fig. 5.** Three model runs demonstrating the effect of variable salt viscosity on stratal architecture. Chronostratigraphic markers displayed every 20 ka. Filled triangles mark the position of the shelf-slope break; open triangles mark the apex of the peripheral substrate bulge.

load on the mobile substrate, and more pronounced subsidence at the shelf edge. Steeper slopes therefore localize the deformation associated with differential loading and result in more closely spaced shelf edge and peripheral bulge.

## *Comparisons with natural examples of differential loading*

Skeryanc & Kolodny (1983) presented an example from the offshore Miocene of the Texas Gulf Coast which they attributed to differential loading of salt and/or overpressured shales (Fig. 7A). Although the precise location is not indicated, regional constraints suggest that the Miocene section was deposited in deep water beyond the shelf break (Worrall & Snelson 1989; Galloway *et al.* 1991). Over a period of 15 Ma, the lower and middle Miocene section expanded across growth faults to greater than 2.5 times its regional thickness. Lesser expansion occurred during the remaining 5 Ma of the late Miocene. The accommodation space for expansion across these growth faults is interpreted to have been provided by seaward withdrawal of ductile materials at depths greater than 5 km. Collectively, the horizons correlated by Skeryanc & Kolodny (1983) describe an asymmetric depocentre with its greatest expansion near the south side, adjacent to the northern edge of the associated 'growth ridge' or peripheral bulge.

Figures 7B and 7C show modelling results which recreate the primary characteristics of the lower and middle Miocene depocentre. Because the asymmetric depocentre is deepest at its far (southern) end, comparison with the general model runs (Figs 3–6) suggests that some external control on sediment geometry is necessary. The preferred modelling set-up (Fig. 7B) therefore imposes topography on the top of the mobile layer which thickens towards the basin centre (such a geometry might result from an earlier phase of substrate withdrawal and/or gravity spreading). As a consequence, the initial sediment influx progrades across the model, with relatively little substrate deformation. At the base of the middle Miocene, the sediment supply is increased from 7 $m^2$ $a^{-1}$ to 11 $m^2$ $a^{-1}$ to simulate the large expansion of sediments observed at that time. The spatial coincidence of a large sediment influx with the thickest substrate (1 km) allows the greatest sediment expansion to occur approximately 35 km from the sediment input point (left side). Although the precise ratio of sediment expansion is not duplicated, this modelling demonstrates that over appropriate times and distances, and using appropriate salt viscosity values, differential sedimentary loading can produce stratal architectures of realistic scale and form.

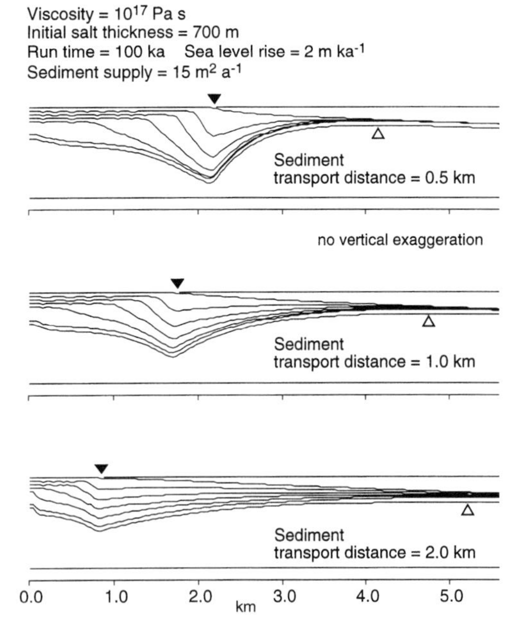


**Fig. 6.** Three model runs demonstrating the effect of variable sediment transport distance on stratal architecture. Chronostratigraphic markers displayed every 20 ka. Filled triangles mark the position of the shelf-slope break; open triangles mark the apex of the peripheral substrate bulge.

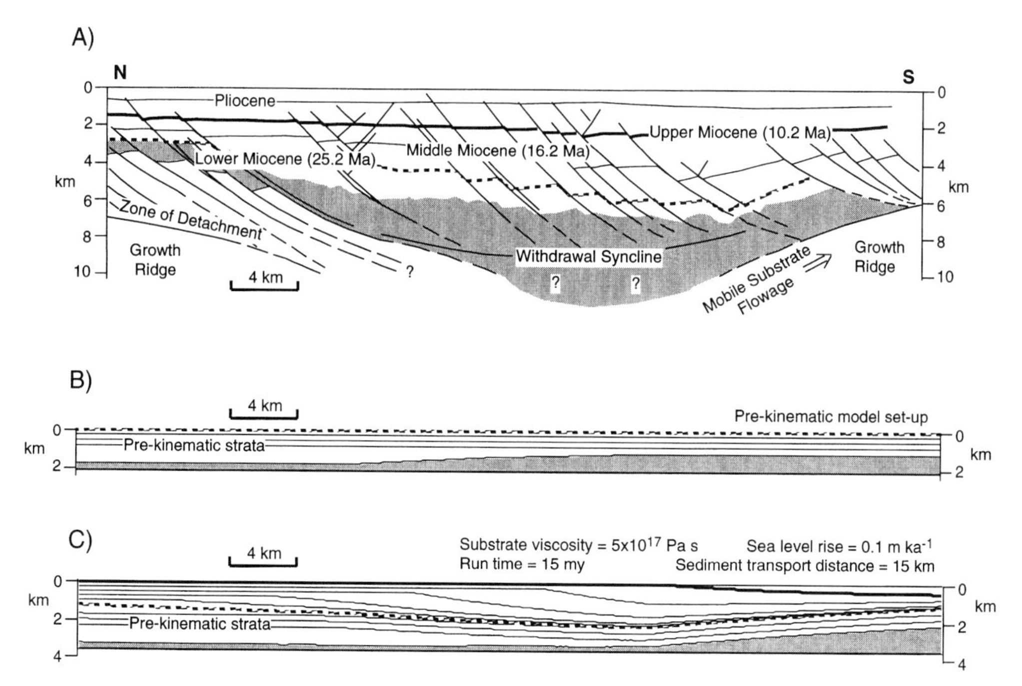


**Fig. 7.** (**A**) Depth-converted seismic line and interpretation as withdrawal syncline, after Skeryanc & Kolodny (1983). This example, from the northwest Gulf of Mexico, is constrained only through the Miocene and Pliocene. Because the identity and geometry of older strata are unclear, precise depth and thickness of ductile layers are ambiguous. (**B**) Model set-up for favoured run. Thickness of the mobile substrate increases from 400 m to 1000 m. Prekinematic strata overlying the substrate are strengthless, but contribute to a more realistic final geometry. (**C**) Results of 15 Ma model run intended to duplicate the lower and middle Miocene stratal architecture depicted in Fig. 7A. Sediment supply increased over time, with the largest pulse coinciding with the base of the middle Miocene ($3\ m^2\ a^{-1}$ at 0–4 Ma, $4\ m^2\ a^{-1}$ at 4 –6 Ma, $5\ m^2\ a^{-1}$ at 6–8 Ma, $7\ m^2\ a^{-1}$ at 8–9 Ma, and $11\ m^2\ a^{-1}$ at 9–15 Ma).

Second-order features attributed to differential loading of salt and/or shale are down-to-basin normal faults associated with deltaic sedimentation (Fig. 7A). Although our model does not incorporate the effects of faulting (overburden is assumed to have zero strength), it can be used to investigate how accommodation space might be created for hangingwall subsidence on growth faults responding to differential loading. Dailly (1976) presented a conceptual model to explain the outward-stepping succession of normal faults observed in the Texas Gulf Coast (Miocene). In his model, the mobile substratum is removed sequentially from beneath seaward-stepping growth faults creating successively younger depocentres (Fig. 8A). It is logical to assume that when accommodation space is created by salt or shale withdrawal, then the fault will cease to slip following complete withdrawal or formation of a weld (terminology of Jackson & Cramez 1989). With a simple model set-up such as depicted in Fig. 3, such successive depocentres are difficult to duplicate, however. Two phenomena interact to create this situation. (1) Substrate welds are difficult to create: because of the flow rate dependence upon mobile layer thickness, the rate of substrate withdrawal decreases with decreasing thickness, thereby making complete withdrawal nearly impossible. (2) Bulge migration outpaces shelf-slope break progradation: as previously described, a constant sediment influx across a widening shelf results in a decreasing progradation rate. Because the slope angle controls bulge migration, with a constant sedimentation rate greater accommodation space is continually created in front of the slope than is filled with sediment. This prevents sediments overrunning the bulge and initiating a new depocentre.

As a consequence, it is impossible to create Dailly's (1976) type of successive depocentres solely by differential loading with a constant sediment supply and/or constant subsidence rate. In Fig. 8b, successive depocentres were created by forcing regressions with intermittent sea-level stillstands. These stillstands increased sediment flux across the shelf, pushed the locus of sedimentation beyond the active bulge, and created a secondary depocentre. Renewed transgression created a third depocentre

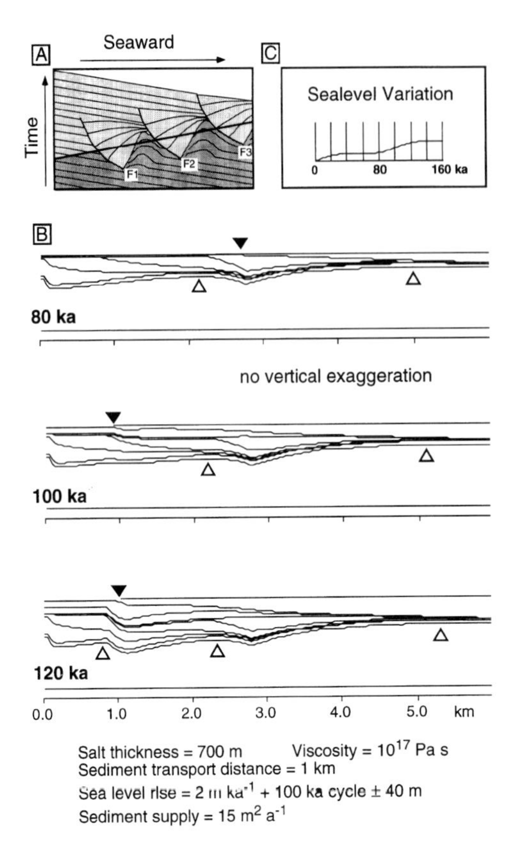


**Fig. 8.** (**A**) Conceptual model of seaward-stepping growth faults (F1–F3) (after Dailly 1976). (**B**) Numerical model results demonstrating how seaward-stepping depocentres can be formed by sea-level still-stands that force sediment beyond the currently active depocentre. Inset (**C**) depicts sea level curve for the model runs. Chronostratigraphic markers are displayed every 20 ka. Filled triangles mark the position of the shelf-slope break; open triangles mark the apex of peripheral substrate bulges. See text for discussion.

intermediate in position between the older two. These results again suggest that a combination of differential loading and external factors such as sea-level and sediment flux variations can account for observed geological structures.

## Discussion and conclusions

It is clear from this modelling that differential loading resulting from laterally varying topography is potentially important in determining the early architecture of deltas and their sedimentary facies distribution. Although our model does not consider the strength of overburden, faulting or density effects, the overall stratal geometries derived from a simple set of assumptions are consistent with natural examples. Importantly, the results show that, with reasonable values of salt thickness and viscosity, rates of salt withdrawal are rapid enough to allow the development of realistic depocentres over geological time scales.

Specifically, this modelling confirms that density inversions are not required to create large-scale movement of salt. The rate of salt movement and overall stratal architecture are dependent upon the salt viscosity, its initial thickness, and the slope of overlying sediments. A migrating bulge of substrate is produced in front of a prograding slope which, with a constant sediment supply and no lateral constraints, migrates faster than the slope itself. Upper slopes may steepen due to substrate withdrawal and lower slopes may become shallower due to substrate inflow. Sedimentary facies will therefore tend to be vertically stacked or backstep landward of a migrating bulge. Additional consequences of this bulge might include compressive structures and gravitational sliding of sediments beyond the base of slope. Buoyancy effects were not considered in this model, but would be expected to enhance the effects of differential loading, perhaps amplifying the forebulge and initiating active diapirism in regions of thin cover.

The authors wish to thank the Fault Dynamics Project (funded by ARCO British, BRASOIL UK Ltd, BP Exploration, Conoco (UK) Limited, Mobil North Sea Limited and Sun Oil Britain) for financial support during this study. HAC gratefully acknowledges additional support from Sun Oil Britain. D. Waltham and I. Davison are thanked for their thoughtful contributions. M. Jackson and I. Lerche are thanked for constructive reviews.

## References

BRUCE, C. H. 1973. Pressured shale and related sediment deformation: Mechanism for development of regional contemporaneous faults. *AAPG Bulletin*, **57**, 878–886.

DAILLY, G. C. 1976. A possible mechanism relating progradation, growth faulting, clay diapirism and overthrusting in a regressive sequence of sediments. *Bulletin of Canadian Petroleum Geology*, **24**, 92–116.

EVAMY, B. D., HAREMBOURE, J., KAMERLING, P., KNAPP, W. A., MOLLOY, F. A. & ROWLANDS, P. H. 1978. Hydrocarbon habitat of the Tertiary Niger delta. *AAPG Bulletin*, **62**, 1–39.

GALLOWAY, W. E., BEBOUT, D. G., FISHER, W. L., DUNLAP, J. B. JR., CABRERA-CASTRO, R. , LUGO-RIVERA, J. E. & SCOTT, T. M. 1991. Cenozoic. *In*: SALVADOR, A. (ed.) *The Gulf of Mexico Basin, The Geology of North America, vol. J*. Geological Society of America, 245–324.

HARDY, S. & WALTHAM, D. 1992. Computer modelling of tectonics, eustacy and sedimentation using the Macintosh. *Geobyte*, **7**, 42–52.

JACKSON, M. P. A. & CRAMEZ, C. 1989. Seismic recognition of salt welds in salt tectonics regimes. *In*: *Gulf of Mexico Salt Tectonics, Associated Processes and Exploration Potential*. Tenth Annual Research Conference, Gulf Coast Section, Society of Economic Paleontologists and Mineralogists (SEPM), 66–71.

—— & TALBOT, C. J. 1986. External shapes, strain rates, and dynamics of salt structures. *Bulletin of the Geological Society of America*, **97**, 305–323.

—— & VENDEVILLE, B. C. 1994. Regional extension as a geologic trigger for diapirism. *Bulletin of the Geological Society of America*, **106**, 57–73.

JONES, M. 1994. Mechanical principles of sediment deformation. *In*: MALTMAN, A. (ed.) *The Geological Deformation of Sediments*. Chapman and Hall, London, 37–71.

MANDL, G. 1988. *Mechanics of Tectonic Faulting*. Elsevier, Amsterdam.

NELSON, T. H. 1991. Salt tectonics and listric-normal faulting. *In*: SALVADOR, A. (ed.) *The Gulf of Mexico Basin. The Geology of North America, Vol J*. Geological Society of America, 73–89.

RIEKE, H. H. & CHILINGARIAN, G. V. 1974. *Compaction of Argillaceous Sediments*. Elsevier, Amsterdam.

SKERYANC, A. J. & KOLODNY, C. R. 1983. Syndepositional structures on mobile substrates. *In*: BALLY, A. (ed.) *Seismic Expression of Structural Styles*. American Association of Petroleum Geologists, Studies in Geology Series 15, 2.3.1-41–42.

VAN KEKEN, P. E., SPIERS, C. J., VAN DEN BERG, A. P. & MUYZERT, E. J. 1993. The effective viscosity of rocksalt: implementation of steady-state creep laws in numerical models of salt diapirism. *Tectonophysics*, **225**, 457–476.

VENDEVILLE, B. C. & JACKSON, M. P. A. 1992a. The fall of diapirs during thin-skinned extension. *Marine and Petroleum Geology*, **9**, 354-371.

—— 1992b. The rise of diapirs during thin-skinned extension. *Marine and Petroleum Geology*, **9**, 331–353.

WALTHAM, D. 1992. Mathematical modelling of sedimentary basin processes. *Marine and Petroleum Geology*, **9**, 265–273.

WEIJERMARS, R., JACKSON, M. P. A. & VENDEVILLE, B. 1993. Rheological and tectonic modeling of salt provinces. *Tectonophysics*, **217**, 143–174.

WORRALL, D. M. & SNELSON, S. 1989. Evolution of the northern Gulf of Mexico, with emphasis on Cenozoic growth faulting and the role of salt. *In*: BALLY, A. & PALMER, A. R. (eds) *The Geology of North America – An Overview*. Geological Society of America, 97–137.

# Temperature dependence of thermal anomalies near evolving salt structures: importance for reducing exploration risk

K. PETERSEN & I. LERCHE

*Department of Geological Sciences, University of South Carolina, Columbia, SC 29208, USA*

**Abstract:** The temperature distribution through time in the subsurface influences hydrocarbon generation. It is therefore of interest to model the spatial variation of temperature through time and the causes of the variation.

The thermal conductivity of salt is a factor of two to three times higher than that of typical sediments. Salt structures often display large vertical relief and so provide low thermal resistance paths for heat conduction to the surface, with focusing through salt at the expense of surrounding sediments. The focused heat re-enters the sediments near the salt structure apex so that local sediments are warmer than remote sediments at equal depth. Sediments close to salt base, and in secondary rim synclines, are cooler than remote sediments at equal depth. Around salt sheets, supra-salt (sub-salt) sediments are warmer (cooler) than remote sediments at equal depth.

Decreasing salt thermal conductivities with increasing temperature are shown to influence thermal anomaly patterns around salt bodies, while temperature anomalies around a laterally moving salt sheet are calculated to exemplify the relative effects of salt shape, depth of burial and heat flow. The spatial temperature history is vital for modelling timing and location of earlier onset of oil generation and maturation above salt structures, and of delay in conversion of trapped oil to gas in deeper sediments compared to the regional regime. The importance of modelling the thermal history in order to reduce exploration risk near salt structures is emphasized.

The presence of salt structures in sedimentary basins impacts structural and stratigraphic basin evolution. The evolution of rim synclines, collapse grabens, growth faults, unconformities and overpressure are some of the phenomena often associated with both autochthonous and allochthonous salt. The structural and stratigraphic features associated with salt movements may significantly influence hydrocarbon migration and entrapment. The thermal features of salt may influence hydrocarbon generation and maturation because the higher thermal conductivity of salt relative to typical sedimentary rocks changes the heat transfer in the vicinity of salt structures.

Thermal anomalies associated with salt have been measured in many different salt provinces. Vizgirda *et al.* (1985) reported positive thermal anomalies of up to 30°C at depths of about 4500 m on the western flank of the West Bay salt structure, offshore Louisiana. The maximum positive anomaly over the central crest can thus be expected to exceed 30°C. Above the Lulu salt structure in the Danish Central Graben there is an anomaly of about +30°C at a depth of 3600 m, becoming insignificant at the shallower depth of 1700 m (Thomsen & Lerche, 1991). Other salt structures in the Danish Central Graben exhibit thermal defocusing, resulting in positive thermal anomalies of 5–15°C (Jensen 1983*a,b*), depending on salt geometry and depth of burial. Downhole temperature measurements deep in the rim synclines are not so frequently reported. However, in the North Caspian depression, subsalt temperatures are recorded to be 20–30°C less than where salt is absent (Proshlyakov *et al.* 1986). The positive thermal anomalies above salt and negative anomalies below salt may alter the timing of initiation of oil generation and of conversion of oil to gas. Four factors are responsible for the observed thermal anomalies:

- shallow salt exhibits a thermal conductivity which is two to three times higher than that of the surrounding sedimentary rocks, causing an increased thermal gradient above salt features. The contrast in thermal conductivities allows for focusing of heat from the subsalt, and surrounding, sediments into salt, resulting in lower temperatures near the salt stem base and below salt than in the regional domain far from the salt;
- the geometry of the salt structure;
- the depth of burial of the salt structure;
- the basement heat flow (Nettleton 1955; Berner *et al.* 1972; Talbot 1978; Bishop 1978; O'Brien *et al.* 1993; Lerche and O'Brien 1994).

Mathematical modelling studies have been performed with various assumptions due to the complexity of calculating the thermal regime around a highly conductive body (Selig and Wallick 1966; Geertsma 1971; Von Herzen *et al.* 1972; Vizgirda *et al.* 1985; Jensen 1983b, 1990; O'Brien and Lerche 1987; Mello *et al.* 1994). The mathematical models are often limited by the salt geometries allowed in the calculations. A quantitative

*From* ALSOP, G. I., BLUNDELL, D. J. & DAVISON, I. (eds), 1996, *Salt Tectonics*,
Geological Society Special Publication No. 100, pp. 275–290.

procedure for modelling the combined evolution of salt and sediments self-consistently (Petersen and Lerche 1993) also allows for modelling the steady-state temperature pattern around the evolving salt, irrespective of the complexity of the salt shapes. This paper will discuss the method used and present some examples to illustrate the range of thermal anomalies associated with different salt shapes and their depths of burial.

## Calculation of thermal focusing and defocusing

The following assumptions are used for modelling the focusing and defocusing of heat:

- steady-state heat flow has been established and heat is transported only by conduction. Conductive heat transfer is a diffusive process where kinetic energy is transmitted between molecules through collision or, in the case of a solid, by phonon transfer. Convective heat transfer requires movement of a medium. If a hot fluid flows into a cold overlying region, it will heat the region. The hot fluid is less dense than the colder, overlying fluid. This situation is gravitationally unstable, implying that cool fluid tends to sink and hot fluid tends to rise. In addition, the main fluid flow is expected to be dominantly vertical due to compaction. Because salt defocuses heat near its top (resulting in a positive temperature anomaly) and focuses heat from the rim syncline sediments (resulting in a negative temperature anomaly), the gravitational instability is reduced. The contribution of convective heat transfer is considered of lesser importance than conduction and is not incorporated in the model;
- the thermal conductivities in the sediments are constant, whereas the thermal conductivity of salt will be allowed to vary with temperature (thus with depth);
- the effects of igneous intrusives and of radioactive decay within the sediments are excluded;
- the sediment surface constitutes an isotherm;
- an horizontal isotherm exists at depth, determined by a multiple of the basement depth (usually around three times the depth to the deepest part of the basement). The basement is taken to be a surface directly below base of salt. Across the depth of the deep horizontal isotherm a spatially uniform heat flux is considered. This constraint is honoured by taking the material between the basement depth and the isotherm depth to have a constant thermal conductivity;
- the temperature is continuous across the salt-sediment, basement-salt and basement-sediment interfaces. The above assumptions imply that vertical heat flux is expected in the regional domain far from the salt, because the influence of the salt is local.

In order to calculate the temperature anomaly in the vicinity of changing salt shapes embedded in a source-free medium the geometric shapes of salt and sediment lithologies must be specified at each instant of time. Furthermore, the thermal conductivities of salt and sediment lithologies, the heat flux evolution with space and time, and the sediment surface temperature with time must be defined. The calculation procedure involves solving the steady-state two-dimensional heat transfer equation by using the method of images (Morse & Feshbach 1953) where the boundary conditions are implicitly involved through the use of the mirror-symmetric Green's function. The details are found in Petersen & Lerche (1994).

## Temperature and heat flow dependence of thermal anomalies

The thermal conductivities of salt and sediments are not constant with depth because thermal conductivities alter with compaction, lithification processes and temperature; salt conductivity is expected to decrease by a factor of about a third (Kappelmeyer & Haenel 1974; Lerche 1991) over a temperature range of 0°C to 130°C. This temperature range covers approximately 4–5 km of sediments. Shale and sand thermal conductivities do not change significantly over that depth range. The reduction in salt thermal conductivity with increasing temperature (therefore with increasing depth of burial) implies that the contrast in salt-sediment thermal conductivity will decrease with depth and thus cause a reduced thermal anomaly pattern around a salt structure. The temperature dependence of thermal conductivity can be expressed as follows.

Let the salt thermal conductivity $\kappa$ vary with temperature as

$$\kappa = \kappa_o(1 + \alpha T)^{-1}. \tag{1}$$

Then specify $\kappa = \kappa_1$ on $T = T_1$ and $\kappa = \kappa_2$ on $T = T_2$ so that

$$\kappa_o/\kappa_1 = (1 + \alpha T_1) \tag{2}$$
$$\kappa_o/\kappa_2 = (1 + \alpha T_2). \tag{3}$$

Then

$$\kappa_1(1 + \alpha T_1) = \kappa_2(1 + \alpha T_2) \tag{4}$$

so that

$$\alpha = (\kappa_2 - \kappa_1)/(\kappa_1 T_1 - \kappa_2 T_2) \tag{5}$$

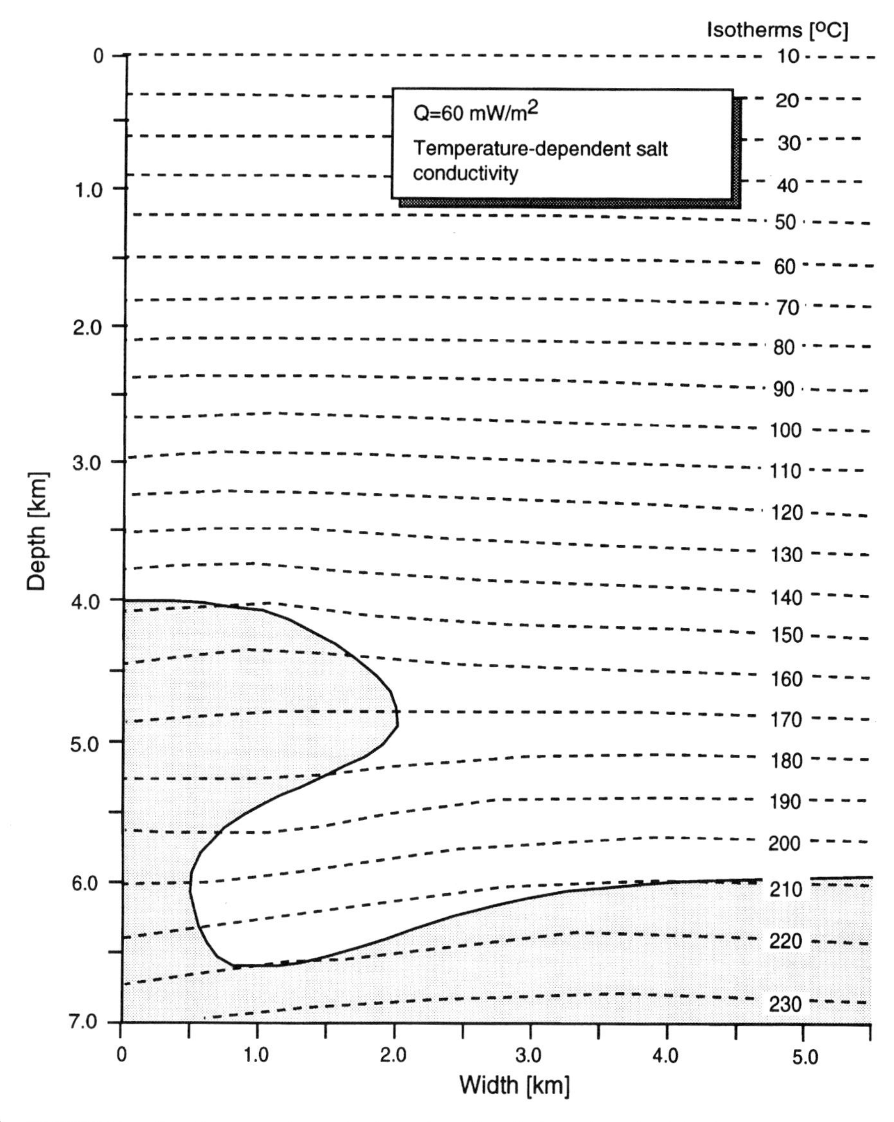


**Fig. 1.** Temperature pattern for an overhang structure where the salt thermal conductivity is allowed to decrease with increasing temperature, thereby providing a decrease in the temperature conductivity contrast between salt and sediments. The temperature gradient increases with depth in the salt and the focusing of heat into the salt is reduced. The temperature anomaly is shown in Fig. 3a.

For any given temperature, $T$, equation (1) can now be used to calculate the corresponding thermal conductivity $\kappa$. The conductivity contrast between salt, sediment and basement at $T = T_{\text{surface}}$ is 3:1:2 with the salt conductivity being 6.0 W/m K. The salt conductivity is reduced from 6.0 W/m K at the surface ($T = T_{\text{surface}}$) to 4.0 W/m K at $T = 130°C$. Figure 1 shows the temperature distribution for a structure which is buried by about 4000 m of sediments. For contrast Fig. 2 shows the temperature distribution when the salt conductivity is held constant ($\alpha = 0$). Clearly the isotherm pattern is different when including the temperature dependence of the salt conductivity. The positive anomaly at the salt crest is virtually unchanged while the negative anomaly in the rim syncline is reduced by more than 50% (compare Figs 3*a* and 3*b*). The temperature increase in the rim syncline is due to the reduction in focusing of heat from the rim syncline sediments into the salt structure. Obviously the temperature dependence of salt thermal conductivity must be taken into consideration when

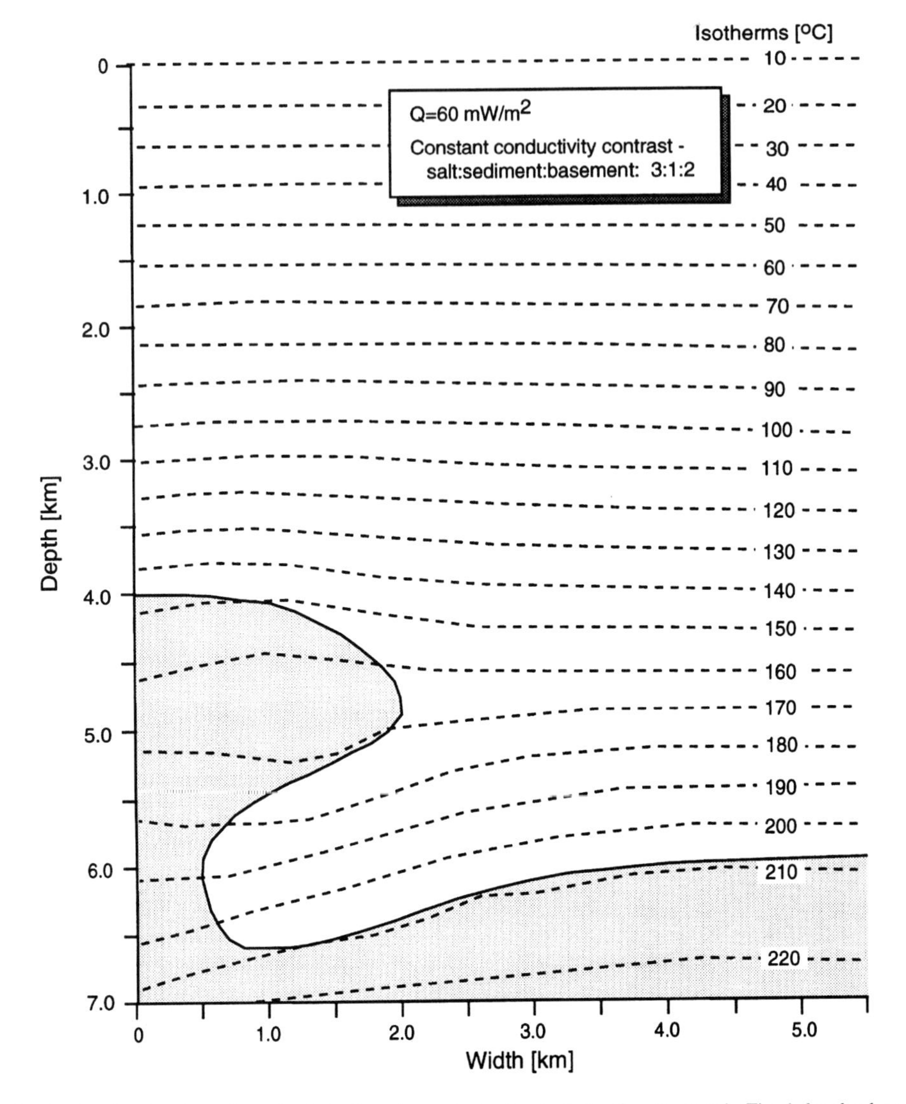


**Fig. 2.** By keeping the thermal conductivity for salt constant with depth for the structure in Fig. 1 the absolute temperature distribution is altered. The temperature anomaly is shown in Fig. 3b.

modelling the focusing and defocusing of heat for salt structures, of significance when estimating the depth for initiation of oil generation and for conversion of oil to gas.

Figure 4 shows the temperature distribution for the salt structure in Fig. 1 when the basement heat flow is reduced from $Q$ = 60 mW/m$^2$ to $Q$ = 40 mW/m$^2$. As expected, the temperature pattern is changed because of the reduction in the regional thermal gradient, but, because the change of contrast in thermal conductivities for salt to sediments is also reduced with depth, only a small change of the thermal anomalies is observed. The lowering of the heat flow, and thus of the thermal gradient, implies that a larger contrast in salt-sediment conductivity is found at the same depth compared to the situation with a higher heat flow. The importance is clear of estimating not only present-day heat flow but also the heat flow variation with time when modelling the temperature history experienced by the sediments. While heat flow in the past can be modelled using thermal indicators (such as vitrinite reflectance) that is not the purpose of this paper; rather the emphasis here is to point out the effects of varying some of the input parameters when modelling the thermal pattern around anomalously conducting shapes.

Salt sheets in the Gulf of Mexico have claimed

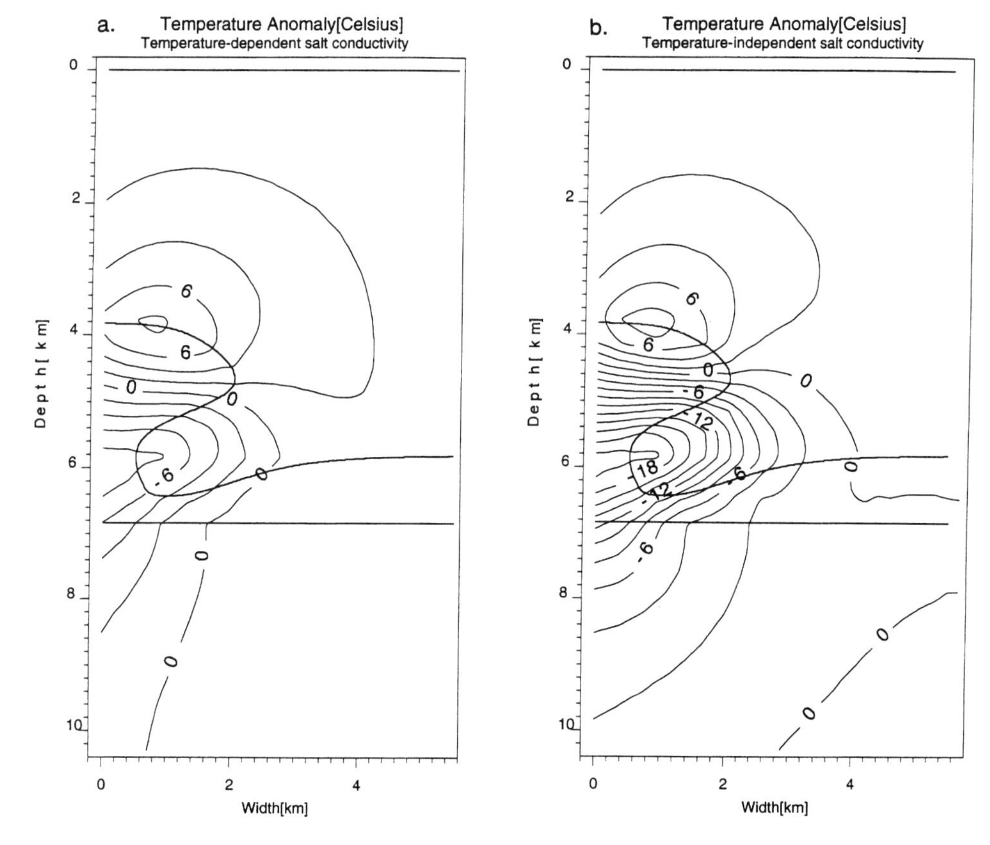


**Fig. 3.** (**a**) The temperature anomalies around a salt structure when allowing the salt thermal conductivity to change with temperature. (**b**) The temperature anomalies when keeping the salt thermal conductivity constant. The defocusing of heat has resulted in similar positive anomalies above the crests (**a**,**b**) whereas the negative anomaly in (**a**) is significantly smaller than in (**b**).

renewed interest (*AAPG Explorer,* January 1994) after the subsalt Mahogany discovery at Ship Shoal 349. The subsalt play is expected to possess a great hydrocarbon potential because:

- the salt acts as a perfect seal;
- subsalt sediments are often highly fractured;
- overpressure buildup results in preservation of porosities higher than in the regional domain;
- the focusing of heat may delay the conversion of oil to gas.

The temperature patterns around salt sheets have been modelled previously by assuming salt bodies of infinite or semi-infinite length (O'Brien & Lerche 1988). When the length-to-thickness ratio is less than about ten, the boundary effects of the feeder stock and the nose of the salt become significant. Figure 5a illustrates a two-dimensional model of a salt sheet/tongue with a lateral extent of about 7.5 km and an average thickness of about 1200 m. The geometry of the isotherms in the lateral part of the sheet structure shows an interesting effect of the focusing and defocusing of heat. Heat focused into the stem will seek to stay in the highly conductive region, if possible, and is therefore focused towards the salt sheet from the stem, as indicated by the slope of the isotherms (see Fig. 5a). In addition, heat is focused by the nose of the salt sheet. Part of the heat is quickly defocused into the overlying sediments, while part of the heat is guided towards the central part of the sheet. The lateral components of heat, flowing from both the right and left ends of the salt sheet, 'widen' the isotherms in the central portion of the sheet, causing a further lowering of the thermal gradient in that area. The sediments below the central portion of the sheet thus experience an even lower temperature than sediments close to the salt stem.

The focusing of heat influences the thermal gradient about 2 km laterally ahead of the salt nose. The maximum positive thermal anomaly is found above the salt stem and is about 16°C (Fig. 5b). The negative anomaly immediately below the central

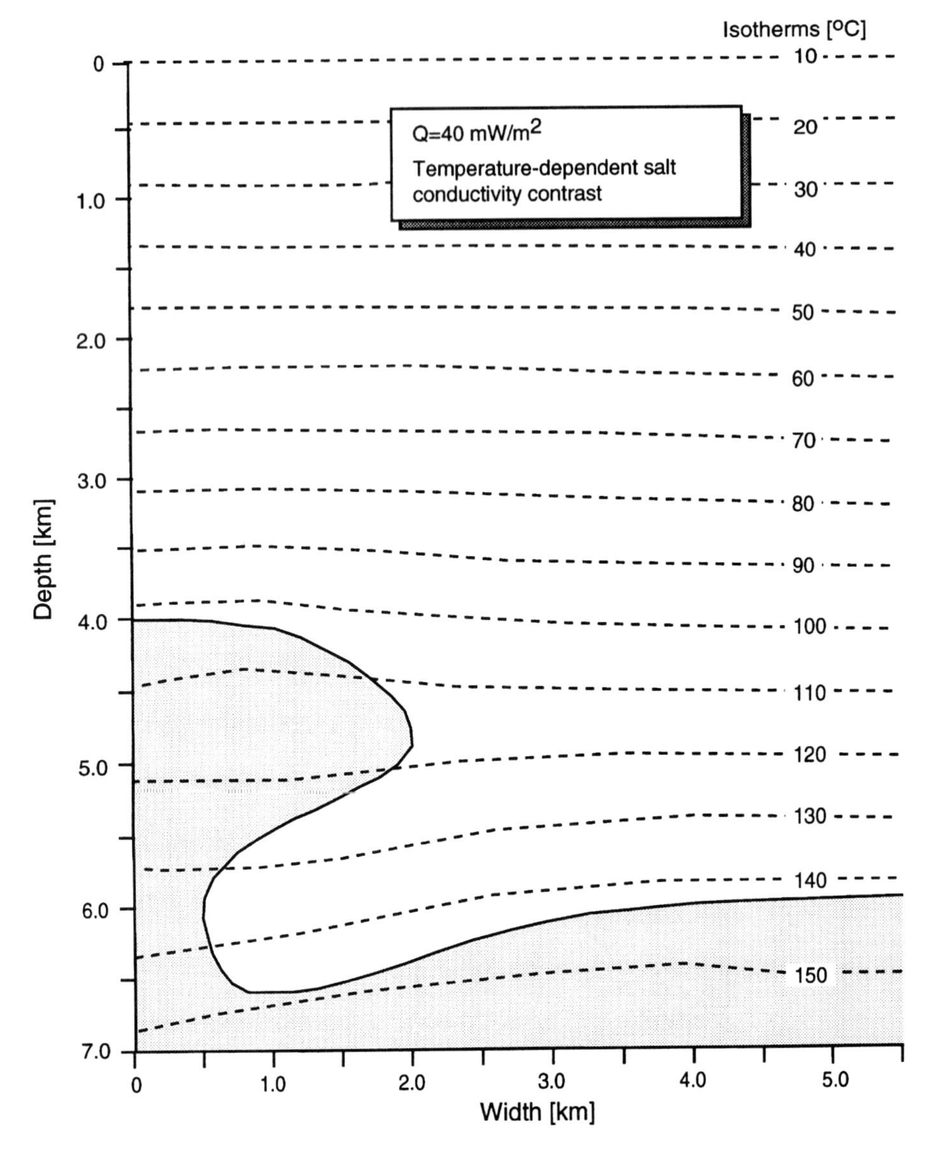


**Fig. 4.** The absolute temperature pattern is changed when the heat flow is reduced from 60 mW/m$^2$ (Fig. 1) to 40 mW/m$^2$ and a higher contrast in thermal conductivity is preserved with depth.

part of the sheet is about 10°C, which is significant for the preservation of trapped hydrocarbons below salt sheets. For instance, when oil is converting to gas far from the salt, oil at the same depth below the salt sheet will be able to undergo about 300–400 m further burial (in the current example) before the conversion will be initiated. A thicker salt sheet will amplify these features.

When the same structure (Fig. 6*a*) experiences a heat flow of 40 mW/m$^2$ (compared to 60 mW/m$^2$ in Fig. 5) the regional temperature gradient and the thermal anomalies are reduced. The anomaly above the salt stem is reduced from a maximum of about +16°C (Fig. 5b) to a maximum of about +10°C (Fig. 6b). The reduction is essentially proportional to the change in heat flow. This result is useful because a quick estimate can be made of the thermal effects of varying the heat flow, once the thermal pattern is calculated for one heat flow value. The rim syncline anomaly is less sensitive to the change in heat flow (see also discussion of Fig. 3).

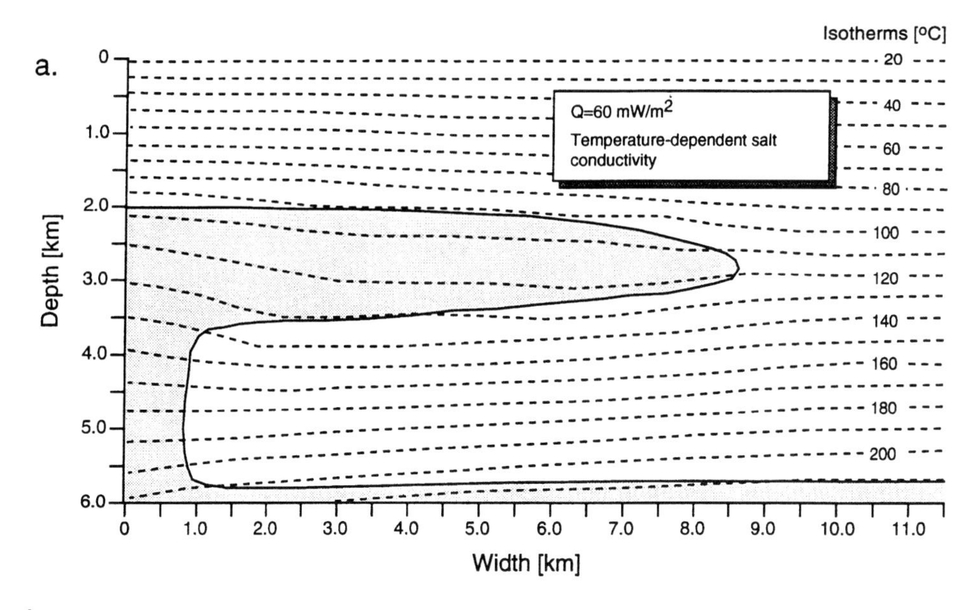


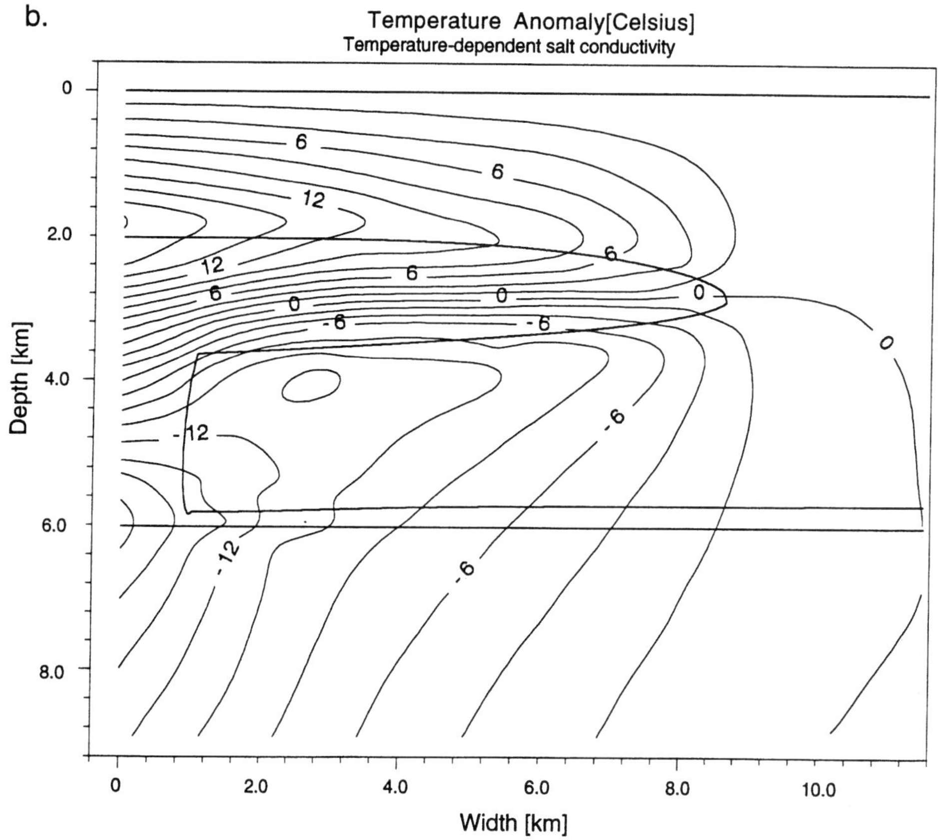


**Fig. 5.** The salt structure resembles the salt sheets often observed in the Gulf of Mexico. (**a**) The temperature interior and exterior to the salt sheet reflects the complicated pattern of focusing and defocusing of heat. (**b**) The thermal anomalies show maximum defocusing above the maximum thickness of salt and maximum focusing near the base of the salt stem. See text for discussion.

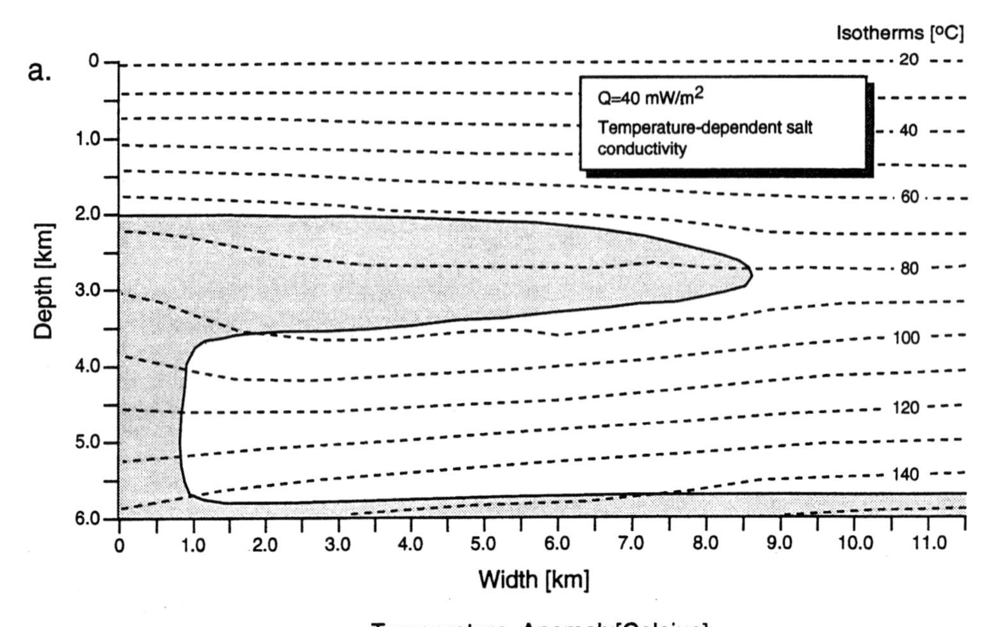


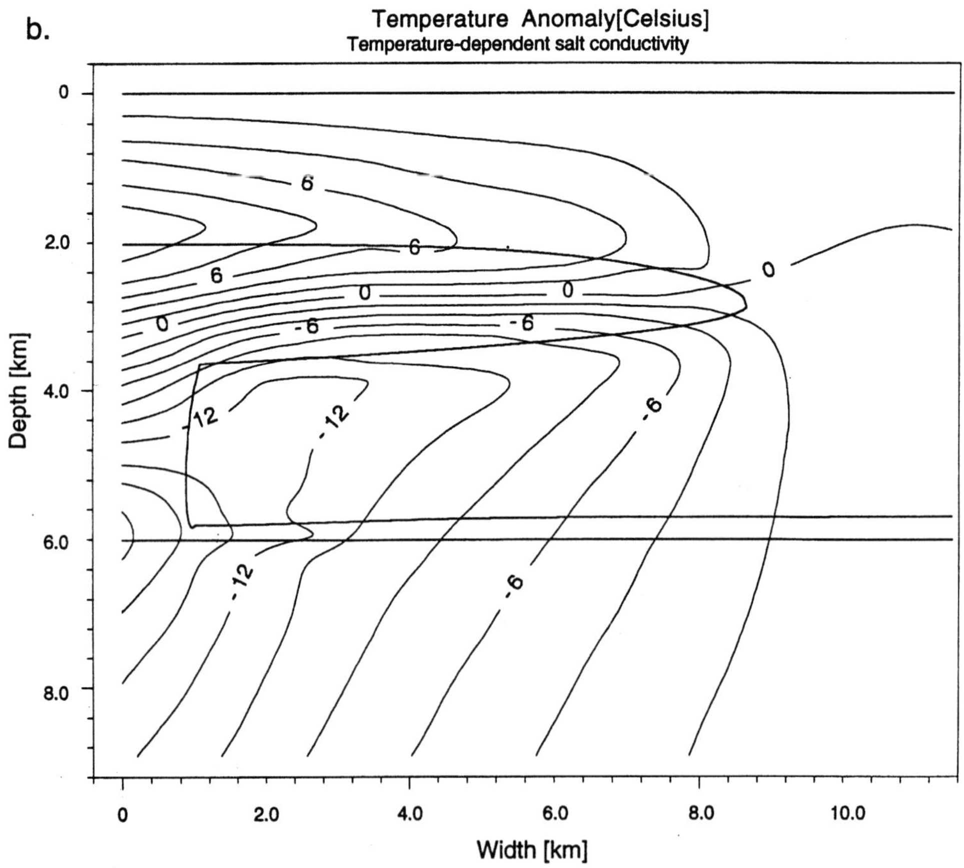


**Fig. 6.** (**a**) A reduction of heat flow from 60 mW/m$^2$ (Fig. 5) to 40 mW/m$^2$ results in (**b**) a reduction of the recorded thermal anomalies, essentially proportional to the change in heat flow.

## Thermal histories around a moving salt sheet

When salt rises through, and past, sedimentary formations the sediments will experience a temperature history different from sediments far from the salt. Depending on depth of burial, the dimensions of a salt sheet and the speed with which a salt sheet moves, the sediments above and below the salt sheet will be exposed to anomalous temperatures for shorter or longer periods of time. The effects of burial, and of the spatial variation of the salt-sediment contrast in thermal conductivity, superpose and are the main controls, together with the heat flow, in determining the temperature history. Figures 7*a–d* show the evolution of an overhang structure to a sheet-like structure and the associated temperature pattern with time (from $t = t_o$ to $t = t_3$) when the sedimentation rate is set equal to the lateral growth rate of the salt sheet. In order to estimate the temperature history for individual sediment particles the deformation history of the sedimentary beds must be modelled, which provides information on the positions of the sediment volumes around the salt sheet in time. Instead of recording the temperature history for different sediment parcels here we examine the temperature history as experienced at different *locations* (Figs 7 & 8), fixed with respect to basement.

The salt overhang moves laterally at a constant rate from times $t = t_o$ to $t = t_3$. Figures 7*a–d* show four locations (fixed with respect to the base of salt) that are followed in time with respect to their temperatures. Initially (Fig. 7*a*) locations A1 and B1 are in the domain of salt influence and locations A2 and B2 are chosen where the thermal gradient is close to the regional gradient far from the salt. As the salt sheet extends laterally with time, the salt nose moves closer to locations A2 and B2 (Figs 7*a–b*). Thereafter the sediments between locations A2 and B2 are replaced by highly conductive salt, resulting in a decrease in the thermal gradient between the two locations (Figs 7*c–d*). At present-day ($t = t_3$) locations A1 and A2 are positioned above the salt sheet whereas locations B1 and B2 are situated below the salt sheet (Fig. 7d). The relation between the temperatures at locations A1, A2, B1, and B2 and the regional temperature at the corresponding depths through time is displayed in Fig. 8 where the absolute temperature anomaly at each location is plotted against a timescale with time $t$ normalized as $(t - t_o)/(t_3 - t_o)$. Figure 8 shows that as the salt sheet increases in lateral extent, the heat flux from the salt stem deviates more towards the salt sheet, causing an increase in the defocusing of heat near the position of location A1, resulting in an increase of the absolute temperature anomaly $\Delta T = T_{obs} - T_{reg}$ with time. Secondly, at location A2, $\Delta T$ increases with time due to the increase in the defocusing of heat caused by the insertion of the salt sheet. Thirdly, $\Delta T$ decreases with time for both locations B1 and B2. Location B2 was already influenced by the salt at $t = t_o$ (Fig. 7*a*) because the extent of the salt was sufficiently large to cause a lateral component of heat flow towards the salt from that location (B2). The increased burial of the salt implies a reduction in the salt–sediment contrast in thermal conductivity and thus a reduced focusing of heat with time for locations B1 and B2. The negligible present-day anomaly beneath the salt sheet would not have been expected if the thermal conductivity of salt had been held constant, which would have resulted in an incorrect negative thermal anomaly of about 30°C.

In order to evaluate the effect of depth of burial on the thermal anomalies the same salt sheet evolution is modelled when the sedimentation rate is about 1/4 of the lateral growth rate of the salt sheet. Figures 9*a–d* show the salt and temperature evolution from times $t = t_o$ to $t = t_3$. Comparison with Figs 7*a–d* shows that the temperatures experienced by the sediments around the salt (Figs 9*a–d*) are increasingly lower through time due to the lesser depth of burial. Figure 10 shows the absolute temperature anomaly with time for locations A1, A2, B1, and B2 (see Fig. 9a for initial locations). Figure 10 displays a similar evolution pattern for locations A1 and A2 as in Fig. 8 whereas the evolution pattern is different for locations B1 and B2. Because the salt sheet in Fig. 9 is shallower than the salt sheet in Fig. 7, the focusing effect of the salt with time remains larger in Fig. 9, as indicated by the magnitude of the negative temperature anomaly with time for locations B1 and B2. The depth of burial, and thus the temperature gradient in the salt, does not cause the same reduction in the thermal conductivity of salt.

A higher rate of sedimentation thus causes a decrease in the focusing of heat below the salt, while a lower rate of sedimentation allows for a relative increase in the negative temperature anomaly below the salt sheet. Depending on the spatial geometry and evolution of the salt sheet, and on the sedimentation history, the sediments underlying the salt sheet are able to keep trapped oil from being converted to gas to a greater depth than sediments far from the salt. The effect decreases with increased burial depth and/or increased heat flow values. Thus, sediments below salt sheets may possess greater hydrocarbon potential due to the temperature and burial history and, in addition, also because the low permeability of salt permits a more perfect seal for potential traps below the salt.

The expected positive anomalies over salt structures increase the maturation level. Thomsen & Lerche (1991) report excess vitrinite reflectance

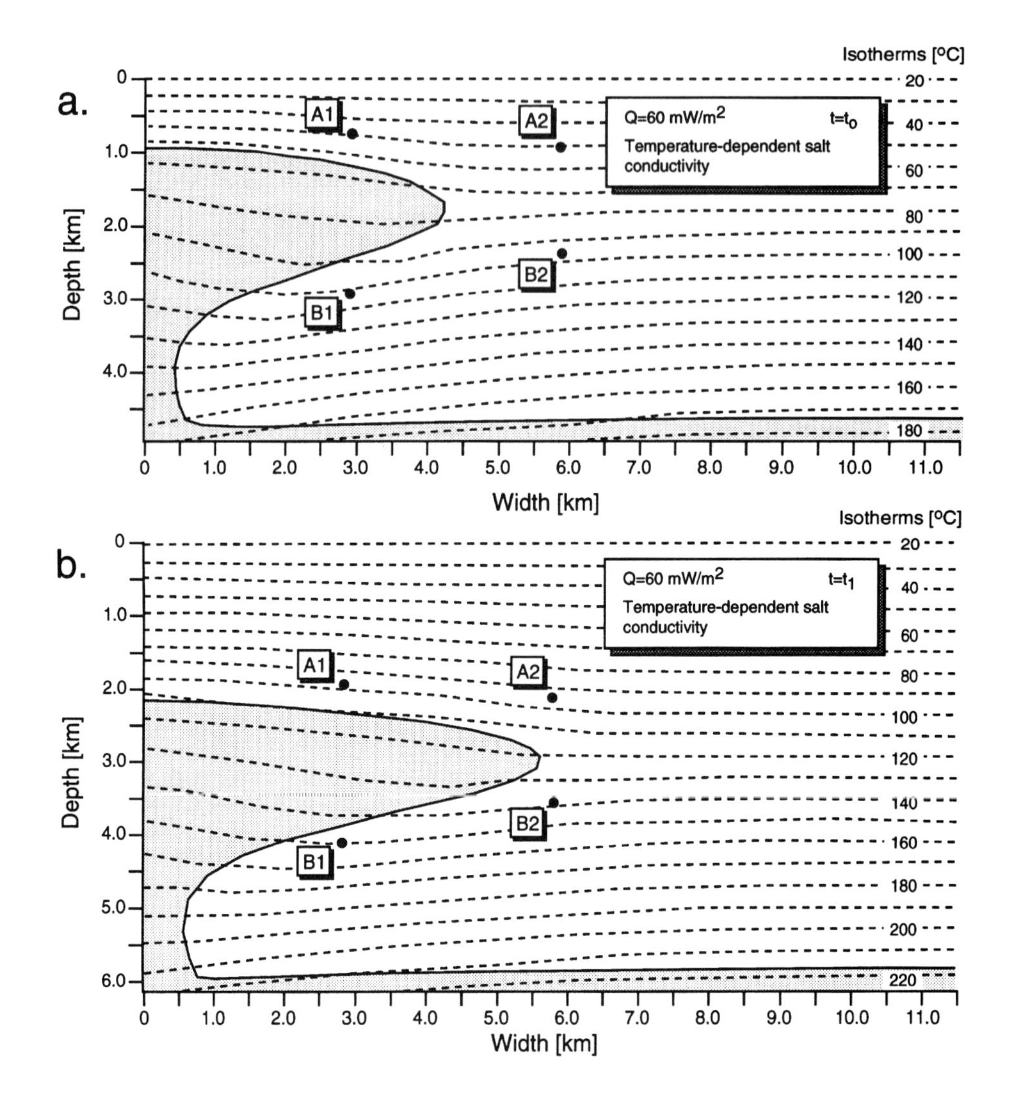


maturity of 0.2 at 3600 m depth decreasing to zero at 3000 m above the Lulu salt structure in the Danish Central Graben. Based on geochemical parameters, Rashid & McAlary (1977) discuss the higher maturity of organic matter above salt compared to far from the salt at comparable depths. The expected negative temperature anomalies beneath salt overhangs and near the salt base must in turn restrain the maturation level by lowering the position of the base of the oil window. Figure 11 shows the oil window for the salt sheet in Fig. 5, where the onset of oil generation is set at the depth where $T = 90$°C, and the conversion of oil to gas at $T = 140$°C. The oil window is raised about 400 metres above the salt and lowered about 400 metres below the salt compared to the regional situation. The hydrocarbon maturity around salt structures can be evaluated by applying the source rock richness and a kinetic breakdown model for kerogen converting to oil and gas.

In the examples presented only the temperature-dependence of the thermal conductivity of salt has been considered. However, the thermal conductivities of sand and shale do not remain constant with temperature. For instance, sand matrix thermal conductivity decreases by about 30% over a range of 0°C to 300°C. Conversely, the thermal conductivities of the sediments increase with compaction due to the loss of low conductivity pore fluid. Including the temperature and compactional effects, the thermal conductivity for sand will change from about 2.5 W/m K at the surface to about 3.1 W/m K at 5 km, assuming a heat flow of 60 mW/m$^2$. The thermal conductivity of shale will change from

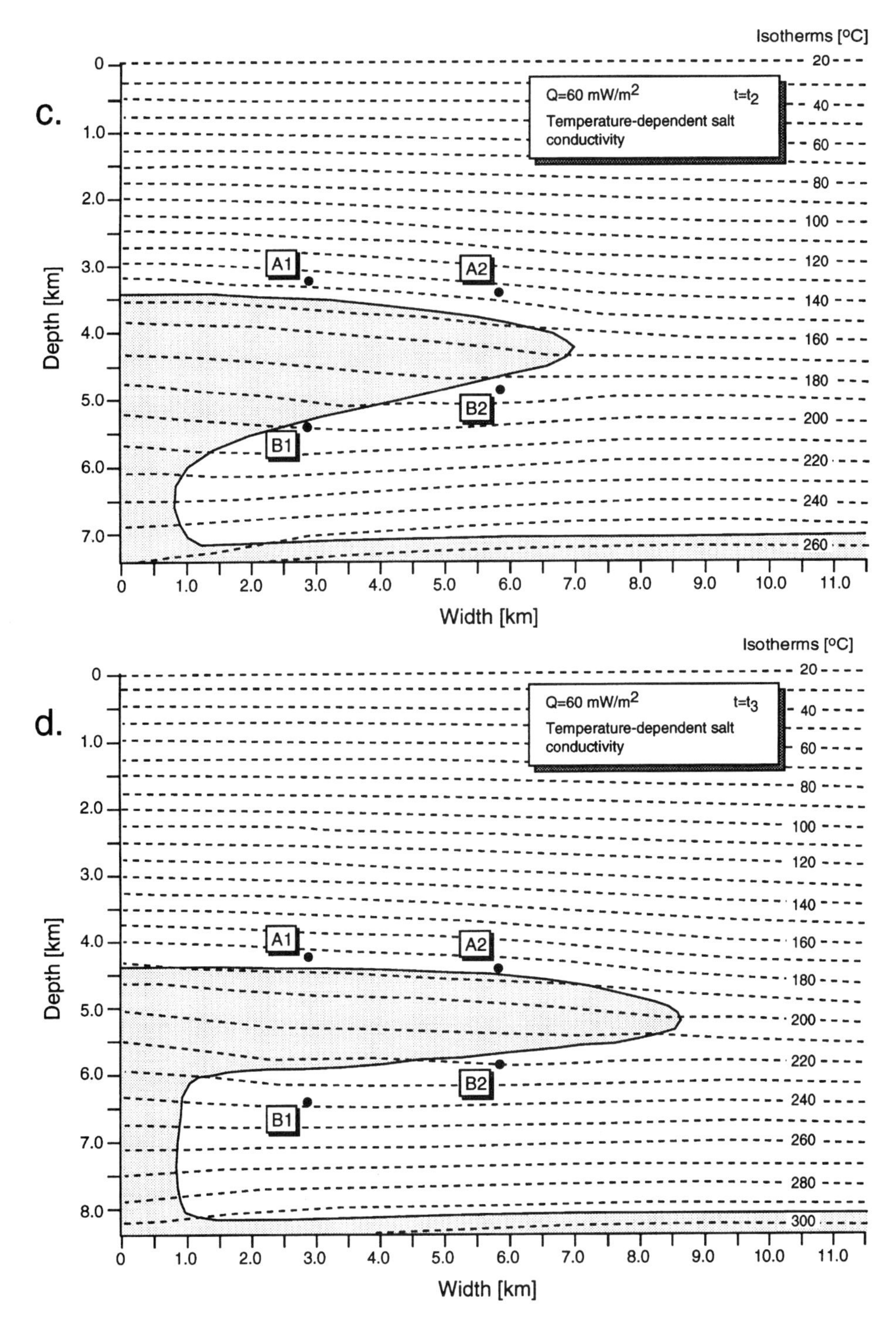


**Fig. 7.** (**a**)–(**d**) show the evolution of a salt sheet from times $t = t_0$ to $t = t_3$ with the sedimentation rate equal to the lateral growth rate of the sheet. The temperature anomaly histories through time of locations A1, A2, B1, and B2 are shown in Fig. 8.

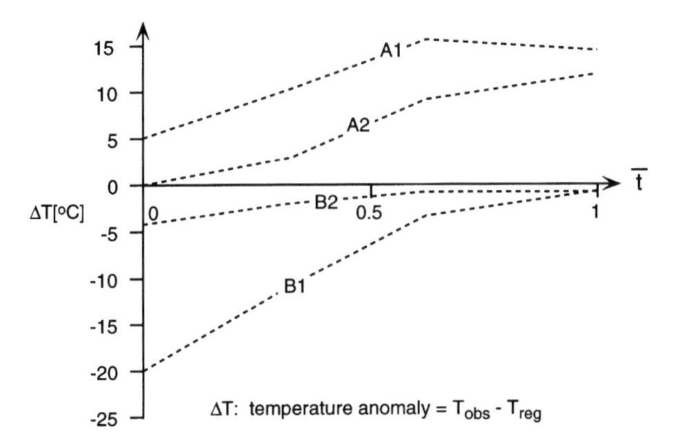


**Fig. 8.** Temperature history at locations A1, A2, B1 and B2 (Fig. 7). The absolute temperature anomaly at each location is plotted against time, $t$, normalized as $(t - t_o)/(t_3 - t_o)$. See text for discussion.

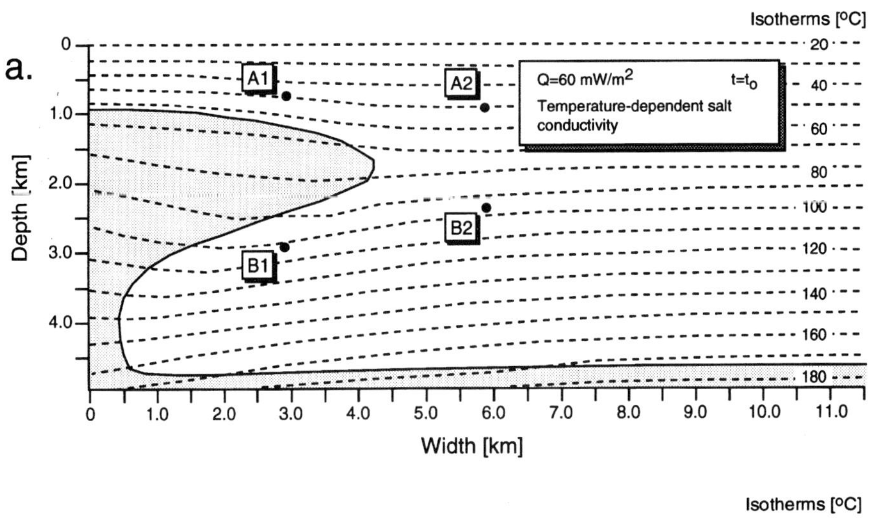


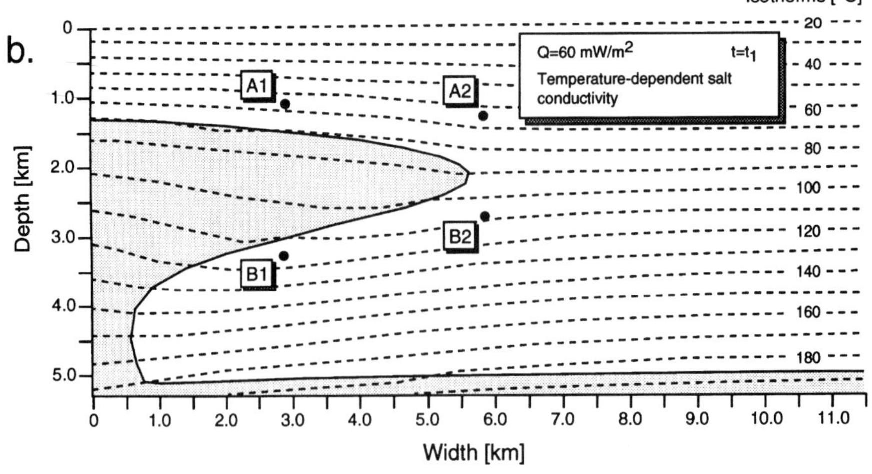

about 1.0 W/m K at the surface to 1.9 W/m K at 5 km depth for the same heat flow value. So the change in thermal conductivity due to temperature is not significant compared to the effect of compaction. Depending on the stratigraphy, higher or lower temperature anomalies can thus be expected; largest in shale-dominated basins.

Because salt acts as an impermeable barrier, overpressure may be expected below salt sheets, the magnitude depending on the subsalt bed geometry and spatial permeability variation (see e.g. Proshlyakov *et al.* 1986; O'Brien *et al.* 1993). Overpressure development implies that larger porosities can be expected below salt sheets compared to the porosities at the same depth in the regional domain. Higher porosity lowers the increase of thermal conductivity over the depth range of overpressure, implying a higher contrast in thermal conductivity between salt and sediments than would otherwise be expected. The effect of the higher porosity has two effects on the temperature anomaly: a lowering of the sediment thermal conductivity leads to an increase in the thermal gradient compared to the regional domain, which

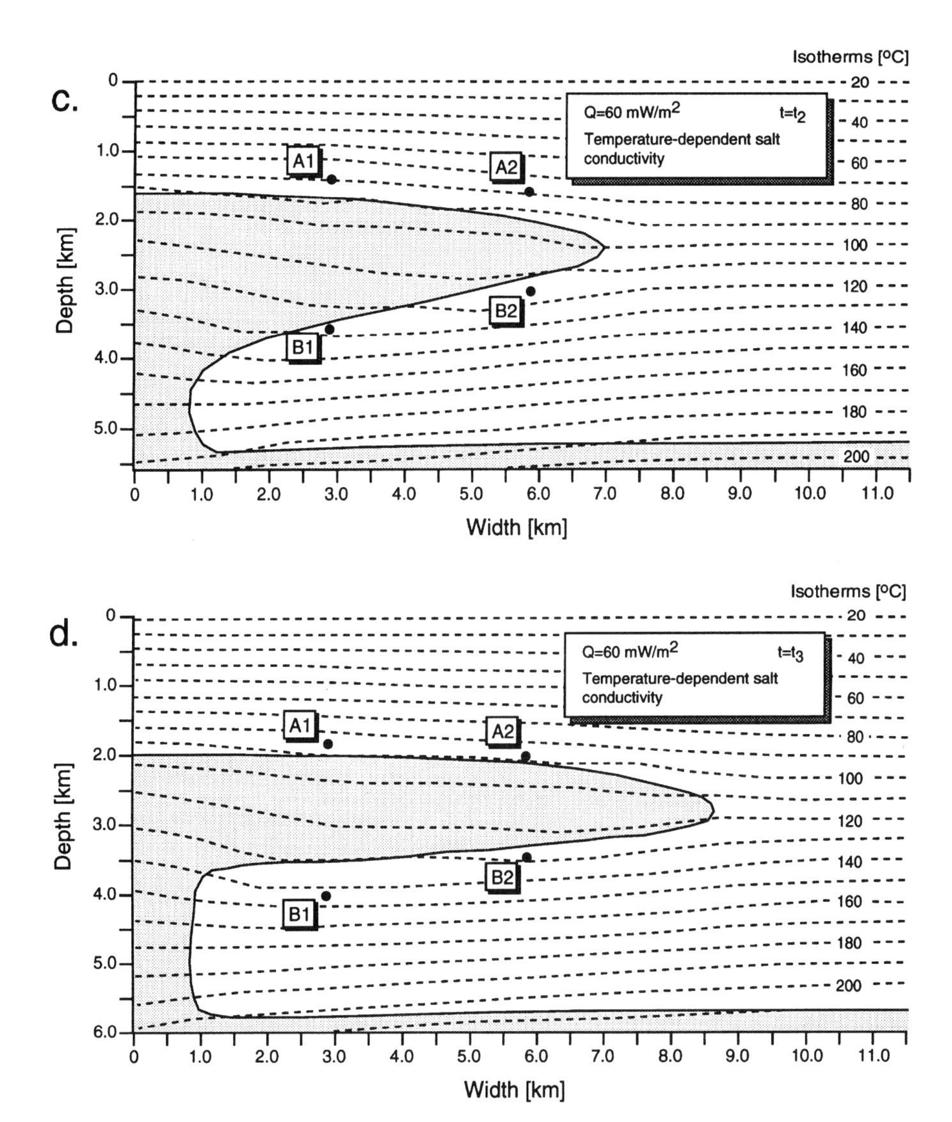


**Fig. 9.** (**a**)–(**d**) show the evolution of a salt sheet at times $t = t_0$ to $t = t_3$ with the sedimentation rate being 1/4 of the lateral growth rate of the sheet. The resulting temperature anomaly histories through time of locations A1, A2, B1, and B2 are shown in Fig. 10.

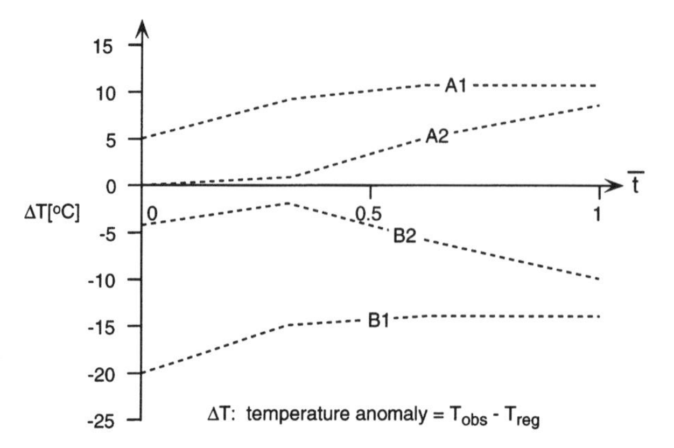


**Fig. 10.** Temperature history at locations A1, A2, B1 and B2 (Fig. 9). The absolute temperature anomaly at each location is plotted against time, $t$, normalized as $(t - t_o)/(t_3 - t_o)$. See text for discussion.

will reduce the negative anomaly; the increase in thermal conductivity contrast between salt and the underlying overpressured sediments will seek to increase the thermal focusing, resulting in an increase in the negative temperature anomaly. Whether the superposition of these two effects results in an increase or decrease in the thermal focusing depends on the vertical range of the overpressured zone, thickness of salt, and depth of burial.

Applications of the model to case histories with real data have been presented in detail in a variety of settings. For instance, in the case of the Lulu-1 salt diapir, Thomsen & Lerche (1993) used both downhole temperature data as well as vitrinite reflectance measurements with depth, relative to the corresponding measurements in a borehole (the Q-1 well) far from the salt, to infer both the thermal focusing and vertical motion of the diapir. In the case of the West Bay, Louisiana salt diapir, Vizgirda *et al.* (1985) used downhole temperature measurements from six boreholes, at increasing lateral distances from the diapir, to demonstrate that the observed spatial variation of temperature was in

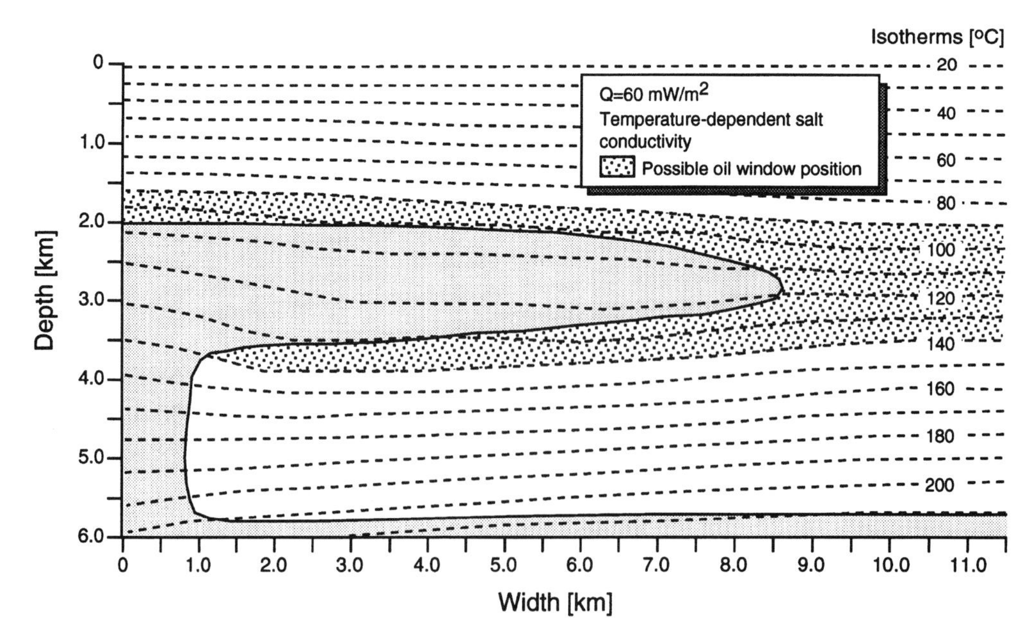


**Fig. 11.** The oil window is raised and lowered above and below the salt, respectively, for the salt sheet in Fig. 5. See text for discussion.

accord with theoretical predictions of thermal focusing. In the case of four wells that penetrated salt sheets in the Gulf of Mexico, O'Brien & Lerche (1988) used measured overpressure profiles, seismic velocities, and downhole density logs to demonstrate that predicted sediment properties were in accord with observations. For the North Caspian depression Proshlyakov *et al.* (1986) have shown that the observed temperatures below salt sheets are between 20–30° lower than at the same sub-surface depth far from salt sheets, precisely in accord with the theoretical predictions.

In short, not only is there an abundance of observational data in a variety of observational settings which has been examined in other communications, but the model predictions for thermal focusing by both diapiric and sheet-like structures are in accord with the observations. In addition, dynamical model predictions of overpressure, porosity retention, and sediment fracturing are also in accord with observed behaviours.

There is little need to re-develop the detailed applications here; they have been adequately addressed elsewhere. The point of this paper has been to bring together in one place the basic physical principles upon which rest the individual applications. Readers interested in specific applications may wish to consult the above references or, indeed, the recent book on salt and sediment dynamics by Lerche & Petersen (1995), which contains copious real data applications of the basic physics spelled out here.

## Conclusions

The examples of thermal focusing by salt presented in this paper demonstrate the following:

(1) The often made assumption that the contrast in thermal conductivities between salt and sediments is constant with depth is justified only when shallow salt structures are considered. But, because salt thermal conductivity decreases with increasing temperature, the focusing effect of salt decreases with increasing depth of burial. As a result: (a) the vertical temperature gradient increases with depth in the salt; (b) the rim syncline sediments experience a smaller negative temperature anomaly with depth than would have been predicted by assuming a constant salt thermal conductivity with depth.

(2) A higher temperature can be maintained above salt structures due to thermal defocusing. The thermal focusing of heat from the rim syncline sediments into the salt structure, and the associated establishment of a negative temperature anomaly compared to the regional domain, depends on the lateral extent of the overhang.

(3) Salt sheets exhibit defocusing of heat at the top of the structures, with the magnitude depending on the thickness of the salt together with the geometry of the salt sheet, which result in positive temperature anomalies in the overlying sediments. The negative temperature anomaly found in the underlying sediments attains its maximum in an area between the salt stem and the central part of the salt sheet, an effect of the lateral components of heat flow within the salt. The focusing of heat into the salt sheet is felt out to a distance in the sediments of about 1.5 times the thickness of the salt sheet.

(4) The horizontal insertion of salt into the sedimentary section changes the thermal history of the sediments around the salt. Sediments which at present occupy positions above a salt sheet have experienced an increase in their temperatures through time, unlike sediments at the same depth in the regional domain. If the salt sheet is buried deeply the sediments above the salt may thus have entered and stayed in the oil window for a longer period of time. Conversely, sediments which at present are located below the salt sheet have been exposed to lower temperatures than sediments at the same depth in the regional domain. A potential source rock far from the salt may have left the oil window while at the same depth near the salt the source rock can remain in the oil window to a greater depth, i.e. subsalt hydrocarbon prospects may exist at a greater depth below the salt than is possible far from the salt. Because overpressure often prevails in subsalt sediments, the thermal conductivities in these sediments will be lower than otherwise anticipated, and may further increase the negative temperature anomaly.

(5) Assessment of the present-day heat flow is confounded because of the uncertainties in measuring the temperature profile with depth properly and in measuring the thermal conductivities of the sediments with depth. Modelling the heat flow history is thus even more uncertain because it relies on the present-day estimated heat flow value and the use of thermal indicators. The examples show that the temperature anomaly and the absolute temperature pattern are both reduced (enhanced), essentially proportional to the decrease (increase) in heat flow. Therefore, the effect of different heat flow values on the temperature pattern can easily be tested and compared with observations.

(6) By calculating thermal anomalies with time, the continued evolution of hydrocarbon potential can be addressed in relation to the dynamical development of the salt, evolving geometry of sedimentary beds, rim syncline evolution, and the expected faulting and fracturing pattern of sediments. Thus an integrated picture can be drawn of the total likelihood of dynamically evolving traps, and their anticipated failure or non-failure, in relation to the generation, migration, and accumulation of hydrocarbons in salt-related traps.

The work reported here was supported by the Industrial Associates of the Basin Modeling Group at the University of South Carolina.

## References

BERNER, H., RAMBERG, H. & STEPHANSSON, O. 1972. Diapirism in theory and experiment. *Tectonophysics*, **15**, 197–218.

BISHOP, R. S. 1978. Mechanism for emplacement of piercement diapirs. *AAPG Bulletin*, **62**, 1561–1583.

GEERTSMA, J. 1971. Finite element analysis of shallow temperature anomalies. *Geophysical Prospecting*, **19**, 662–681.

JENSEN, P. K. 1983*a*. Formation temperatures in the Danish Central Graben. *Danmarks Geologiske Undersøgelse, Årbog 1982*, 91–106.

—— 1983*b*. Calculations on the thermal conditions around a salt diapir. *Geophysical Prospecting*, **31**, 481–489.

—— 1990. Analysis of the temperature field around salt diapirs. *Geothermics*, **19**, 273–283.

KAPPELMEYER, O. & HAENEL, R. 1974. Geothermics: with special reference to applications. *Geoexploration Monograph 4*, Borntraeger, Berlin.

LERCHE, I. 1991. Temperature dependence of thermal conductivity and its impact on assessments of heat flux. *Pure and Applied Geophysics*, **136**, 1–11.

—— & O'BRIEN, J. J. 1994. Understanding subsalt overpressure may reduce drilling risks. *Oil and Gas Journal*, 24 January, 28–34.

—— & PETERSEN, K. 1995, *Salt and Sediment Dynamics*. CRC Press, Boca Raton.

MELLO, U. T., ANDERSON, R. T. & KARNER, G. D. 1994. Salt restrains maturation in subsalt plays. *Oil and Gas Journal*, 31 January, 101–107.

MORSE, P. M. & FESHBACH, H. 1953. *Methods of Theoretical Physics*, **1**, McGraw-Hill Book Co., New York.

NETTLETON, L. L., 1955. History of concepts of Gulf Coast salt dome formation. *AAPG Bulletin*, **39**, 2373-2383.

O'BRIEN, J. J. & LERCHE, I. 1987. Heat flow and thermal maturation near salt diapirs. *In*: LERCHE, I. & O'BRIEN, J. J. (eds) *Dynamical Geology of Salt and Related Structures*. Academic Press, Orlando, 711–750.

—— & —— 1988. *Impact of Heat Flux around Salt Diapirs and Salt Sheets in the Gulf Coast on Hydrocarbon Maturity: Models and Observations*. Gulf Coast Association of Geological Societies Transactions, **38**, 231–243.

——, —— & YU, Z. 1993. Measurements and models under salt sheets in the Gulf of Mexico. *In*: ARMENTROUT, J., BLOCK, R., OLSON, M. C. & PERKINS, B. F. (eds) *Rates of Geologic Processes*. GCSSEPM Foundation 14th Annual Research Conference, 155–163.

PETERSEN, K. & LERCHE, I. 1993. Interactive salt and sediment evolution: self-consistent quantitative models. *In*: COBBOLD, P. (ed.) *New Insights Into Salt Tectonics*. Tectonophysics Special Issue, **228**, 211–238.

—— & —— 1994. Quantification of thermal anomalies around salt structures. *Geothermics* (in press).

PROSHLYAKOV, B. K. *et al.* 1986. Sub-salt sediments of the North Caspian depression – a unique complex of natural oil-gas reservoir. *Journal of Petroleum Geology*, **22**, 430–432.

RASHID, M. A. & MCALARY, J. D. 1977. Early maturation of organic matter and genesis of hydrocarbons as a result of heat from a shallow piercement salt dome. *Journal of Geochemical Exploration*, **8**, 549–569.

SELIG, F. & WALLICK, G. C. 1966. Temperature distribution in salt domes and surrounding sediments. *Geophysics*, **31**, 346–361.

TALBOT, C. J., 1978. Halokinesis and thermal convection, *Nature*, **273**, 739–741.

THOMSEN, R. O. & LERCHE, I. 1991. Salt diapir velocity assessment from temperature and thermal indicator anomalies: application to Lulu-1, Danish North Sea. *Terra Nova*, **3**, 500–509.

VIZGIRDA, J., O'BRIEN, J. J. & LERCHE, I. 1985. Thermal Anomalies on the flanks of a salt dome. *Geothermics*, **14**, 553–565.

VON HERTZEN, R. P., HOSKINS, H. & ANDEL, T. 1972. Geophysical studies in the Angola Diapir Field. *Geological Society of America Bulletin*, **83**, 1901–1910.

# Salt diapirism with simultaneous brittle faulting and viscous flow

ALEXEI N. B. POLIAKOV[1,2], YURI YU. PODLADCHIKOV[3], ETHAN CH. DAWSON[4] & CHRISTOPHER J. TALBOT[5]

[1]*Laboratoire de Géophysique et Tectonique (URA 1760), Université Montpellier II, 34095 Montpellier 05, France*

[2]*PMMH (URA 857), ESPCI, 10 rue Vauquelin, 75231 Paris 05, France*

[3]*Department of Sedimentary Geology, Vrije Universiteit, 1081 HV Amsterdam, The Netherlands*

[4]*Department of Civil Engineering, University of Minnesota, Minneapolis, Minnesota 55455, USA*

[5]*Hans Ramberg Tectonic Lab, Uppsala University, Norbyvägen 16, 752 36 Uppsala, Sweden*

**Abstract**: In this paper we show for the first time how sedimentary rocks above a salt diapir can deform in either a brittle or a viscous way depending on the stress state and strain rates. Most existing models for salt tectonics incorporate only one or other of these deformation mechanisms resulting in different conclusions. Taking as an example the controversial problem of buoyant salt diapirism, we demonstrate how diapirs can grow in an overburden which may deform by both creep and faulting. We argue that salt movement can take place under low differential stresses because sediments may deform by pressure-solution creep. These zones of diffuse deformation co-exist with faults where differential stresses are high enough to reach the yield limit at which sediments fail. We demonstrate two conditions under which buoyant diapirs can cause substantial faulting in their overburden without the application of tectonic forces. Buoyant diapirs can fault overburdens with high viscosity contrast ($\mu_{overburden}/\mu_{salt} > 10^2$) if the topographic relief they induce in the surface is removed by rapid erosion.

The mechanisms that generate salt diapirs are not completely understood. There are two points of view depending on the rheology assumed for the overburden of the salt. The classic approach was to ignore the fact that most rocks fault and to simplify the behaviour of the salt overburden to a viscous fluid. In this case, buoyancy forces alone are sufficient to drive diapirs of salt into denser viscous layers (Ramberg 1981; Jackson & Talbot 1986). In contrast, insisting that overburdens that fault are brittle solids leads to the problem that buoyancy forces alone are insufficient to drive diapirs into brittle overburdens. This is because the density contrast between salt and the most likely overburdens results in differential stresses too low to deform truly brittle rocks. This logic has led many recent workers to assume that forces in addition to gravity are necessary to initiate and promote salt diapirs. Mechanisms proposed for providing such external tectonic stresses are: thin-skinned regional extension (Vendeville & Jackson 1992; Weijermars *et al.* 1993, Nalpas & Brun 1993; Daudré & Cloetingh 1994), gravitational gliding (Demercian *et al.* 1993), or differential loading (Jackson & Talbot 1991; Schultz-Ela *et al.* 1993; Davison *et al.* 1993).

This paper combines these two extreme points of view. We explore the possibility that the overburden of a salt diapir can deform by simultaneous brittle faulting and viscous flow. This idea is motivated by the observation that sediments have two major modes of deformation: either a homogeneous distributed flow or localized shear zones or faults. The first is most likely due to a pressure-solution creep mechanism and results in viscous-like behaviour. The latter is due to cataclastic faulting resulting in brittle behaviour.

Because the role of viscous-like deformation in sedimentary rocks is doubted by some authors, we would like to mention some field evidence for this type of deformation; e.g. folds in sedimentary rocks at different scales (Price & Cosgrove 1990; Suppe 1985), periodic spacing of low-amplitude salt pillows related to the salt and overburden thicknesses (Jackson *et al.* 1990; Hughes & Davison 1993, Demercian *et al.* 1993; Rönnlund; 1989) and drag folding on diapir flanks with no observed faults (Ratcliffe *et al.* 1992). Brittle behaviour (i.e. faults) in sedimentary rocks related to salt diapirs is not in doubt because it is easily observed in the field and in seismic profiles. However, the role of faults in salt diapirism is not completely clear.

We see no contradiction in assuming simultaneous viscous and brittle rheology for sedimentary rocks. There are situations when diapirs are

*From* Alsop, G. I., Blundell, D. J. & Davison, I. (eds), 1996, *Salt Tectonics*, Geological Society Special Publication No. 100, pp. 291–302.

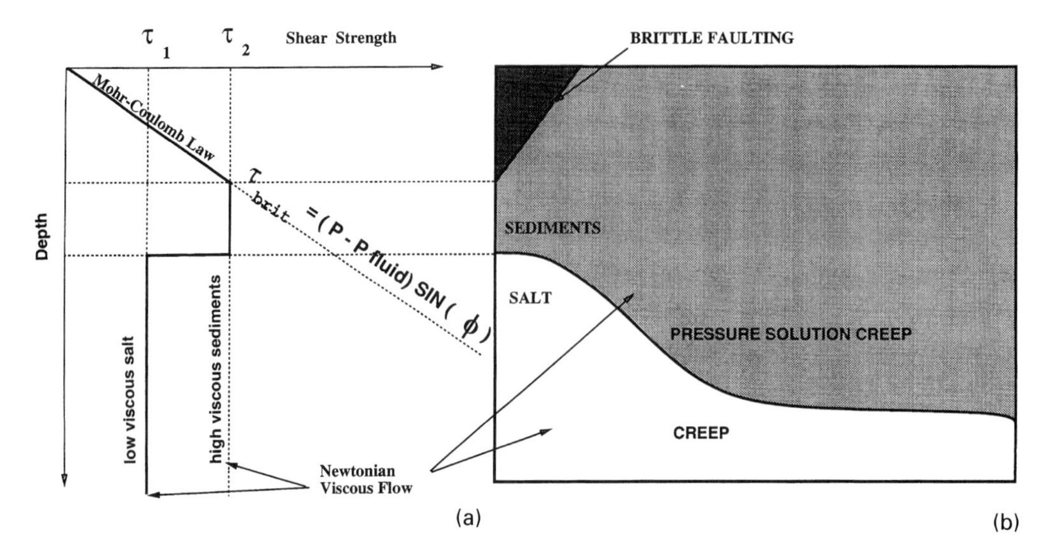


**Fig. 1.** Simple model of the strength of the sediments and buoyant salt diapir for a constant strain rate. Rheology of salt is assumed to be linearly viscous. Sediments may fail in a brittle manner (black zone in the upper left corner of (**b**) or may deform homogeneously by pressure-solution creep (grey area in (**b**). Rocks deform by the mechanism requiring the lowest differential stress. Note that there are no applied external stresses.

controlled by purely brittle deformation in their overburden and other situations where they are controlled by purely viscous deformation of the overburden. However, there is a whole spectrum of intermediate situations which combine both diffused and localized deformation in patterns that depend upon strain rates and stresses.

This paper presents a single example in which these modes of deformation coexist. We model numerically the evolution of a diapir in a brittle–viscous overburden. In order to simplify the problem, we consider only the classic case of buoyantly driven diapirs (i.e. without any externally applied loads). We show that two effects can concentrate the relatively weak gravitational forces of buoyancy sufficiently to produce faulting in the overburden. The two conditions necessary to produce significant brittle deformation are a high-viscosity contrast between the salt and its overburden, and the continual erosion and redeposition of the topographic relief induced in the surface above a rising diapir. Both of these conditions are reasonably realistic. The primary goal of this paper is to answer the longstanding question: can salt diapirs rise without the application of external forces?

## Estimation of rheological behaviour

The results of any numerical modelling exercise depend strongly on the underlying assumptions. It has been shown that the growth of salt diapirs is relatively insensitive to the rheology of the diapir itself (van Keken *et al.* 1993; Weijermars *et al.* 1993). The diapir material can be treated as a simple linear viscous fluid. On the other hand, diapir growth turns out to be very sensitive to the rheology of the overburden. The overburden rheology is crucial and divides diapir models into two camps with completely different interpretations of field observations (Cobbold 1993). Here, instead of imposing either a viscous or a brittle rheology on the overburden, we include both rheologies, and allow the system to choose its own mode of deformation depending on stress and strain rates. We feel that this fairly simple and understandable approach is closer to the true nature of the problem.

Figure 1B shows the geometry of the two-layer system we consider, with salt at the bottom and sediments at the top.The density of the salt is taken as 2200 kg $m^{-3}$ and the density of the sediments increases exponentially with depth, varying between 1800 kg $m^{-3}$ at the surface and 2500 kg $m^{-3}$ at a depth of 4 km, due to compaction (as in Biot & Ode 1965). The viscosity of the salt is assumed to be $10^{17}$ Pa s. Let us estimate the viscosity of the overburden layer. Sedimentary overburdens are almost always saturated with water. When differential stresses are applied, mineral grains are dissolved in regions of high stress and are redeposited in regions of low stress. The result is a deformation with a behaviour similar to that of a Newtonian fluid and known as 'pressure solution creep' (Turcotte & Schubert 1982; Rutter 1976).

There is persuasive petrographic evidence for this type of mechanism in the field (Gratier 1987; Kirby & McCormick 1990; Aharonov & Rothman in press). However, pressure solution mechanisms are difficult to reproduce in laboratory experiments at low stresses and temperatures ($T < 400^{\circ}$C), therefore the viscosity is uncertain and remains a free parameter. There are probably other rate-dependent mechanisms of deformation in sediments, such as compaction, grain boundary sliding, fatigue and twinning in calcite, but they are even more uncertain and less evident than the pressure solution creep. In order to estimate the viscosity of the sediments we have performed simulations varying many parameters (including the rate of erosion and sedimentation). We have found that the qualitative results do not depend on the absolute values of the viscosities but instead depend on a few dimensionless parameters. The dimensionless parameters involve material properties, the geometry of the problem and the rate of sedimentation (Poliakov *et al* 1993*b*, Podladchikov *et al.* 1993). One important parameter, is the ratio of the viscosity of the sediments to the viscosity of the diapir. Varying this parameter, we approximately match observed shapes of diapirs and their growth times using viscosity contrasts between $10^2$ and $10^4$. The same values are reported from studies of the relationship between the salt structure wavelength and the thickness ratio between salt and overburden (Rönnlund 1989; Hughes & Davison 1993).

Let us roughly predict the stress state and rheological behaviour of the overburden for the system shown in Fig. 1B. We assume that the shear strain-rate $\dot{\epsilon}$ is constant along the left vertical boundary in Fig. 1B. We can then estimate the shear stresses along the same boundary. Shear stresses due to creep are simply calculated as $\tau_{creep1,2} = \mu_{1,2}\dot{\epsilon}$, and are shown by dotted lines in Fig. 1A. In this figure the indices 1 and 2 denote salt and overburden, respectively. The shear stress $\tau_{brit}$ needed for the rocks to fail as brittle solids is calculated assuming that the sediments obey a Mohr–Coulomb friction law and, assuming a simple lithostatic water pressure, $p_{\rm water} = \rho_{\rm water} gz$ at depth $z$

$$\tau_{\rm brit} = (\,p - p_{\rm water}\,)\sin\phi$$

where $\tau_{\rm brit}$ is the deviatoric stress, $p$ is the pressure in the rock matrix and $\phi$ is the friction angle, which is approximately 30° for sedimentary rocks. Rocks will deform in a brittle or creeping manner depending on which mechanism requires lower stresses. For a given geometry and strain rate, the resulting stress (solid line in Fig.1a) corresponds to the minimum of stress from the creep law (vertical dotted line, Fig.1a) and stress from brittle faulting (inclined dotted line, Fig.1a). This type of rheological law divides the sediments into creeping and brittle areas as a function of stress and strain rate distribution in the sediments.

## A new modelling strategy: link between viscous and visco-elasto-plastic simulations

Simulation of the rheological behaviour described above is a difficult technical problem. The difficulties involve: tracking of discontinuities between materials with different rheologies (Poliakov & Podladchikov 1992), viscoelastic aspects of diapirism and problems of remeshing the distorted computational grid (Poliakov *et al.* 1993*a*; Braun & Sambridge 1994; Hassani 1994), sedimentation and erosion (Podladchikov *et al.* 1993), and localization in a pressure-dependent (brittle) material with a non-associated plastic flow rule (Poliakov *et al.* 1993*c*; Poliakov & Herrmann 1994). Simultaneous modelling of all of these aspects of diapirism is extremely challenging. However, the problem can be made tractable by combining different numerical methods and by making reasonable physical simplifications.

The method we apply in this paper consists of combining the results from two independent numerical programs. Each program has its own advantages and disadvantages. These two programs are: (1) an implicit Eulerian–Lagrangian finite-element code PPFEM that permits simulation of long histories of diapiric evolution that include erosion and sedimentation but allow only viscous deformations (Podladchikov *et al.* 1993); (2) an explicit Lagrangian finite-element code PAROVOZ (based on the FLAC technique, (Cundall 1989)) which can simulate complex (visco-elasto-plastic) rheological behaviour but is complicated at large deformations because of remeshing problems and limitations on the critical time step. This time step is controlled by the rate of elastic wave propagation and by inertial effects (Poliakov *et al.* 1993*a*).

Comparing simulations from these two different codes we found conditions under which the results are sufficiently close that it is reasonable to transmit information from one code to the other. Here we will briefly explain the loose coupling procedure we have used. For visco-elastic flow there is a 'viscous' characteristic time scale

$$T_{\rm visc} = \frac{\mu}{S} \qquad (1)$$

and the Maxwell relaxation time

$$T_{\rm relax} = \frac{\mu}{G} \qquad (2)$$

where $S$ is the characteristic differential stress and $G$

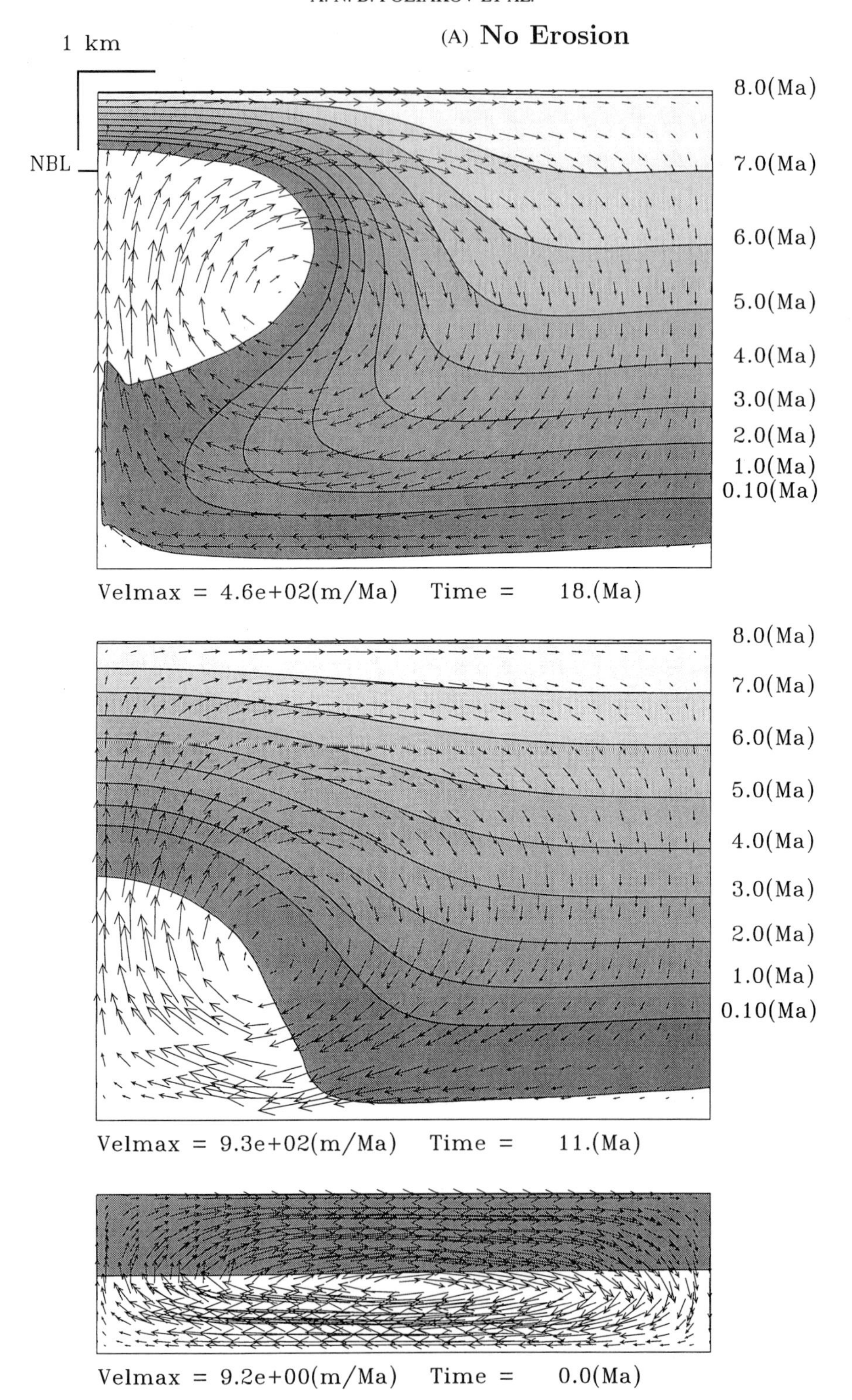


**Fig. 2.** Evolution of bedding markers, and velocity field for a buoyant diapir in the model without erosion (**A**) and with fast erosion (**B**) at the rate of sedimentation of 500 m per Ma. Rheology of the salt diapir and overburden is assumed to be linearly viscous, where $\mu_{overburden}/\mu_{salt} = 10^3$. All other parameters are the same as in the model ner500

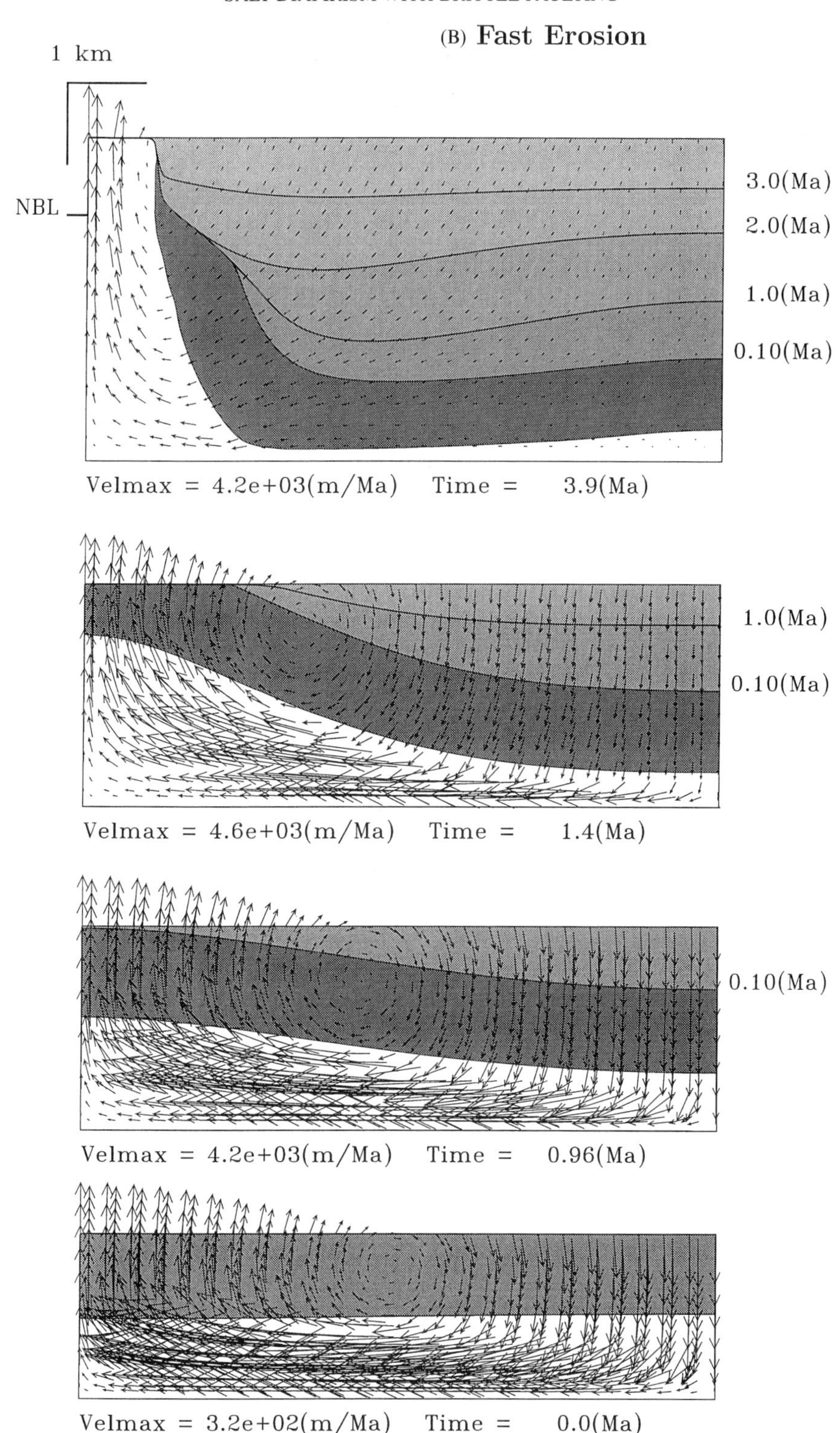


and er500 from Poliakov *et al.* (1993*b*). NBL is the neutral buoyancy level at which the densities of the salt and compacting sediments are equal. Note the strong acceleration in the diapir growth rate, change in the shape and growth through and above the neutral buoyancy level in the case with fast erosion compared to the case without erosion.

is an elastic shear modulus. For our case we assume that the buoyancy forces are the only source of differential stress, which can be roughly estimated as $S = \Delta\rho g L$, where $\Delta\rho$ is a characteristic density contrast, $L$ is the scale of the problem and $g$ is the acceleration due to gravity. The ratio $De$ between the Maxwell and 'viscous' time scales is called the Deborah number and controls visco-elastic effects

$$De = \frac{T_{relax}}{T_{visc}} = \frac{S}{G} = \frac{\Delta\rho g L}{G} \quad (3)$$

Thus, one can see that visco-elastic behaviour of materials with constant viscosity is independent of the absolute value of viscosity but depends only on the ratio of the characteristic shear stresses to the elastic modulus (the viscosity controls the time scale). At a small value of $De$ ($<10^{-2}$) the materials behave as almost purely viscous fluids. At a high value of $De$ the materials behave as almost purely elastic solids. We have shown numerically and analytically that an increase in the value of $De$ accelerates diapir growth for the isoviscous case (Poliakov *et al.* 1993*a*). High $De$ relies may be important for large-scale mantle flows.

Let us roughly estimate the characteristic buoyancy force $S$ and $De$ for diapirism in sedimentary basins. We can use the same definition for $De$ as above if the viscosity in the overburden is constant. This viscosity controls the 'viscous' time scale of diapiric growth and the maximum Maxwell relaxation time. Thus, the viscosity of the salt does not play a role in this analysis because the relaxation time in the salt is much shorter than that of the overburden. Assuming that $\Delta\rho = 300$ kg m$^{-3}$, $g = 10$ ms$^{-2}$ and $L = 3 \times 10^3$ m, we obtain

$$S \approx 9 \times 10^6 \text{ Pa} \quad (4)$$

which is close to the estimates of Davison *et al.* (1996). Assuming that compacted sediments have an elastic shear modulus $G = 3 \times 10^{10}$ Pa, we find that

$$De \approx 3 \times 10^{-4} \quad (5)$$

Thus, for salt diapirism, the influence of visco-elasticity is insignificant, at least for the isoviscous case (Poliakov *et al.* 1993*a*). Note that this upper limit estimation of deviatoric stress does not give us the spatial distribution of stresses and we cannot make conclusions about faulting from this simple analysis.

These results indicate that stresses in a visco-elastic material at low Deborah number are close to the stresses in a purely viscous material (i.e. stresses due to elastic deformation are nearly relaxed). This means that simulations with a purely viscous rheology which are numerically inexpensive will not differ much from those obtained during expensive visco-elastic simulations which are much more difficult technically. In turn, we assume that stresses for a visco-elastic and for a visco-elasto-plastic material are close to each other as well if the zones of plastic deformation are relatively small compared with the dimension of the whole problem. Thus, calculations with a purely viscous code are a good simplification of the real problem, giving approximately correct stress fields, velocities and diapir shapes. However, 'viscous' simulations cannot simulate faults approximated by localized zones of shear zones in an elasto-plastic material. At the same time, these 'viscous' stresses are a good initial stress approximation for simulations with visco-elasto-plastic rheology. Following this concept, we make the following proposal.

First, we calculate the whole diapiric evolution using our fast 'viscous' code (e.g. Fig. 2). These simulations reveal fairly well the main 'bulk' effects of buoyant diapirism: diapir shapes, growth times, and interactions with erosion and sedimentation. To see faulting requires calculations at a finer scale and it is quite reasonable to use the geometry of the diapir and sediments obtained from the viscous PPFEM code and input them into the PAROVOZ code as initial conditions. Note that the geometry of a diapir and sediments (i.e. the distribution of viscosities and densities) is sufficient to define the stress field for a viscous rheology in a steady-state problem (the stress in a viscous fluid depends only on the strain rate and the viscosity, but not on the history of loading). Thus, we can then continue the simulation with the PAROVOZ code using the full visco-elasto-plastic rheology (Figs 3–5). In this way, we can compute brittle (plastic) zones which will be formed in zones where stresses reach the yield condition. Of course, by this method we do not have feed-back, i.e. the brittle zones have no influence on the viscous evolution. However, as we will demonstrate below, for many problems this approach is well justified: zones of plastic deformation appear to be relatively narrow compared to the size of the problem (Fig. 5b) and they do not significantly change the velocity field (i.e. the shapes of the salt diapirs).

In other words, we propose to estimate *a posteriori* fault zones associated with a viscous diapir with a low Deborah number (where the overburden is nearly viscous because its relaxation time is fast compared to the characteristic time of viscous flow). We assume that the relatively small brittle zones do not significantly affect the evolution of the diapir. This last assumption is supported by seismic observations that suggest that the offsets along faults around many salt diapirs are insignificant and

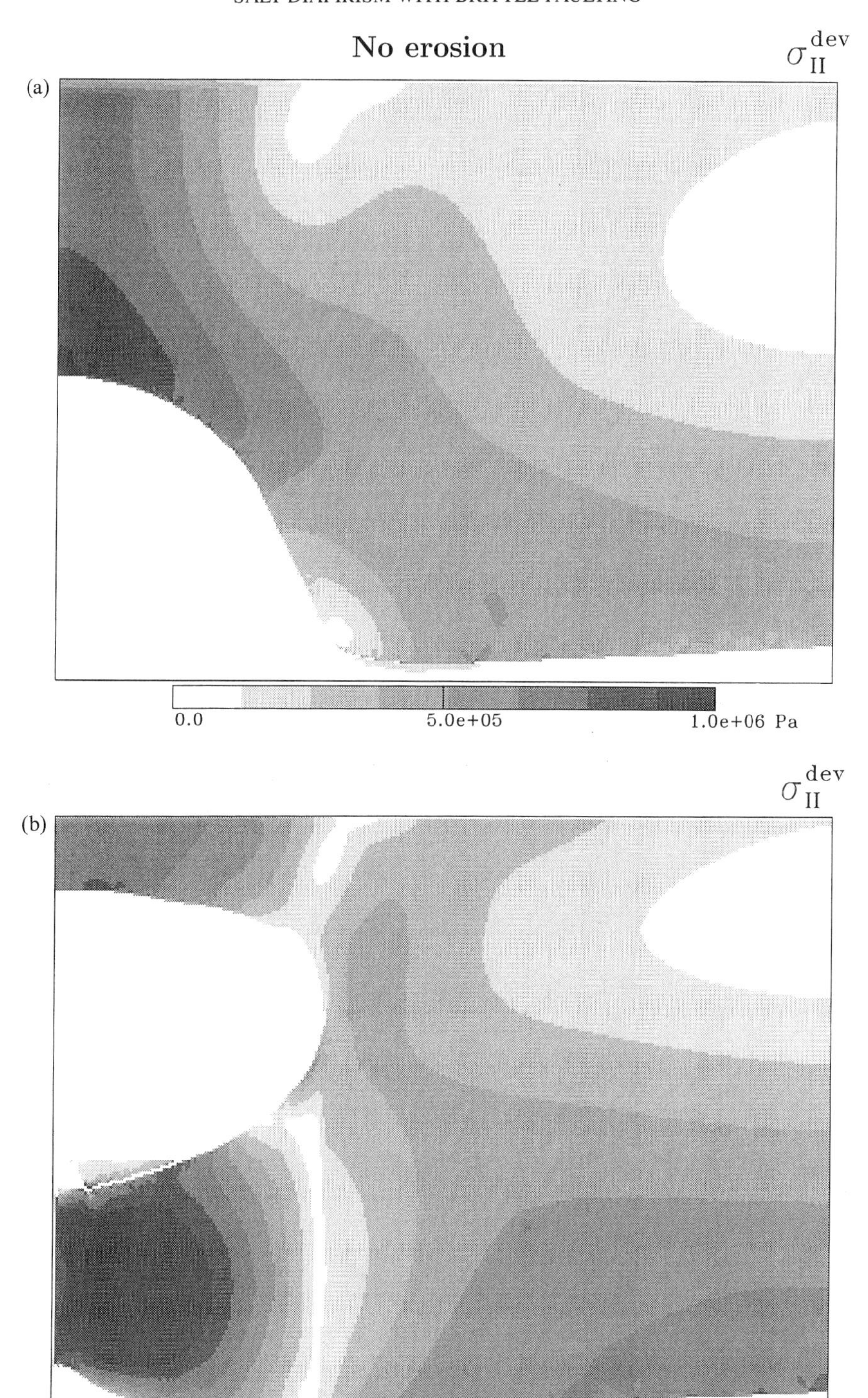


**Fig. 3.** Maximum differential stress (second invariant of deviatoric stress) values are shaded in the model without erosion (model ner500, Fig. 2A) at the times of 11 Ma (**a**) and 18 Ma (**b**). Rheology of the overburden is visco-elasto-plastic and calculations are performed by code PAROVOZ. There are no significant zones of plastic deformation because of low differential stresses.

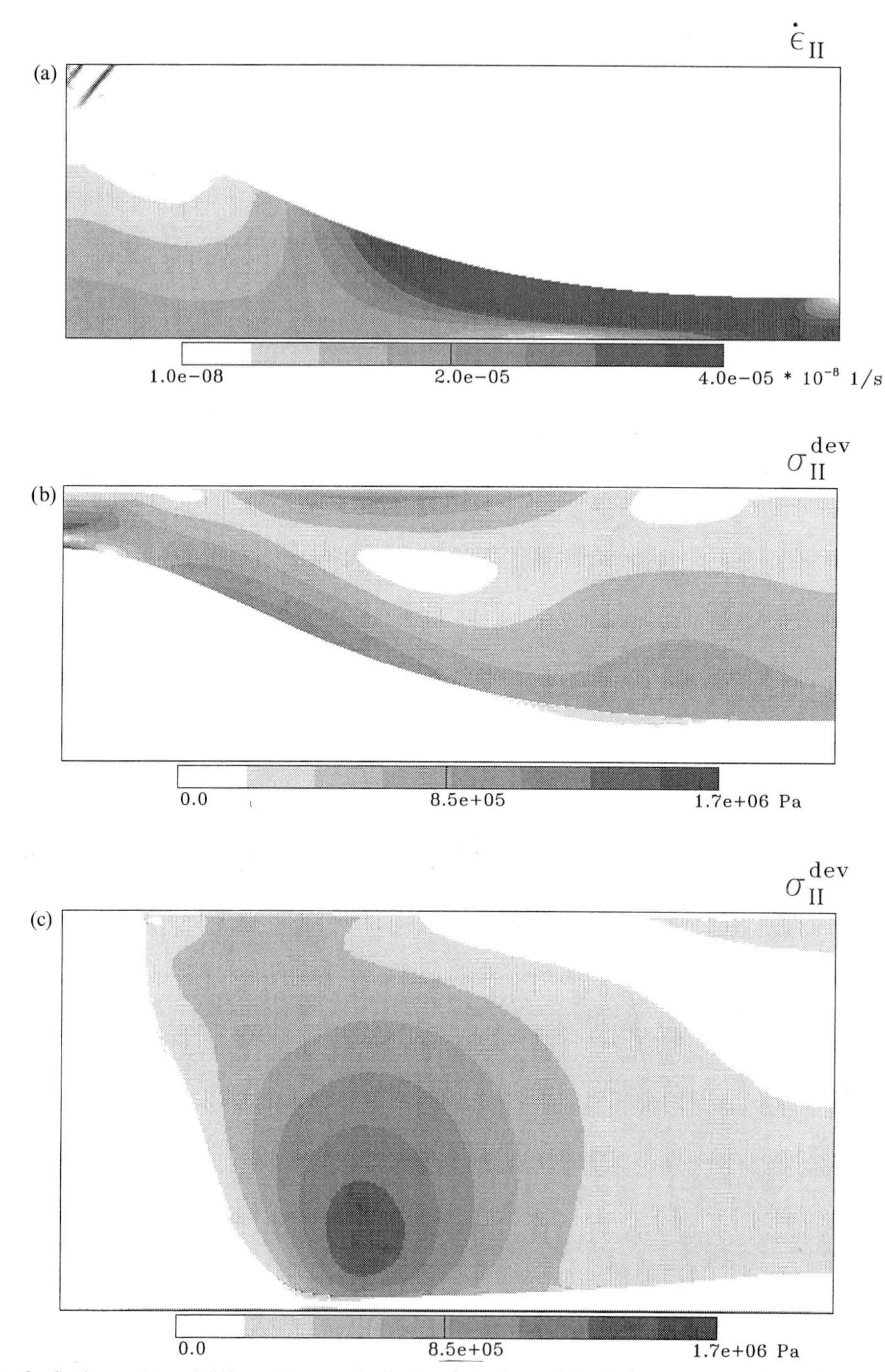


**Fig. 4.** Strain rate (**a**) and differential stress distribution (**b**) in the model with fast erosion (model er500, Fig. 2B) at the time of 1.4 Ma, and differential stress distribution at the time of 3.9 Ma (**c**). Rheology of the overburden is visco-elasto-plastic. Stress distribution at the non-linear stage before extrusion is most favourable for faulting (shear localization) above the top of the diapir (**a**). This effect is the result of high strain rates (effect of erosion), redistribution and concentration of stresses (effect of viscosity contrast between salt and overburden). There is almost no faulting at the stage of extrusion. The diapir is not sealed and does not create high deviatoric stresses (**c**).

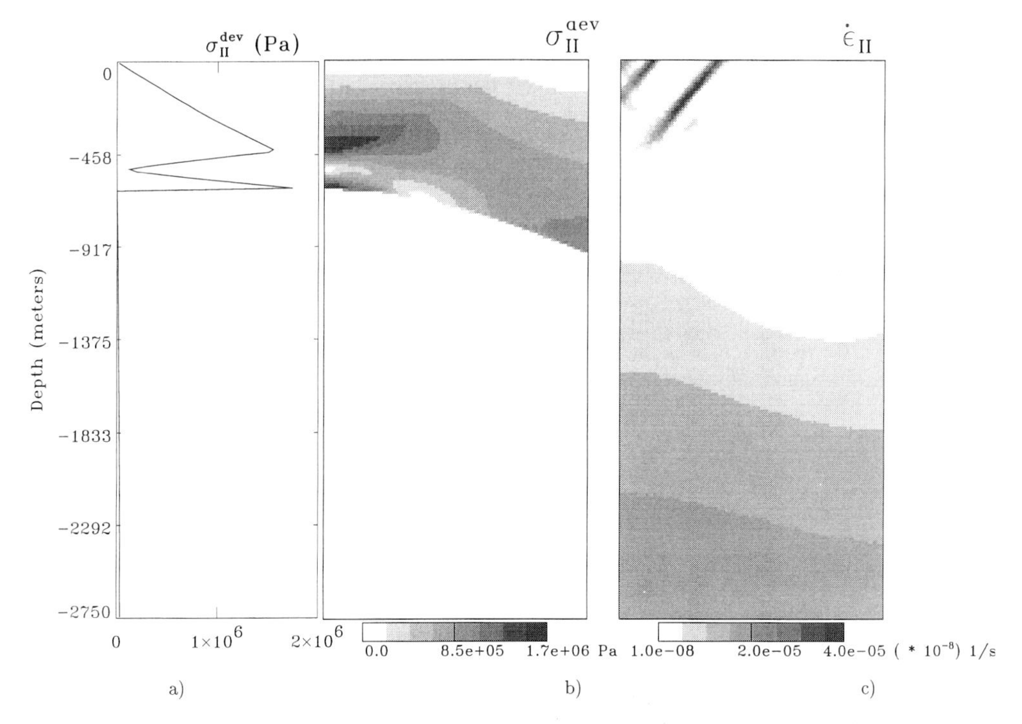


**Fig. 5.** Vertical profile of deviatoric stress (**a**) at the left boundary in Fig. 4b. Detailed view of maximum shear stress (**b**) and strain rate field (**c**) in the left part of the Fig. 4a,b. The upper part of the strength profile (**a**) and the upper left corner in the stress field (**b**) show a plastic zone where stress is relaxed to the Mohr–Coulomb yield surface. Strain rate field shows zones of localized shear corresponding to faults (**c**).

that faults do not change the general rounded shapes of diapirs.

## Effects on the stress distribution: erosion and viscosity contrast

Figure 1A shows that the depth of the transition between brittle Mohr–Coulomb behaviour and creep depends on the differential stress in the upper layer. Where the strain rates (and therefore the creep stresses) are low during the initial stages of diapirism, any brittle zones that develop are small and very shallow (as shown in seismic interpretations by Hughes & Davison (1993)). Activity along fault zones is most significant during intermediate diapiric stages, when growth is highly non-linear. The strain rates and thus deviatoric stresses are highest during diapiric growth. During the final stages, the mature diapir either spreads at its level of neutral buoyancy, and the strain rate falls to zero, or extrudes onto the surface to establish a channel flow. In both cases, deviatoric stresses decrease and new faulting becomes insignificant.

In order to maximize shear stresses in the upper layer and see the strongest interaction between diapiric growth and faulting, one needs either to increase the strain rate in the upper layer without decreasing its viscosity, or to increase the viscosity of the upper layer without significantly decreasing the strain rate.

The first case can occur due to surface erosion and resedimentation of a topographic bulge induced above the rising diapir. This redistribution of overburden unbalances the vertical stresses and considerably speeds-up diapirism by being equivalent to an additional force pulling the diapir upwards (Biot & Ode 1965). We have found numerically (Poliakov *et al.* 1993*b*; Podladchikov *et al.* 1993) that erosion is a major factor in diapiric evolution, with the following feed-back effects:

- strain rates are increased by up to 50 times;
- the spherical or 'mushroom' shapes typical of diapirs without erosion (Figs 2A & 3) are changed to columnar or 'finger-like' structures with erosion (Figs 2B & 4);
- diapirs are extruded onto the top surface.

We will show below that all of these effects make erosion an important parameter influencing faulting and creep around a salt diapir. Another important parameter controlling stresses in the overburden turns out to be the viscosity contrast between the whole overburden and the salt. Note that the change of the absolute value of viscosity does not change the stress magnitude and its distribution because the system is driven only by internal buoyancy forces. Any increase in viscosity is matched by a corresponding decrease in strain rate because the driving buoyancy force remains the same. However, changing the viscosity contrast results in a non-linear redistribution of stresses. This can lead to large stress concentrations high in the shallow overburden (Poliakov *et al.* 1993*a*). Thus, a high viscosity contrast between the overburden and the salt produces more faulting than a low viscosity contrast, given the same buoyancy forces. If the viscosity of the overburden changes with depth as a function of temperature, faulting can be even more significant. However, in this paper we show only calculations with a constant viscosity for the upper layer.

## Simultaneous faulting and creep

Assuming that erosion has a strong impact on diapiric evolution, we explore the influence of this parameter on faulting and creep. In order to show the most distinctive situation, we use the same model parameters as in Poliakov *et al.* (1993*b*). Figure 2 shows the results of simulations with the PPFEM viscous code of the evolution of a two-layered system of salt and sedimentary overburden accumulating at a sedimentation rate of 500 m per Ma. A sinusoidal perturbation is imposed on the boundary between the salt and overburden with a wavelength estimated from an analysis following Biot & Ode (1965). In Fig. 2 layers of sediments are distinguished by markers with ages shown on the right. Free-slip boundary conditions were chosen for the lateral and bottom boundaries. The arrows represent the velocity field and are scaled for every figure independently, with a maximum value shown under each picture. The case with no erosion is shown in Fig. 2A (model ner500) and the case with fast erosion in Fig. 2B (model er500). In order to simulate these two cases we simply changed the upper boundary conditions from free-slip (no erosion) to free-stress, where the upper boundary is kept flat to simulate fast erosion. We use the term 'fast erosion' in the sense that material is redistributed so fast that the upper surface is always flat. This is a good approximation of erosion rates with a coefficient of diffusion higher than $10^{11}$ $m^2$ per Ma.

It is easy to see the differences between these two cases in terms of the shapes of the diapirs, velocity of growth and the geometry of the sedimentary layers. Following the philosophy described in the section on modelling strategy, we import the results from the purely viscous PPFEM code shown in Fig. 2 into the visco-elastic-plastic code PAROVOZ to make *a posteriori* estimation of fault zones. Recall that this procedure is allowable when the visco-elastic stresses are relaxed (*De* is low) and plastic zones are not large relative to the size of the problem. Both of these conditions are met in our simulations.

To see the most non-trivial stages of diapiric evolution, we have chosen two stages from each of the models ner500 and er500 and recalculated them with the visco-elastic-plastic code. Figure 3 shows the distribution of the deviatoric stress for two non-linear stages in Fig. 2A. At both time steps the shear stresses are not high enough to generate zones of brittle damage, except in the very thin uppermost overburden where the strength is close to zero. For the case without erosion, we conclude that the differential stresses remain low and, therefore, that diapirs are only likely to grow if their overburden deforms by a distributed creep mechanism without significant faulting.

Figure 4 gives the strain rate (a) and the stress fields (b) and (c) for the case of fast erosion in the model er500 (Fig. 2B). The velocity and strain rates in the model with erosion are much higher than in the model without erosion. Erosion concentrates shear stresses so much that they reach the yield strength in the overburden directly above the diapir (Fig. 4b). This leads to a significant zone of faulting (Fig. 4a). Figure 5 focuses on the distribution of shear stress and the shear strain rate (c) for the left of Fig. 4a. One can clearly see the zone of plastic deformation in Fig. 5a (linear dependence of stress with depth) and in Fig. 5b (the sharp change in the distribution of shear stress in the horizontal direction) and the dark zones of localized deformation above the diapir which correspond to faults in Fig. 5c.

At later stages (Fig. 4c), when the diapir extrudes onto the surface, shear stresses decrease and there is almost no new faulting. In this case, the salt can be understood to be driven along a channel by the difference in hydrostatic pressures at the bottom of the diapir and the column of overburden.

## Conclusions

In this paper we demonstrate that the same overburden above a salt diapir can deform in two ways: either as a brittle solid or as a viscous fluid, depending on the stress state and strain rates. Theoretical and experimental difficulties mean that most other models of salt tectonics can incorporate one or the other of these deformation mechanisms, but seldom both. Model results with brittle overburdens differ considerably from models with viscous overburdens.

Taking into account both types of deformation mechanism increases the realism of models of some problems. An example is the controversial problem of buoyant salt diapirism. We have demonstrated here how diapirs can grow due to gravity forces alone in overburdens which deform by both creep and faulting. There is nothing controversial in advocating penetrative strains in overburdens deforming by pressure-solution creep under low differential stresses. Zones of homogeneous deformation can easily co-exist with faults where differential stresses reach the yield strength of the overburden.

Because purely buoyant diapirism produces very low differential stresses, most of the deformation will be accommodated by creep. Nevertheless, buoyant diapirism can still involve significant faulting during the fastest non-linear stages of growth, where a high viscosity contrast between salt and overburden leads to high stress concentrations, and where the rate of diapirism is accelerated by fast erosion and redeposition of the topography induced in the surface above a rising diapir. Both of these conditions result in regions with high shear stresses developing faults between blocks of overburden which undergo slow creeping deformation.

This paper was inspired by Ian Davison. Thanks are also due to Peter Cundall, Hans Herrmann, David Yuen, Jean Chery, Francis Lucazeau, Geoffrey King, Marc Daigneres, Einat Aharonov, Euripides Papamichos and David Mainprice for the interest in our research which was a major driving force of this work. Alexei Poliakov is very grateful to Swedish NFR for support during his work in ESPCI, Paris. This paper is publication no. 950315 of the Netherlands Research School of Sedimentary Geology. HLRZ (Höchstleistungsrechenzentrum), KFA-Jüelich and MSI (Minnesota Supercomputer Institute) are gratefully thanked for crucial computational resources.

## References


AHARONOV E. & ROTHMAN D. (in press) Growth of correlated pore-scale structures. A dynamical model. *Journal of Geophysical Research.*

BIOT, M. A. & ODE, H. 1965. Theory of gravity instability with variable overburden and compaction, *Geophysics*, **30**, 215–227.

BRAUN, J. & SAMBRIDGE, M. 1994. Dynamical Lagrangian Remeshing (DLR): A new algorithm for solving large strain deformation problems and its application to fault-propagation folding. *Earth and Planetary Science Letters*, **124**, 211–220.

COBBOLD, P. R. (ed.) 1993. New insights into salt tectonics. *Tectonophysics*, **228**, 1–448.

CUNDALL, P. A. 1989. Numerical experiments on localization in frictional materials, *Ingenieur-Archiv*, **59**, 148–159.

DAUDRÉ, B. & CLOETINGH, S. 1994. Numerical modelling of salt diapirism: influence of the tectonic regime. *Tectonophysics*, **240**, 225–274.

DAVISON, I., BOSENCE D., ALSOP, G. I. &AL-AAWAH, M. H. 1996. Deformation and sedimentation around active Miocene salt diapirs on the Tihama Plain, northwest Yemen. *This volume.*

——, INSLEY M., HARPER, M., WESTON, P., BLUNDELL, D., MCCLAY, K. & QUALLINGTON, A. 1993. Physical modelling of overburden deformation around salt diapirs. *Tectonophysics*, **228**, 225–274.

DEMERCIAN, S., SZATMARI, P. & COBBOLD, P. R. 1993. Style and pattern of salt diapirs due to thin-skinned gravitational gliding, Campos and Santos basins, offshore Brasil. *Tectonophysics*, **228**, 393–433.

GRATIER, J. P. 1987. Pressure solution-deposition creep and associated tectonic differentation. *In*: JONES, M. E. & PRESTON R. M. (eds) *Deformation of Sediments and Sedimentary Rocks*. Geological Society, London, Special Publication, **29**, 25–38.

HASSANI, R. 1994. *Modelisation Numerique de la Deformation des Systemes Geologiques*. PhD Thesis. University Montpellier II.

HUGHES, M. & DAVISON, I. 1993. Geometry and growth kinematics of salt pillows in the southern North Sea. *Tectonophysics*, **228**, 239–254.

JACKSON, M. P. A. & TALBOT, C. J. 1986. External shapes, strain rates, and dynamics of salt structures. *Bulletin of the Geological Society of America*, **97**, 305–323.

—— & —— 1991. *A Glossary of Salt Tectonics*. Bureau of Economic Geology, The University of Texas at Austin, Geological Circular, **91-4**.

——, CORNELIUS, R. R., CRAIG, C. H., GANSSER, A., STOCKLIN, J. & TALBOT, C. J. 1990. *Salt Diapirs of the Great Kavir, Central Iran*, Geological Society of America, Memoir, **177**, 139.

KIRBY, S. H. & MCCORMICK, J. W. 1990. Inelastic properties of rocks and minerals: strength and rheology. *In*: CARMICHAEL R. S. (ed.) *Practical Handbook of Physical Properties of Rocks and Minerals.*

NALPAS, T. & BRUN, J.-P. 1993. Salt flow and diapirism related to extension at crustal scale. *Tectonophysics*, **228**, 349–362.

PODLADCHIKOV, Y., TALBOT C. & POLIAKOV A. N. B. 1993. Numerical models of complex diapirs. *Tectonophysics*, **228**, 189–198.

POLIAKOV, A. N. B. & HERRMANN H. J. 1994. Self-organized criticality in plastic shear bands. *Geophysical Research Letters*, **21**, 2143–2146.

——, ——, PODLADCHIKOV, Y. Y. & ROUX, S. 1994. Fractal plastic shear bands. *Fractals*, **2**, 567–581.

—— & PODLADCHIKOV, Y. Y. 1992. Diapirism and topography. *Geophysical Journal International*, **109**, 553–564.

——, CUNDALL, P., PODLADCHIKOV, Y. & LAYKHOVSKY, V. 1993*a*. An explicit inertial method for the simulation of viscoelastic flow: An evaluation of elastic effects on diapiric flow in two- and three-layers models. *In*: STONE, D. B. & RUNCORN, S. K. (eds.), *Flow and Creep in the Solar System: Observations, Modeling and Theory*. Kluwer, Dordrecht, 175–95.

——, PODLADCHIKOV, Y. & TALBOT C. 1993*c*. Initiation of salt diapirs with frictional overburden: numerical experiments. *Tectonophysics*, **228**, 199–210.

——, VAN BALEN, R., PODLADCHIKOV, Y. Y., DAUDRE, B., CLOETINGH, S. & TALBOT, C. J. 1993*b*. Numerical analysis of how sedimentation and redistribution of surficial sediments affects salt diapirism. *Tectonophysics*, **226**, 199–216.

PRICE, N. J. & COSGROVE, J. W. 1990. *Analysis of Geological Structures*. Cambridge University Press. Cambridge.

RAMBERG, H. 1981. *Gravity, Deformation and the Earth's Crust.* 2nd edn. Academic, London.

RATCLIFFE, D. W., GRAY, S. H. & WHITMORE, JR, N. D. 1992. Seismic imaging of salt structures in the Gulf Coast of Mexico. *Leading Edge*, **11**, 15–34.

RÖNNLUND, P. 1989. Viscosity ratio estimates from natural Rayleigh–Taylor instabilities. *Terra Nova*, **1**, 344–348.

RUTTER, E. H. 1976. The kinetics of rock deformation by pressure solution. *Philosophical Transactions of the Royal Society of London*, **283**, 203–219.

SUPPE, J. 1985. *Principles of Structural Geology*. Prentice-Hall, Englewood Cliffs, New Jersey.

SCHULTZ-ELA, D. D., JACKSON, M. P. A. & VENDEVILLE, B. C. 1993. Mechanics of active salt diapirism. *Tectonophysics*, **228**, 275–315.

TURCOTTE, D. L. & SCHUBERT, G. 1982. *Geodynamics, Applications of Continuum Physics to Geological Problems*. Wiley, New York.

VAN KEKEN, P. E., SPIERS, C. J., VAN DEN BERG, A. P. & MUYZERT, E. J. 1993. The effective viscosity of rock-salt: Implementation of steady state creep laws in numerical models of salt diapirism. *Tectonophysics*, **225**, 457–476.

VENDEVILLE, B. C. & JACKSON, M. P. A. 1992. The rise of diapirs during thin-skinned extension. *Marine and Petroleum Geology*, **9**, 331–353.

WEIJERMARS, R., JACKSON, M. P. A. & VENDEVILLE, B. 1993. Rheological and tectonic modelling of salt provinces. *Tectonophysics*, **217**, 143–174.

# Index

References to figures and tables are given in *italics*